THE FINITE ELEMENT METHOD WITH HEAT TRANSFER AND FLUID MECHANICS APPLICATIONS

This book is intended for advanced undergraduate and graduate students. The first four chapters are devoted to introduction of the finite element concept. The focus of the book then covers two essential areas – heat transfer and fluid mechanics – topics with different finite element formulations. The heat transfer applications begin with the classical one-dimensional thin-rod problem, followed by a discussion of the two-dimensional heat transfer problem, including a variety of boundary conditions. Finally, a complicated-geometry three-dimensional problem involving a cooled radial turbine rotor is presented, with the cooling passages treated as "heat sinks" in the finite element analysis. For fluid mechanics, the concept of "nodeless" degrees of freedom is introduced, with real-life fluid flow applications. The time-dependent finite element analysis topic is addressed through the problem of unsteady stator/rotor flow interaction within a turbomachinery stage. Finally, the concept of "virtually deformable finite elements," as it relates to the problem of fluid-induced vibration, is explained in detail with many practical applications.

Erian A. Baskharone is Professor Emeritus of Mechanical and Aerospace Engineering at Texas A&M University. He is a member of the ASME Turbomachinery Executive Committee. Dr. Baskharone was a senior engineer with Allied-Signal Corporation, responsible for the aerothermodynamic design of various turbofan and turboprop engines. His research covered a wide spectrum of turbomachinery topics, including the unsteady stator/rotor flow interaction and the fluid-induced vibration of the space shuttle main engine turbopumps. His finite element–based perturbation approach to the problem of turbomachinery fluid-induced vibration is well known. At Texas A&M, he received the General Dynamics Award of Excellence in Engineering Teaching (1991) and the Amoco Foundation Award for Distinguished Teaching (1992). He is the author of *Principles of Turbomachinery in Air-Breathing Engines* (Cambridge University Press, 2006) and *Thermal Science: Essentials of Thermodynamics, Fluid Mechanics and Heat Transfer* (2012).

The Finite Element Method with Heat Transfer and Fluid Mechanics Applications

Erian A. Baskharone
Texas A&M University

CAMBRIDGE
UNIVERSITY PRESS

University Printing House, Cambridge CB2 8BS, United Kingdom

One Liberty Plaza, 20th Floor, New York, NY 10006, USA

477 Williamstown Road, Port Melbourne, VIC 3207, Australia

314-321, 3rd Floor, Plot 3, Splendor Forum, Jasola District Centre, New Delhi - 110025, India

103 Penang Road, #05-06/07, Visioncrest Commercial, Singapore 238467

Cambridge University Press is part of the University of Cambridge.

It furthers the University's mission by disseminating knowledge in the pursuit of education, learning and research at the highest international levels of excellence.

www.cambridge.org
Information on this title: www.cambridge.org/9781107039810

First published 2014

A catalogue record for this publication is available from the British Library

Library of Congress Cataloging in Publication data
Baskharone, Erian A., 1947–
The finite element method with heat transfer and fluid mechanics applications / Erian A. Baskharone.
pages cm
Includes bibliographical references and index.
ISBN 978-1-107-03981-0 (hardback)
1. Heat–Transmission. 2. Fluid mechanics. 3. Finite element method. I. Title.
QC320.22.F56B37 2013
536′.200151825–dc23 2013009965

ISBN 978-1-107-03981-0 Hardback

Brief Contents

1 **The Finite Element Method: Introductory Remarks** 1
2 **Some Methods for Solving Continuum Problems** 23
3 **Variational Approach** 27
4 **Requirements for the Interpolation Functions** 33
5 **Heat Transfer Applications** 35
6 **One-Dimensional Steady-State Problems** 42
7 **The Two-Dimensional Heat-Conduction Problem** 51
8 **Three-Dimensional Heat-Conduction Applications with Convection and Internal Heat Absorption** 66
9 **One-Dimensional Transient Problems** 95
10 **Fluid Mechanics Finite Element Applications** 106
11 **Use of Nodeless Degrees of Freedom** 114
12 **Finite Element Analysis in Curvilinear Coordinate** 149
13 **Finite Element Modeling of Flow in Annular Axisymmetric Passages** 176
14 **Extracting the Finite Element Domain from a Larger Flow System** 189
15 **Finite Element Application to Unsteady Flow Problems** 201
16 **Finite Element–Based Perturbation Approach to Unsteady Flow Problems** 237

Appendix A. Natural Coordinates for Three-Dimensional Surface Elements 321
Appendix B. Classification and Finite Element Formulation of Viscous Flow Problems 324
Appendix C. Numerical Integration 331
Appendix D. Finite Element–Based Perturbation Analysis: Formulation of the Zeroth-Order Flow Field 335
Appendix E. Displaced-Rotor Operation: Perturbation Analysis 344
Appendix F. Rigorous Adaptation to Compressible-Flow Problems 355

Detailed Contents

Preface *page* xv

1 **The Finite Element Method: Introductory Remarks** 1

The Mathematical Approach: A Variational Interpretation 2
Continuum Problems 3
Terminology and Preliminary Considerations: Types of Nodes 4
Degrees of Freedom 4
Interpolation Functions: Polynomials 6
One Independent Variable 6
Two Independent Variables 7
Three Independent Variables 7
Deriving Interpolation Functions 8
Natural Coordinates 10
Natural Coordinates in One Dimension 11
Natural Coordinates in Two Dimensions 12
Natural Coordinates in Three Dimensions 14
Curve-Sided Isoparametric Elements 16
Coordinate Transformation 17
Evaluation of Elemental Matrices 20
References 22

2 **Some Methods for Solving Continuum Problems** 23

Overview 23
The Ritz Method 24
Example: The Ritz Method 24
The Finite Element Method: Relation to the Ritz Method 26

3 **Variational Approach** 27

Example of Piecewise Approximation 28
Elemental Equations from a Variational Principle 30

4 **Requirements for the Interpolation Functions** 33

5 **Heat Transfer Applications** 35

Variational Approach 35
Example 37
Approximation of Integrals 38
One-Dimensional Steady-State Problems 39
Finite Element Formulation 40

6 **One-Dimensional Steady-State Problems** 42

Variational Statement 42
Finite Element Formulation 44
Numerical Results 48
Problems 49

7 **The Two-Dimensional Heat-Conduction Problem** 51

Variational Statement 51
Finite Element Formulation 52
Numerical Solution 59
Numerical Results 63

8 **Three-Dimensional Heat-Conduction Applications with Convection and Internal Heat Absorption** 66

The Problem of Cooling a Radial Turbine Rotor: Overview 66
Governing Equations 67
Finite Element Variational Formulation 68
Euler's Theorem of Variational Calculus 68
Derivation of the Variational Statement 69
Discretization of the Continuum 71
Evaluation of $dI_k/d\{T\}$ 73
Evaluation of $dI_g/d\{T\}$ 78
Evaluation of $dI_h/d\{T\}$ 79
Evaluation of $dI_q/d\{T\}$ 82
The Final Set of Equations 83
Turbine Rotor Configuration and Cooling Techniques 83
Determination of the Hot Turbine Boundary Conditions 84
Rotor Blade 84
Rotor Disk Backside 85
Rotor Hub 85
Cooled Turbine Rotor Calculations 85
Rotor Disk Cooling 86
Blade Cooling through a Slot 86
Blade Cooling through Radially Drilled Holes 86
Numerical Results 87
Remark 93

Problems 93
References 94

9 **One-Dimensional Transient Problems** 95
Variational Statement 95
Finite Element Formulation 96
Numerical Solutions 101
Euler's Method 101
Crank-Nicolson Method 103
Purely Implicit Method 103
References 105

10 **Fluid Mechanics Finite Element Applications** 106
Introduction 106
Inviscid Incompressible Flows 107
Problem Statement 107
Velocity Potential and Stream-Function Formulations 109
Flow around Multiple Bodies by Superposition 111
References 113

11 **Use of Nodeless Degrees of Freedom** 114
Overview 114
Flow-Governing Equations 120
Boundary Conditions 122
Flow Inlet Station $(B - C)$ 122
Flow Exit Station $(D - A)$ 122
Periodic Boundaries $(A - B$ and $D - C)$ 123
Domain-Splitting Boundaries $(E - F$ and $G - H)$ 123
Airfoil Suction and Pressure Surfaces 123
Finite Element Analysis 124
Galerkin's Weighted-Residual Approach 127
Applications 128
Flow Analysis in a Rectilinear-Airfoil Cascade 128
Field Discretization Model 128
Computational Results and Accuracy Assessment 128
Periodicity Conditions in Radial Cascades 130
Flow Investigation in a Radial-Turbine Scroll 130
Finite Element Analysis 132
Introduction of a Velocity Potential Discontinuity 134
Computational Results 134
Proposed Analysis Upgrades 135
Domains with High Degrees of Multiconnectivity 135
Axial-Flow Stator with a Spanwise Circulation Variation 138
Problems 138
References 147

12 **Finite Element Analysis in Curvilinear Coordinate** 149
Introduction 149
Analysis Guidelines and Limitations 153
Flow-Governing Equations 154
Continuity Equation 154
Through-Flow Momentum Equation 154
Tangential Momentum Equation 154
Boundary Conditions 156
Finite Element Formulation 158
Continuity Equation 160
Through-Flow Momentum Equation 160
Tangential Momentum Equation 160
Iterative Solution Procedure 161
Application Examples 162
Example 1: Second-Stage Stator of a Gas Turbine 164
Example 2: Low-Aspect-Ratio Turbine Stator 165
Proposed Analysis Upgrades 167
Adaptation to a Rotating-Blade Cascade 167
Inclusion of the Flow Turbulence Aspect 169
Problems 169
References 174

13 **Finite Element Modeling of Flow in Annular Axisymmetric Passages** 176
Introduction 176
Analysis 177
Flow-Governing Equations 177
Turbulence Closure 178
Boundary Conditions 180
Finite Element Formulation 181
Method of Numerical Solution 183
Numerical Results 183
Grid Dependency of the Flow Field 184
Diffuser Flow Field and Off-Design Performance 184
References 187

14 **Extracting the Finite Element Domain from a Larger Flow System** 189
Introduction 189
Analysis 191
Selection Options of the Computational Domain 191
Flow-Governing Equations 192
Boundary Conditions 193
Stage Inlet Station 193
Impeller Inlet and Exit Stations 193
Stage Exit Station 193

Solid Boundary Segments 193
Finite Element Formulation 194
Numerical Results 194
References 199

15 **Finite Element Application to Unsteady Flow Problems** 201
Introduction 201
Example 201
Flow-Governing Equations 204
Continuity Equation 204
Radial Momentum Equation 204
Tangential Momentum Equation 205
Axial Momentum Equation 205
Boundary Conditions 205
Finite Element Formulation 207
Time-Integration Algorithm 210
Numerical Procedure 211
Computational Results 211
Proposed Analysis Upgrades 219
Bidirectional Transfer of Boundary Conditions 219
Starting Point 220
Two-Way Stator/Rotor Exchange of Boundary Conditions 220
Continuity of the Variables' Normal Derivatives through Implicit Means 223
Methodology 223
Analysis 224
Problems 226
References 235

16 **Finite Element–Based Perturbation Approach to Unsteady Flow Problems** 237
Overview 237
Foundation of the Finite Element–Based Perturbation Approach 238
Definition of the Force-Related Rotordynamic Coefficients 240
Computational Development: Analysis of the Centered-Rotor Flow Field 242
Flow-Governing Equations 242
Continuity Equation 242
Axial Momentum Equation 242
Radial Momentum Equation 243
Tangential Momentum Equation 243
Boundary Conditions 243
Flow Inlet Station 243
Flow Exit Station 244
Solid Boundary Segments 244
Introduction of the Upwinding Technique 244

Finite Element Formulation 245
Method of Numerical Solution 248
Assessment of the Centered-Rotor Flow Field 249
Computational Development: Building the Zeroth-Order Flow Model 249
Strategy 251
Transition to an Alternate Frame of Reference 252
Adaptation of the Axisymmetric Flow Solution 253
Flow-Governing Equations in the Rotating Frame of Reference 255
Continuity Equation 255
x-Momentum Equation 255
y-Momentum Equation 255
z-Momentum Equation 255
Calculation of the Force-Related Rotordynamic Coefficients 256
Applications: Benchmark Test Case—Comparison with Cal Tech's Experimental Work 258
Background 258
Features of the Centered-Rotor Flow Field 259
Assessment of the Fluid-Induced Force Components 260
Applications: Perturbed Flow Structure due to Synchronous Whirl 262
Overview 262
Grid Dependency Investigation 264
Samples of the Computational Results 265
Comparison with Experimental Data 267
Applications: Rotordynamic Analysis of Labyrinth Seals 273
Literature Survey 273
Centered-Rotor Flow Field 275
Investigation of the Grid Dependency 278
Fluid-Induced Forces and Rotordynamic Coefficients 280
Applications: Rotordynamic Behavior of a Shrouded Pump Impeller 283
Centered-Impeller Subproblem: Contouring the Flow Domain 284
Centered-Impeller Subproblem: Boundary Conditions 284
Stage Inlet Station 284
Impeller Inlet and Exit Stations 285
Stage Exit Station 285
Solid Boundary Segments 285
Flow Structure 285
Simulation of the Impeller Subdomain Effects 291
Worthiness of Simulating the Impeller Subdomain 292
Results of the Perturbation Analysis 294
Assessment of the Single-Harmonic Perturbation Assumption 296
Applications: Investigation of Annular Seals under Conical Whirl 298
Rotordynamic Analysis of the Fluid/Rotor Interaction System 299
Computational Results 301
Applications: Interrelated Effects of the Cylindrical/Conical Rotor Whirl 303

Expanded Rotordynamic Analysis 304
Computational Results 306
Applications: Compressible-Flow Gas Seals Using a Simplified Adiabatic-Flow Approach 307
Computational Results 308
Comment 308
Proposed Upgrades of the Perturbation Analysis 309
Inclusion of the Shear-Stress Perturbations in Computing the Fluid-Induced Forces 309
Rigorous Adaptation to Compressible-Flow Problems 311
Relevant Remarks 311
Problems 313
References 317

Appendix A. Natural Coordinates for Three-Dimensional Surface Elements 321
Appendix B. Classification and Finite Element Formulation of Viscous Flow Problems 324
Appendix C. Numerical Integration 331
Appendix D. Finite Element–Based Perturbation Analysis: Formulation of the Zeroth-Order Flow Field 335
Appendix E. Displaced-Rotor Operation: Perturbation Analysis 344
Appendix F. Rigorous Adaptation to Compressible-Flow Problems 355
Index 369

Preface

This textbook complements but goes beyond many existing texts that introduce the concept of the finite element method and its numerous applications in virtually all engineering disciplines. The focus of this book branches out to two essential thermal science areas. These are the heat transfer and fluid flow disciplines. First, there is a total of four introductory chapters, which (combined) introduce the general finite element concept. Next, the finite element heat transfer applications are discussed. The heat transfer applications vary from the simple one-dimensional thin-rod problem all the way up to a fully three-dimensional heat-conduction problem with various boundary conditions and internal heat absorption. In the fluid mechanics part of the book, each chapter is focused on a single flow problem category varying in complexity from a simple incompressible potential flow field to the full-scale three-dimensional turbulent flow structure in complex-geometry computational domains. Complex problems such as the perturbation technique (within a finite element framework) are also addressed. Among these, the common factor is in the means with which the finite element model is uniquely adapted to overcome a historically persisting analytical challenge in this and other similar problems under the same classification. While emphasizing the problem's fluid mechanics foundation, including such aspects as the extraction of the computational domain from a larger system and the choice of sound engineering boundary conditions, an equal emphasis is placed on handling the inherent analytical challenge. Also explained in each case is the choice of a suitable finite element formulation, particularly in problems involving convection-dominated flow fields.

In the heat transfer part of the book, all finite element equations are derived from scratch. Emphasized in this part is the importance of matrix algebra, a tool that simplifies both building the element's contributions and transmitting them to the so-called global system. The book doesn't shy away from such complicating factors as internal heat generation and varying thermal conductivity. Use of the method in three-dimensional applications is illustrated through the problem of cooling a radial-inflow turbine rotor, where the secondary (or cooling) flow stream is simulated through a sequence of heat sinks. The numerical results stand to attest to the efficiency and accuracy of the finite element method, at least in this case.

The optimal situation in an analytical heat transfer course is for the entire class to have the necessary mathematical background to solve the advanced problems that are being discussed. The instructor can then devote his or her efforts to the engineering aspects of problem solving. It has been the instructor's experience, however, to be faced with a class that has not yet had the necessary mathematical background or, at least, is currently enrolled in a course that may provide some of the information they need. This usually means that a heat transfer course must teach mathematics as well as engineering. Of course, this is not unique to the analytical heat transfer part of this book but extends to the other thermal-science discipline, namely, fluid mechanics.

A major part of this book's worthiness lies in the carefully written segments with which each chapter in the fluid mechanics part of the book is concluded. This is where extensions of the finite element approach from the analytical or application viewpoints are proposed. While a rather compact version of each flow model has already been published, it was an absolute priority to present the model details in a more comprehensible format as well as to thoroughly discuss the computational results as related to the analytical choices made. It is, however, each chapter's concluding segment that thoroughly explains how the flow analysis can be upgraded in a research-type environment, promoting the student's masterfulness of the finite element technique in the process. The provided yet uninvestigated upgrades here take the traditional role of limited-purpose exercises in other engineering textbooks covering the same engineering discipline.

In the interest of fairness, the majority of the above-cited concluding segments go well beyond what may be viewed as natural or obvious extensions of the different finite element models developed in this book. These largely involve the very analytical core of each flow model. Above all, these yet unpublished segments are written to leave the student with virtually no questions about the purpose of each upgrade or how to implement it.

Through four carefully selected flow problems with ascending difficulty levels, this books aims, in part, to illustrate how a conventional finite element model can be altered to accommodate some features that are unique to each of these problem categories. Of these, the first problem is that of a simple potential flow through an airfoil cascade. The *chronic* problem in the majority of existing analytical models in this area has to do with the need to externally impose the magnitude of circulation around the airfoil, either explicitly or through a fictitiously prescribed exit flow angle. Such an action, in the author's opinion, overspecifies the problem and, in effect, implies (at least theoretically) a flow field that is purely of the analyst's own choosing. Still insufficient to analytically close the problem, some of the more accurate cascade-flow models use an iterative procedure to determine the circulation magnitude that, in the end, satisfies some acceptable but not precise trailing-edge flow behavior. The current finite element analysis is based on recognition of the circulation as a *nodeless* degree of freedom with a unique magnitude that is determined, noniteratively in the final solution.

The second problem essentially deals with a similar problem category, with the exception that the convergence or divergence of the endwalls (hub and casing) is now simulated through a variable-thickness stream filament in what is referred to as *quasi*-three-dimensional cascade-flow analysis, with the flow viscosity aspects added this time. Over the filament surfaces, the acting forces (e.g., those due to shear stress) have been ignored historically. The offered finite element analysis accounts for these effects through what is loosely termed *source* terms in deriving the finite element equivalent of the flow-governing equations. The analysis goes further in elevating the solution accuracy through a unique implicit approach within the finite element flow model to implement the so-termed cascade periodicity conditions without having to sacrifice the finite element equations over the periodic boundaries.

Next comes a more involved problem, where two subdomains, namely, those of the stator and rotor, within a typical turbomachinery stage, are combined together. In this case, the relative motion between the two subdomains, coupled with the close proximity of the two cascades, rules out any attempt to eliminate the time-dependency features of the stage flow field. The importance of this problem stems from the fact that the unsteady mutual flow interaction between these two subdomains, which is dictated by such aspects as the stator/rotor spacing and the relative blade counts, may lead to a catastrophic premature mechanical failure in one or both blade rows. The task, in this large-scale problem, is to devise a stable time-marching approach with the boundary conditions over the stator/rotor interface surface that is, meanwhile, updated every step of the way.

The last fluid mechanics application is where the core of the finite element model is examined and then manipulated to give a novel finite element–based perturbation model for a special class of time-dependent flow fields. At the heart of this approach is the idea of treating the assembly of finite elements as that of "virtually" deformable subdomains. The petrturbation model here is applicable to a wide range of problems where a rotor undergoes a "whirling" motion that is composed of a lateral eccentricity coupled with a precession (or whirl) frequency that is generally less than, equal to, or even higher than the rotor speed. Application of the perturbation model to a wide variety of problems featuring fluid-induced forces (as well as moments resulting from the so-called canonical whirl) is thoughtfully presented, together with such aspects as the solution-grid dependency.

1 The Finite Element Method: Introductory Remarks

The finite element method is a numerical technique for obtaining approximate solutions to a wide spectrum of engineering problems. Although originally developed to study the stresses in complex airframe structures, it has since been extended and applied to a broad field in continuum mechanics. Because of its diversity and flexibility as an analysis tool, this particular technique is receiving much attention in both academia and industry fields.

Although this brief comment on the finite element method answers the question posed by the section heading, it does not give us the operational definition we need to apply the method to a particular problem. Such an operational definition, along with a description of the method fundamentals, requires considerably more than just one paragraph to develop. Hence, the first segment of this book is devoted to basic concepts and fundamental theory. Before discussing more aspects of the finite element method, we should first consider some of the circumstances leading to its inception, and we should briefly contrast it with other numerical techniques.

In more and more engineering situations today, we find that it is necessary to obtain approximate numerical solutions to problems rather than exact closed-form solutions. For example, we may want to find the load capacity of a plate that has several stiffeners and odd-shaped holes, the concentration of pollutants during nonuniform atmospheric conditions, or the rate of fluid flow through a passage of arbitrary shape. Without too much effort, we can write down the governing equations and boundary conditions for these problems, but we immediately see that no simple analytical solution can be found. The difficulty in these examples lies in the fact that either the geometry or some other features of the problem are irregular. Analytical closed-form solutions to problems of this type seldom exist, yet those are the kinds of problems that engineers are called on to solve.

The resourcefulness of the analyst usually comes to the rescue, providing several alternatives to overcome the dilemma. One possibility is to make meaningful assumptions, to ignore the difficulties, and to reduce the problem to one that can be handled. Sometimes this procedure works. However, more often than not, it leads to serious inaccuracies or totally wrong answers. Now that large-scale digital computers are widely available, a more viable alternative is to retain the complexities of the problem and try to find an approximate numerical solution.

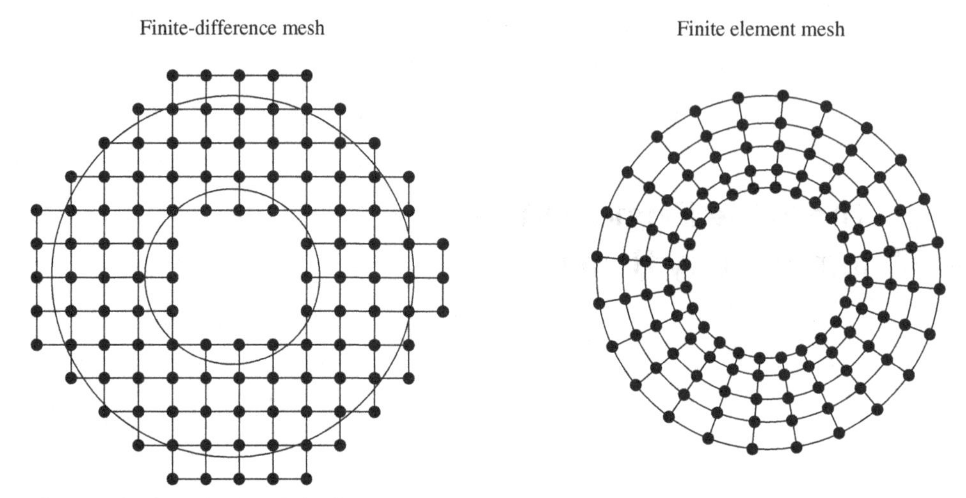

Figure 1.1. Superiority of the finite element method in simulating curved-domain boundary segments.

Several approximate numerical analysis methods have evolved over the years, with the most commonly used method being the finite-difference technique. The familiar finite-difference model of a problem gives pointwise approximation to the governing equations. The model, formed by writing difference equations for an array of grid points, is improved as more points are used. With finite-difference techniques, we can treat some fairly difficult problems, but, for example, when we encounter irregular geometries or unusual specification of boundary conditions, we find that finite-difference techniques become too hard to use.

In addition to the finite-difference method, another, more recent numerical method, known as the *finite element method*, has emerged. Unlike the finite-difference method, which envisions the solution domain as an array of grid points, the finite element method envisions the solution region as a buildup of small, nonoverlapping and interconnected subregions termed *elements*. A finite element model of a problem gives a *piecewise* approximation to the governing equations. The basic premise of the finite element method is that a solution domain can be modeled or approximated analytically by replacing it with an assemblage of discrete elements. Since these elements can be put together in a variety of ways, they can be used to represent exceedingly complex shapes.

As an example of how finite-difference and finite element models might be used to represent a complex geometric shape, consider the annular passage in Figure 1.1. Note the superiority of the finite element method in handling such problems because the elements can be shaped in such a way that matches the solution-domain curved boundary segments. Figure 1.1 is meant to simply illustrate the finite element model in contrast to a typical finite-difference mesh of the same annulus.

The Mathematical Approach: A Variational Interpretation

This approach is helpful in gaining an understanding of the finite element method, but instrumental difficulties arise when we try to apply it to complex problems. In this section, we take a broader view and interpret the finite element method as an approximate means of solving variational problems. At this point, however, we cannot faithfully discuss the many specific techniques that are useful for particular types of problems. These specialized aspects of the finite element method will be introduced later in this and succeeding chapters.

To set the stage for the introduction of the mathematical concepts and to give them a place in the overall collection of solution techniques, we begin with a general discussion of the continuum problems of mathematical physics.

After briefly mentioning some of the more popular solution techniques for different classes of problems, we establish the necessary technology and definitions to show how variational problems and the finite element method are related. The variational basis of the finite element method dictates the criteria to be satisfied by the so-called element interpolation functions and enables us to make definitive statements about the convergence of results as we use an ever-increasing number of smaller and smaller elements.

After a discussion of the variational approach to the formulation of element equations, we consider a detailed example. The last segment of this chapter contains the problem of how to find variational principles for use in the finite element method.

Continuum Problems

Problems in engineering and science fall into two fundamentally different categories depending on which point of view we adopt. One point of view is that all matter consists of single particles that retain their identity and nature as they move through space. Their position in space at any instant is given by the coordinates in some reference frame, and these coordinates are functions of time – the only independent variable for any process. This viewpoint, known as the *Lagrangian* viewpoint, is the basis for *Newtonian* particle mechanics.

The other viewpoint, the one we will use, stems from the *continuum* rather than the molecular or particle approach to nature. In the continuum (sometimes termed *Eulerian*) viewpoint, we say that all processes are characterized by field quantities that are defined at every point in space. The independent variables in continuum problems are the coordinates of space and time. The Eulerian viewpoint allows us to focus our attention on one point in space and then observe the phenomena occurring there.

Continuum problems are concerned with fields of temperature, stress, mass concentration, displacement, electromagnetic, and acoustic potentials, to name just a few examples. These problems arise from phenomena in nature that are approximately characterized by partial differential equations and their boundary conditions.

We will briefly discuss the nature of continuum problems typically encountered and some of the possible means of solution. Then we will return to the topic of solving these problems using the finite element method. Continuum problems of mathematical physics are often referred to as *boundary-value problems* because their solution is sought in some domain defined by a given boundary, on which certain constraints termed *boundary conditions* are specified. Except for free-boundary problems, the boundary shape and its location are always known. Sometimes, however, it is the analyst's responsiblity to extract it from a bigger system. The boundary may be defined by a curve or a surface of n-dimensional space, and the domain it defines may be finite or infinite depending on the extremities of the boundary. The boundary is said to be *closed* if conditions affecting the solution of the problem are specified everywhere on the boundary (even though part of the boundary may extend to infinity) and *open* if part of the boundary extends to infinity and no boundary conditions are specified on the part at infinity. It is important to note that our definition of a boundary value problem departs from the usual one. The usual definition distinguishes between boundary value problems and initial value problems, where time is an independent variable. Because of our definition of the domain boundary, we may describe all partial differential equations and their boundary conditions as boundary value problems.

Terminology and Preliminary Considerations: Types of Nodes

In Figures 1.2 through 1.6, the number of basic element shapes and typical locations of nodes assigned to these elements are illustrated. We allude to the exterior and interior nodes, but now we can be more specific. Nodes are classified as either exterior or interior depending on their location relative to the geometry of an element. Exterior nodes lie on the boundary of an element, and they represent the points of connection between bordering elements. Nodes positioned at the corners of elements, along the edges, or on the surfaces are all exterior nodes. For one-dimensional elements such as those in Figure 1.2, there are only two exterior nodes because only the ends of the element connect to other one-dimensional elements. In contrast to exterior nodes, interior nodes are those that do not connect with neighboring elements.

Degrees of Freedom

Two other features, in addition to shape, characterize a particular element type: (1) the number of nodes assigned to the element and (2) the number and type of nodal variables chosen for it. Often the nodal variables associated with the element are referred to as the *degrees of freedom* of the element. This terminology, which we will adopt, is a spin-off from the solid mechanics field, where the nodal variables are usually nodal displacements and sometimes derivatives of displacements. Nodal

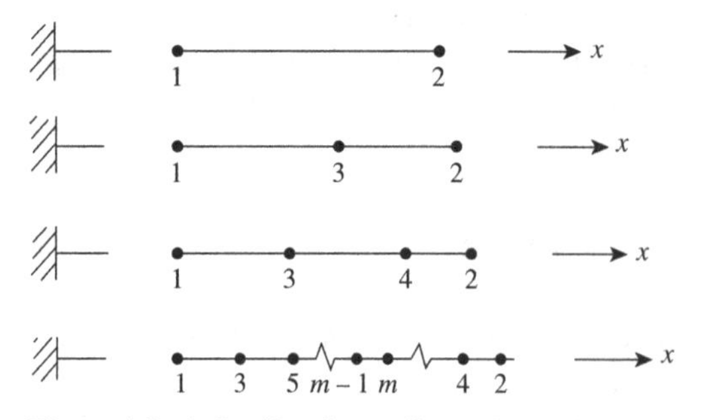

Figure 1.2. A family of one-dimensional line elements.

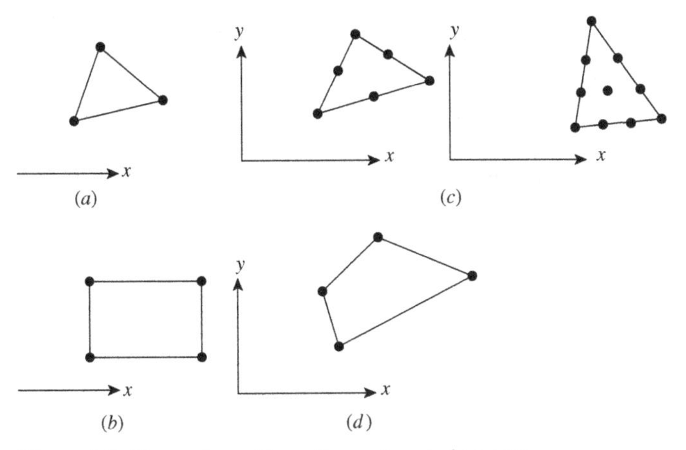

Figure 1.3. Examples of two-dimensional elements.

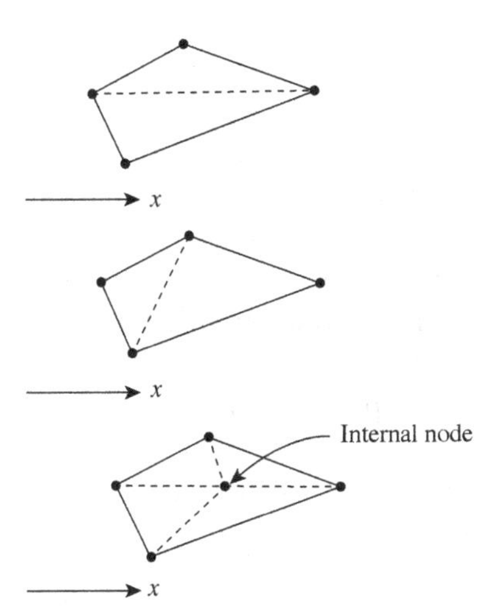

Figure 1.4. The quadriteral element formed by combining triangles.

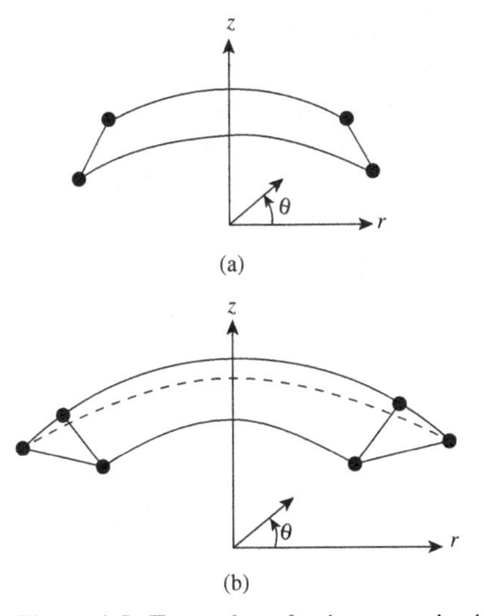

Figure 1.5. Examples of axisymmetric-ring elements.

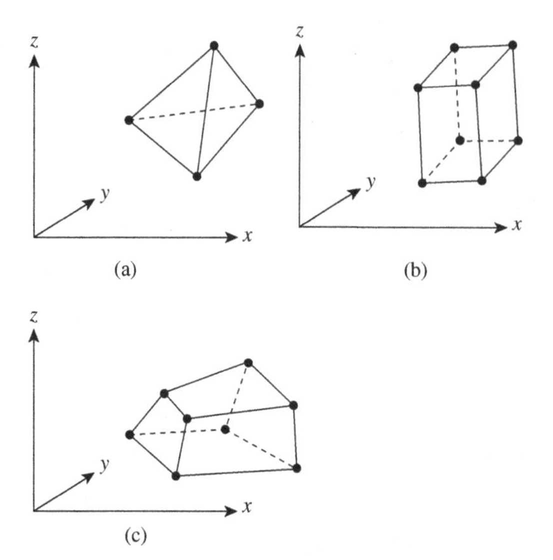

Figure 1.6. Three-dimensional elements.

degrees of freedom can be interior or exterior in relation to the element boundaries, depending on whether they are assigned to interior or exterior nodes.

In the fluid mechanics section of this book, one of the first few problems (Chapter 11) has to do with the fluid flow in multiply-connected solution domains. In this case, the field variable is the velocity potential ϕ. In addition, however, the circulation around the lifting body is added to a specific group of computational points as a *nodeless* degree of freedom. With the aid of this chapter's contents, therefore, the student is encouraged to generalize the term *degree of freedom* in a manner that is consistent with the problem physics.

Interpolation Functions: Polynomials

In the finite element literature, the functions used to represent the behavior of a field variable within an element are called *interpolation functions*, *shape functions*, or *approximation functions*. We have used and will continue to use only the first term in this text. Although it is conceivable that many types of functions could serve as interpolation functions, only polynomials have received widespread use. The reason is that polynomials are relatively easy to manipulate mathematically. In other words, they can be integrated or differentiated without difficulty. Trigonometric functions also possess this property, but they are seldom used. Here we will employ only polynomials of various types and orders to generate suitable interpolation functions. The polynomials we will consider follow.

One Independent Variable

In one dimension, a general complete nth-order polynomial may be written as follows:

$$P_n(x) = \sum_{i=0}^{T_n^{(1)}} \alpha_i x^{(i)}$$

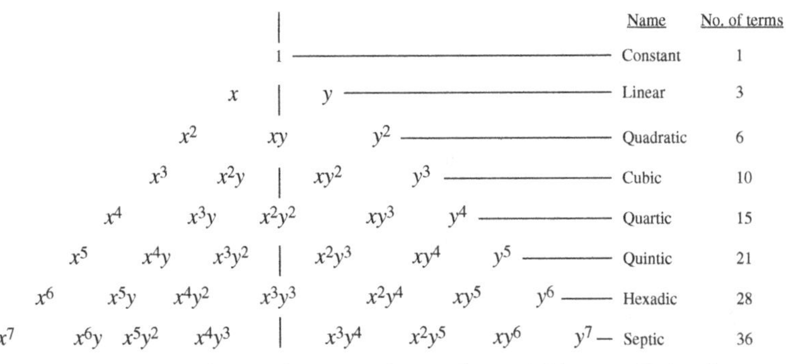

Figure 1.7. Array of terms in a complete polynomial in two dimensions.

where the number of terms in the polynomial is $T_n^{(1)} = n+1$. For $n = 1$, $T_1^{(1)} = 2$, and $P_1(x) = \alpha_{)} + \alpha_1 x$. For $n = 2$, $T_2^{(1)} = 3$, and $p_2(x) = \alpha_0 + \alpha_1 x + \alpha_2 x^2, \ldots$, and so on.

Two Independent Variables

In two dimensions, a complete nth-order polynomial may be written as follows:

$$P_n(x,y) = \sum_{k=1}^{T_n^{(2)}} \alpha_k x^i y^j, \qquad i+j \leq n$$

where the number of terms in the polynomial is $T_n^{(2)} = [(n+1)(n+2)]/2$. For $n = 1$, $T_1^{(2)} = 3$, and $P_1(x,y) = \alpha_1 + \alpha_2 x + \alpha_3 y$. For $n = 2$, $T_2^{(2)} = 6$, and $P_2(x,y) = \alpha_1 + \alpha_2 x + \alpha_3 y + \alpha_4 xy + \alpha_5 x^2 + \alpha_6 y^2, \ldots$, and so on.

Gallagher [1] suggested a convenient way to illustrate the terms in a complete two-dimensional polynomial. If the terms are placed in a triangular array of ascending order, we obtain an arrangement similar to the Pascal triangle (Figure 1.7). We note that the sum of exponents of any term in this triangular array is the corresponding number in the well-known Pascal triangle of binomial coefficients.

Three Independent Variables

In three dimensions, a complete nth-order polynomial may be written as follows:

$$P_n(x,y,z) = \sum_{l=1}^{T_n^{(3)}} \alpha_l x^i y^j z^k \qquad i+j+k \leq n$$

where the number of terms in the polynomial is

$$T_n^{(3)} = \frac{(n+1)(n+2)(n+3)}{6}$$

For $n = 1$, $T_1^{(3)} = 4$, and $P_1(x,y,z) = \alpha_1 + \alpha_2 x + \alpha_3 y + \alpha_4 z$. For $n = 2$, $T_2^{(3)} = 10$, and $P_2(x,y,z) = \alpha_1 + \alpha_2 x + \alpha_3 y + \alpha_4 z + \alpha_5 xy + \alpha_6 xz + \alpha_7 yz + \alpha_8 x^2 + \alpha_9 y^2 + \alpha_{10} z^2, \ldots$, and so on.

The terms in a complete three-dimensional polynomial may also be arrayed in a manner that is analogous to the triangular array in two dimensions. The array

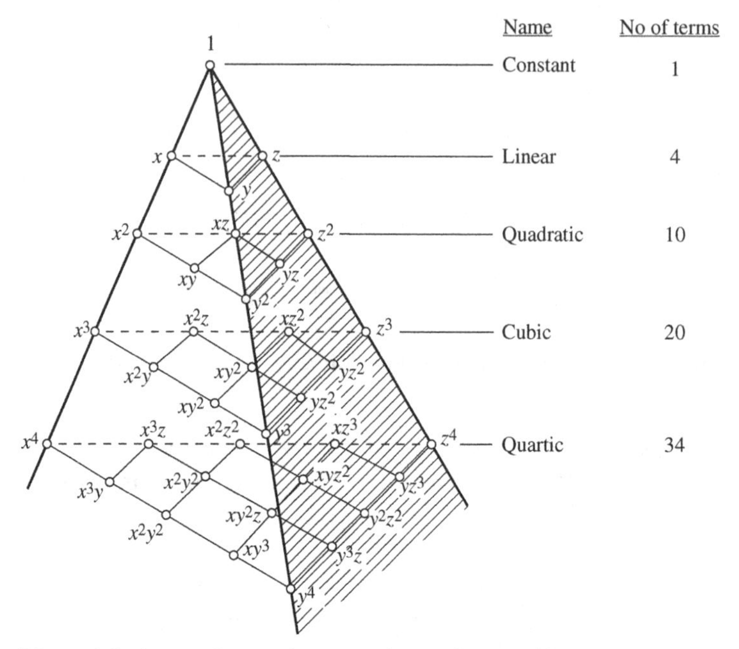

Figure 1.8. Array of terms in a complete polynomial in three dimensions.

becomes a tetrahedron, with the various terms placed at different planar levels, as shown in Figure 1.8.

Deriving Interpolation Functions

Thus far we have seen how a field variable can be represented within an element as a polynomial series whose coefficients are the generalized coordinates. In this section we will see how the interpolation functions for the physical degree of freedom are derived. These interpolation functions emerge from the basic procedure for expressing the generalized coordinates in terms of the nodal degrees of freedom.

The basic ideas can be illustrated through a simple example in two dimensions. Suppose that we wish to construct a rectangular element with nodes positioned at the element corners (Figure 1.9). If we assign one value of ϕ to each node, the element, then, will have four degrees of freedom, and we may select, as an interpolation model, a four-term polynomial such as

$$\phi(x,y) = \alpha_1 + \alpha_2 x + \alpha_3 y + \alpha_4 xy$$

The generalized coordinates may now be found by evaluating this interpolation function at each of the four nodes and then inverting the resulting set of simultaneous equations. Thus we may write

$$\phi_1 = \alpha_1 + \alpha_2 x_1 + \alpha_3 y_1 + \alpha_4 x_1 y_1$$

$$\phi_2 = \alpha_1 + \alpha_2 x_2 + \alpha_3 y_2 + \alpha_4 x_2 y_2$$

$$\phi_3 = \alpha_1 + \alpha_2 x_3 + \alpha_3 y_3 + \alpha_4 x_3 y_3$$

$$\phi_4 = \alpha_1 + \alpha_2 x_4 + \alpha_3 y_4 + \alpha_4 x_4 y_4$$

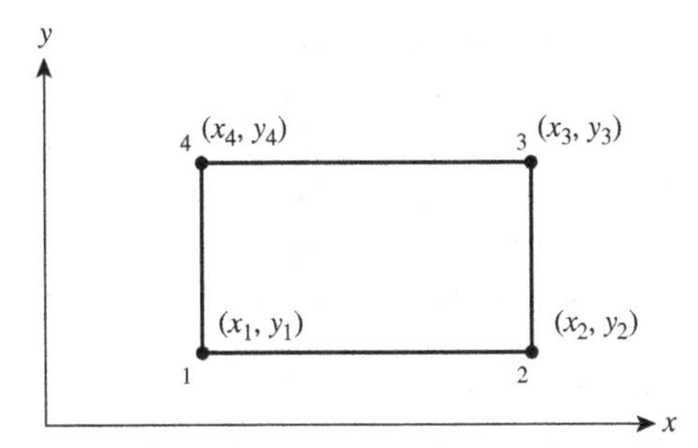

Figure 1.9. A rectangular element with sides parallel to the axes of the global coordinate system.

or, in matrix notation,

$$\{\phi\} = [G]\{\alpha\}$$

where the preceding column vectors and matrix are defined as follows:

$$\{\phi\} = \begin{Bmatrix} \phi_1 \\ \phi_2 \\ \phi_3 \\ \phi_4 \end{Bmatrix}$$

$$[G] = \begin{bmatrix} 1 & x_1 & y_1 & x_1y_1 \\ 1 & x_2 & y_2 & x_2y_2 \\ 1 & x_3 & y_3 & x_3y_3 \\ 1 & x_4 & y_4 & x_4y_4 \end{bmatrix}$$

$$\{\alpha\} = \begin{Bmatrix} \alpha_1 \\ \alpha_2 \\ \alpha_3 \\ \alpha_4 \end{Bmatrix}$$

In principle, then, we can express the generalized coordinates as the solution of the preceding matrix equation for $\{\alpha\}$, that is,

$$\{\alpha\} = [G]^{-1}\{\phi\}$$

Expressing the terms of the interpolation polynomial in the original expression for $\phi(x, y)$ as a product of a row vector and a column vector, we can write

$$\phi = [P]\{\alpha\}$$

where $$[P] = [1 \text{ x y xy}]$$

Thus, through simple substitution, we get

$$\phi = [P][G]^{-1}\{\phi\} = [N]\{\phi\}$$

with $$[N] = [P][G]^{-1}$$

This expression for the field variable ϕ, though obtained for one case, is generally applicable to all straight-sided elements. The original interpolation polynomial $[P]\{\alpha\}$ should not be confused with the interpolants N_i associated with the nodal degree of freedom. The distinction to note here is that $[P]\{\alpha\}$ is an interpolation function that applies to the whole element and expresses the field-variable behavior in terms of the generalized coordinates, whereas the interpolants N_i refer to individual nodes and individual degrees of freedom, and they represent the field-variable behavior. It is easy to see from the last expression of ϕ that the function N_i referring to node i takes on unit value at node i and zero value at all other nodes of the element.

The procedure for expressing the generalized coordinates in terms of the nodal degrees of freedom is actually the method commonly used to derive the nodal interpolation functions N_i. The procedure is straightforward and may be carried out easily, but sometimes difficulties are encountered. For some types of elements models, the inverse of the matrix $[G]$ may not exist for all orientations of the element in the *global* coordinate system. If an explicit expression for $[G]^{-1}$ is obtained algebraically, it may be possible to see under what conditions $[G]^{-1}$ does not exist and then try to avoid those circumstances when constructing the element mesh. Such an approach, however, is seldom recommended. Another disadvantage stems from the computational effort required to obtain $[G]^{-1}$ when it exists. For a large number of elements with many degrees of freedom, the computational cost can be prohibitive.

These reasons have motivated many researchers to try to obtain the nodal interpolation functions N_i by inspection, often relying on the use of special coordinate systems called *natural coordinates*. This particular topic is discussed separately in Appendix A for the general three-dimensional heat-conduction problem.

Throughout the remainder of this book, many elements with different shapes in two- and three-dimensional applications, including curve-sided elements, will be used. In each case, the shape functions will be defined and used in solving a variety of heat transfer and fluid mechanics problems.

Natural Coordinates

A local coordinate system that relies on the element geometry for its definition and whose coordinates range between zero and unity within the element is known as a *natural coordinate system*. Such systems have the property that one particular coordinate has a unity value at one node of the element and zero value at the other node(s); its variation between nodes is linear. We may construct natural coordinate systems for two-node line elements, three-node triangular elements, four-node quadrilateral elements, four-node tetrahedral elements, and so on.

The use of natural coordinates in deriving interpolation functions is particularly advantageous because special closed-form integration formulas can often be used to evaluate the integrals in the element equations. Natural coordinates also play a crucial role in the development of curve-sided elements.

The basic purpose of a natural coordinate system is to describe the location of a point inside an element in terms of the coordinates associated with the nodes of the element. We denote the natural coordinates as L_i, $(i = 1, 2, \ldots, n)$, where n is the number of external nodes of the element. One coordinate is associated with node i and has unit value there. It will become evident that the natural coordinates are functions of the global Cartesian coordinate system in which the element is defined.

Natural Coordinates in One Dimension

Figure 1.10 shows a line element in which we desire to define a natural coordinate system. If we select L_1 and L_2 as the natural coordinates, the location of the point x_P may be expressed as a linear combination of the nodal coordinates x_1 and x_2, that is,

$$x_P = L_1 x_1 + L_2 x_2$$

Because x_P can be any point on the line element, we can drop the subscript P for convenience. The coordinates L_1 and L_2 may be interpreted as weighting functions relating the coordinates of the end nodes to the coordinate of any interior point. Clearly, the weighting functions are not independent because we, in the end, will have

$$L_1 + L_2 = 1$$

The preceding two equations may be solved simultaneously for the functions L_1 and L_2 with the following result:

$$L_1(x) = \frac{x_2 - x}{x_2 - x_1}$$

$$L_2(x) = \frac{x - x_1}{x_2 - x_1}$$

The functions L_1 and L_2 are seen to be simply ratios of length and are often referred to as *length coordinates*. The variation of L_1 is shown in Figure 1.11. The linear interpolation used for the field variable ϕ can be written directly as follows:

$$\phi(x) = \phi_1 L_1 + \phi_2 L_2$$

If ϕ is taken to be a function of L_1 and L_2, differentiation of ϕ follows the chain-rule formula:

$$\frac{d\phi}{dx} = \frac{\partial\phi}{\partial L_1}\frac{\partial L_1}{\partial x} + \frac{\partial\phi}{\partial L_2}\frac{\partial L_2}{\partial x}$$

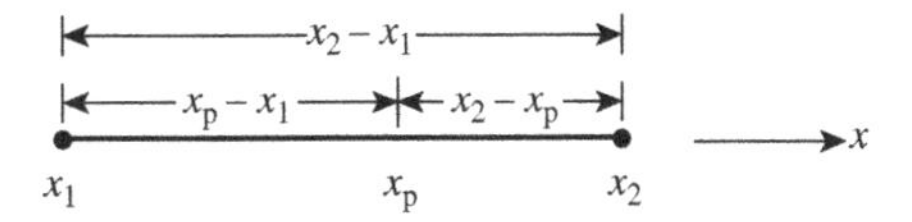

Figure 1.10. Two-noded line element.

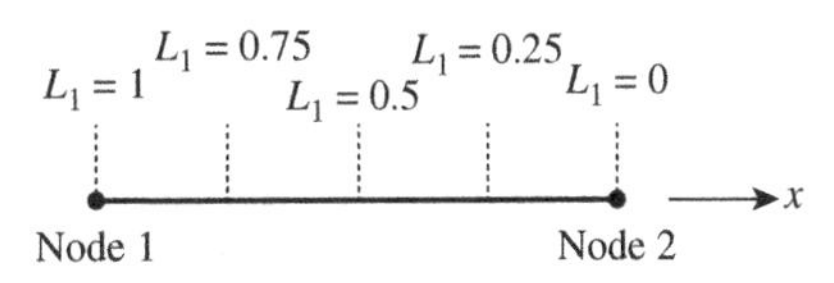

Figure 1.11. Variation of a length coordinate within an element.

where
$$\frac{\partial L_1}{\partial x} = \frac{-1}{x_2 - x_1}$$
$$\frac{\partial L_2}{\partial x} = \frac{1}{x_2 - x_1}$$

Natural Coordinates in Two Dimensions

The development of natural coordinates for triangular elements follows the same procedure we used for the one-dimensional case. Again, the goal is to choose coordinates L_1, L_2, and L_3 to describe the location of any point x_P within the element or on its boundary (Figure 1.12). The Cartesian coordinates of a point in the element should be linearly related to the new coordinates by the following equations:

$$x = L_1 x_1 + L_2 x_2 + L_3 x_3$$
$$y = L_1 y_1 + L_2 y_2 + L_3 y_3$$

In additions to these equations, we impose a third condition requiring that the weighting functions sum to be equal to unity, that is,

$$L_1 + L_2 + L_3 = 1$$

From this relationship, it is clear that only two of the natural coordinates can be independent, just as with the *length* coordinates presented earlier.

Inversion of the x and y expressions (earlier) gives the natural coordinates in terms of the Cartesian coordinates:

$$L_1(x,y) = \frac{1}{2\delta}(a_1 + b_1 x + c_1 y)$$
$$L_2(x,y) = \frac{1}{2\delta}(a_2 + b_2 x + c_2 y)$$
$$L_3(x,y) = \frac{1}{2\delta}(a_3 + b_3 x + c_3 y)$$

where:
$$2\Delta = \begin{vmatrix} 1 & x_1 & y_1 \\ 1 & x_2 & y_2 \\ 1 & x_3 & y_3 \end{vmatrix}$$

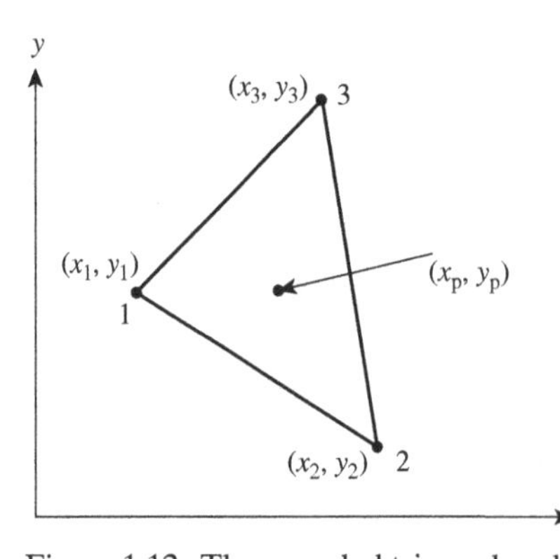

Figure 1.12. Three-noded triangular element.

which is equal to twice the area of the triangle (1-2-3) in Figure 1.12. Also,

$$a_1 = x_2y_3 - x_3y_2$$
$$b_1 = y_2 - y_3$$
$$c_1 = x_3 - x_2$$

The other coefficients are obtained by cyclically permuting the subscripts.

The natural coordinates L_1, L_2, and L_3 are precisely the interpolation functions for linear interpolation over a triangle; that is, $N_i \equiv L_i$ for the linear triangle. A little algebraic manipulation will reveal that the natural coordinates for a triangle have an interpretation analogous to that of length coordinates for a line. Just as $L_1(x)$ for the line element is a ratio of lengths, $L_1(x,y)$ for the triangular element is a ratio of areas. Figure 1.13 shows how the natural coordinates, often called *area coordinates*, are related to areas. As shown in Figure 1.13, when the point (x_P, y_P) is located on the element's boundary, one of the area segments vanishes, and hence the appropriate area coordinate along that boundary is identically zero. For example, if (x_P, y_P) is on line 1-3, then

$$L_2 = \frac{A_2}{\Delta} = 0 \quad \text{because } A_2 = 0$$

If we interpret the field variable ϕ as a function of L_1, L_2, and L_3 instead of x and y, differentiation becomes

$$\frac{\partial\phi}{\partial x} = \frac{\partial\phi}{\partial L_1}\frac{\partial L_1}{\partial x} + \frac{\partial\phi}{\partial L_2}\frac{\partial L_2}{\partial x} + \frac{\partial\phi}{\partial L_3}\frac{\partial L_3}{\partial x}$$

$$\frac{\partial\phi}{\partial y} = \frac{\partial\phi}{\partial L_1}\frac{\partial L_1}{\partial y} + \frac{\partial\phi}{\partial L_2}\frac{\partial L_2}{\partial y} + \frac{\partial\phi}{\partial L_3}\frac{\partial L_3}{\partial y}$$

where

$$\frac{\partial L_i}{\partial x} = \frac{b_i}{2\Delta}, \frac{\partial L_i}{\partial y} = \frac{c_i}{2\Delta} \qquad i = 1,2,3$$

Another type of natural coordinate system can be established for a four-node quadrilateral element in two dimensions. Figure 1.14 shows a general quadrilateral element in the global Cartesian coordinate system and a local natural coordinate system. In the natural coordinate system whose origin exists at the centroid, the

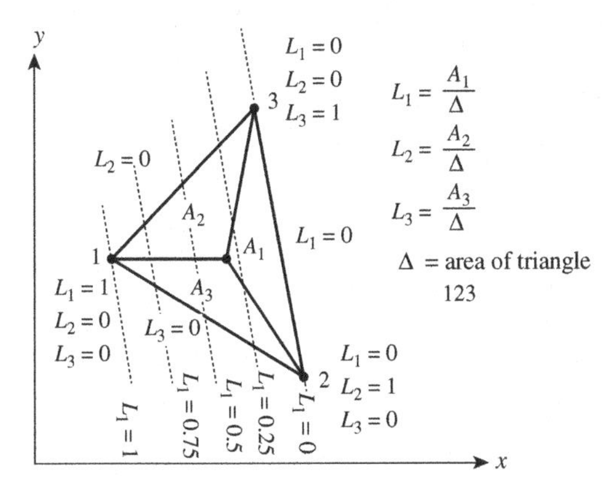

Figure 1.13. Area coordinates for a triangle.

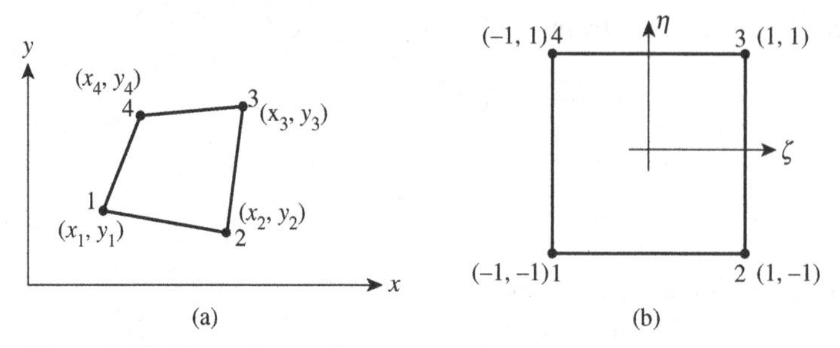

Figure 1.14. Natural coordinates for a general quadrilateral element: (*a*) Cartesian coordinates; (*b*) Natural coordinates.

quadrilateral element is a square with sides extending to $\xi = \pm 1$, $\eta = \pm 1$. The local and global coordinates are related by the following expressions:

$$x = \frac{1}{4}[(1-\xi)(1-\eta)x_1 + (1+\xi)(1-\eta)x_2 + (1+\xi)(1+\eta)x_3 + (1-\xi)(1+\eta)x_4]$$

$$y = \frac{1}{4}[(1-\xi)(1-\eta)y_1 + (1+\xi)(1-\eta)y_2 + (1+\xi)(1+\eta)y_3 + (1-\xi)(1+\eta)y_4]$$

Rather than trying to solve the preceding two expressions for ξ and η in terms of x and y, the nodal coordinates then proceed as before, using numerical methods to carry out differentiation and integration. We will postpone discussion of numerical procedures until we consider the isoparametric element concepts later in this chapter.

Natural Coordinates in Three Dimensions

We can define natural coordinates for the four-node tetrahedron in a manner that is analogous to the procedure for the three-node triangle. The result, as the reader may expect, is a set of *volume* coordinates whose physical interpretation turns out to be a ratio of volumes in the tetrahedron. Figure 1.15 shows a typical element and defines the node numbering scheme.

The global Cartesian coordinates and the local natural coordinates are related as follows:

$$x = L_1x_1 + L_2x_2 + L_3x_3 + L_4y_4$$

$$y = L_1y_1 + L_2y_2 + L_3y_3 + L_4y_4$$

$$z = L_1z_1 + L_2z_2 + L_3z_3 + L_4z_4$$

$$L_1 + L_2 + L_3 + L_4 = 1$$

These equations can be inverted to give

$$L_1 = \frac{1}{6V}(a_i + b_ix + c_iy + d_iz) \qquad i = 1, 2, 3 \text{ and } 4$$

$$6V = \begin{vmatrix} 1 & x_1 & y_1 & z_1 \\ 1 & x_2 & y_2 & z_2 \\ 1 & x_3 & y_3 & z_3 \\ 1 & x_4 & y_4 & z_4 \end{vmatrix}$$

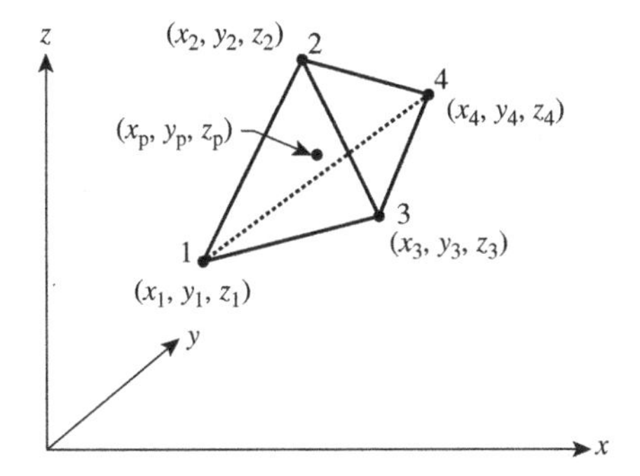

Figure 1.15. A tetrahedral element whose nodes are numbered according to the right-hand rule.

which is equal to six times the tetrahedron volume defined by nodes 1, 2, 3, and 4. Besides,

$$a_1 = \begin{vmatrix} x_2 & y_2 & z_2 \\ x_3 & y_3 & z_3 \\ x_4 & y_4 & z_4 \end{vmatrix}$$

$$b_1 = -\begin{vmatrix} 1 & y_2 & z_2 \\ 1 & y_3 & z_3 \\ 1 & y_4 & z_4 \end{vmatrix}$$

$$c_1 = -\begin{vmatrix} x_2 & 1 & z_2 \\ x_3 & 1 & z_3 \\ x_4 & 1 & z_4 \end{vmatrix}$$

$$d_1 = -\begin{vmatrix} x_2 & y_2 & 1 \\ x_3 & y_3 & 1 \\ x_4 & y_4 & 1 \end{vmatrix}$$

The other constants are obtained through a cyclic permutation of subscripts 1, 2, 3, and 4. Because the constants are the cofactors of the determinant in the expression for $6V$, attention must be given to the appropriate sign. If the tetrahedron is defined in a right-handed Cartesian coordinate system, the preceding set of expressions is valid only when the nodes are numbered so that nodes 1, 2, and 3 are ordered counterclockwise when viewed from node 4.

Figure 1.16 illustrates the physical interpretation of natural coordinates for a tetrahedron. The appropriate differentiation formulas are as follows:

$$\frac{\partial \phi}{\partial x} = \sum_{i=1}^{4} \frac{\partial \phi}{\partial L_i} \frac{\partial L_i}{\partial x}$$

$$\frac{\partial \phi}{\partial y} = \sum_{i=1}^{4} \frac{\partial \phi}{\partial L_i} \frac{\partial L_i}{\partial y}$$

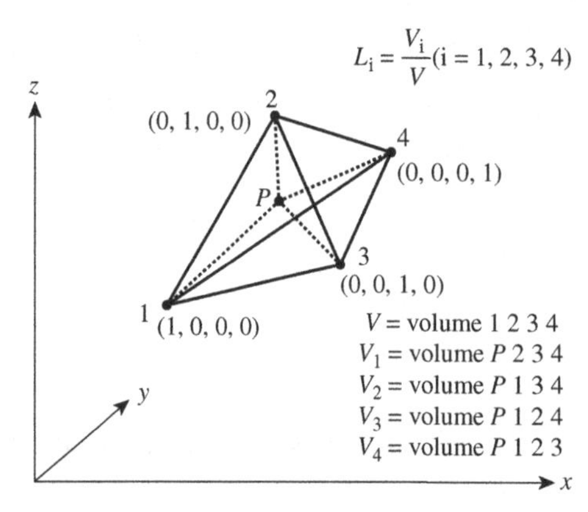

Figure 1.16. Volume coordinates.

$$\frac{\partial \phi}{\partial z} = \sum_{i=1}^{L} \frac{\partial \phi}{\partial L_i} \frac{\partial L_i}{\partial z}$$

where

$$\frac{\partial L_i}{\partial x} = \frac{b_i}{6V}$$

$$\frac{\partial L_i}{\partial y} = \frac{c_i}{6V}$$

$$\frac{\partial L_i}{\partial z} = \frac{d_i}{6V}$$

Natural coordinates can also be established for general hexahedral elements in three dimensions. For the eight-node hexahedron in Figure 1.17, the equations relating the Cartesian to the natural coordinates are

$$x = \sum_{i=1}^{8} x_i L_i$$

$$y = \sum_{i=1}^{8} y_i L_i$$

$$z = \sum_{i=1}^{8} z_i L_i$$

where $$L_i = \frac{1}{8}(1 + \zeta \zeta_1)(1 + \eta \eta_1)(1 + \xi \xi_i) \qquad i = 1, 2, \ldots, 8$$

Again, inversion of the preceding two relationships is not possible, so we must use numerical methods to carry out differentiation and integration. This will be clearer next.

Curve-Sided Isoparametric Elements

Fitting a curved boundary with straight-sided elements often leads to a satisfactory representation of the boundary, but better fitting would be possible if curve-sided

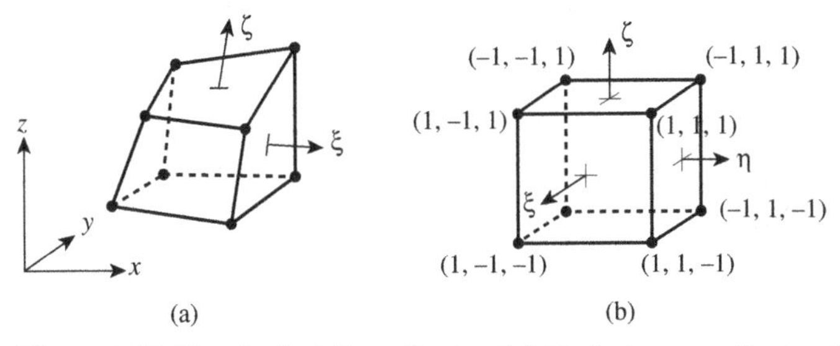

Figure 1.17. Hexahedral Coordinates: (*a*) Cartesian coordinates; (*b*) Natural coordinates.

elements could be formulated for the task. If curve-sided elements were available, it would be permissible to use a smaller number of larger elements and still achieve a close boundary representation. Also, in practical three-dimensional problems, where the great number of degrees of freedom can overburden even the largest computers, it is sometimes essential to have a means of reducing the problem size by using fewer elements.

Coordinate Transformation

The essential idea underlying the development of curve-sided elements centers around mapping or transforming simple geometric shapes in some local coordinate system into distorted shapes in the global Caretisian coordinate system and then evaluating the element equations for the curve-sided elements that come out. An example will help to clarify the concepts. For the purpose of discussion, we will restrict our example to two dimensions, but all concepts extend immediately to one and three dimensions.

Suppose that we wish to represent a solution domain in the x-y Cartesian coordinates by a network of curve-sided quadrilateral elements, and furthermore, we desire the field variable ϕ to have a quadratic variation within each element. According to our previous discussion, if we choose the nodal values of ϕ as degrees of freedom, three nodes must be associated with each side of the element. The solution domain and the desired finite element model might appear as shown in Figure 1.18. To construct one typical element of this assemblage, we focus our attention on the simpler "parent" element in the $\xi - \eta$ local coordinate system shown in Figure 1.19. This element is the second member of the so-called serendipity family of rectangular elements, and the quadratic variation of ϕ within the element may be expressed as follows:

$$\phi(\zeta,\eta) = \sum_{i=1}^{8} N_i(\zeta,\eta)\phi_i$$

where N_i are the serendipity functions given as follows:

For nodes at $\zeta \pm 1$ and $\eta \pm 1$,

$$N_i(\zeta,\eta) = \frac{1}{4}(1+\xi\xi_i)(1+\eta\eta_i)(\zeta\zeta_i+\eta\eta_i-1)$$

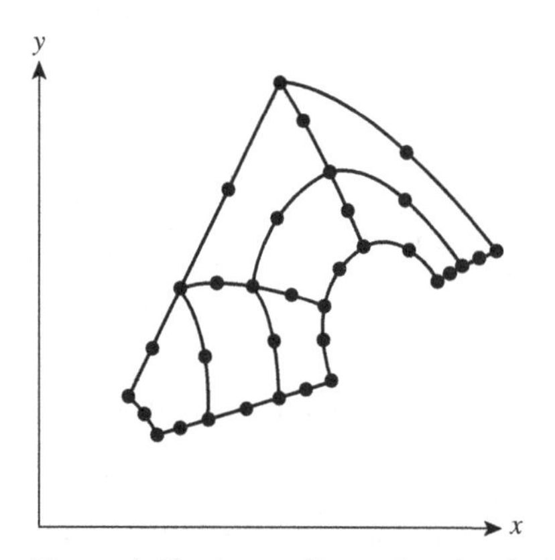

Figure 1.18. A two-dimensional solution domain represented by curved quadrilateral elements.

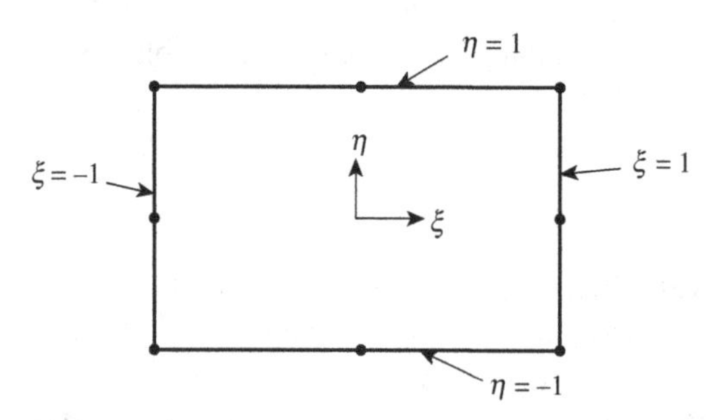

Figure 1.19. Parent rectangular element in local coordinates.

For nodes at $\zeta = 0$ and $\eta \pm 1$,

$$N_i(\zeta, \eta) = \frac{1}{2}(1 - \xi^2)(1 + \eta\eta_i)$$

For nodes at $\zeta \pm 1$ and $\eta = 0$,

$$N_i(\zeta, \eta) = \frac{1}{2}(1 + \xi\xi_1)(1 - \eta^2)$$

The nodes in the ξ-η plane may be mapped into corresponding nodes in the x-y plane by defining the following two relationships:

$$x = \sum_{i=1}^{8} F_i(\xi, \eta)x_i$$

$$y = \sum_{i=1}^{8} F_i(\xi, \eta)y_i$$

Extension of this mapping procedure to elements with a different number of nodes and elements in other dimensions is obvious.

For this example, the mapping functions F_i must be quadratic because the curved boundaries of the element in the x-y plane need three points for their unique specification, and the F_i should take on the proper values of unity and zero when

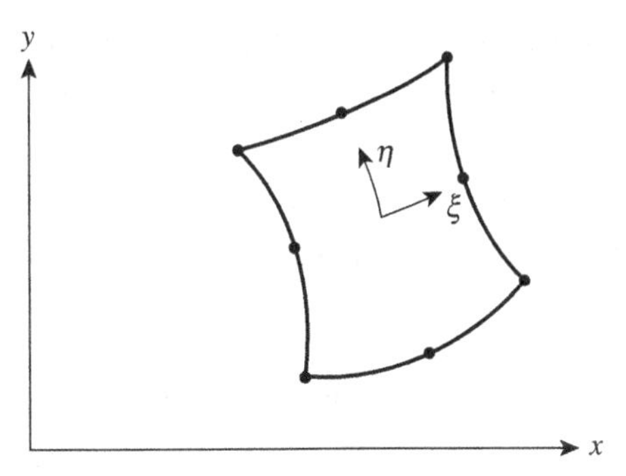

Figure 1.20. A curve-sided quadrilateral element resulting from mapping the rectangular parent element.

evaluated at the nodes in the ξ-η plane. Functions meeting these requirements are precisely the quadratic serendipity interpolation functions presented earlier. Hence we can write

$$x = \sum_{i=1}^{8} N_i(\zeta, \eta) x_i$$

$$y = \sum_{i=1}^{8} N_i(\zeta, \eta) y_i$$

where the N_i's have already been defined. The mapping defined in the preceding x and y expressions results in a curve-sided quadratic element of the type shown in Figure 1.20. For this particular element, the functional representation of the field variable and the functional representation of its curved boundaries are expressed by interpolation functions of the same order. Curve-sided elements formulated in this way are called *isoparametric elements*. Different terminology is used to describe the curve-sided elements whose geometry and field-variable representations are described by polynomials of different order. The number of nodes used to define a curved element may be different from the number at which the element's degrees of freedom are specified. In contrast to isoparametric elements, we define *subparametric elements* as those whose geometry is described by a polynomial of an order that is lower than that used for the field variable and *superparametric elements* as those whose geometry is described by a higher-order polynomial. Of the three categories of curve-sided elements, isoparametric elements are the most commonly used, but some forms of the other two are employed, for instance, in the stress analysis of shell structures. In this discussion, we will confine our attention to isoparametric elements.

When writing equations such as that used to define both x and y earlier, we assume that indeed the transformation between the local ξ-η coordinates and the global x-y coordinates is unique; that is, we assume that each point in one system has a corresponding point in the other system. If the transformation is nonunique, then we can expect violent and undesirable distortions in the x-y system that may fold the curved element back on itself.

An important consideration in the construction of curve-sided elements is the preservation of continuity conditions in the global coordinate system. In this regard, Zienkiewicz [2] advanced two useful guidelines:

- If two adjacent curve-sided elements are generated from parent elements whose interpolation functions satisfy interelement continuity, these curved elements will be continuous.
- If the interpolation functions are given in the local coordinate system, and they ensure the continuity of ϕ in the parent element, then ϕ will also be continuous in the curve-sided element.

Evaluation of Elemental Matrices

Having described the element shape, we face the task of evaluating the element equations by carrying out the usual integrations appearing in them. In general, the terms in the element equations will contain integrals of the following form:

$$\int_{A^{(e)}} f\left(\phi, \frac{\partial \phi}{\partial x}, \frac{\partial \phi}{\partial y}\right) dx\, dy$$

where $A^{(e)}$ is the area of the curve-sided element. Because ϕ is expressed as a function of the local coordinates ξ and η, it is necessary to express $\partial\phi/\partial x$, $\partial\phi/\partial y$, and $dx\, dy$ in terms of ξ and η as well. This can be done as follows:

$$\frac{\partial \phi}{\partial x} = \sum_{i=1}^{8} \frac{\partial N_i}{\partial x} \phi_i$$

$$\frac{\partial \phi}{\partial y} = \sum_{i=1}^{8} \frac{\partial N_i}{\partial y} \phi_i$$

Hence we must express $\partial N_i/\partial x$ and $\partial N_i/\partial y$ in terms of ζ and η. Because of the inverse form of the expressions for x and y (earlier), we write, by the chain rule of differentiation,

$$\frac{\partial N_i}{\partial \xi} = \frac{\partial N_i}{\partial x}\frac{\partial x}{\partial \xi} + \frac{\partial N_i}{\partial y}\frac{\partial y}{\partial \xi}$$

$$\frac{\partial N_i}{\partial \eta} = \frac{\partial N_i}{\partial x}\frac{\partial x}{\partial \eta} + \frac{\partial N_i}{\partial y}\frac{\partial y}{\partial \eta}$$

or

$$\begin{Bmatrix} \dfrac{\partial N_i}{\partial \xi} \\ \dfrac{\partial N_i}{\partial \eta} \end{Bmatrix} = \begin{bmatrix} \dfrac{\partial x}{\partial \xi} & \dfrac{\partial y}{\partial \xi} \\ \dfrac{\partial x}{\partial \eta} & \dfrac{\partial y}{\partial \eta} \end{bmatrix} \begin{Bmatrix} \dfrac{\partial N_i}{\partial x} \\ \dfrac{\partial N_i}{\partial y} \end{Bmatrix} = [J] \begin{Bmatrix} \dfrac{\partial N_i}{\partial x} \\ \dfrac{\partial N_i}{\partial y} \end{Bmatrix}$$

where $[J]$ is the Jacobian matrix,

$$[J] = \begin{bmatrix} \dfrac{\partial x}{\partial \xi} & \dfrac{\partial y}{\partial \xi} \\ \dfrac{\partial x}{\partial \eta} & \dfrac{\partial y}{\partial \eta} \end{bmatrix}$$

We evaluate an element Jacobian matrix using the coordinate-transformation equation (earlier). For the eight-node quadratic element,

$$[J(\xi,\eta)] = \begin{bmatrix} \sum_{i=1}^{8} \frac{\partial N_i}{\partial \xi}(\xi,\eta)x_i & \sum_{i=1}^{8} \frac{\partial N_i}{\partial \xi}(\xi,\eta)y_i \\ \sum_{i=1}^{8} \frac{\partial N_i}{\partial \eta}(\xi,\eta)x_i & \sum_{i=1}^{8} \frac{\partial N_i}{\partial \eta}(\xi,\eta)y_i \end{bmatrix}$$

To derive the desired derivative of ϕ, we must invert the equations defining $\partial\phi/\partial x$ and $\partial\phi/\partial y$ (earlier), and this involves finding the inverse of the Jacobian matrix as follows:

$$\begin{Bmatrix} \frac{\partial N_i}{\partial x} \\ \frac{\partial N_i}{\partial y} \end{Bmatrix} = [J]^{-1} \begin{Bmatrix} \frac{\partial N_i}{\partial \xi} \\ \frac{\partial N_i}{\partial \eta} \end{Bmatrix}$$

where i varies between 1 and 8

Now,

$$\begin{Bmatrix} \frac{\partial \phi}{\partial x} \\ \frac{\partial \phi}{\partial y} \end{Bmatrix} = \begin{bmatrix} \frac{\partial N_1}{\partial x} & \frac{\partial N_2}{\partial x} & \cdots & \frac{\partial N_8}{\partial x} \\ \frac{\partial N_1}{\partial y} & \frac{\partial N_2}{\partial y} & \cdots & \frac{\partial N_8}{\partial y} \end{bmatrix} \begin{Bmatrix} \phi_1 \\ \phi_2 \\ \cdot \\ \cdot \\ \cdot \\ \phi_8 \end{Bmatrix}$$

Thus

$$\begin{Bmatrix} \frac{\partial \phi}{\partial x} \\ \frac{\partial \phi}{\partial y} \end{Bmatrix} = [J]^{-1} \begin{bmatrix} \frac{\partial N_1}{\partial \xi} & \frac{\partial N_2}{\partial \xi} & \cdots & \frac{\partial N_8}{\partial \xi} \\ \frac{\partial N_1}{\partial \eta} & \frac{\partial N_2}{\partial \eta} & \cdots & \frac{\partial N_8}{\partial \eta} \end{bmatrix} \begin{Bmatrix} \phi_1 \\ \phi_2 \\ \cdot \\ \cdot \\ \cdot \\ \phi_8 \end{Bmatrix}$$

To complete the evaluation of the integral, we need to express the element of the incremental area $dx\,dy$ in terms of $d\zeta\ d\eta$. Texts on advanced calculus show that

$$dx\,dy = |J|\,d\xi\ d\eta$$

where $|J|$ is the determinant of $[J]$. With these transformations, integrals such as $\int_{A^e} f(\phi, \partial\phi/\partial x, \partial\phi/\partial y)dx\,dy$ reduce to the following form:

$$\int_{-1}^{+1}\int_{-1}^{+1} f'(\xi\eta)d\xi\ d\eta$$

where f' is the transformed function f.

Although the integration limits are now those of the simple parent element, the transformed integrand f' is not a simple function that permits closed-form integration. For this reason, it is necessary to resort to numerical integration, and this poses no particular difficulty. Always associated with the procedure of numerical integration is the question of how accurately the integration needs to be done to ensure convergence and to guard against making the resulting system matrix singular.

Irons [3, 4] provides the following guideline for isoparametric element formulation: "Convergence of the finite element process should occur if the numerical integration is accurate enough to evaluate the area or volume of the curve-sided element exactly."

The sample isoparametric element we have considered is just one of many possibilities. Actually, we can start with the basic elements in any dimension, elements that may be described by local coordinates ζ, η, and ξ or natural coordinates L_1, L_2, L_3, and so on, and transform these into curve-sided elements. When the parent element such as a triangle or a tetrahedron is expressed in terms of natural coordinates, it is necessary to express one of the L's in terms of the others because not all the natural coordinates are independent. Also, in this case, the limits of integration must be changed to correspond to the boundaries of the element. In any event, numerical integration is still necessary.

REFERENCES

[1] Gallagher, R. H., "Analysis of Plate and Shell Structures," *Proceedings of Symposium on Application of Finite Element Methods in Civil Engineering*, ASCE–Vanderbilt University, Nashville, TN, November 1969.

[2] Z. Zienkiewicz, O. C., *The Finite-Element Method*, McGraw-Hill, New York, 1974.

[3] Irons, B. M., "Engineering Application of Numerical Integration in Stiffness Method, *AIAA Journal*, Vol. 14, 1966, pp. 2035–7.

[4] Egatoudis, J., Irons, B, and Zienkiewicz, O. C., "Curved Isoparametric Quadrilateral Elements for Finite Element Analysis, *Int. J. Solids Struct.*, Vol. 4, 1968, pp. 31–42.

2 Some Methods for Solving Continuum Problems

Overview

There are many approaches to the solution of linear and nonlinear boundary-value problems, and they range from completely analytical to completely numerical. Of these, the following deserve attention:

- Direct integration (exact solution):
  - Separation of variables
  - Similarity solutions
  - Fourier and Laplace transformations
- Approximate solutions
  - Perturbation
  - Power series
  - Probability schemes
  - Method of weighted residuals (MWR)
  - Finite difference techniques
  - Ritz method
  - Finite element method

For a few problems, it is possible to obtain an exact solution by direct integration of the governing differential equation. This is accomplished, sometimes, by an obvious separation of variables or by applying a transformation that makes the variables separable and leads to a similarity solution. Occasionally, a Fourier or Laplace transformation of the differential equation leads to an exact solution. However, the number of problems with exact solutions is severely limited, and most of these have already been solved.

Because regular and singular perturbation methods are primarily applicable when the nonlinear terms in the equation are small in relation to the linear terms, their usefulness is limited. The power-series method is powerful and has been employed with some success, but because the method requires generation of a coefficient for each term in the series, it is relatively tedious. It is also difficult, if not impossible, to demonstrate that the series converges.

The probability schemes, usually classified under the heading of *Monte Carlo methods*, are used for obtaining a statistical estimate of a desired quantity by random sampling. These methods work best when the desired quantity is a statistical parameter, and sampling is done from a selective population.

With the advent of high-speed digital computers, it appears that the three currently outstanding methods for obtaining approximate solutions of high accuracy are the method of weighted residuals, the finite-difference method, and the finite element method. These methods are related, as we will see, and in some cases the finite-difference and finite element methods can be shown to be special cases of the method of weighted residuals.

The Ritz Method

This method consists of *assuming* the form of the unknown solution in terms of known functions (trial functions) with unknown adjustable parameters (the trial functions are sometimes called *coordinate functions*). From the family of trial functions, we select the function that renders a specific functional stationary. The procedure is to substitute the trial functions into the functional, thereby expressing the functional in terms of the adjustable parameters. The functional is then differentiated with respect to each parameter, and the resulting equation is set equal to zero. If there are n unknown parameters, there will be n simultaneous equations to be solved for these parameters. By this means, the approximate solution is chosen from the family of assumed solutions.

The procedure does nothing more than give us the *best* solution from the family of assumed solutions. Clearly, then, the accuracy of the approximate solution depends on the choice of trial functions. We require that the trial functions be defined over the whole solution domain and that they satisfy, at least, some and usually all of the boundary conditions. Sometimes, if we know the general nature of the desired solution, we can improve the approximation by choosing the trial functions to reflect this very nature. Generally, the approximation improves as the family of trial functions and the number of adjustable parameters increase. If the trial functions are part of an infinite set of functions that are capable of representing the unknown function to any degree of accuracy, the process of including more and more trial functions leads to a series of approximate solutions that converges to the true solution. Often a family of trial functions is constructed from polynomials of successively increasing degree, but in certain cases, other kinds of functions may offer advantages.

Example: The Ritz Method

To illustrate the Ritz method, we will consider a simple example. Suppose that we want the function $\phi(x)$ to satisfy the following partial differential equation:

$$\frac{\partial^2 \phi}{\partial x^2} = -f(x)$$

with the following boundary conditions:

$$\phi(a) = A$$
$$\phi(b) = B$$

We assume that $f(x)$ is a continuous function in the closed interval $[a,b]$. This problem is equivalent to finding the function $\phi(x)$ that minimizes the following functional:

$$J(\phi) = \int_a^b [\frac{1}{2}\frac{d\phi^2}{dx} - f(x)\phi(x)]\,dx$$

We will ignore the fact that this problem has an exact solution and proceed to find an approximate solution. According to the Ritz method, we assume that the desired solution can be approximately represented in $[a,b]$ by a combination of selected trial functions of the form

$$\phi(x) = C_1\psi_1(x) + C_2\psi_2(x) + \cdots + C_n\psi_n(x) \qquad a<b$$

where the n constants and C_i are adjustable parameters to be determined. The trial functions should be selected so that the expression $\phi(x)$ satisfies the boundary conditions regardless of the choice of the constants C_i. Using polynomials is a simple and convenient way to construct the trial functions. Thus, if $A = B = 0$, for example, we can write

$$\phi(x) \approx (x-a)(x-b)(C_1 + C_2 x + C_3 x^2 + \cdots + C_n x^{n-1})$$

as a possible series of trial functions. When we substitute this approximate expression for $\phi(x)$ into the functional to be minimized, we obtain, after carrying out the integration, the following result:

$$J = J(C_1, C_2, \cdots, C_n)$$

Now we require that the constants C_i be so chosen to minimize J. Hence, from differential calculus, we have

$$\frac{\partial J}{\partial C_i} = 0 \qquad i \leq 1 \leq n$$

When these n equations are solved for the n parameters C_i, the approximate solution to $\phi^n = -f$ is obtained, as shown in Figure 2.1. The accuracy of the approximate

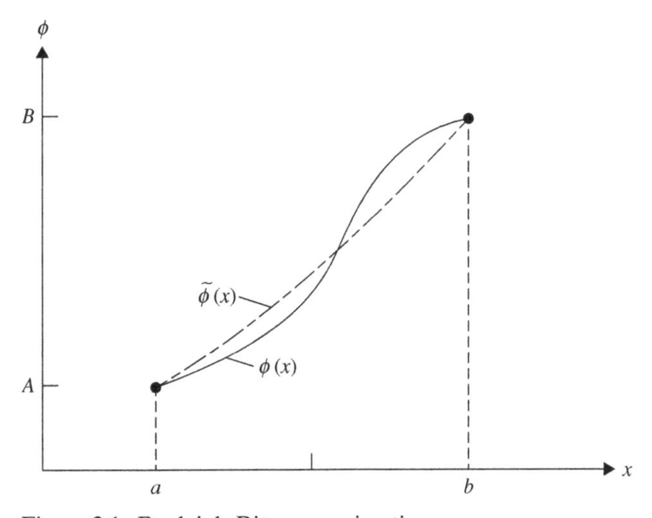

Figure 2.1. Rayleigh-Ritz approximation.

solution depends on the number of C's used in the trial function. Generally, as n increases, the accuracy improves. To assess the improvement in accuracy as more C's are used, we can repeatedly solve the problem by taking successively more terms in the approximation; that is, we can use

$$\phi_1(x) \approx (x-a)(x-b)C_1$$

$$\phi_2(x) \equiv (x-a)(x-b)(C_1 + C_2 x)$$

$$\phi_3(x) \approx (x-a)(x-b)(C_1 + C_2 x + C_3 x^2)$$

and so on. By comparing the results at the end of each calculation, we can estimate the effect of adding more terms on accuracy.

The Finite Element Method: Relation to the Ritz Method

The finite element method and the Ritz method are essentially equivalent. Each method uses a set of trial functions as the starting point for obtaining an approximate solution, and both methods take a linear combination of trial functions that makes a given functional stationary. The major difference between these two methods is that the assumed trial functions in the finite element method are not defined over the entire solution domain, and they have to satisfy no boundary conditions but only certain continuity conditions and then only sometimes. Because the Ritz method uses functions defined over the whole domain, it can only be used for domains of relatively simple geometric shapes. In the finite element method, the same geometric limitations exist, but only over the elements. Because elements with simple shapes can be assembled to represent exceedingly complex geometries, the finite element method is far more versatile than the Ritz method. From a strictly mathematical standpoint, the finite element method is a special case of the Ritz method only when the piecewise trial functions abide by certain continuity and completeness conditions. These are discussed separately in Chapter 4.

3 Variational Approach

The basic idea of the finite element method is to divide the solution domain into a finite number of subdomains that are termed *elements*. These elements are connected only at nodal points in the domain and on the element boundaries. In this way, the solution domain is discretized and represented as a patchwork of elements. Frequently, the finite element boundaries are straight lines or planes, so if the solution domain has curved boundaries, these are approximated by a series of straight, flat segments.

The mathematical interpretation of the finite element method requires us to generalize our definition of an element and to think of elements in less physical terms. Instead of viewing an element as a physical part of the system, we view it as part of the solution domain where the phenomena of interest are occurring. We imagine the solution domain to be sectioned by lines (or general planes in n dimensions) that define the boundaries of an element. The elements are interconnected only at imaginary nodal points on the boundaries or surfaces of the elements. For solid-mechanics problems, we no longer need to imagine that the elements deform or change shape; rather, we define them as regions of space where a displacement field exists. The nodes of an element are then simply located in space where the displacement and, possibly, its derivatives are known or sought. Similarly, for fluid mechanics problems, the elements are regions over which a pressure field exists and through which the fluid is flowing. The mathematical interpretation of a finite element mesh is that it is a spatial subdivision rather than a material subdivision. The broader interpretation of an element allows us to carry over many of the basic ideas from one problem area to another.

In the finite element procedure, once the element mesh for the solution domain has been decided, the behavior of the unknown field variable over each element is approximated by continuous functions expressed in terms of the nodal values of the field variable and, sometimes, the nodal values of its derivatives up to a certain order. The functions defined over each finite element are collectively called *interpolation functions*, *shape functions*, or *field-variable models*. The collection of interpolation functions for the whole solution domain provides a piecewise approximation to the field variable.

Example of Piecewise Approximation

To illustrate the nature of this piecewise approximation, we consider the representation of a two-dimensional field variable $\phi(x, y)$. We will show that the nodal values of ϕ can uniquely and continuously define $\phi(x, y)$ throughout the domain of interest in the $x = y$ plane, and we will introduce the notation for an interpolation function.

Suppose that we have the domain shown in Figure 3.1, and we section it into triangular elements with nodes at the vertices of the triangle. With this type of domain discretization, we can allow ϕ to vary linearly over each element (Figure 3.2). The plane passing through the three nodal values of ϕ associated with element (e) is described as follows:

$$\phi^{(e)}(x, y) = \beta_1{}^{(e)} + \beta_2{}^{(e)}x + \beta_3{}^{(e)}y$$

We can express the constants $\beta_1{}^{(e)}$, $\beta_2{}^{(e)}$, and $\beta_3{}^{(e)}$ in terms of the coordinates of the element nodes and the nodal values of ϕ by evaluating the preceding expression at

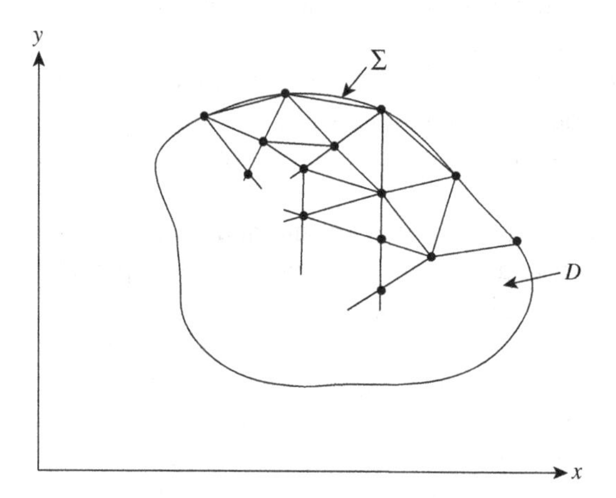

Figure 3.1. Two-dimensional domain divided into triangular elements.

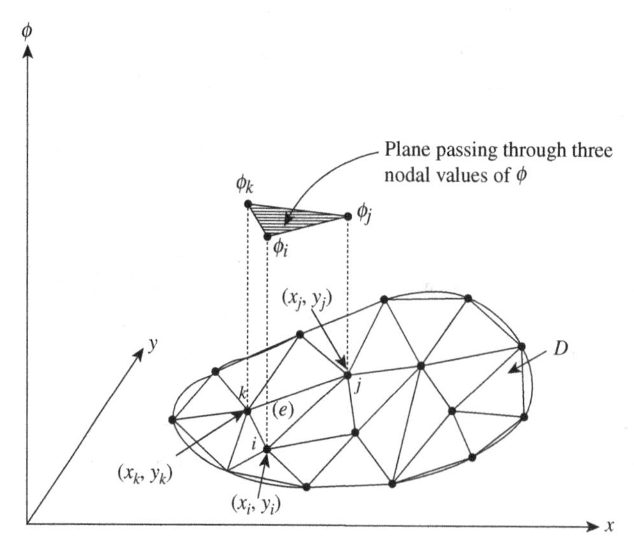

Figure 3.2. Subdivided domain D and a piecewise linear solution surface.

each node as follows:

$$\phi_1 = \beta_1{}^{(e)} + \beta_2{}^{(e)} x_i + \beta_3{}^{(e)} y_i$$
$$\phi_2 = \beta_1{}^{(e)} + \beta_2{}^{(e)} x_j + \beta_3{}^{(e)} y_j$$
$$\phi_3 = \beta_1{}^{(e)} + \beta_2{}^{(e)} x_k + \beta_3{}^{(e)} y_k$$

Solving these three equations yields:

$$\beta_1{}^{(e)} = \frac{\phi_i(x_j y_k - y_j x_k) + \phi_j(x_k y_i - x_i y_k) + \phi_k(x_i y_j - x_j y_i)}{2\Delta}$$
$$\beta_2{}^{(e)} = \frac{\phi_i(y_j - y_k) + \phi_j(y_k - y_i) + \phi_k(y_i - y_j)}{2\Delta}$$
$$\beta_3{}^{(e)} = \frac{\phi_i(x_k - x_j) + \phi_j(x_i - x_k) + \phi_k(x_j - x_i)}{2\Delta}$$

where $$\Delta = (x_j y_k - x_k y_j) + (x_k y_i - x_i y_k) + (x_i y_j - x_j y_i)$$

The preceding is the area of the triangle whose vertices are i, j, and k.

Substituting and rearranging terms, we get

$$\phi^{(e)}(x,y) = \frac{a_i + b_i x + c_i y}{2\Delta}\phi_i + \frac{a_j + b_j x + c_j y}{2\Delta}\phi_j + \frac{a_k + b_k x + c_k y}{2\Delta}\phi_k$$

where
$$a_i = x_j y_k - x_k y_j$$
$$b_i = y_j - y_k$$
$$c_i = x_k - x_j$$

and the other coefficients are obtained through a cyclic permutation of the subscripts i, j, and k. In general, the functions $N^{(e)}$ are called *shape functions* or *interpolation functions*, and they play a most important role in all finite element analyses. In matrix notation, we can write

$$\phi^{(e)}(x,y) = N_i\phi_i + N_j\phi_j + N_k\phi_k$$

If the domain contains M elements, the complete representation of the field variable over the whole domain is given as follows:

$$\phi(x,y) = \sum_{e=1}^{M} \phi^{(e)}(x,y) = \sum_{e=1}^{M} [N^{(e)}]\{\phi^{(e)}\}$$

Examination of this equation reveals that if the nodal values of ϕ are known, we can represent the whole solution surface $\phi(x,y)$ as a series of interconnected triangular planes. This many-faceted surface has no discontinuities or *gaps* at interelement boundaries because the values of ϕ at any two nodes defining an element boundary uniquely determine the linear variation of ϕ along that boundary.

Although we obtained these equations for a particular interpolation function (linear) and a particular element type (three-noded linear triangle), these equations have general validity. For more complex interpolation functions and more element types, such as those discussed later in this text, the form of the preceding two expressions remains the same; only the number of terms in the row and column matrices will be different. Hence, if a solution domain is subdivided into elements, we can represent the unknown field variable in each element as follows:

$$\phi^{(e)} = [N^{(e)}]\{\phi\}$$

where $[N^{(e)}]$ is the row vector of interpolation functions that are functions of the nodal coordinates, and $\{\phi\}^{(e)}$ is the column vector that is the collection of r discrete values consisting of nodal values of ϕ that are associated with the element and, perhaps, some other parameters that characterize the element and those are not identified with any node. Such *nodeless* variables may appear when constraints are imposed on the field variable or when parameters are assigned to an element (or a group of elements) via modes of interpolation functions that vanish on the element boundaries.

We will now assume that we have the field variable ϕ completely represented in the solution domain in terms of a collection of nodal values of ϕ. Under this assumption, if these discrete values are known, then the problem of finding an approximation of ϕ is solved.

Elemental Equations from a Variational Principle

The finite element solution here involves picking the nodal values of ϕ so as to make the functional $I(\phi)$ stationary. This functional is unique to each problem and will be discussed in detail throughout the remainder of the first (heat transfer) part of this text. To achieve this with respect to the nodal values of ϕ, we assume that

$$\delta I(\phi) = \sum_{i=1}^{n} \frac{\partial I}{\partial \phi_i} \delta\phi_i = 0$$

where n is the nodal number of discrete values of ϕ assigned to the solution domain. Because the $\delta\phi_i$'s are independent, the preceding equation can hold only if

$$\frac{\partial I}{\partial \phi_i} = 0 \qquad i = 1,2,3,\ldots,n$$

If the interpolation functions giving our piecewise approximation of ϕ obey certain continuity and compatibility conditions, which are discussed in Chapter 4, the functional $I(\phi)$ may be written as a sum of individual functionals defined for all elements of the assemblage, that is,

$$I(\phi) = \sum_{e=1}^{M} I^{(e)}\left(\phi^{(e)}\right)$$

where M is the total number of elements and the superscript (e) denotes a typical element. Hence, instead of working with the functional defined over the whole solution region, we may focus our attention on the functionals defined for the individual elements. We now have the following relationship:

$$\delta I = \sum_{e=1}^{M} \delta I^{(e)} = 0$$

where the variation of $I^{(e)}$ is taken only with respect to the nodal values associated with element (e). The last equation implies that

$$\left\{ \frac{\partial I^{(e)}}{\partial \phi} \right\} = \frac{\partial I^{(e)}}{\partial \phi_j} \qquad j = 1, 2, \ldots, r$$

where r is the number of nodes assigned to the typical element (e). The last set of equations comprise a set of r equations that characterize the behavior of element (e). The fact that we can represent the functional for the assemblage of elements as the sum of the functionals for all individual elements provides the key to formulating individual element equations from a variational principle. If, for example, the governing differential equations and boundary conditions for a problem are linear and self-adjoint, the corresponding variational statement of the problem involves a quadratic functional. When ϕ is quadratic, $I^{(e)}(\phi)$ is also quadratic, and the last equation for element (e) can always be written as follows:

$$\left\{ \frac{\partial I^{(e)}}{\partial \phi} \right\} = [K]^{(e)}\{\phi\}^{(e)} - \{F\}^{(e)} = \{0\}$$

where $[K]$ is a square matrix of "stiffness" coefficients, $\{\phi\}^{(e)}$ is the column vector of nodal values, and $\{F\}$ is the vector of resulting nodal actions (load vector). The complete set of finite element equations for the problem is assembled by adding all the derivatives of I for all elements. Symbolically, we can write the complete set of equations as follows:

$$\frac{\partial I}{\partial \phi_i} = \sum_{e=1}^{M} \frac{\partial I^{(e)}}{\partial \phi_i} = 0 \qquad i = 1, 2, \ldots, n$$

or

$$\left\{ \frac{\partial I}{\partial \phi} \right\} = \{0\}$$

Our problem is solved when the preceding set of equations is solved simultaneously for the nodal values of ϕ. If there are q nodes in the solution domain where ϕ is specified by boundary conditions, there will be $n - q$ equations to be solved for the $n - q$ unknowns. Note that the summation indicated in the preceding equation contains many zero terms because only elements sharing node i will contribute to $\partial I / \partial \phi$. If node i does not belong to element (e), then $\partial I / \partial \phi_i = 0$. This fact is manifested in the "narrow-band" and "sparseness" properties of the resulting matrix of stiffness coefficients.

We have now established the means for formulating individual finite element equations from a variational principle. The procedure can be summarized as follows: If the functional for a given problem can be expressed as the sum of functionals evaluated for all elements, we may focus our attention on an isolated element without regard for its eventual location in the assemblage. To derive the equations governing the element's behavior, we first use interpolation functions to define the unknown field variable ϕ in terms of its nodal values associated with the element; then we evaluate the functional $I^{(e)}$ by substituting the assumed form for $\phi^{(e)}$ and its derivatives and carry out the integration over the domain defined by the element's boundaries. Finally, we perform the differentiations, and the result is the set of equations defining the element behavior. Because the differentiations are with respect to the discrete nodal values, we only employ calculus, not calculus of variations.

4 Requirements for the Interpolation Functions

Our procedure for formulating the individual element equations from a variational principle and our privilege to assemble these equations to obtain the system's (global) equations rely on the assumption that the interpolation functions satisfy certain requirements. The requirements we place on the choice of interpolation functions stem from the need to ensure that our approximate solution converges to the correct one when we use an increasing number of smaller elements, that is, when we refine the element mesh. Mathematical proofs of convergence assume that the process of mesh refinement occurs in a regular fashion as follows:

- The elements must be made smaller in such a way that every point of the solution domain can always be within an element, regardless of how small the element might be.
- All previous meshes must be contained in the refined meshes.
- The form of interpolation functions must remain unchanged during mesh refinement.

These three conditions are illustrated in Figure 4.1, where a simple two-dimensional solution domain in the form of an equilateral triangle is discretized with an increasing number of three-noded triangles. We note that when elements with straight boundaries are used to model solution domains with curved boundaries, the first two conditions are not satisfied, and rigorous mathematical proofs of convergence may not be obtainable. Despite this limitation, many applications of the finite element method to problems with non–polygonal solution domains yield acceptable engineering solutions.

To generate monotonic convergence in the sense just described, and to make the assembly of individual element equations meaningful, we require that the interpolation functions $N^{(e)}$ in the expressions

$$\phi^{(e)} = [N^{(e)}]\{\phi\}^{(e)} \qquad e = 1, 2, \ldots, M$$

be chosen so that the following general requirements are met:

- At elemental interfaces (boundaries), the field variable ϕ and any of its partial derivatives up to one order less than the highest-order derivative appearing in $I(\phi)$ must be continuous.

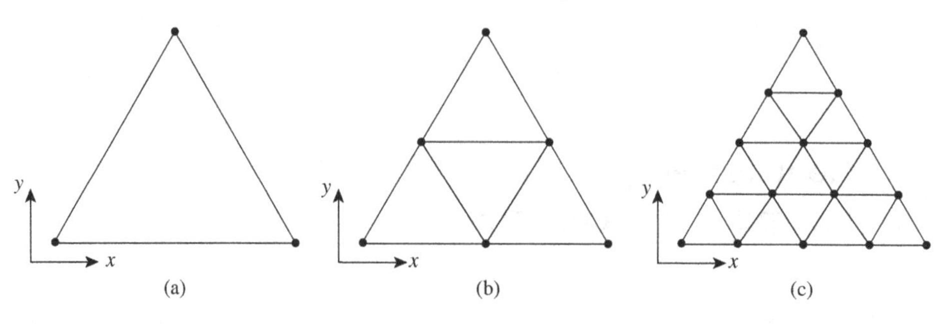

Figure 4.1. Example of successive mesh refinement.

- All uniform states of ϕ and its partial derivatives up to the highest order appearing in $I(\phi)$ should have representation in $\phi^{(e)}$ when, in the limit, the element size shrinks to zero.

The first of these requirements is known as the *compatibility* requirement and the second as the *completeness* requirement. Elements whose interpolation functions satisfy the first requirement are called *compatible* or *conforming* elements, whereas those satisfying the second requirement are termed *complete* elements. The definition of an *incompatible* element is obvious.

The compatibility requirement helps to ensure that the integral $I(\phi)$ is well defined. Without interelement continuity, we cannot be sure that the integral in $I(\phi)$ is unique. Uncertain contributions may arise from the "gaps" between elements. If the compatibility requirement is violated, it is sometimes possible to add special boundary integrals for compensation. It is always desirable, when carrying out a finite element analysis, to be sure that mesh refinement will lead to answers that are converging to the correct solution. For some problems, however, choosing interpolation functions that meet all requirements may be difficult and may involve excessive numerical computations. For this reason, some investigators have ventured to formulate interpolation functions for elements that do not meet all the compatibility and completeness requirements. In some instances, acceptable convergence has been obtained, whereas, in others, convergence to an incorrect solution has occurred. Some applications have shown that the compatibility condition does not always lead to the most rapid convergence. However, when extending the finite element method for use in other areas where far less experience is available, the safest approach is to pick functions and formulate elements that satisfy both requirements.

5 Heat Transfer Applications

Variational Approach

A thorough understanding of the calculus of variations is not necessary to use the finite element method. However, an introduction will be valuable because the starting point of the finite element method has a much different form than we have been exposed to. The calculus of variations provides the bridge between what we are already familiar with and the statement of the problem that is required for the finite element formulation.

A simple problem in the theory of variational calculus seeks to find a function $u(x)$ that minimizes the following integral:

$$I = \int_{x=0}^{L} F\left[x, u(x), \frac{du}{dx}\right] dx \tag{5.1}$$

Observe that F is a function of both $u(x)$ and its derivative. In this simple problem, we will also specify the values of $u(x)$ at $x = 0$ and $x = \text{L}$. That is, we will require that

$$u(o) = u_0 \tag{5.2a}$$

$$u(L) = u_L \tag{5.2b}$$

To find $u(x)$, we are going to consider every possible function that satisfies Equation (5.1). From all these possible functions, we will be seeking the one that gives I its minimum value. This set of possible functions may be represented simply by $\bar{u}(x, \epsilon)$, where

$$\bar{u}(x, \epsilon) = u(x) + \epsilon\eta(x) \tag{5.3}$$

The function $u(x)$ is the desired function that will minimize I, and $\epsilon\eta(x)$ is called a *variation* of this function. The functions $\bar{u}$ and u are shown in Figure 5.1. The function $\eta(x)$ is a completely arbitrary function of x, except that we must insist that

$$\eta_0 = 0 \tag{5.4a}$$

$$\eta(L) = 0 \tag{5.4b}$$

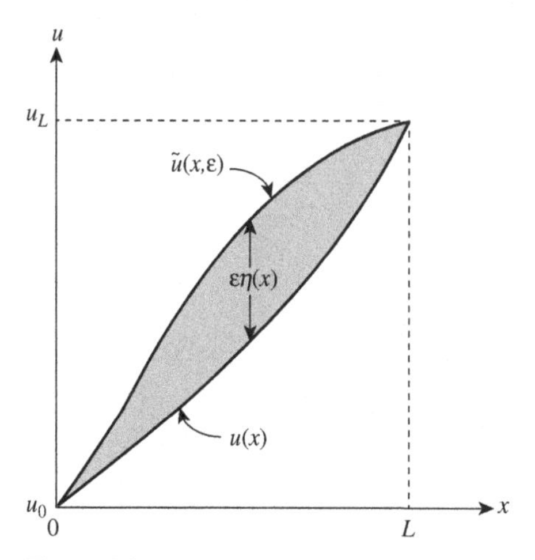

Figure 5.1. Desired variational solution $u(x)$ and a trial function.

We will have the correct values at $x=0$ and $x=L$. That is, we want the two following conditions satisfied:

$$\bar{u}(0,\epsilon) = u(0) \tag{5.5a}$$

$$\bar{u}(L,\epsilon) = u(L) \tag{5.5b}$$

for Equations (5.1) and (5.2) to be satisfied.

Notice that the function we are seeking is included in the complete set of possible functions given by Equation (5.3). The desired function $u(x)$ is the one in the complete set that has $\epsilon = 0$. That is, from Equation (5.3),

$$\bar{u}(x,\epsilon) = u(x) \tag{5.6}$$

We will also need to consider the derivative of $\bar{u}(x,\epsilon)$ with respect to x. This may be obtained by differentiating Equation (5.3) to give

$$\bar{u}(x,\epsilon) = u'(x) + \epsilon\eta'(x) \tag{5.7}$$

Now let us consider the integral that is obtained by replacing $u(x)$ and du/dx in Equation (5.1) with $\bar{u}(x,\epsilon)$. That is, let us consider

$$I(\epsilon) = \int_{x=0}^{L} F[x, \bar{u}(x,\epsilon), \bar{u}'(x,\epsilon)]\, dx \tag{5.8}$$

Observe that this integral is a function of ϵ because it will still appear as a parameter after the integration over x is carried out. Also notice that when $\epsilon = 0$, the integral in Equation (5.8) reduces to the integral we started with [Equation (5.1)] because $\bar{u}(x,\epsilon) = u(x)$ when $\epsilon = 0$, as given by Equation (5.6). This means that we want $I(\epsilon)$ to have a minimum value when $\epsilon = 0$.

To find a minimum of $I(\epsilon)$, we must first differentiate it with respect to ϵ. Because the limits of integration are not functions of ϵ, we may use the Leibnitz rule to write

$$\frac{dI(\epsilon)}{d\epsilon} = \int_{x=0}^{L} \frac{\partial}{\partial\epsilon} F[x, \bar{u}(x,\epsilon)\bar{u}'(x,\epsilon)]\, dx$$

The chain rule of calculus now may be employed to give

$$\frac{dI(\epsilon)}{d\epsilon}=\int_{x=0}^{L}\left[\frac{\partial F}{\partial u}\frac{\partial u}{\partial \epsilon}+\frac{\partial F}{\partial u'}\frac{\partial u'}{\partial \epsilon}\right]dx \tag{5.9}$$

From Equations (5.6) and (5.7), it follows that

$$\begin{aligned}\frac{\partial u}{\partial \epsilon}&=\eta(x)\\ \frac{\partial u'}{\partial \epsilon}&=\eta'(x)=\frac{du(x)}{dx}\end{aligned} \tag{5.9}$$

Therefore, Equation (5.9) may be rewritten as follows:

$$\frac{dI(\epsilon)}{d\epsilon}=\int_{x=0}^{L}\left[\frac{\partial F}{\partial u}\eta(x)+\frac{\partial u'}{du(x)}\right]dx \tag{5.10}$$

The second term on the right-hand side may be integrated by parts to give

$$\frac{dI(\epsilon)}{d\epsilon}=\int_{x=0}^{L}\frac{\partial F}{\partial u}\eta(x)dx+\frac{\partial F}{\partial u}[\eta(x)]_{x=0}^{L}-\int_{x=0}^{L}\eta(x)\frac{d}{dx}\left[\frac{\partial F}{\partial u'}\right]dx \tag{5.11}$$

From Equation (5.4), we see that the integrated term vanishes at both its upper and lower limits. Thus, on recombining the two integrals, Equation (5.11) reduces to

$$\frac{dI(\epsilon)}{d\epsilon}_{\epsilon=0}=\int_{x=0}^{L}\eta(x)\left[\frac{\partial F}{\partial u}-\frac{d}{dx}\left(\frac{\partial F}{\partial u'}\right)\right]dx$$

We will now insist that this be equal to zero when $\epsilon=0$ so that $I(0)$ will attain its minimum value. Thus

$$\frac{dI}{d\epsilon}_{\epsilon=0}=\int_{x=0}^{L}\eta(x)\left[\frac{\partial F}{\partial u}-\frac{d}{dx}\left(\frac{\partial F}{\partial u'}\right)\right]dx=0$$

Observe, from Equation (5.3), that $\bar{u}$ has become u because ϵ has been set equal to zero. Therefore, for I to be a minimum, the following condition must be satisfied:

$$\frac{\partial F}{\partial u}-\frac{d}{dx}\left(\frac{\partial F}{\partial u'}\right)=0 \tag{5.12}$$

This differential equation is called the *Euler-Lagrange equation*. Its boundary conditions in this simple problem are given by Equation (5.2). The solution $u(x)$ to this differential equation will be the function that minimizes the original integral.

Example

A proper example here is to find the function $u(x)$ that minimizes the following integral:

$$I=\frac{1}{2}\int_{0}^{\pi/2}(u'^2-U^2)dx \tag{5.13}$$

subject to the following boundary conditions:

$$u(0)=0 \tag{5.14a}$$

$$u\left(\frac{\pi}{2}\right)=1 \tag{5.14b}$$

By comparison with Equation (5.1), the function F is given by

$$F = \frac{1}{2}(u'^2 - u^2)$$

The partial derivatives of F with respect to u and u' are then found to be

$$\frac{\partial F}{\partial u} = -u$$
$$\frac{\partial F}{\partial u'} = u$$

These may now be substituted into Equation (5.12) to give

$$-u - \frac{d}{dx}(u') = 0$$

which, on carrying out the indicated differentiation, becomes

$$u'' + u = 0 \tag{5.15}$$

Thus the function $u(x)$ that minimizes the integral in Equation (5.13) is the function that satisfies the differential equation and the boundary conditions given by Equation (5.14).

The solution to this problem is

$$u(x) = \sin x$$

One could then compute the integral to determine the minimum value. Thus

$$I = \frac{1}{2}\int_0^{\pi/2}\left(\cos^2 x - \sin^2 x\right) dx$$

On carrying out the integration, it turns out that $I = 0$. Any other form for $u(x)$ that satisfies only the boundary conditions would give a larger value for this integral.

Approximation of Integrals

In the finite element method, the problem will be cast as an integral to be minimized, and we will use a numerical approximation of the integral (see Appendix C) to obtain the solution.

Figure 5.2 shows a function $F(x)$ for which we would like to find

$$I = \int F(x)dx$$

The integral may be broken up into a number of subintegrals (over finite-elements) as shown in Figure 5.2. The subintegral $I^{(e)}$ between x_i and x_j may be approximated by assuming a linear variation of $F(x)$ between these two points. Thus

$$I^{(e)} = \frac{F_i + F_j}{2}(x_j - x_i)$$

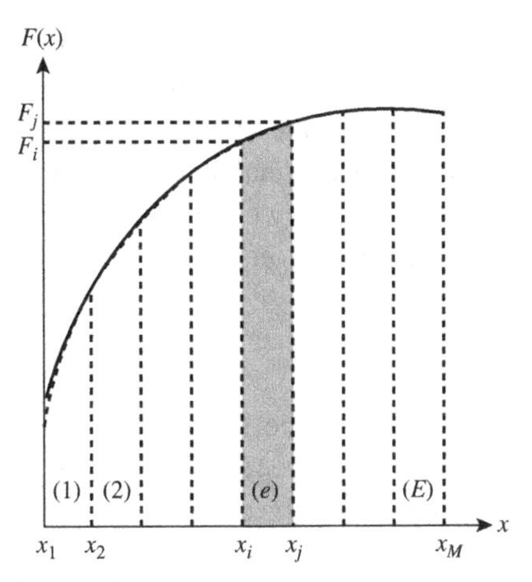

Figure 5.2. Approximation of integrals.

This is the trapezoidal rule for approximating the area under a curve. The entire integral would then be the sum of all integrals over all these elements. That is,

$$I = \sum_{e=1}^{E} I^{(e)}$$

If each of the increments $\Delta x^{(e)} = x_j - x_i$ is of the same size Δx, the preceding expression may be rewritten as follows:

$$I = \frac{\Delta x}{2} \sum_{e=1}^{E} (F_i + F_j)^{(e)}$$

For $E = 4$, for instance, we get

$$I = \frac{\Delta x}{2}[F_1 + 2F_2 + 2F_3 + 2F_4 + F_5]$$

Note that the integral is now approximated in terms of a finite number of values of the integrand. This is the *composite trapezoidal rule* for equal intervals.

One-Dimensional Steady-State Problems

As an illustration of a one-dimensional steady-state problem, let us consider a thin rod for an example. The governing differential equation and boundary conditions may be stated as follows:

$$kA\frac{d^2T}{dx^2} - hp(T - T_O) = 0$$

$$T(0) = T_O$$

$$\left(\frac{dT}{dx}\right)_{x=L} = 0$$

The well-established finite-difference method "attacks" the differential equation directly by replacing derivatives with differences. The finite element method, on the other hand, begins with a variational statement (which is one available option) of the problem. Therefore, we will first have to find the variational statement that corresponds to the preceding differential equation and boundary conditions. Once this is done, we can begin the problem's finite element formulation.

Finite Element Formulation

For the purpose of generality, let us consider the solution to be time-dependent. Our solution domain Ω is divided into M elements of r nodes each. By the usual procedure, we express the temperature and its gradients within each element as follows:

$$T^{(e)}(x,y,z,t) = \sum_{i=1}^{r} N_i(x,y,z)T_i(t)$$

$$\frac{\partial T^{(e)}}{\partial x}(x,y,z,t) = \sum_{i=1}^{r} \frac{\partial N_i}{\partial x}(x,y,z)T_i(t)$$

$$\frac{\partial T^{(e)}}{\partial y}(x,y,z,t) = \sum_{i=1}^{r} \frac{\partial N_i}{\partial y}(x,y,z)T_i(t)$$

$$\frac{\partial T^{(e)}}{\partial z}(x,y,z,t) = \sum_{i=1}^{r} \frac{\partial N_i}{\partial z}(x,y,z)T_i(t)$$

or in matrix notation as

$$\{A\} = [N][B]T^{(e)}$$

$$\text{where } \{A\} = \begin{Bmatrix} \frac{\partial T}{\partial x}(x,y,z,t) \\ \frac{\partial T}{\partial y}(x,y,z,t) \\ \frac{\partial T}{\partial z}(x,y,z,t) \end{Bmatrix}$$

$$[N(x,y,z)] = [N_1 N_2,\ldots,N_r]$$

$$[B(x,y,z)] = \begin{Bmatrix} \frac{\partial N_1}{\partial x} & \frac{\partial N_2}{\partial x} & \cdots & \frac{\partial N_r}{\partial x} \\ \frac{\partial N_1}{\partial y} & \frac{\partial N_2}{\partial y} & \cdots & \frac{\partial N_r}{\partial y} \\ \frac{\partial N_1}{\partial z} & \frac{\partial N_2}{\partial z} & \cdots & \frac{\partial N_r}{\partial z} \end{Bmatrix}$$

$T_i(t)$ is the value of T at each node, and $[T(t)]$ is the vector of the element nodal temperatures. The second-order heat-conduction equation requires only C^0 continuity (meaning continuity of the field variable across the element's interfaces), and we may use temperature as the only nodal unknown. We focus on a single element and, for simplicity, omit the superscript (e). The method of weighted residuals then may be used to derive the element equation starting with the energy equation. A review of this method follows in Chapter 6.

6 One-Dimensional Steady-State Problems

As an illustration of a one-dimensional steady-state problem (Figure 6.1), let us again consider the famous thin-rod problem. The governing differential equation and boundary conditions may be stated as follows:

$$kA\frac{d^2T}{dx^2} - hp(T - T_O) = 0 \tag{6.1}$$

$$T(0) = T_0 \tag{6.2a}$$

$$\left(\frac{dT}{dx}\right)_{x=L} = 0 \tag{6.2b}$$

Variational Statement

The process through which one might deduce the variational statement from the differential equation begins by rewriting the differential equation in the following form:

$$hp(T - T_O) - \frac{d}{dx}(kAT') = 0$$

First, the function F must be determined. Here T replaces u and $T' = dT/dx$ replaces u' (refer to Chapter 5). With that, we arrive at the following:

$$\frac{\partial F}{\partial T} = hp(T - T_O)$$

$$\frac{\partial F}{\partial T'} = kAT'$$

Recalling that T and T' are to be treated as independent functions, each of these expressions may be partially integrated to give the following:

$$F = \frac{1}{2}hp(T - T_O)^2 + f(T')$$

$$F = \frac{1}{2}KAT'^2 + g(T)$$

Therefore,

$$F = \frac{1}{2}\left[hp(T - T_O)^2 + kAT'^2\right]$$

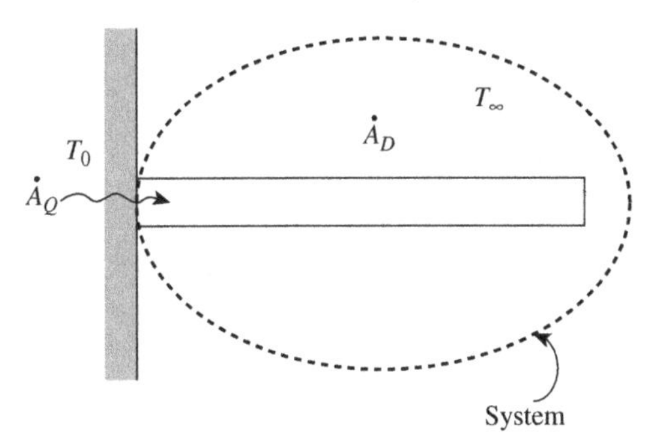

Figure 6.1. Thin-rod heat-conduction problem.

There could be an additive constant in F, but the same function that minimizes the integral of F would also minimize the integral of F plus a constant. Therefore, this option is omitted.

An equivalent problem, therefore, to the solution of Equation (6.1), with boundary conditions given by Equation (6.2), is to find the function $T(x)$ that satisfies the same boundary conditions and minimizes the integral

$$I = \frac{1}{2}\int_{x=0}^{L}\left[kA\left(\frac{dT}{dx}\right)^2 + hp(T - T_\infty)^2\right] dx$$

Before moving on to the finite element approximation, it is of interest to attach some physical significance to the integral I that we are trying to minimize. The expression for I may be rewritten using absolute temperatures as follows:

$$I = \frac{1}{2}\int_{x=0}^{L}\left\{kA\left[\frac{d(T - T_\infty)}{dx}\right]^2 + hp(T - T_\infty)^2\right\} dx$$

In writing this expression, we have used the fact that T_O is a constant and can therefore be included in the derivative without changing anything.

The first term may next be integrated by parts to give

$$I = \frac{1}{2}\left[kA\frac{d(T - T_\infty)}{dx}(T - T_\infty)\right]_{x=0}^{L} - \frac{1}{2}\int_0^L (TT_\infty)\frac{d}{dx}\left[kA\frac{d}{dx}(T - T_\infty)\right] dx$$
$$+\frac{1}{2}\int_0^L hp(T - T_\infty)^2\, dx$$

The integrated term vanishes at the upper limit because the derivative is zero there. Thus the preceding expression reduces to

$$I = -\frac{1}{2}kA\left(\frac{dT}{dx}\right)_{x=0}(T_O - T_\infty) + \frac{1}{2}\int_0^L (T - T_\infty)\left[hp(T - T_\infty) - \frac{d}{dx}\left(kA\frac{dT}{dx}\right)\right] dx$$

where we have now removed T_∞ from the derivatives.

The integral in the preceding expression is identically zero because the term in brackets is zero. Because the heat transfer rate at $x = 0$ is given by

$$q_O = -kA\left(\frac{dT}{dx}\right)_{x=0}$$

the expression for I reduces to

$$I = \frac{1}{2} q_O (T_O - T_\infty)$$

This result may be divided by $T_O/2$ to give

$$\frac{2I}{T_O} = q_O \left(1 - \frac{T_\infty}{T_O}\right)$$

Finite element Formulation

Now that we have the variational statement of the problem, we are ready to begin the finite element formulation to obtain the approximate solution for the temperature as a function of x.

Taking $T_\infty = 0$, the integral to be minimized may be written as follows:

$$I = \frac{1}{2} \int_0^L \left[kA \left(\frac{dT}{dx} \right)^2 + hpT^2 \right] dx$$

The finite element method will play the same role in finding the approximate temperature profile to minimize this integral as the well-established finite-difference method played in finding approximate solutions to differential equations.

The first step is to specify nodal-point locations along the interval $x = 0$ to $x = L$, as shown in Figure 6.2. These nodal points will be numbered from 1 to M. The interval between two adjacent nodes is called an *element* in this one-dimensional application. These elements are numbered from 1 to E. A typical element (e) is, typically, between the nodes i and j, as shown in Figure 6.3. This numbering scheme is slightly different from that in the finite-difference method, where the nodal-point numbers are also used to designate the region surrounding the nodal point. In the finite element method, the numbering of nodal points is entirely separate from the numbering of elements.

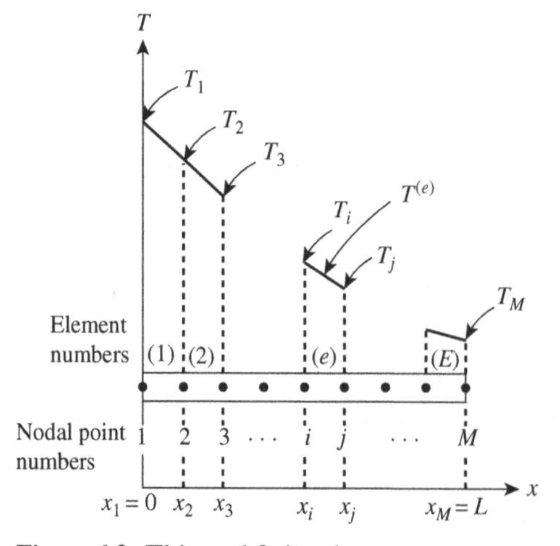

Figure 6.2. Thin-rod finite element arrangement.

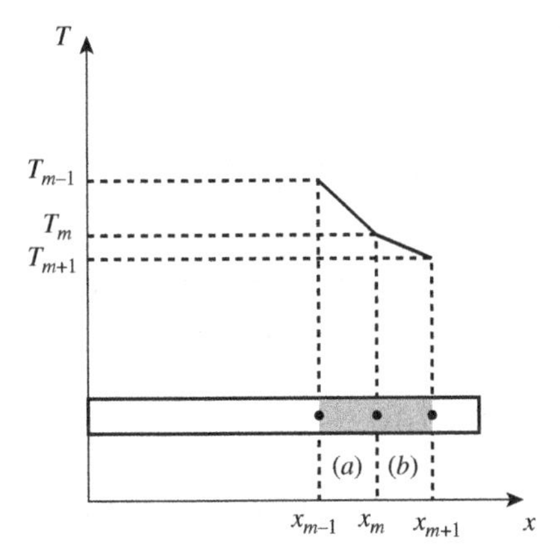

Figure 6.3. Finite elements adjacent to node m.

To evaluate the preceding integral, we will break it up into E subintegrals over each of the E elements. That is, we will consider the following:

$$I = I^{(1)} + I^{(2)} + \cdots + I^{(e)} + \cdots + I^{(e)} = \sum_{e=1}^{E} I^{(e)}$$

where the integral $I^{(e)}$ over the typical element (e) is given by

$$I^{(e)} = \frac{1}{2}\int_{x_i}^{x_j}\left[kA\left(\frac{dT}{dx}\right)^2 + hpT^2\right]$$

To evaluate this elemental integral, we will need to know, or assume, something about the temperature distribution within the finite element. The easiest assumption to make is that it varies linearly over each individual element, as indicated in Figure 6.3. That is, we will assume a linear variation between T_i and T_j. These nodal temperatures are as yet unknown. The whole idea of the finite element method here is to pick these temperatures $(T_1, T_2, \ldots, T_i, T_j, \ldots, T_M)$ in such a way as to make the complete integral as small as possible.

We will assume that the temperature $T^{(e)}$ within the element (e) is given as follows:

$$T^{(e)} = C_1{}^{(e)} + C_2{}^{(e)}x$$

The constants C_1 and C_2 have a superscript (e) on them because they will, in general, be different for each element. These constants may be determined by evaluating the preceding temperature-distribution assumption at x_i and x_j, where the temperatures are T_i and T_j, respectively. Thus

$$T_i = C_1{}^{(e)} + C_2{}^{(e)}x_i$$

$$T_j = C_1{}^{(e)} + C_2{}^{(e)}x_j$$

Solving these two equations simultaneously, we get the following result:

$$C_1{}^{(e)} = \frac{x_jT_i - x_iT_j}{x_j - x_i}$$

$$C_2^{(e)} = \frac{T_j - T_i}{x_j - x_i}$$

The superscript (e) could have been written on x_i, x_j, T_i and T_j because these also refer to the element (e). They have all been omitted, however, because they are rather cumbersome, and the subscripts i and j could very well serve to remind us that we are dealing with the small element (e). The temperature distribution within the typical element (e) can now be found by simple substitution as follows:

$$T^{(e)} = \frac{1}{x_j - x_i}\left[(x_j T_i - x_i T_j) + (T_j - T_i)x\right]$$

The temperature derivative within the element is then obtained by simple differentiation as follows:

$$\frac{dT^{(e)}}{dx} = \frac{T_j - T_i}{x_j - x_i}$$

Next, we can substitute this assumed temperature distribution into the original integral to get the following result:

$$I^{(e)} = \frac{1}{2}\int_{x_i}^{x_j}\left\{k^{(e)}A^{(e)}\frac{(T_j - T_i)^2}{(x_j - x_i)^2} + \frac{h^{(e)}p^{(e)}}{(x_j - x_i)^2}\left[(x_j T_i - x_i T_j) + (T_j - T_i)x\right]^2\right\} dx$$

At this point it should be observed that after integration over x is carried out, $I^{(e)}$ will be a function of of T_i and T_j. All the other symbols that appear are known.

To minimize I, we will have to find the derivative of $I^{(e)}$ with respect to both T_i and T_j. We therefore have two types of operations, namely, integration over x and differentiation with respect to T_i and T_j. These can be performed in any order. Let us, arbitrarily, do the differentiation first and then the integration.

The differentiation with respect to T_i, in its first stage, gives rise to the following:

$$\frac{\partial I^{(e)}}{\partial T_i} = \frac{k^{(e)}A^{(e)}}{x_j - x_i}(T_i - T_j) + \frac{h^{(e)}p^{(e)}}{(x_j - x_i)^2}\left[\left(x_j^3 - 3x_i^2 - x_i^3\right)\left(\frac{1}{3}T_i + \frac{1}{6}T_j\right)\right]$$

Knowing the algebraic fact that

$$x_j^3 - 3x_j^2 x_i + 3x_j x_i^2 - x_i^3 = (x_j - x_i)^3$$

then the preceding simplifies to

$$\frac{\partial I^{(e)}}{\partial T_i} = \frac{k^{(e)}A^{(e)}}{x_j - x_i}(T_i - T_j) + \frac{h^{(e)}p^{(e)}(x_j - x_i)}{6}(2T_i + T_j)$$

Similarly, we arrive at the second differentiation result as follows:

$$\frac{\partial I^{(e)}}{\partial T_j} = \frac{k^{(e)}A^{(e)}}{x_j - x_i}(T_j - T_i) + \frac{h^{(e)}p^{(e)}(x_j - x_i)}{6}(T_i + 2T_j)$$

It is interesting to note, in passing, that the quantity $(k^{(e)}A^{(e)})/(x_j - x_i)$ that appears in these expressions is the reciprocal of the conduction thermal resistance of the element. The term $h^{(e)}p^{(e)}(x_j - x_i)$ is the reciprocal of the convection resistance of the element.

We are now prepared to go about finding a minimum of I. As we observed in the discussion of the original integral, the subintegral $I^{(e)}$ is a function of its two nodal temperatures T_i and T_j. This means that the complete integral will be a function of the complete set of unknown nodal temperatures. That is,

$$I = I(T_1, T_2, \ldots, T_m, \ldots, T_M)$$

To find a minimum of I, we will have to differentiate it with respect to each of the nodal temperatures and set each derivative equal to zero. Thus, if there are M nodal temperatures, there will be M equations that will be obtained.

Differentiation with respect to a typical nodal temperature T_m gives rise to the following:

$$\frac{\partial I}{\partial T_m} = \frac{\partial I^{(1)}}{\partial T_m} + \frac{\partial I^{(2)}}{\partial T_m} + \cdots + \frac{\partial I^{(a)}}{\partial T_m} + \frac{\partial I^{(b)}}{\partial T_m} + \cdots + \frac{\partial I^{(E)}}{\partial T_m}$$

As shown in Figure 6.3, elements (a) and (b) are the elements on both sides of node m. Thus this expression can be rewritten as follows:

$$\frac{\partial I}{\partial T_m} = \frac{\partial I^{(a)}}{\partial T_m} + \frac{\partial I^{(b)}}{\partial T_m}$$

because all other partial derivatives are zero.

Each of the two derivatives on the right-hand side of the preceding equation may be evaluated using the general expressions developed for the typical element (e). This step will yield the following:

$$\frac{\partial I^{(a)}}{\partial T_m} = \frac{kA}{\Delta x}(T_m - T_{m-1}) + \frac{hp\Delta x}{6}(T_{m-1} + 2T_m)$$

We have replaced $(x_j - x_i)$ by simply Δx, and we will, for simplicity, assume it to be the same for every element for the present discussion. Then, for element (b), and by setting $i = m$ and $i = m + 1$, we find the following:

$$\frac{\partial I^{(b)}}{\partial T_m} = \frac{kA}{\Delta x}(T_m - T_{m+1}) + \frac{hp\Delta x}{6}(2T_m + T_{m+1})$$

The preceding two expressions may now be added together to give

$$\frac{\partial I}{\partial T_m} = \frac{kA}{\Delta x}(-T_{m-1} + 2T_m - T_{m+1}) + \frac{hp\Delta x}{6}(T_{m-1} + 4T_m + T_{m+1})$$

This expression is now set equal to zero to provide one of the conditions for minimizing I. Thus

$$\frac{kA}{\Delta x}(-T_{m-1} + 2T_m - T_{m+1}) + \frac{hp\Delta x}{6}(T_{m-1} + 4T_m + T_{m+1}) = 0$$

Multiplying by $6\Delta x/kA$ and rearranging terms, we get

$$-\left[6 - m^2(\Delta x)^2\right] T_{m-1} + \left[12 + 4m^2(\Delta x)^2\right] T_m - \left[6 - m^2(\Delta x)^2\right] T_{m+1} = 0$$

where $m^2 = hp/kA$.

Finally, the equation may be rewritten as follows:

$$-T_{m-1}+DT_m-T_{m+1}=0$$

where
$$D=\frac{12+4m^2(\Delta x)^2}{6-m^2(\Delta x)^2}$$

This equation holds true for any of the interior nodes. For the first interior node, however, $m = 1$, it must be recognized that $T_{m-1} = T_\infty$ is the given boundary condition.

For the node M, we must find the partial derivative of I with respect to T_M. In this case, however, the temperature T_M appears only in element (E). Thus

$$\frac{\partial I}{\partial T_M}=\frac{\partial I^{(E)}}{\partial T_M}$$

Now, setting $i=M-1$ and $j=M$, we get the following:

$$\frac{\partial I}{\partial T_M}=\frac{kA}{\Delta x}(T_M-T_{M-1})+\frac{hp\Delta x}{6}(T_{M-1}+2T_M)$$

This may be set equal to zero (as the second boundary condition) and rearranged to give the following equation:

$$-2T_{M-1}+DT_M=0$$

Now, assuming that $M=4$, the system of equations, in matrix form, can be cast as follows:

$$\begin{bmatrix} D & -1 & & \\ -1 & D & -1 & \\ & -1 & D & -1 \\ & & -2 & D \end{bmatrix}\begin{Bmatrix} T_1 \\ T_2 \\ T_3 \\ T_4 \end{Bmatrix}=\begin{Bmatrix} T_\infty \\ 0 \\ 0 \\ 0 \end{Bmatrix}$$

where
$$D=\frac{12+4(mL)^2(\Delta x)^2}{6-(mL)^2(\Delta x)^2}$$

Numerical Results

The numerical results for $mL=2$ are shown in Figure 6.4, which shows the improvement in the solution that can be obtained by increasing the number of elements. Another important computation that should be made is the heat transfer rate from the thin rod. This may be obtained by integrating the convection-related heat losses over the entire length of the rod:

$$q_O=q_C=\int_\Omega hpTdx$$

This integral may be evaluated by breaking it up into subintegrals over all elements. That is,

$$q_O=q_C^{(1)}+q_C^{(2)}+\cdots+q_C^{(e)}+\cdots+q_C^{(E)}$$

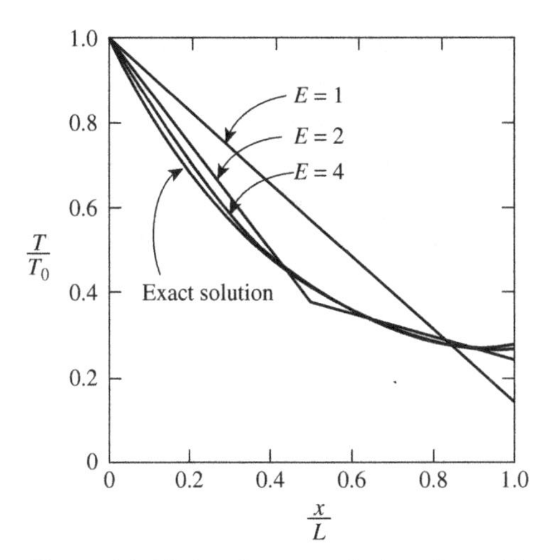

Figure 6.4. Finite element solutions for a thin rod with $mL = 2$.

The convection heat loss from element (e) may then be written as follows:

$$q_C{}^{(e)} = \int_{x_i}^{x_j} hpT\,dx$$

The expression for $T^{(e)}$ (earlier) may then be substituted for the elemental temperature variation. Thus

$$q_C{}^{(e)} = \int_{x_i}^{x_j} \frac{h^{(e)}p^{(e)}}{x_j - x_i}[(x_jT_i - x_iT_j) + (T_j - T_i)x]\,dx$$

On integration, substituting limits, and simplifying, this expression reduces to

$$q_C{}^{(e)} = h^{(e)}p^{(e)}(x_j - x_i)\frac{T_i + T_j}{2}$$

This simply states that the arithmetic mean temperature should be used to evaluate the convection-related heat loss from the element. For the case of four elements, the normalized result for the heat transfer rate is given as follows:

$$\frac{q_O}{hpLT_O} = \frac{1}{8}(1 + 2T_1 + 2T_2 + 2T_3 + T_4)$$

where h, p, and $(x_j - x_i)$ have been fixed as constants

PROBLEMS

P.1 The governing differential equation and boundary conditions for uniform energy generation in a plane wall are given as follows:

$$k\frac{d^2T}{dx^2} + g''' = 0$$

$$T(0) = 0$$

$$T'(L) = 0$$

a. Determine an equivalent variational formulation.
b. Using matrix representation, develop the finite element set of equations.
c. Obtain the finite element solutions for $x_{ij} = L, L/2$, and $L/4$.

P.2 Go back to the thin-rod problem, with the solution given in Figure 6.4. Now repeat the problem solution with $E = 6$. Again, show that the numerical solution approaches the exact solution as the number of finite-elements is increased.

7 The Two-Dimensional Heat-Conduction Problem

As an illustration of the use of the finite element method for the solution of two-dimensional steady-state problems, this section is concerned with the geometry shown in Figure 7.1 and Figure 7.2. We must first obtain an equivalent variational statement for the problem.

The governing partial differential equation and boundary conditions are as follows:

$$k\left(\frac{\partial^2 T}{\partial x^2}+\frac{\partial^2 T}{\partial y^2}\right)+g'''=0 \tag{7.1}$$

$$\left(\frac{\partial T}{\partial x}\right)_{x=0}=0 \tag{7.2}$$

$$\left(\frac{\partial T}{\partial y}\right)_{y=0}=0 \tag{7.3}$$

$$T(L,y)=T_\infty \tag{7.4}$$

$$T(x,L)=T_\infty \tag{7.5}$$

where g''' is the strength (in J/m^3) of a heat *source* or *sink* inside the body at hand.

Variational Statement

The variational statement for a two-dimensional problem involves the minimization of a double integral. The ideas of variational calculus were presented earlier in the introductory chapters and can now be extended to the problem of finding a minimum for the integral

$$I=\int\int F(x,y,u,u_x,u_y)\,dx\,dy \tag{7.6}$$

The resulting Euler-Lagrange equation for the case of specified value of u (Chapter 5) on the boundaries turns out to be

$$\frac{\partial F}{\partial u}-\frac{\partial}{\partial x}\left(\frac{\partial F}{\partial u_x}\right)-\frac{\partial}{\partial y}\left(\frac{\partial F}{\partial u_y}\right)=0 \tag{7.7}$$

We could then work backward from the differential equation (7.1) and find the variational statement for the problem in the same way we did for the thin-rod problem. Because the process is not of immediate concern to us at this moment, we will simply state that the following integral must be minimized:

$$I = \frac{1}{2}\int_{y=0}^{L}\int_{x=0}^{L}\left[k\left(\frac{\partial T}{\partial x}\right)^2 + k\left(\frac{\partial T}{\partial y}\right)^2 - 2g'''T\right]dx\,dy \tag{7.8}$$

Finite Element Formulation

Now that we have been given the function that must be minimized [Equation (7.8)], we can begin the finite element formulation to find an approximate solution to the problem of heat transfer through the two-dimensional problem at hand. We could go about the integration and differentiation processes just as we did in Chapter 6 without ever mentioning the word *matrix*. However, because matrices are used extensively in the finite element literature and lend themselves to computer applications, we will adopt the matrix approach, also used in Chapter 6. Many of the steps will be repeated, but this is done extensively to reinforce your understanding of this approach. You will undoubtedly find it helpful to refer to Chapter 6 as you study the corresponding steps in the finite element formulation in this section. It should be emphasized, however, that the use of matrices is not essential to the method (only convenient and computer-oriented). The use of matrices allows us to develop the formulation in a much more general way with no additional difficulty and will be immediately applicable to more advanced problems.

The underlying problem is identical to the thin-rod problem discussed in Chapter 6. We have a function $I(T)$ that we want to minimize. This will be done by setting its derivative with respect to each of the nodal temperatures T equal to zero. That is,

$$\frac{dI}{dT} = 0 \tag{7.9}$$

It will be convenient to separate I into two parts by rewriting Equation (7.8) as follows:

$$I_k = \frac{1}{2}\int\int_A \left[k\left(\frac{\partial T}{\partial x}\right)^2 + k\left(\frac{\partial T}{\partial y}\right)^2\right]dx\,dy \tag{7.10}$$

$$I_g = \int\int_A g'''T\,dx\,dy \tag{7.11}$$

Observe that in writing Equations (7.10) and (7.11), we have generalized the problem somewhat by considering an arbitrary region of interest A. This is perfectly valid, and it can be carried out with no additional difficulties. We will later return to the original problem we set out to solve.

Equation (7.1) may now be substituted into Equation (7.9) to give

$$\frac{dI}{dT} = \frac{dI_k}{dT} - \frac{dI_g}{dT} = 0 \tag{7.12}$$

The region of interest A, a square slab in this case (see Figure 7.2), will be broken up into E finite-elements so that the integrals in Equation (7.12) can be computed

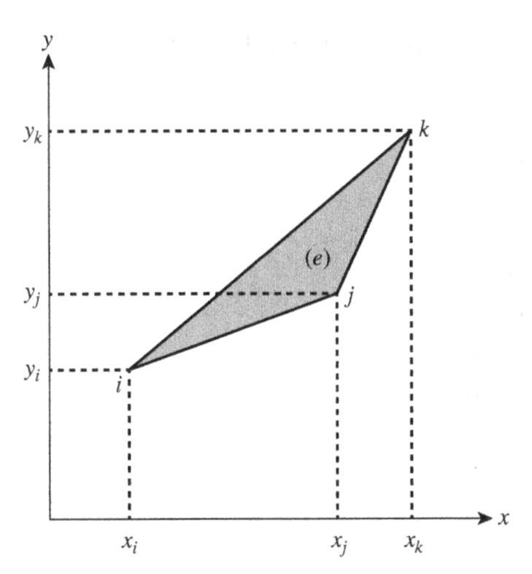

Figure 7.1. Finite element triangular element for two-dimensional conduction.

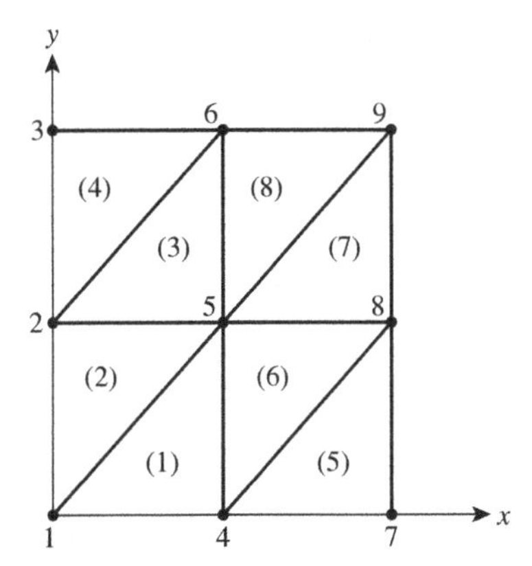

Figure 7.2. Finite element solution for square generation (Case 1)

as a sum over each of the small elements. That is, we will write

$$\frac{dI_k}{dT} = \sum_{e=1}^{E} \frac{dI_k{}^{(e)}}{dT} \tag{7.13}$$

$$\frac{dI_g}{dT} = \sum_{e=1}^{E} \frac{dI_g{}^{(e)}}{dT} \tag{7.14}$$

Equations (7.13) and (7.14) will be computed separately and then substituted into Equation (7.12).

In this two-dimensional conduction problem, we will use a triangular finite element (see Figure 7.1). This means that $I_k{}^{(e)}$ and $I_g{}^{(e)}$ will be functions of the three corner temperatures T_i, T_j, and T_k. Consequently, the partial derivatives with respect to all other nodal temperatures will be zero. Following the same line of reasoning discussed in Chapter 6, we can write

$$\frac{dI_k{}^{(e)}}{dT^{(e)}} = D^{(e)} \frac{dI_k{}^{(e)}}{dT^{(e)}} \tag{7.15}$$

where the "mapping" matrix $D^{(e)}$ is an $M \times 3$ matrix that is defined as follows:

$$D^{(e)} = \begin{bmatrix} 0 & 0 & 0 \\ 0 & 0 & 0 \\ 1 & 0 & 0 \\ 0 & 0 & 0 \\ 0 & 0 & 0 \\ 0 & 1 & 0 \\ 0 & 0 & 0 \\ 0 & 0 & 0 \\ 0 & 0 & 1 \\ 0 & 0 & 0 \\ 0 & 0 & 0 \end{bmatrix}$$

where the rows with 1 in them are the ith, jth, and kth rows. Also,

$$\frac{dI_k{}^{(e)}}{dT^{(e)}} = \begin{Bmatrix} \dfrac{dI^{k(e)}}{dT_i} \\ \dfrac{dI^{k(e)}}{dT_i} \\ \dfrac{dI^{k(e)}}{dT_i} \end{Bmatrix}$$

Equation (7.15) through the last expressions (7.18) applies equally well to $I_g{}^{(e)}$ if we simply replace k by g. Thus we have now reduced the problem to one of considering each element separately and then adding up all the individual results.

Within each element, we will assume that the temperature varies linearly between the three corner temperatures. This may be expressed as follows:

$$T^{(e)} = C_1{}^{(e)} + C_2{}^{(e)}x + C_3{}^{(e)}y \tag{7.16}$$

or

$$T^{(e)} = p^T C^{(e)} \tag{7.17}$$

where

$$p^T = [1\ x\ y] \tag{7.18}$$

and

$$C^{(e)} = \begin{Bmatrix} C_1 \\ C_2 \\ C_3 \end{Bmatrix} \tag{7.19}$$

where the last expression is (7.19)

The coefficients C_1, C_2, and C_3 may be evaluated by insisting that $T^{(e)}$ be equal to T_i, T_j, and T_k at the element corner nodes in Figure 7.1. Thus we end up with the following:

$$T^{(e)} = P^{(e)} C^{(e)} \tag{7.20}$$

where

$$T^{(e)} = \begin{Bmatrix} T_i \\ T_j \\ T_k \end{Bmatrix}$$

and

$$P^{(e)} = \begin{bmatrix} 1 & x_i & y_i \\ 1 & x_j & y_j \\ 1 & x_k & y_k \end{bmatrix}$$

The constants $C^{(e)}$ may be found by multiplying Equation (7.20) by the inverse of $P^{(e)}$, $R^{(e)} = P^{(e)^{-1}}$:

$$R^{(e)} T^{(e)} = P^{(e)^{-1}} P^{(e)} C^{(e)} \tag{7.21}$$

Because $P^{(e)^{-1}} P^{(e)} = I$ (unit matrix) and $IC^{(e)} = C^{(e)}$, Equation (7.21) reduces to

$$C^{(e)} = R^{(e)} T^{(e)} \tag{7.22}$$

This result can be substituted into Equation (7.20) to arrive at the following expression for the temperature distribution within the element:

$$T^{(e)} = p^T R^{(e)} T^{(e)} \tag{7.23}$$

The matrix $R^{(e)}$, which is the inverse of $P^{(e)}$, has been worked out and is given as follows:

$$R^{(e)} = \frac{1}{x_{ij} y_{jk} - x_{jk} y_{ij}} \begin{bmatrix} x_j y_k - x_k y_j & x_k y_i - x_i y_k & x_i y_j - x_j y_i \\ -y_{jk} & y_{ik} & -y_{ij} \\ x_{jk} & -x_{ik} & x_{ij} \end{bmatrix}$$

We are now ready to direct our attention to the function $I_k{}^{(e)}$, which, from Equation (7.10), may be written for the element (e) as follows:

$$I_k{}^{(e)} = \frac{k^{(e)}}{2} \int\int_{A^{(e)}} \left[(p_x{}^T R^{(e)} T^{(e)})^2 (p_y{}^T R^{(e)} T^{(e)})^2 \right] dx\,dy$$

This can now be differentiated with respect to each of the three corner temperatures at the same time. Note that in deriving the preceding expression, we have assumed that the element is sufficiently small that the thermal conductivity is constant throughout the entire element. Now

$$\frac{dI_k{}^{(e)}}{dT^{(e)}} = \frac{k^e}{2} \int\int_{A^{(e)}} \left[2(p_x{}^T R^{(e)} T^{(e)})^T + 2(p_y{}^T R^{(e)} T^{(e)}) \right.$$
$$\left. +2(p_y{}^T R^{(e)} T^{(e)})(p_y{}^T R^{(e)} T^{(e)})(p_y{}^T R^{(e)})^T \right] dx\,dy \tag{7.24}$$

In arriving at this expression, we have assumed that the thermal conductivity of the element is independent of the temperature in the differentiation process.

Following the same line of reasoning used in Chapter 6, we may interchange the order of multiplication in Equation (7.24) and then take the transposes of $p_x{}^T R^{(e)}$ and $p_y{}^T R^{(e)}$ to arrive at the following expression:

$$\frac{dI_k{}^{(e)}}{dT^{(e)}} = k^{(e)} \int\int_{A^{(e)}} b \left[R^{(e)^T} p_x p_x{}^T R^{(e)} T^{(e)} + R^{(e)^T} p_y p_y{}^T R^{(e)} T^{(e)} \right] dx\,dy$$

One of the rules of matrix algebra is that $AB + AC = A(B + C)$ and, similarly, $AC + BC = (A + B)C$. We will invoke this rule. From Equations (7.20) and (7.24), we observe that $T^{(e)}$ and $R^{(e)}$ are independent of x and y. Therefore, they may be factored out of the integral (being careful to maintain the order of matrix multiplication) to give the following:

$$\frac{dI_k{}^{(e)}}{dT^{(e)}} = k^{(e)}R^{(e)T} \int\int_{A^{(e)}} (p_x p_x{}^T p_y p_y{}^T)\,dx\,dy\,R^{(e)}T^{(e)} \tag{7.25}$$

We are finally ready to carry out the integration. The required partial derivatives of p^T may be found by differentiating Equation (7.18) to give

$$p_x{}^T = [0\ 1\ 0]$$

$$p_y{}^T = [1\ 0\ 0]$$

Then, because $\alpha^{T^T} = \alpha$, it follows that

$$p_x = \begin{Bmatrix} 0 \\ 1 \\ 0 \end{Bmatrix}$$

and

$$p_y = \begin{Bmatrix} 0 \\ 0 \\ 1 \end{Bmatrix}$$

The integrand in Equation (7.25) may now be expressed as follows:

$$p_x p_x{}^T + p_y p_y{}^T = \begin{Bmatrix} 0 \\ 1 \\ 0 \end{Bmatrix} [0\ 1\ 0] + \begin{Bmatrix} 0 \\ 0 \\ 1 \end{Bmatrix} [0\ 0\ 1] = \begin{bmatrix} 0 & 0 & 0 \\ 0 & 1 & 0 \\ 0 & 0 & 1 \end{bmatrix} + \begin{bmatrix} 0 & 0 & 0 \\ 0 & 0 & 0 \\ 0 & 0 & 1 \end{bmatrix}$$

$$= \begin{bmatrix} 0 & 0 & 0 \\ 0 & 1 & 0 \\ 0 & 0 & 1 \end{bmatrix}$$

We may then write the integral as follows:

$$\int_{A^{(e)}} (p_x P_x{}^T + P_y P_y{}^T)dA = \int_{A^{(e)}} \begin{bmatrix} 0 & 0 & 0 \\ 0 & 1 & 0 \\ 0 & 0 & 1 \end{bmatrix} dA$$

Because the matrix is constant, it may be taken outside the integral sign. The result is recognized as the area of the triangular element. Thus the foregoing expression reduces to

$$\int\int_{A^{(e)}} (p_x p_x{}^T + p_y p_y{}^T)\,dx\,dy = A^{(e)} \begin{bmatrix} 0 & 0 & 0 \\ 0 & 1 & 0 \\ 0 & 0 & 1 \end{bmatrix}$$

This integral can now be premultiplied by $R^{(e)^T}$ and postmultiplied by $R^{(e)}$. The algebraic details are analogous to the steps previously taken in Chapter 6, but they become a little long-winded and will therefore be omitted here. The final result is

$$\frac{dI_k{}^{(e)}}{dT^{(e)}} = K^{(e)}T^{(e)} \tag{7.26}$$

where the element conduction matrix is given as follows:

$$\begin{bmatrix} x_{jk}^2 + y_{jk}^2 & -(x_{ik}x_{jk} + y_{ik}y_{jk}) & x_{ij}x_{jk} + y_{ij}x_{jk} + y_{ij}y_{jk} \\ \cdot & x_{ik}^2 + y_{ik}^2 & -(x_{ij}x_{ik} + y_{ij}y_{jk}) \\ \cdot & \cdot & x_{ij}^2 + y_{ij}^2 \end{bmatrix} \tag{7.27}$$

where the dots simply are meant to signify a symmetric matrix. We will be able to save on computer storage and CPU time consumption.

The area of the triangle is easily computed in terms of the nodal positions. It can be shown that

$$A^{(e)} = \frac{1}{2} \mid x_{ij}y_{jk}y_{ij} \mid$$

It should be noted that the algebraic steps required to arrive at Equation (7.27) are time-consuming and might even destroy the usefulness of the finite element method if they had to be carried out every time one wants to solve a new problem. There are, nevertheless, two saving factors. First, with the widespread use of computers, general-purpose computer programs (collectively referred to as *canned* programs) are being developed to handle conduction problems. The finite element method is ideally suited for such general-purpose programs. Thus the algebra is done only once, and the results are incorporated into a general computer program. The second factor is that if you don't want to carry out the algebraic steps by hand, you can always ask the computer to do the matrix multiplication for you. Because this requires more computational time and must be repeated for each problem, it is cheaper to do the algebra whenever possible if you are aiming at a general-purpose computer program.

Equation (7.26) can now be substituted into Equation (7.15) and the result substituted into Equation (7.14) to give

$$\frac{dI_k}{dT} = \sum_{e=1}^{E} D^{(e)} K^{(e)} D^{(e)^T} T \tag{7.28}$$

where the matrix $D^{(e)}$ was defined earlier. Thus

$$\frac{dI_k}{dT} = \sum_{e=1}^{E} D^{(e)} K^{(e)} D^{(e)^T}$$

It is now convenient to define a global conduction matrix as follows:

$$K = \sum_{e=1}^{E} D^{(e)} K^{(e)} D^{(e)^T} \tag{7.29}$$

Then Equation (7.28) may be simplified to

$$\frac{dI_k}{dT} = KT \tag{7.30}$$

This is the final result we need for Equation (7.12).

Now let us consider I_g, which may be written for an element (e) from Equation (7.11) as follows:

$$I_g^{(e)} = g''' \int\int_{A^{(e)}} p^T R^{(e)} T^{(e)} \tag{7.31}$$

In writing this expression, we have assumed that the element is sufficiently small that the generation rate g''' may be taken as uniform and moved outside the integral sign. It can be different for each element, however. Substituting Equation (7.23) into Equation (7.31) gives rise to

$$I_g{}^{(e)} = g'''^{(e)} \int\int_{A^{(e)}} p^T R^{(e)} T^{(e)}\, dx\, dy$$

This may be differentiated with respect to $T^{(e)}$ to give

$$\frac{dI_g{}^{(e)}}{dT^{(e)}} = g'''^{(e)} \int\int_{A^{(e)}} \left(p^T R^{(e)}\right)^T dx\,dy$$

Transposing $p^T R^{(e)}$ and recognizing that $R^{(e)}$ is independent of x and y, the following relationship is obtained:

$$\frac{dI_g{}^{(e)}}{dT^{(e)}} = g'''^{(e)} R^{(e)T} \int\int_{A^{(e)}} p\, dx\,dy$$

From Equation (7.18), we can then rewrite this equation as follows:

$$\frac{dI_g{}^{(e)}}{dT^{(e)}} = g''' R^{(e)T} \int\int_{A^{(e)}} \begin{Bmatrix} 1 \\ x \\ y \end{Bmatrix} dx\,dy$$

Each term in the column matrix must be individually integrated. The first of these three integrals is recognized as the area of the triangular element. The second integral is the first moment of the triangle around the y axis. By definition of the centroid (c) of the area x_c, this integral may be written as $x_c{}^{(e)} A^{(e)}$. Similarly, the third integral would be given by $y_c{}^{(e)} A^{(e)}$. Thus, after factoring out the common $A^{(e)}$, we arrive at the following:

$$\frac{dI_g{}^{(e)}}{dT^{(e)}} = g'''^{(e)} A^{(e)} R^{(e)T} \begin{Bmatrix} 1 \\ x_c \\ y_c \end{Bmatrix}$$

or

$$\frac{dI_g{}^{(e)}}{dT^{(e)}} = g^{(e)(e)} A^{(e)} R^{(e)T} p_c{}^{(e)} \tag{7.32}$$

where

$$p_c{}^{(e)} = \begin{Bmatrix} 1 \\ x_c \\ y_c \end{Bmatrix}$$

The centroid of a triangle is easily computed in terms of the corner positions as follows:

$$x_c{}^{(e)} = \frac{x_i + x_j + x_k}{3} \tag{7.33a}$$

$$y_c{}^{(e)} = \frac{y_i + y_j + y_k}{3} \tag{7.33b}$$

If the matrix multiplication in Equation (7.32) is carried out in detail, the following result is obtained:

$$\frac{dI_g{}^{(e)}}{dT^{(e)}} = g^{(e)} \tag{7.34}$$

where the element heat-generation matrix is given as follows:

$$g^{(e)} = \frac{g'''^{(e)}}{A}3\begin{Bmatrix}1\\1\\1\end{Bmatrix}$$

We can now write the following:

$$\frac{dI_g}{dT} = \sum_{e=1}^{E} D^{(e)}g^{(e)} \tag{7.35}$$

If we define a global generation matrix as

$$g = \sum_{e=1}^{E} D^{(e)}g^{(e)} \tag{7.36}$$

Equation (7.35) will then, reduce to

$$\frac{dI_g}{dT} = g \tag{7.37}$$

Numerical Solution

Let us now apply the finite element formulation we have derived in this chapter to the example we originally set out to solve. Figure 7.1 shows one possible selection of nodal points and triangular elements. The input data describing the geometry and physical properties of the finite element system are given in Table 7.1. For convenience in storing the symmetric matrix, i, j, and k have been so assigned that $i < j < k$. In a real situation, numerical values would be specified for the physical quantities rather than letter symbols. Notice how easy it would be to specify any set of nodal positions and different values of $k^{(e)}$ and $g'''^{(e)}$ for each element. This is one of the beauties of the finite element method. Also note that there will be, at most, four off-diagonal rows above the main diagonal because the maximum difference between i, j, and k is now 4.

The computer, then, starts with element (1) and begins to build up the matrix $[K]$. It would first make the following calculations for element (1):

$$x_{ij} = x_j - x_i = \frac{L}{2} - 0 = \frac{L}{2}$$

$$x_{ik} = x_k - x_i = \frac{L}{2}$$

$$x_{jk} = x_k - x_j = \frac{L}{2} - \frac{L}{2} = 0$$

$$y_{ij} = 0$$

$$y_{ik} = y_k - y_i = \frac{L}{2}$$

$$y_{jk} = \frac{L}{2}$$

$$A^{(e)} = \frac{1}{2}\left|x_{ij}y_{ik} - x_{ik}y_{ij}\right| = \frac{L^2}{8}$$

Table 7.1. *Finite element solution for square generation*

Nodal coordinates			Element information					
i	x_i	y_i	(e)	i	j	k	k	g'''
1	0	0	1	1	4	5	k	g'''
2	0	L_2	2	1	2	5	k	g'''
3	0	L	3	2	5	6	k	g'''
4	L_2	0	4	2	3	6	k	g'''
5	L_2	L_2	5	4	7	8	k	g'''
6	L_2	L	6	4	5	8	k	g'''
7	L	0	7	5	8	9	k	g'''
8	L	L_2	8	5	6	9	k	g'''
9	L	L						

Because there are eight equal triangles in this example, the result that $A^{(e)} = L^2/8$ is obviously correct. The magnitude of $A^{(e)}$ will appear as a constant multiplier in the element conduction matrix (7.27). Thus the computer would then evaluate

$$\frac{k^{(1)}A^{(1)}}{(x_{ij}y_{jk} - x_{jk}y_{ij})^2} = \frac{kA^{(e)}}{\left(2A^{(e)}\right)^2} = \frac{k}{4A^{(e)}} = \frac{2k}{L^2}$$

Then, from Equation (7.27), the components of the element's conduction matrix $K^{(1)}$ can be computed as follows:

$$K_{11}{}^{(1)} = \frac{k}{2}$$

$$K_{12}{}^{(1)} = -\frac{k}{2}$$

$$K_{13}{}^{(1)} = 0$$

$$K_{21}{}^{(1)} = K_{12}{}^{(1)} \qquad \text{(Matrix is symmetric.)}$$

$$K_{22}{}^{(1)} = k$$

$$K_{23}{}^{(1)} = -\frac{k}{2}$$

$$K_{31}{}^{(1)} = K_{13}{}^{(1)}$$

$$K_{32}{}^{(1)} = K_{23}{}^{(1)}$$

$$K_{33}{}^{(1)} = \frac{k}{2}$$

The element matrix $K^{(1)}$ may then be written as follows:

$$K^{(1)} = \frac{k}{2}\begin{bmatrix} 1 & -1 & 0 \\ . & 2 & -1 \\ . & . & 1 \end{bmatrix}$$

with the dots simply meaning that the matrix is symmetric. After both K and g have been properly stored, we have the following equations to solve:

$$\begin{bmatrix} 2 & -1 & 0 & -1 & & & & & \\ -1 & 4 & -1 & 0 & -2 & & & & \\ 0 & -1 & 2 & 0 & 0 & -1 & & & \\ -1 & 0 & 0 & 4 & -2 & 0 & -1 & & \\ & -2 & 0 & -2 & 8 & -2 & 0 & -2 & \\ & & -1 & 0 & -2 & 4 & 0 & 0 & -1 \\ & & & -1 & 0 & 0 & 2 & -1 & 0 \\ & & & & -2 & 0 & -1 & 4 & -1 \\ & & & & & -1 & 0 & -1 & 2 \end{bmatrix} \begin{Bmatrix} T_1 \\ T_2 \\ T_3 \\ T_4 \\ T_5 \\ T_6 \\ T_7 \\ T_8 \\ T_9 \end{Bmatrix} = \frac{g'''L^2}{12k}\begin{Bmatrix} 2 \\ 3 \\ 1 \\ 3 \\ 6 \\ 3 \\ 1 \\ 3 \\ 2 \end{Bmatrix}$$

The $k/2$ factor has been moved to the right-hand side of the equation for convenience. The terms below the main diagonal ordinarily would not be stored because the matrix is symmetric. They have been shown here because this will help in understanding the next step to be taken.

Before solving these equations, they must be modified to incorporate the specified-temperature boundary condition without having to destroy the symmetry of the matrix on the left-hand side. In this example, nodes 3 and 6 through 9 have specified-temperature boundary conditions to have a temperature of T_O. To handle node 3, we would first transfer all the off-diagonal terms that multiply T_3 over to the right-hand side of the matrix equation. The components of these terms are then set equal to zero on the left-hand side. Then, in the third equation, the off-diagonal terms are set equal to zero, the diagonal terms are set to unity, and the third components of the right-hand side are set equal to T_O (the given value for T_3). After these modifications have been made, the preceding matrix equation will now appear as follows:

$$\begin{bmatrix} 2 & -1 & 0 & -1 & & & & & \\ -1 & 4 & 0 & 0 & -2 & & & & \\ 0 & 0 & 1 & 0 & 0 & 0 & & & \\ -1 & 0 & 0 & 4 & -2 & 0 & -1 & & \\ & -2 & 0 & -2 & 8 & -2 & 0 & -2 & \\ & & 0 & 0 & -2 & 4 & 0 & 0 & -1 \\ & & & -1 & 0 & 0 & 2 & -1 & 0 \\ & & & & -2 & 0 & -1 & 4 & -1 \\ & & & & & -1 & 0 & -1 & 2 \end{bmatrix} \begin{Bmatrix} T_1 \\ T_2 \\ T_3 \\ T_4 \\ T_5 \\ T_6 \\ T_7 \\ T_8 \\ T_9 \end{Bmatrix} = \frac{g'''L^2}{12k}\begin{Bmatrix} 2 \\ 3 \\ 0 \\ 3 \\ 6 \\ 3 \\ 1 \\ 3 \\ 3 \end{Bmatrix} - T_O\begin{Bmatrix} 0 \\ -1 \\ -1 \\ 0 \\ 0 \\ -1 \\ 0 \\ 0 \\ 0 \end{Bmatrix}$$

The subtractions from the right-hand side have been shown in a symbolic form. In the computer, the numerical values would actually be used, and there would only be a one-column matrix. It should be observed that the column matrix multiplying T_O contains the components in the third column of the original matrix in the preceding

$$
\begin{bmatrix}
4 & -2 & 0 & 0 & -2 & & & & & & & & & & & \\
-1 & 4 & -1 & 0 & 0 & -2 & & & & & & & & & & \\
0 & -1 & 4 & -1 & 0 & 0 & -2 & & & & & & & & & \\
0 & 0 & -1 & 4 & 0 & 0 & 0 & -2 & & & & & & & & \\
-1 & 0 & 0 & 0 & 4 & -2 & 0 & 0 & -1 & & & & & & & \\
 & -1 & 0 & 0 & -1 & 4 & -1 & 0 & 0 & -1 & & & & & & \\
 & & -1 & 0 & 0 & -1 & 4 & -1 & 0 & 0 & -1 & & & & & \\
 & & & -1 & 0 & 0 & -1 & 4 & 0 & 0 & 0 & -1 & & & & \\
 & & & & -1 & 0 & 0 & 0 & 4 & -2 & 0 & 0 & -1 & & & \\
 & & & & & -1 & 0 & 0 & -1 & 4 & -1 & 0 & 0 & -1 & & \\
 & & & & & & -1 & 0 & 0 & -1 & 4 & -1 & 0 & 0 & -1 & \\
 & & & & & & & -1 & 0 & 0 & -1 & 4 & 0 & 0 & 0 & -1 \\
 & & & & & & & & -1 & 0 & 0 & 0 & 4 & -2 & 0 & 0 \\
 & & & & & & & & & -1 & 0 & 0 & -1 & 4 & -1 & 0 \\
 & & & & & & & & & & -1 & 0 & 0 & -1 & 4 & -1 \\
 & & & & & & & & & & & -1 & 0 & 0 & -1 & 4
\end{bmatrix}
\begin{bmatrix} u_1 \\ u_2 \\ u_3 \\ u_4 \\ u_6 \\ u_7 \\ u_8 \\ u_9 \\ u_{11} \\ u_{12} \\ u_{13} \\ u_{14} \\ u_{16} \\ u_{17} \\ u_{18} \\ u_{19} \end{bmatrix}
=
\begin{bmatrix} \frac{4}{3}(\frac{1}{16}) \\ \frac{1}{16} \\ \frac{1}{16} \\ \frac{1}{16} \\ \frac{1}{16} \\ \frac{1}{16} \\ \frac{1}{16} \\ \frac{1}{16} \\ \frac{1}{16} \\ \frac{1}{16} \\ \frac{1}{16} \\ \frac{1}{16} \\ \frac{1}{16} \\ \frac{1}{16} \\ \frac{1}{16} \\ \frac{1}{16} \end{bmatrix}
$$

Figure 7.3. Assembled set of finite element equations.

matrix equation, except for the diagonal term. Also notice that the third equation has now been decoupled from the rest of the equations and has the solution $T_3 = T_O$ as specified by the boundary condition.

We would then have to consider nodes 6 through 9 in the same manner. The final result is as follows:

$$
\begin{bmatrix}
2 & -1 & 0 & -1 & & & & & \\
-1 & 4 & 0 & 0 & -2 & & & & \\
0 & 0 & 1 & 0 & 0 & 0 & & & \\
-1 & 0 & 0 & 4 & -2 & 0 & 0 & & \\
 & -2 & 0 & -2 & 8 & 0 & 0 & 0 & \\
 & & 0 & 0 & 1 & 0 & 0 & 0 & \\
 & & & 0 & 0 & 1 & 0 & 0 & 0 \\
 & & & & 0 & 0 & 0 & 1 & 0 \\
 & & & & & 0 & 0 & 0 & 1
\end{bmatrix}
\begin{Bmatrix} T_1 \\ T_2 \\ T_3 \\ T_4 \\ T_5 \\ T_6 \\ T_7 \\ T_8 \\ T_9 \end{Bmatrix}
=
\begin{Bmatrix} \left(\dfrac{g'''L^2}{6k}\right) \\ \left(\dfrac{g'''L^2}{4k}\right) + T_O \\ T_O \\ \left(\dfrac{g'''L^2}{2k}\right) + 4\,T_O \\ T_O \\ T_O \\ T_O \end{Bmatrix}
$$

This seems like a major overhaul of the matrix equation in Figure 7.3 in order to handle the specified-temperature boundary condition. Indeed, it is major for this example, in which five of the nine nodes have specified temperatures. In a more practical case, however, there will be many more interior nodes relative to the boundary nodes. Had we cut the spacing in half, there would have been only 9 of 25 nodes that would have a specified temperature. Smaller nodal spacings would reduce the percentage even further.

We now have a set of simultaneous algebraic equations to solve. Several methods can be used for solution. In this case, however, because the matrix is symmetric (and we have taken great pains to keep it so when incorporating the specified boundary temperatures), it can be handled more efficiently.

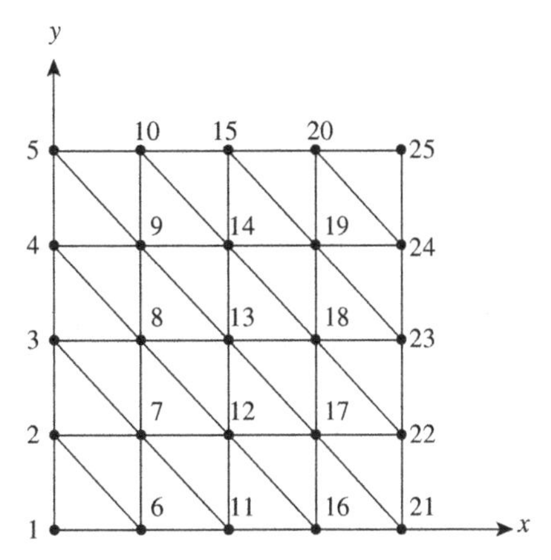

Figure 7.4. Finite element model for square generation (case 1).

Node	Exact	Finite elements	Error, exact-approx.	Error, % of $u(0,0)$
1	0.2947	0.3013	–0.0066	–2.24
2	0.2789	0.2805	–0.0016	–0.54
3	0.2293	0.2292	0.0001	0.03
4	0.1397	0.1392	0.0005	0.17
6	0.2789	0.2805	–0.0016	–0.54
7	0.2642	0.2645	–0.0003	–0.10
8	0.2178	0.2172	0.0006	0.20
9	0.1333	0.1327	0.0006	0.20
11	0.2293	0.2292	0.0001	0.03
12	0.2178	0.2172	0.0006	0.20
13	0.1811	0.1801	0.0010	0.34
14	0.1127	0.1117	0.0010	0.34
16	0.1397	0.1392	0.0005	0.17
17	0.1333	0.1327	0.0006	0.20
18	0.1127	0.1117	0.0010	0.34
19	0.0728	0.0715	0.0013	0.44

Figure 7.5. Finite element solutions for square generation.

Numerical Results

The numerical results for the example where $\Delta x = L/4$ (Figure 7.4) considered in the preceding section are shown in Figure 7.5. This figure shows how generally low the error percentages are throughout the entire slab.

In case 2 (Figure 7.6), we would arrive at a different set of algebraic equations, but only slightly different. Just the first term in the column matrix on the right-hand side of the preceding matrix equation would change. It would become $2/3(1/16)$ instead of $4/3(1/16)$ for case 1. It is interesting to note, in passing, that the finite-difference case is the arithmetic average of these two cases.

The solution for case 2 is compared with that for case 1 in Figure 7.7 for $y/L = 0$. The finite-difference solution is still the arithmetic average of these two solutions. The exact solution is conspicuously missing because in any practical solution, you would not know it. We now have the task of deciding which of these two solutions

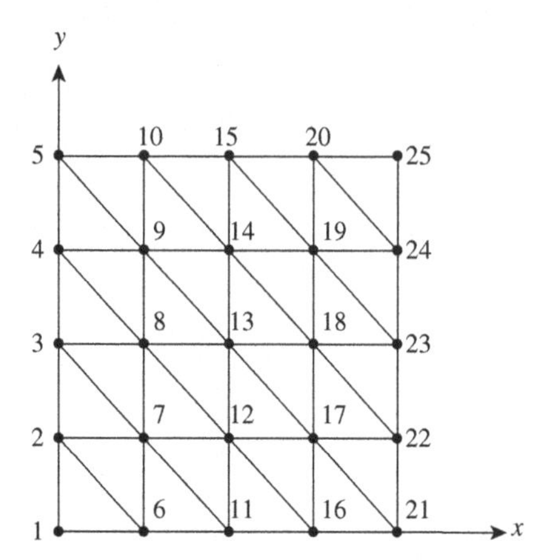

Figure 7.6. Finite element arrangement for square generation (case 2).

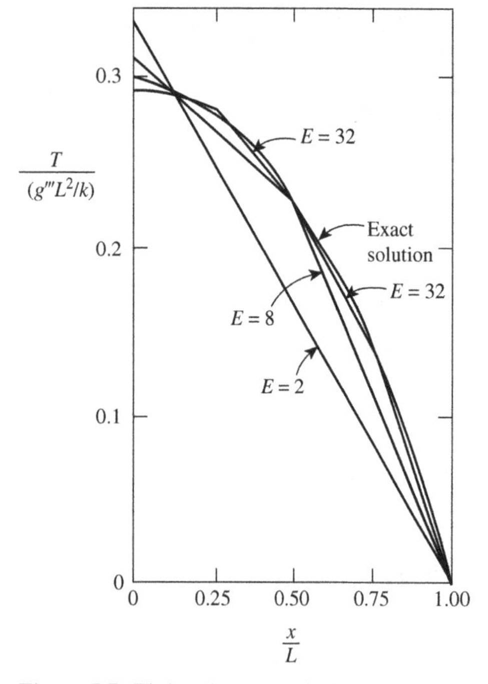

Figure 7.7. Finite element solution for square generation along $y = 0$.

(or three solutions, if you include the finite-difference solution, which is not shown in the figure) is the best.

The variational statement of the problem gives us a way of determining which solution is the "best." The variational statement indicates that a certain integral must be minimized. The best solution may be said to be the one that gives the minimum value for this integral. Figure 7.8 shows an interesting comparison of these solutions for 1, 4, and 16 nodal points. Note that for a specified number of nodal temperatures, the finite element (case 1) solution is the best solution. It should also be noted that the integral becomes smaller (i.e., the solution becomes better) as the number of nodal points is increased. The three solutions also group closer

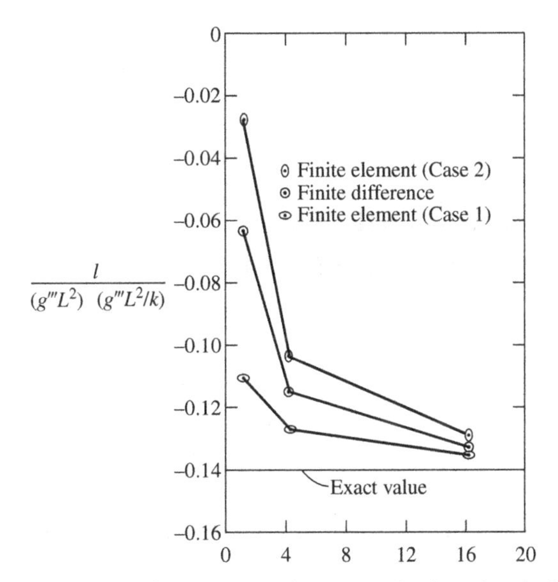

Figure 7.8. Comparison of computed values for the integral I.

together as the nodal points increase so that for a large number of nodal points the three solutions will be indistinguishable (and identical to the exact solution). Nevertheless, note that the superiority of the finite element method, as the most versatile numerical method there is, in handling general irregular-boundary boundary conditions.

8 Three-Dimensional Heat-Conduction Applications with Convection and Internal Heat Absorption

The problem statement, in general, was discussed earlier with two-dimensional heat-conduction modeling. There is no need, therefore, to repeat the analysis, except to point out that the problem at hand is much more involved. Computerwise, the problem requires a much greater amount of memory and CPU time consumption. In fact, the mere completion of the solution-domain discretization model, using one of the finite element categories, such as the curved-boundary isoparametric element, is perhaps half the work there.

It is the author's opinion that exploring a complex-geometry three-dimensional heat-conduction problem would serve students much more than just casting some bulky matrix equations. The chosen problem, in this chapter, has to do with, perhaps, one of the most complicated bodies there can be. The conduction "body" here is a "slice" in a radial inflow turbine rotor, one that includes one blade and the corresponding segment of the center body (Figure 8.1). The variational approach is, once again, used, and the simplest three-dimensional finite element, meaning the tetrahedron, is used to discretize the problem domain. Naturally, source (actually sink) terms will have to appear in the governing equations because they represent the heat loss due to the passing cooling air, wherever that applies in the blade unit. This in no way eliminates the heat convection through the blade unit. The only difference, in fact, is that heat convection will be the means by which the flowing hot gases interact with the rotor, whereas heat sinks model the effect of the cooling air. Note that in calculating the heat-convection coefficient h, the relative velocity between the hot gases and the rotating body will, and should be, used.

In the following, the governing equations and different boundary conditions are stated. Next, the finite element model is derived. Note that it is the relative flow thermophysical properties that will count here (meaning the total relative pressure and total relative temperature). For example, it is the flow velocity relative to the spinning rotor that interferes with the calculation of the local heat-convection coefficient.

The Problem of Cooling a Radial Turbine Rotor: Overview

The problem of studying heat transfer in a radial-inflow turbine rotor is mainly one of heat conduction through this three-dimensional body, subject to different

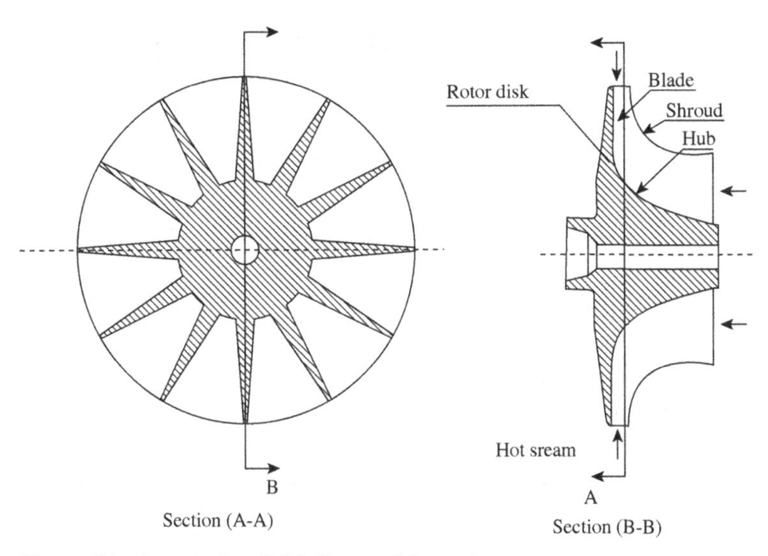

Figure 8.1. A typical radial-inflow turbine rotor.

categories of boundary conditions over all its surface segments, as well as heat sinks representing the internal cooling effect inside special interior ducts. Referring to Figure 8.1, it can be seen that a large portion of the rotor surface is exposed to the main hot flow (assumed to be predominantly air). Another portion is exposed to either internal or external cooling flow streams (also air). Heat that is transferred by convection to or from these surface portions is one type of these boundary conditions. Another type of boundary conditions involves a specified magnitude of heat flux through the rotor at some parts of its surfaces. This type includes thermally insulated surfaces, which, in reality, allow a negligible amount of heat flux.

Governing Equations

The differential equation governing the heat conduction through a three-dimensional body, through which the thermal conductivity is a variable, can be stated as follows:

$$\frac{\partial}{\partial x}(kT_x)+\frac{\partial}{\partial y}(kT_y)+\frac{\partial}{\partial z}(kT_z)+g=0$$

where

- $T_x \equiv \partial T/\partial x$.
- $T_y \equiv \partial T/\partial y$.
- $T_z \equiv \partial T/\partial z$.
- k is the body-material thermal conductivity.
- g represents the strength, generally, of the heat sources (or sinks) that may exist inside the body and is numerically equal to the rate of heat energy generated (or absorbed) per unit volume of the solid body. Of course, g is positive for sources and negative for sinks, with the latter being the case here.

For convenience, the groups of surface segments through which heat is transferred by convection will be referred to by group S. On the other hand, the segments on which the heat flux is specified will be denoted by C. The boundary conditions that may be assigned to the body surface, therefore, can be expressed as follows:

$$k\frac{\partial T}{\partial x}l+k\frac{\partial T}{\partial y}m+\frac{\partial T}{\partial z}n+h(T-T_\infty)=0 \qquad \text{on convection surfaces } S$$

$$k\frac{\partial T}{\partial x}l+k\frac{\partial T}{\partial y}m+k\frac{\partial T}{\partial z}n+q=0 \qquad \text{on the group } C$$

where

- h is the local convection heat transfer coefficient.
- T_∞ is the local stream temperature.
- q is the locally specified heat flux.
- l, m, and n are the direction cosines of the local outward unit vector normal to the surface under consideration.

Finite Element Variational Formulation

As a reminder, continuum problems often have two conceptually equivalent formulations: a differential formulation and a variational one. In the differential formulation, the problem is to integrate a differential equation or a system of such equations, subject to given boundary conditions. In the variational formulation, however, the problem is to find the unknown function or functions that extremize or make stationary a functional or a system of functionals, subject to the same given boundary conditions.

The equivalence of these two formulations is apparent from the calculus of variations, which shows that the functionals are extremized only when one or more of the Euler equations and their boundary conditions are satisfied. These equations are precisely the governing differential equations of the problem. In the following, the variational formulation of the problem will be deduced using Euler's theorem of variational calculus.

Euler's Theorem of Variational Calculus

Let $I(T)$ denote a functional defined as follows:

$$I(T)=\int_V f(x,y,z,T,T_x,T_y,T_z)dV+\int_S\left(\frac{aT^2}{2}-bT\right)dS+\int_C eT\,dC$$

where

- V represents the domain volume.
- S and C represent portions of the boundaries of that volume on which physical boundary conditions, other than specified values of the field variable T, are prescribed. These include convection heat transfer and the effect of specified heat flux.
- a, b, and c are physical constants.

If the functional $I(T)$ is to be extremized $[\delta I(T) = 0]$, then the necessary and sufficient conditions to be satisfied are

$$\frac{\partial}{\partial x}\left(\frac{\partial I}{\partial T_x}\right) + \frac{\partial}{\partial y}\left(\frac{\partial I}{\partial T_y}\right) + \frac{\partial}{\partial z}\left(\frac{\partial I}{\partial T_z}\right) - \frac{\partial I}{\partial T} = 0 \qquad \text{within } V$$

$$\frac{\partial I}{\partial T_x} l + \frac{\partial I}{\partial T_y} m + \frac{\partial I}{\partial T_z} n + aT - b = 0 \qquad \text{on the surface } S \text{ (convection-related)}$$

$$\frac{\partial I}{\partial T_x} l + \frac{\partial I}{\partial T_y} m + \frac{\partial I}{\partial T_z} n + c = 0 \qquad \text{on the surface } C \text{ (specified-heat-flux-related)}$$

The preceding three expressions are referred to as *Euler-Lagrange conditions.*

Derivation of the Variational Statement

The variational statement of the present problem can be directly derived through comparing the partial differential equation (earlier) and boundary conditions with the Euler-Lagrange conditions. Doing so, we get

$$\frac{\partial f}{\partial T_x} = kT_x \qquad \text{or} \qquad f = \frac{1}{2} k T_x^2 + F_1(T, T_y, T_z)$$

$$\frac{\partial f}{\partial T_y} = kT_y \qquad \text{or} \qquad f = \frac{1}{2} k T_y^2 + F_2(T, T_x, T_z)$$

$$\frac{\partial f}{\partial T_z} = kT_z \qquad \text{or} \qquad f = \frac{1}{2} k T_z^2 + F_3(T, T_x, T_y)$$

$$\frac{\partial I}{\partial T} = -g \qquad \text{or} \qquad f = -gT + F_4(T_x, T_y, T_z)$$

where T, T_x, T_y, and T_z are functions of x, y, and z. From the last four equations, the following expression for $f(T, T_x, T_y, T_z)$ can be easily obtained:

$$f = \frac{k}{2}\left[\left(\frac{\partial T}{\partial x}\right)^2 + \left(\frac{\partial T}{\partial y}\right)^2 + \left(\frac{\partial T}{\partial z}\right)^2\right] - gT$$

with the negative sign of the heat-source term implying that the assumed heat generation here, through a heat source, is actually a case of heat absorption through heat sinks.

Now Euler-Lagrange conditions can be rewritten as follows:

$$k\frac{\partial T}{\partial x} l + k\frac{\partial T}{\partial y} m + k\frac{\partial T}{\partial z} n + aT - b = 0 \qquad \text{on surface } S$$

$$k\frac{\partial T}{\partial x} l + k\frac{\partial T}{\partial y} m + k\frac{\partial T}{\partial z} n + e = 0 \qquad \text{on surface } C$$

At this point, we are prepared to draw the following correspondences:

$$a \equiv h$$

$$b \equiv hT_\infty$$

$$c \equiv q$$

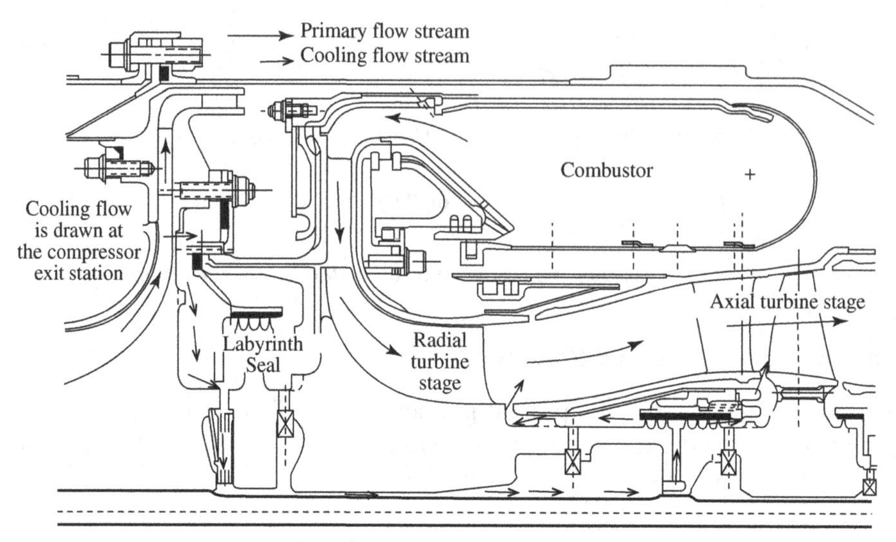

Figure 8.2. Typical path of the cooling air across the gas generator.

Finally, we arrive at the following expression for $I(T)$:

$$I(T) = \frac{1}{2}\int_V \left[k\left(\frac{\partial T}{\partial x}\right)^2 + k\left(\frac{\partial T}{\partial y}\right)^2 + k\left(\frac{\partial T}{\partial z}\right)^2 - 2gT \right] dV + \frac{1}{2}\int_S h(T^2 - 2TT_\infty)dS - \int_C qT\,dC$$

Again, solution of the conduction-governing equation (earlier) and the associated boundary conditions are equivalent to the determination of the temperature distribution $T(x, y, z)$ that makes stationary the functional $I(T)$

Figure 8.2 shows the "coolant" flow path from the last compressor stage all the way up to our first-stage turbine where it is being discharged. The coolant here is naturally a relatively low-temperature stream of air that is unburned and possesses a pressure with a magnitude that is roughly equal to that at the first-stage-turbine inlet station.

Figure 8.3 shows the rotor "slice," or unit, in which the temperature study is to be conducted. The slice here, by reference to Figure 8.3, is composed of one of the rotor blades together with its corresponding hub segment.

Figure 8.4 illustrates the internal cooling techniques that will be investigated in this section. The final temperature distributions associated with these cooling categories are also shown in the same figure and will be discussed later in this section. Note that the temperature magnitudes in this figure are in degrees Fahrenheit.

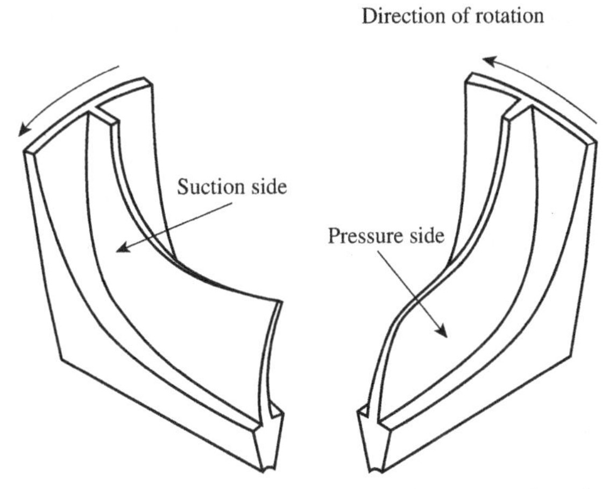

Figure 8.3. A "slice" of the rotor unit to be analyzed.

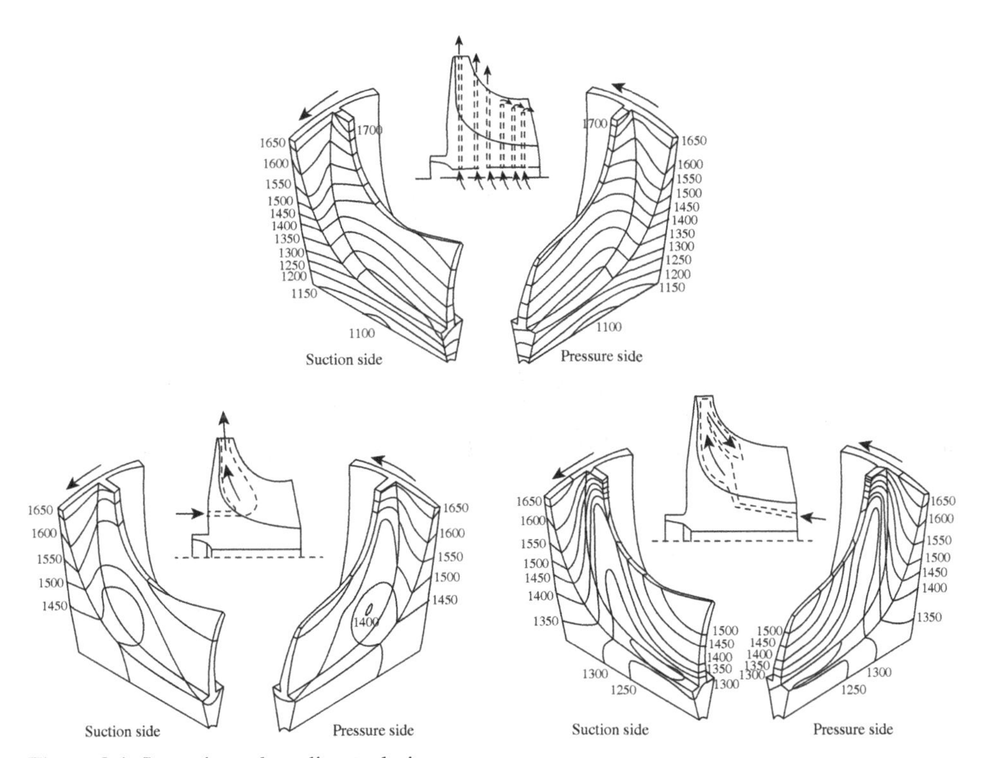

Figure 8.4. Investigated cooling techniques.

Discretization of the Continuum

The temperature distributions in Figure 8.5 are but samples of the finite-element results. To arrive at those, the computational domain was first discretized using tetrahedrons as the discretization units as shown in Figure 8.5. Figure 8.6 shows the two mechanisms used to externally cool the rotor backdisk. These lead to a coolant air stream that is directed radially inwards then radially outwards. In this, and similar cases, the necessary condition for $I(T)$ to have an extremum becomes:

$$\frac{dI}{d\{T\}} = \{0\}$$

where $\{T\}$ is the column vector containing all temperatures at the nodal points, that is,

$$\left\{ \begin{array}{c} \{T\} = T_1 \\ T_2 \\ \cdot \\ \cdot \\ \cdot \\ T_M \end{array} \right\}$$

Consequently,

$$\frac{dI}{d\{T\}} = \left\{ \begin{array}{c} \dfrac{\partial I}{\partial T_1} \\ \dfrac{\partial I}{\partial T_2} \\ \cdot \\ \cdot \\ \cdot \\ \dfrac{\partial I}{\partial T_M} \end{array} \right\}$$

Before evaluating the different differentiations (earlier), it is convenient to break the functional $I(T)$ up into four separate integrals as follows:

$$I = I_k + I_g + I_h - I_q$$

where

$$I_k = \frac{1}{2}\int_V \left[k\left(\frac{\partial T}{\partial x}\right)^2 + k\left(\frac{\partial T}{\partial y}\right)^2 + k\left(\frac{\partial T}{\partial z}\right)^2 \right] dV$$

$$I_g = \int_V gT\,dV$$

$$I_h = \frac{1}{2}\int_S h(T^2 - 2TT_\infty)\,dS$$

$$I_q = \int_C qT\,dC$$

which yields

$$\frac{dI_k}{d\{T\}} - \frac{dI_g}{d\{T\}} + \frac{dI_h}{d\{T\}} - \frac{dI_q}{d\{T\}} = \{0\}$$

The next step is to evaluate each of the four terms appearing in the preceding equation. In each case, the field of integration will be discretized into finite-elements of varying volumes. An interpolation function will be assumed to approximate the field variable (meaning the temperature T) within each element. Choices of the element's shape and interpolation order are done in such a way that convergence of the numerical results is achieved as the element sizes are progressively reduced down to zero.

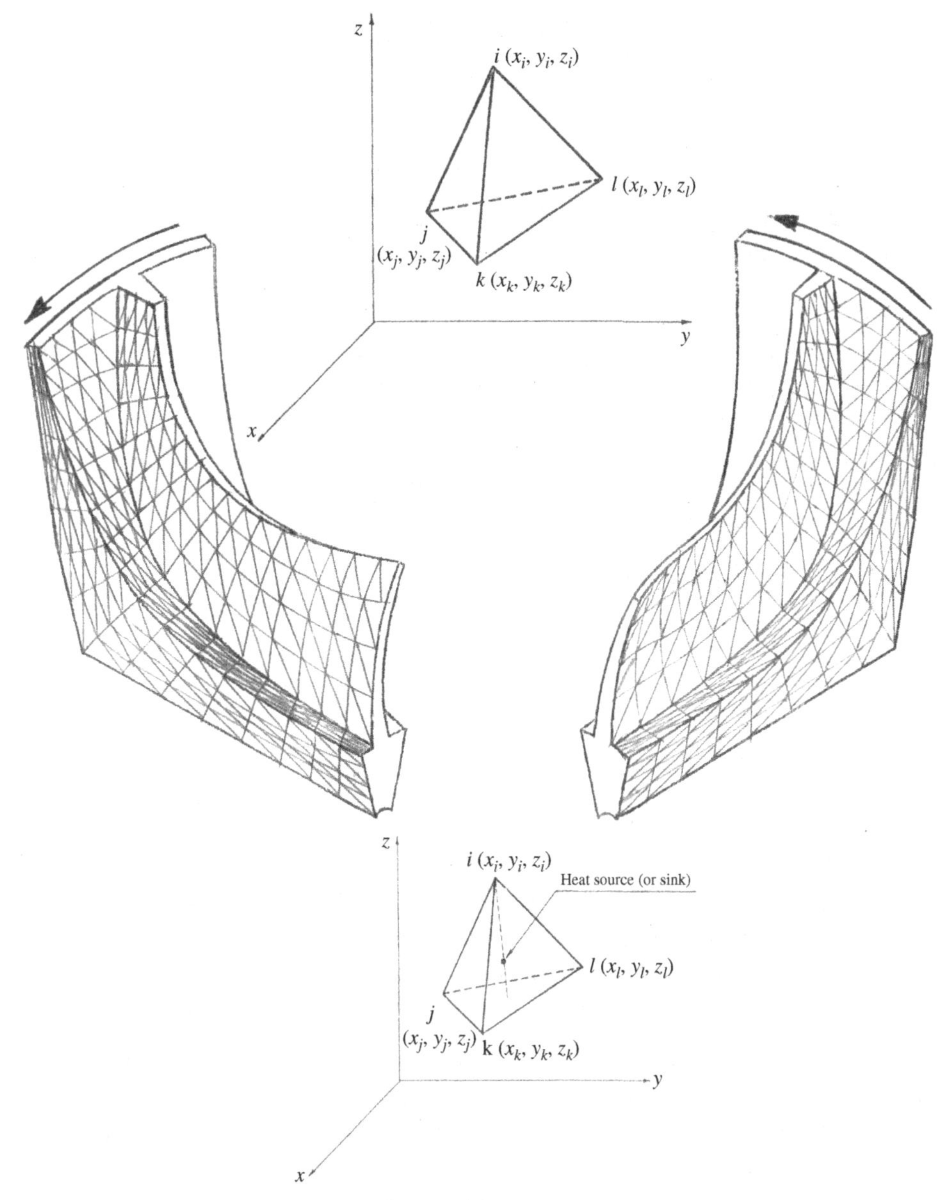

Figure 8.5. Finite element discretization model.

Evaluation of $dI_k/d\{T\}$

Because the field integration of I_k is the three-dimensional body volume V, the discretization of this field must take the form of three-dimensional elements. These "finite" elements are, in this section, chosen to be four-noded tetrahedrons (Figure 8.5) occupying the field of integration (V) and having the M nodal points as their vertices. This type of element is advantageous over cubic elements and triangular prisms because it can match the body curved boundary segments better. A typical tetrahedral element (e) is shown in Figure 8.5, whereas Figure 8.6 shows the nodal points' numbering scheme on the meridional rotor segment projection.

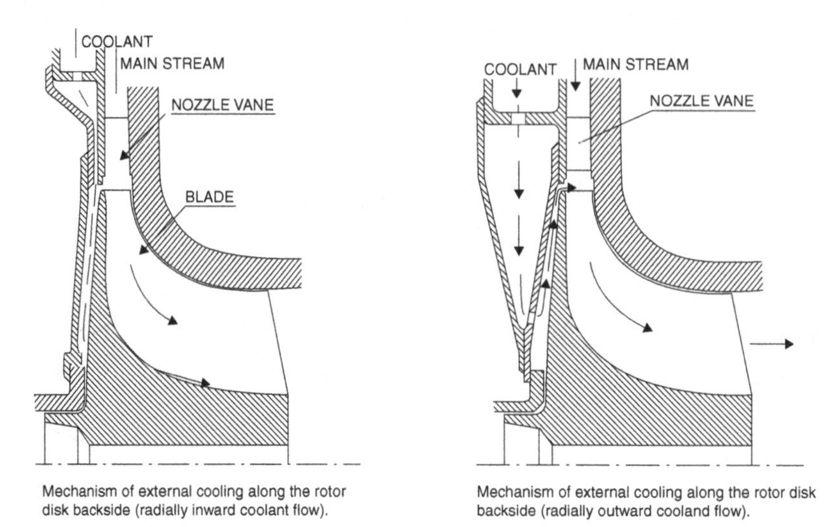

Figure 8.6. Mechanisms of External Cooling.

The discretized form of I_k takes the following form:

$$I_k = \sum_{e=1}^{M} I_k{}^{(e)}$$

where

- (e) refers to a typical tetrahedral element.
- E is the total number of elements occupying the volume V.
- $I_k{}^{(e)} = \frac{1}{2}\int_{V^{(e)}} \left[k^{(e)}(\partial T^{(e)}/\partial x)^2 + k^{(e)}(\partial T^{(e)}/\partial y)^2 + k^{(e)}(\partial T^{(e)}/\partial z)^2\right] dV$.
- $V^{(e)}$ represents the volume of the typical element (e).
- $T^{(e)}$ represents the temperature within this element.
- $k^{(e)}$ represents the average thermal conductivity within the element.

Now an $M \times 4$ matrix $[D^{(e)}]$ is assigned to each element such that, for a typical element (e), the rows of this matrix contain zero entries except for four rows, corresponding to the nodal numbers assigned to the tetrahedron's vertices, as follows:

$$\begin{bmatrix} 0 & 0 & 0 & 0 \\ \cdot & \cdot & \cdot & \cdot \\ \cdot & \cdot & \cdot & \cdot \\ 1 & 0 & 0 & 0 \\ \cdot & \cdot & \cdot & \cdot \\ \cdot & \cdot & \cdot & \cdot \\ 0 & 1 & 0 & 0 \\ \cdot & \cdot & \cdot & \cdot \\ \cdot & \cdot & \cdot & \cdot \\ 0 & 0 & 1 & 0 \\ \cdot & \cdot & \cdot & \cdot \\ \cdot & \cdot & \cdot & \cdot \\ 0 & 0 & 0 & 1 \\ \cdot & \cdot & \cdot & \cdot \\ \cdot & \cdot & \cdot & \cdot \end{bmatrix}$$

with the nonzero rows corresponding to the numbers assigned to the tetrahedron's vertices.

This "mapping" matrix $[D^{(e)}]$ helps in placing the contributions of the local (4×4) local matrix of "influence" coefficients appropriately in the so-called global matrix of coefficients. Each row in this matrix containing the digit 1 corresponds to the local number that is assigned to the element in the global assembly of elements.

It is clear that ${I_k}^{(e)}$ depends only on the column vector $\{T\}^{(e)}$, where

$$\{T\}^{(e)} = \begin{Bmatrix} T_1 \\ T_2 \\ T_3 \\ T_4 \end{Bmatrix}$$

Thus, referring to the definition of $[D]^{(e)}$, the following relationship can be obtained:

$$\frac{d{I_k}^{(e)}}{d\{T\}} = [D]^{(e)} \frac{d{I_k}^{(e)}}{d\{T\}^{(e)}}$$

where

$$\frac{d{I_k}^{(e)}}{d\{T\}^{(e)}} = \begin{Bmatrix} \dfrac{d{I_k}^{(e)}}{dT_i} \\ \dfrac{d{I_k}^{(e)}}{dT_j} \\ \dfrac{d{I_k}^{(e)}}{dT_k} \\ \dfrac{d{I_k}^{(e)}}{dT_l} \end{Bmatrix}$$

where i, j, k and l are nodal numbers assigned to the finite-element vertices. Through simple substitution, we get

$$\frac{dI_k}{d\{T\}} = \sum_{e=1}^{E} [D]^{(e)} \frac{d{I_k}^{(e)}}{d\{T\}^{(e)}}$$

In order to evaluate $d{I_k}^{(e)}/d\{T\}$, an interpolation function, which approximates the piecewise temperature variation within the element (e), has to be assumed. Because the functions appearing under the integral sign in the element's matrix equation does not contain derivatives of higher than the first order, the interpolation function must be so selected to ensure, only, the temperature continuity (and not the temperature normal derivatives) across the elements' interfaces. The linear interpolation function, in this case, assumes the following form:

$$T^{(e)} = \{P\}^T \{C\}^{(e)}$$

where

- $\{P\}^T$ represents the row matrix [1 x y z], and

- $\{C\}^{(e)}$ is the element's column vector of constants, defined as follows:

$$\{C\}^{(e)} = \begin{Bmatrix} C_1 \\ C_2 \\ C_3 \\ C_4 \end{Bmatrix}$$

$\{C\}^{(e)}$ is to be evaluated in terms of the coordinates of and temperatures at the tetrahedral element's vertices i, j, k, and l.

Substituting the coordinates and temperatures of the element's vertices, a set of three simultaneous algebraic equations is obtained,

$$\begin{Bmatrix} T_i \\ T_j \\ T_k \\ T_l \end{Bmatrix} = \begin{bmatrix} 1 & x_i & y_i & z_i \\ 1 & x_j & y_j & z_j \\ 1 & x_k & y_k & z_k \\ 1 & x_l & y_l & z_l \end{bmatrix} \begin{Bmatrix} C_1 \\ C_2 \\ C_3 \\ C_4 \end{Bmatrix}$$

which can be rewritten in the following compact form:

$$\{T\}^{(e)} = [P]^{(e)}\{C\}^{(e)}$$

Premultiplying the preceding equation by $[R]^{(e)} = [P^{(e)^{-1}}]$, the following equation is obtained:

$$\{C\}^{(e)} = [R]^{(e)}\{T\}^{(e)}$$

Substituting this expression into the expanded element's matrix equation, the interpolation function takes the following form:

$$T^{(e)} = \{P\}^T[R]^{(e)}\{T\}^{(e)}$$

Differentiating,

$$\frac{\partial T^{(e)}}{\partial x} = \{P_x\}^T[R]^{(e)}\{T\}^{(e)}$$

$$\frac{\partial T^{(e)}}{\partial y} = \{P_y\}^T[R]^{(e)}\{T\}^{(e)}$$

$$\frac{\partial T^{(e)}}{\partial z} = \{P_z\}^T[R]^{(e)}\{T\}^{(e)}$$

where

$$[P_x]^T = [0\ 1\ 0\ 0]$$

$$[P_y]^T = [0\ 0\ 1\ 0]$$

$$[P_z]^T = [0\ 0\ 0\ 1]$$

Now, substituting these expressions into the definition of the elemental functional $I_k{}^{(e)}$, we get

$$I_k{}^{(e)} = \frac{k^{(e)}}{2} \int_{V^{(e)}} [\,(\{P_x\}^T[R]^{(e)}\{T\}^{(e)})^2 + (\{P_y\}^T[R]^{(e)}\{T\}^{(e)})^2 + (\{P_z\}^T[R]^{(e)}\{T\}^{(e)})^2\,]\,dV^{(e)}$$

Making use of the rules governing differentiating the product of matrices, and noting that all terms of the form similar to $\{P_x\}^T[R]^{(e)}\{T\}^{(e)}$ are scalar quantities, the differentiation of $I_k{}^{(e)}$ with respect to the column vector $\{T\}^{(e)}$ gives the following result:

$$\frac{dI_k{}^{(e)}}{d\{T\}^{(e)}} = k^{(e)}[R]^{(e)}\left[\int_{V^{(e)}}(\{P_x\}\{P_x\}^T + \{P_y\}\{P_y\}^T + \{P_z\}\{P_z\}^T)dV^{(e)}\right][R]^{(e)}\{T\}^{(e)}]dV^{(e)}$$

Using the expressions for $\{P_x\}$, $\{P_y\}$, and $\{P_z\}$, the integrand in the preceding equation assumes the following form:

$$\{P_x\}\{P_x\}^T + \{P_y\}\{P_y\}^T + \{P_z\}\{P_z\}^T = \begin{bmatrix} 0 & 0 & 0 & 0 \\ 0 & 1 & 0 & 0 \\ 0 & 0 & 1 & 0 \\ 0 & 0 & 0 & 1 \end{bmatrix}$$

which equals a constant matrix $[E]$.

Through simple substitution, we get

$$\frac{dI_k{}^{(e)}}{d\{T\}^{(e)}} = k^{(e)}V^{(e)}[R]^T[E][R]^{(e)}\{T\}^{(e)} = [K]^{(e)}\{T\}^{(e)}$$

where $$[K]^{(e)} = k^{(e)}V^{(e)}R]^{(e)^T}[E][R]^{(e)}$$

with $V^{(e)}$ being the volume of the typical tetrahedral finite element (e) with vertices i, j, k, and l:

$$V^{(e)} = \frac{1}{6}\begin{vmatrix} 1 & x_i & y_i & z_i \\ 1 & x_j & y_j & z_j \\ 1 & x_k & y_k & z_k \\ 1 & x_l & y_l & z_l \end{vmatrix}$$

The column vector $\{T\}^{(e)}$, containing the element's four nodal temperatures, can be related to the general (or global) column vector $\{T\}$, which contains all the nodal temperatures, through the use of the definition of the (mapping) matrix $[D]^{(e)}$ as follows:

$$\{T\}^{(e)} = [D]^{(e)^T}\{T\}$$

Finally, and through simple substitution, the following result is attained:

$$\frac{dI_k}{d\{T\}} = \left(\sum_{e=1}^{E}[D]^{(e)}[K]^{(e)}[D]^{(e)^T}\right)\{T\}$$

where $$[K]^{(e)} = \sum_{e=1}^{E}[D]^{(e)}[K]^{(e)}[D]^{(e)^T}$$

The matrix $[K]$ is an $M \times M$ matrix and is called the *global conduction matrix*.

Evaluation of $dI_g/d\{T\}$

The discretized form of the functional $I_g(T)$ is

$$I_g = \sum_{n=1}^{N} I_g{}^{(n)}$$

where
$$I_g{}^{(n)} = g^{(n)} \int_{V^{(e)}} T^{(n)}\, dV$$

with the following symbols' interpretations:

- N is the total number of elements where heat is generated (or absorbed).
- (n) refers to a typical element in which a heat source (or sink) exists.
- $g^{(n)}$ is the strength of the heat source (or sink).
- $V^{(n)}$ is the volume of the typical element (n).
- $\{T\}^{(n)}$ is the temperature-interpolating function within the element (n).

Following a procedure similar to that used in evaluating $dI_k/d\{T\}$, the interpolation function takes the following form:

$$T^{(n)} = \{P\}^T [R]^{(n)} \{T\}^{(n)}$$

where
$$\{P\}^T = [1\ x\ y\ z]$$

$$[R]^{(n)} = \begin{bmatrix} 1 & x_i & y_i & z_i \\ 1 & x_j & y_j & z_j \\ 1 & x_k & y_k & z_k \\ 1 & x_l & y_l & z_l \end{bmatrix}$$

$$\{T\}^{(n)} = \begin{Bmatrix} T_i \\ T_j \\ T_k \\ T_l \end{Bmatrix}$$

in which i, j, k, and l represent the numbers assigned to the vertices of the typical element (n), as before.

Now we get, through simple substitution and differentiation, the following:

$$\frac{dI_g{}^{(n)}}{d\{T\}^{(n)}} = g^{(n)} [R]^{(n)T} \int_{V^{(n)}} \{P\}\, dV$$

where
$$P = \begin{Bmatrix} 1 \\ x \\ y \\ z \end{Bmatrix}$$

The preceding equation can be rewritten in the following form:

$$\frac{dI_g{}^{(n)}}{d\{T\}^{(n)}} = g^{(n)}[R]^{(n)T}\begin{Bmatrix} \int_{V^{(n)}} dV \\ \int_{V^{(n)}} x dV \\ \int_{V^{(n)}} y dV \\ \int_{V^{(n)}} z dV \end{Bmatrix} = g^{(n)}V^{(n)}[R]^{(n)^T}\{P\}_C{}^{(n)} = \{G\}^{(n)}$$

where $$\{G\}^{(n)} = g^{(n)}V^{(n)}[R]^{(n)^T}\{P\}_C{}^{(n)}$$

and $$\{P\}_C{}^{(n)} = \begin{Bmatrix} 1 \\ x_c \\ y_c \\ z_c \end{Bmatrix}$$

with x_c, y_c, and z_c being the centroid coordinates of the typical element (n):

$$x_c = \frac{1}{4}(x_i + x_j + x_k + x_l)$$

$$y_c = \frac{1}{4}(y_i + y_j + y_k + y_l)$$

$$z_c = \frac{1}{4}(z_i + z_j + z_k + z_l)$$

Introducing the concept of the "mapping" matrix $[D]$, the following expression is obtained:

$$\frac{dI_g}{d\{T\}} = \sum_{n=1}^{N} [D]^{(n)}\{G\}^{(n)} = [G]$$

The global matrix $[G]$ represents the summation of the properties of all elements within which heat energy is produced (or absorbed).

Evaluation of $dI_h/d\{T\}$

The discretized form of the functional $I_h(T)$ is

$$I_h = \sum_{l=1}^{L} I_h{}^{(l)}$$

where $$I_h{}^{(l)} = \frac{h^{(l)}}{2}\int_{S^{(l)}} 2(T^{(l)2} - 2T_\infty{}^{(l)}\{T\}^{(l)})dS$$

and

- l refers to a typical triangular element on the body surface through which heat is transferred by convection (Figure 8.7). Analysis of this (three-dimensional) surface element, within the natural-coordinate framework, is contained in Appendix A.
- L represents the total number of such triangular elements.
- $S^{(l)}$ is the area of the triangular element (l) on the body's surface.

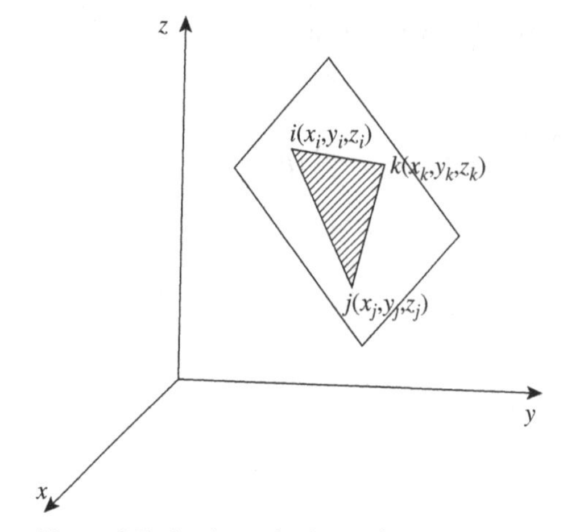

Figure 8.7. A triangular boundary element.

Differentiation of the preceding equation yields:

$$\frac{dI_h}{d\{T\}} = \sum_{l=1}^{L} \frac{dI_h{}^{(l)}}{d\{T\}} \sum_{l=1}^{L} \frac{dI_h{}^{(l)}}{d\{T\}}$$

For each triangular element in this family, an $M \times 3$ matrix $[D]$ is now defined such that for a typical element (l) with the nodal numbers i, j, and k assigned to its vertices, the rows of the matrix $[D]^{(l)}$ contain zero entries except for the ith, jth, and kth rows:

$$D^{(l)} = \begin{bmatrix} 0 & 0 & 0 \\ \cdot & \cdot & \cdot \\ \cdot & \cdot & \cdot \\ 1 & 0 & 0 \\ \cdot & \cdot & \cdot \\ \cdot & \cdot & \cdot \\ 0 & 1 & 0 \\ \cdot & \cdot & \cdot \\ \cdot & \cdot & \cdot \\ 0 & 0 & 1 \\ \cdot & \cdot & \cdot \\ \cdot & \cdot & \cdot \end{bmatrix}$$

where i, j, and k are the numbers assigned to the three nodal points which constitute the vertices of the element (l). With this definition, the preceding equation can be rewritten as follows:

$$\frac{dI_h}{d\{T\}} = \sum_{l=1}^{L} [D]^{(l)} \frac{dI_h{}^{(l)}}{d\{T\}^{(l)}}$$

where L is the total number of triangular surface elements covering the body's boundary where heat is transferred by convection.

Referring to Appendix A, the natural coordinates are now employed to approximate the temperature distribution within the triangular element (l), resulting in the following form of an interpolation function:

$$T^{(l)} = [L_1 L_2 L_3] \begin{Bmatrix} C_1 \\ C_2 \\ C_3 \end{Bmatrix}$$

where:

$$\{P\}^T = [L_1 L_2 L_3]$$

Substituting the coordinates and temperatures of the nodal points i, j, and k into the preceding equation, the following set of simultaneous algebraic equations is obtained:

$$\{T\}^{(l)} = \begin{Bmatrix} T_i \\ T_j \\ T_k \end{Bmatrix} = \begin{bmatrix} L_{1i} & L_{2i} & L_{3i} \\ L_{1j} & L_{2j} & L_{3j} \\ L_{1k} & L_{2k} & L_{3k} \end{bmatrix} \begin{Bmatrix} C_1 \\ C_2 \\ C_3 \end{Bmatrix} = \begin{bmatrix} 1 & 0 & 0 \\ 0 & 1 & 0 \\ 0 & 0 & 1 \end{bmatrix} \begin{Bmatrix} C_1 \\ C_2 \\ C_3 \end{Bmatrix}$$

or $\{C\}^{(l)} = \{T\}^{(l)}$

where $\{T\}^{(l)} = \begin{Bmatrix} T_i \\ T_j \\ T_k \end{Bmatrix}$

Through simple substitution, the following expression is obtained on carrying out the differentiation process:

$$\frac{dI_h{}^{(l)}}{d\{T\}^{(l)}} = h^{(l)} \begin{bmatrix} \int_{S^{(l)}} L_1{}^2 dS & \int_{S^{(l)}} L_1 L_2 dS & \int_{S^{(l)}} L_1 L_3 dS \\ \int_{S^{(l)}} L_1 L_2 dS & \int_{S^{(l)}} L_2{}^2 dS & \int_{S^{(l)}} L_2 L_3 dS \\ \int_{S^{(l)}} L_1 L_3 dS & \int_{S^{(l)}} L_2 L_3 dS & \int_{S^{(l)}} L_3{}^2 dS \end{bmatrix} \begin{Bmatrix} T_i \\ T_j \\ T_k \end{Bmatrix} - h^{(l)} T_\infty{}^{(l)} \begin{Bmatrix} \int_{S^{(l)}} L_1 dS \\ \int_{S^{(l)}} L_2 dS \\ \int_{S^{(l)}} L_3 dS \end{Bmatrix}$$

Each element in the last two matrices can be evaluated by substituting the proper values of natural coordinates' indices (Huebner, K.H., 1975) as follows:

$$\frac{dI_h{}^{(l)}}{d\{T\}^{(l)}} = \frac{1}{12} h^{(l)} T_\infty{}^{(l)} S^{(l)} \begin{Bmatrix} 1 \\ 1 \\ 1 \end{Bmatrix} = [H]^{(l)} \{T\}^{(l)} - \{F\}^{(l)}$$

where $[H]^{(l)} = \frac{1}{12} h^{(l)} S^{(l)} \begin{bmatrix} 2 & 1 & 1 \\ 1 & 2 & 1 \\ 1 & 1 & 2 \end{bmatrix}$

$$\{F\}^{(l)} = \frac{1}{3} h^{(l)} T_{\infty}{}^{(l)} S^{(l)} \begin{Bmatrix} 1 \\ 1 \\ 1 \end{Bmatrix}$$

Now we can derive the following expression:

$$\begin{aligned} \frac{dI_h}{d\{T\}} &= \sum_{l=1}^{L} [D]^{(l)} ([H]^{(l)} \{T\}^{(l)} - \{F\}^{(l)}) \\ &= \left(\sum_{l=1}^{L} [D]^{(l)} [H]^{(l)} [D]^{(l)^T} \right) \{T\} - \sum_{l=1}^{L} [D]^{(l)} \{T\}^{(l)} = [H]\{T\} - \{F\} \end{aligned}$$

where
$$[H] = \sum_{l=1}^{L} [D]^{(l)} [H]^{(l)} [D]^{(l)^T}$$

$$\{F\} = \sum_{l=1}^{L} [D]^{(l)} \{F\}^{(l)}$$

Note that $[H]$ is an $M \times M$ matrix, whereas $\{F\}$ is an $M \times 1$ column vector, where M is the total number of nodes.

Evaluation of $dI_q/d\{T\}$

A derivation similar to that used in evaluating $dI_h/d\{T\}$ can be followed, leading to the following final result:

$$\frac{dI_q}{d\{T\}} = \sum_{r=1}^{R} [D]^{(r)} \{Q\}^{(r)}$$

where
$$\{Q\}^{(r)} = \frac{1}{3} q^{(r)} c^{(r)} \begin{Bmatrix} 1 \\ 1 \\ 1 \end{Bmatrix}$$

and

- (r) refers to a typical triangular element through which a specified heat flux is transferred.
- $c^{(r)}$ is the element's area.
- $q^{(r)}$ is the value of specified heat flux.
- R is the total number of the specified heat-flux elements.

The preceding equation can be rewritten in the following compact form:

$$\frac{dI_q}{d\{T\}} = \{Q\}$$

$$\{Q\} = \sum_{r=1}^{R} [D]^{(r)} \{Q\}^{(r)}$$

The Final Set of Equations

Through simple substitution, we get

$$([K]+[H])\{T\}=\{G\}+\{F\}+\{Q\}$$

where $\{T\}$ is the $M \times 1$ column vector of all nodal temperatures. This equation can be rewritten in the following compact form:

$$[A]\{T\}=\{B\}$$

where

$$[A]=[K]+[H]$$

$$\{B\}=\{G\}+\{F\}+\{Q\}$$

$[A]$ is an $M \times M$ square matrix, whereas $\{B\}$ is an $M \times 1$ column vector. Before inverting the matrix of "influence" coefficients, the final boundary condition type is invoked. This has to do with the specified-temperature nodes. These nodes will overwhelmingly exist on the body surface. This kind of boundary conditions was discussed earlier in this chapter in conjunction with two-dimensional heat conduction applications, and will therefore not be repeated here.

Turbine Rotor Configuration and Cooling Techniques

The task at hand now is to predict is to predict the temperature distribution in a hot radial-inflow turbine rotor (Figure 8.1) and to investigate the effectiveness of different cooling techniques of the rotor blades. In all cases, the coolant flow rate is not to exceed 3 percent of the primary flow rate.

Figure 8.1 shows two views of this rotor. The cooling-air pathway from the exit station of the last compressor stage, which leads to the first-stage rotor (where the coolant is discharged), is shown in Figure 8.2.

A turbine rotor having the same geometry and operating conditions as that of reference [1] was selected. The rotor was divided into 12 identical units, where the number of rotor blades is indeed 12, with each unit consisting of one blade and its corresponding hub segment. The temperature distribution is predicted in only one of these units and is assumed to be the same in all other units. Figure 8.3 shows the geometry of this rotor unit. The turbine operating conditions are as follows:

- Rotor speed = 67,000 rpm.
- Hot-gas mass flow rate = 2 kg/s.
- Relative total temperature at the rotor inlet station = 1283 K.
- Absolute total temperature at the rotor inlet station = 2205 K.
- Absolute total pressure at the rotor inlet station = 17.5 bar.

In the following, the methods used in calculating the boundary conditions for the hot (uncooled) turbine rotor are presented. Different cooling methods will also be discussed, together with the method of calculating the boundary conditions associated with each of them.

Determination of the Hot Turbine Boundary Conditions

Rotor Blade

This is done using the relative velocity distribution of the main gas stream along the shroud, hub and mean lines which are given in the same reference, and through extensive interpolation, we arrive at the nodal-point magnitude of relative velocity to be used in calculating the convection heat transfer coefficient (h). The heat transfer coefficient is calculated along the rotor blade using Von Karman's form of Reynolds' analogy [6], namely,

$$\frac{Nu}{Re^{0.5}} = uPr\left(\frac{c_f Re}{2}\right)^{0.5}\left\{\left(\frac{c_f}{2}\right)^{-0.5}+5\left[\left(Pr-1\right)+\frac{5Pr+1}{6}\right)\right]\right\}^{-1}$$

where
$$Nu = \frac{hc}{k} \quad \text{(Nusselt number)}$$
$$Re = \frac{Wc}{\nu} \quad \text{(Reynolds number)}$$
$$Pr = \frac{c_p \mu}{k} \quad \text{(Prandtl number)}$$

and

- c is a characteristic length, taken as the mean magnitude of the blade true chord.
- h is the convection heat transfer coefficient.
- W is the local relative free-stream velocity.
- μ is the gas dynamic viscosity coefficient.
- ν is the gas kinematic viscosity coefficient.
- c_f is the local friction coefficient.
- $u = W/W_2$, with W_2 being the gas relative velocity far downstream from the rotor exit station.

In the empirical formula (earlier), the local friction coefficient c_f is calculated using the following empirical relationship for turbulent flow over plane surfaces [7]:

$$c_{f_x} = 0.0576\left(\frac{W_x}{\nu}\right)^{-0.2}$$

where x is the distance along the blade measured from the blade leading edge.

For the stagnation condition heat transfer coefficient, the following relationship [4] was used:

$$\frac{Nu}{Re^{0.5}} = 0.57Pr^{0.4}\left(\frac{W_1}{W_2}\cdot\frac{2c}{r_1}\right)^{0.5}$$

where

- r_1 is the blade leading-edge radius.
- W_1 and W_2 are the relative gas velocities far upstream and far downstream from the rotor, respectively.

Rotor Disk Backside

The heat transfer coefficient distribution on the rotor disk backside is calculated using the following two relationships [7]:

$$Nu_l = \frac{hr_O}{k} = 0.35\left(\frac{\omega r_o{}^2}{\nu}\right)^{0.5} \qquad \text{(laminar flow)}$$

$$Nu_t = \frac{hr}{k} = 0.0195\left(\frac{\omega r^2}{\nu}\right)^{0.8} \qquad \text{(turbulent flow)}$$

where

- ω is the rotor speed in rad/s.
- r is the local radius.
- r_O is the outer radius of the disk backside.

The critical rotational Reynolds number $\omega r_{crit}{}^2/\nu$, at which transition from laminar to turbulent flow occurs, was taken to be 250,000.

Rotor Hub

The two expressions relating the Nusselt number to the heat transfer coefficient, together with the main-flow relative velocity distribution along the hub, were used to calculate the heat transfer coefficient distribution along the hub.

Cooled Turbine Rotor Calculations

Figure 8.3 shows two isometric views of the rotor-blade unit under investigation. In external cooling of the rotor-disk backside, the coolant emerges from a circumferential slot located at the trailing edges of the stator blades (actually the correct phrase here should be stator vanes), and is concentric with the turbine rotor disk. The coolant is allowed to flow radially inward over the rotor disk backside and, then, is discharged through holes in the hot-air stream along the rotor hub. In the other type of the rotor disk cooling, the cooling airflow is directed radially outward and terminates at the leading edge (Figure 8.6), where it is discharged. The rotor blade is cooled in two different ways in which the cooling air exists inside the blade through a slot and is then terminated at the blade leading edge (Figure 8.4) or through 1.8-mm-diameter radial holes drilled in the blade and is, finally, injected at the suction side of the blade, where it is needed to provide momentum to what could very well be a region of flow separation and recirculation. Note that this is the optimal side, for the reason just stated.

For all cases considered, it is assumed that the cooling air is to be drawn off the compressor's exit station in a power-plant setting, arriving at the rotor under a pressure magnitude that is almost equal to the total pressure at the turbine inlet station, a temperature of 728 K, and a flow rate that is equal to 1.5 percent of the main (hot) flow rate.

In the following, the boundary conditions associated with each type of cooling will be discussed.

Rotor Disk Cooling

The general procedure followed in each of the two types of rotor-disk cooling is the same. First, the distribution of the heat transfer coefficient along the coolant direction is calculated using the expression

$$\Delta T_c = \frac{h_c A(T_\infty - T_c)}{c_p \dot{m}_c}$$

where

- h_c is the convection heat transfer coefficient.
- A is the area of the ring through which heat is transferred by convection.
- T_m is the average "metal" temperature at this particular ring.
- T_c is the corresponding average of the coolant temperature.
- c_p is the coolant specific heat (under constant pressure).
- $\dot{m}_c$ is the coolant flow rate.

A trial-and-error procedure was needed for the calculation of coolant temperature distribution because the metal temperature distribution along the disk backside is not known in advance.

Blade Cooling through a Slot

For the calculation of convection heat transfer coefficient distribution along the slot (Figure 8.4), the following expression for turbulent flow through ducts [7] is used:

$$h = 0.023 \frac{k}{D_H} \left(\frac{V D_H}{\nu} \right)^{0.8} (Pr)^{0.33}$$

where V represents the coolant velocity, whereas the hydraulic diameter D_H is defined as follows:

$$D_H = 4 \frac{\text{coolant-passage-cross-sectional-area}}{\text{wetted-perimeter}}$$

The temperature distribution of the cooling air as it flows through the slot is calculated using the previously cited heat-balance equation through a trial-and-error procedure.

Blade Cooling through Radially Drilled Holes

The heat transfer coefficient for this type of cooling (Figure 8.4) is calculated using the following expression [5]:

$$h = 0.023 \frac{k}{d} {Re_d}^{0.8} (Pr)^{0.33}$$

where

- d is the hole's diameter, and
- $Re_d = Vd/\nu$, with V being the coolant velocity relative to the (spinning) rotor.

A similar procedure to that used in the last type of cooling is again followed to calculate the cooling air temperature distribution along the radial holes.

In this type of cooling, and due to the relatively small hole diameter (compared with the rotor unit dimensions), the flow of cooling air through these holes is assumed to represent an arrangement of heat sinks located inside the tetrahedral elements into which the body is divided. The strength of each heat sink is equal to the rate of heat energy transferred to the cooling air as it passes through that element.

Numerical Results

Based on the preceding analysis, a Fortran computer program was prepared by the author. The program was then used to predict the temperature distribution in the rotor of the radial-inflow turbine discussed earlier (see Figure 8.1). The program was next used to attain the temperature distribution in the rotor under the different techniques of cooling discussed earlier. The turbine rotor used in this study is that of reference [5], and under the same operating conditions. Figure 8.3 shows the *slice* of the rotor body, including one blade and its corresponding segment of the center body (housing is not shown in the figure).

First, the four cooling methods (discussed in the preceding section) were separately investigated at a cooling-flow/primary-flow mass-flow ratio of 1.5 percent. Increasing this percentage to 3 percent, the rotor temperature distribution corresponding to different combinations of external and internal convection cooling is also computed.

The program input for the cases with no internal cooling through a slot was based on a total of 256 nodal points inside and on the boundary segments of the rotor unit (Figure 8.5). Using these nodes, the body was divided up into 225 elements starting, first, with hexahedral elements, which were then broken up into three tetrahedral elements each. This process produced a final total of 675 tetrahedral elements to work with (Figures 8.5 and 8.6). To account for the different boundary conditions involved, the body surface was covered by 371 surface triangular elements (Figure 8.7). On the other hand, for the cases including internal-convection cooling through a slot in the rotor blade, the total number of nodes came out to be 300, with 837 solid tetrahedral elements and 433 triangular surface elements.

In order to establish a basis for evaluating the effect of the different cooling methods, a parameter that is generically termed *cooling effectiveness* was introduced. This parameter is defined as follows:

$$\tau = \frac{T_{m,O} - T_{m,c}}{T_{T,g} - T_{T_T,c}}$$

where

- τ is the cooling effectiveness.
- $T_{m,O}$ is the metal temperature obtained at a certain point with no cooling.
- $T_{m,c}$ is the metal temperature at the same point with a particular type of cooling.
- $T_{T,g}$ is the total relative temperature (note that we are dealing with a rotating frame of reference that is attached to the rotor) of the main hot gas at the rotor inlet station.

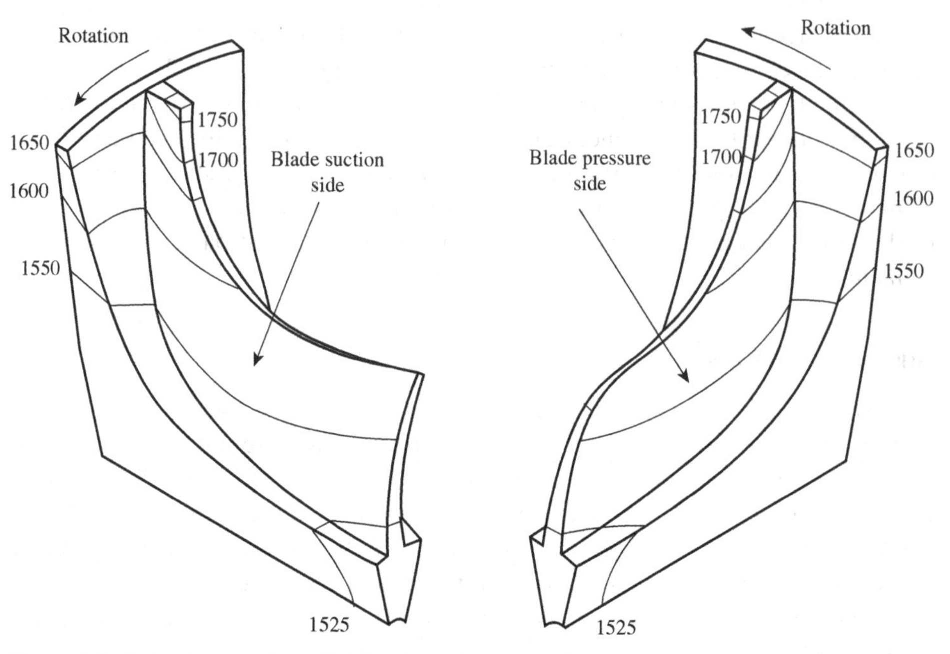

Figure 8.8. Rotor temperature distribution without cooling.

- $T_{T,c}$ is the total temperature of the cooling air at the inlet station of its passage(s).

This definition facilitates comparison of the different cooling techniques for the same coolant/primary mass-flow ratio.

According to reference [5], the region of the blade/hub interface at which the main (or primary) flow changes direction from radial to axial (Figure 8.2) was found to be a region of high stress and stress concentration levels. A metal temperature that is higher than 1400°F (1033 K) in this region would cause metal creep. This fact furnishes another basis for evaluating the different types of cooling. Evidently, from a stress standpoint, the best cooling technique is that which produces the lowest temperature in this critically stressed region.

Figure 8.8 shows the original temperature distribution (in degrees Fahrenheit) in the hot (meaning uncooled) rotor unit. It can be seen that the temperature magnitudes in the segment of radially directed flow are too high for a long-life turbine operation. Furthermore, an average temperature of 1540°F is found to exist in the highly stressed region of the blade/hub interface, which indicates severe mechanical consequences from a stress viewpoint.

Two types of cooling were first applied to the rotor disk. These involve an external flow of the cooling air along the disk backside in the radially outward direction (Figures 8.6 and 8.9). The coolant flow was 1.5 percent of the turbine primary mass-flow rate. The distribution of temperature and cooling effectiveness are shown in Figures 8.9 and 8.10, respectively, for the first case. As these two figures indicate, the effect of this cooling category is almost limited to the rotor disk. The cooling effectiveness is generally low, particularly for the rotor blade.

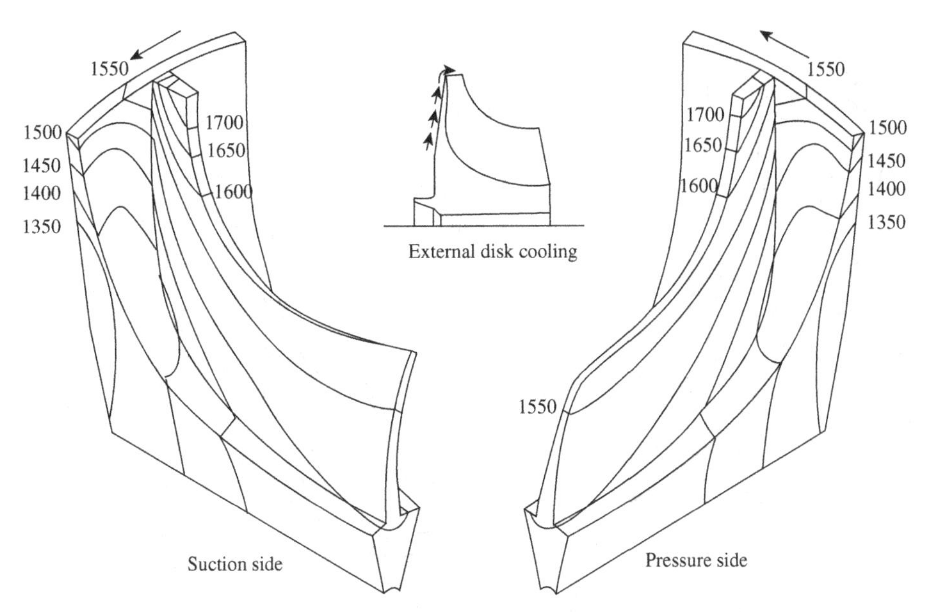

Figure 8.9. Temperature distribution with 1.5 percent external disc cooling.

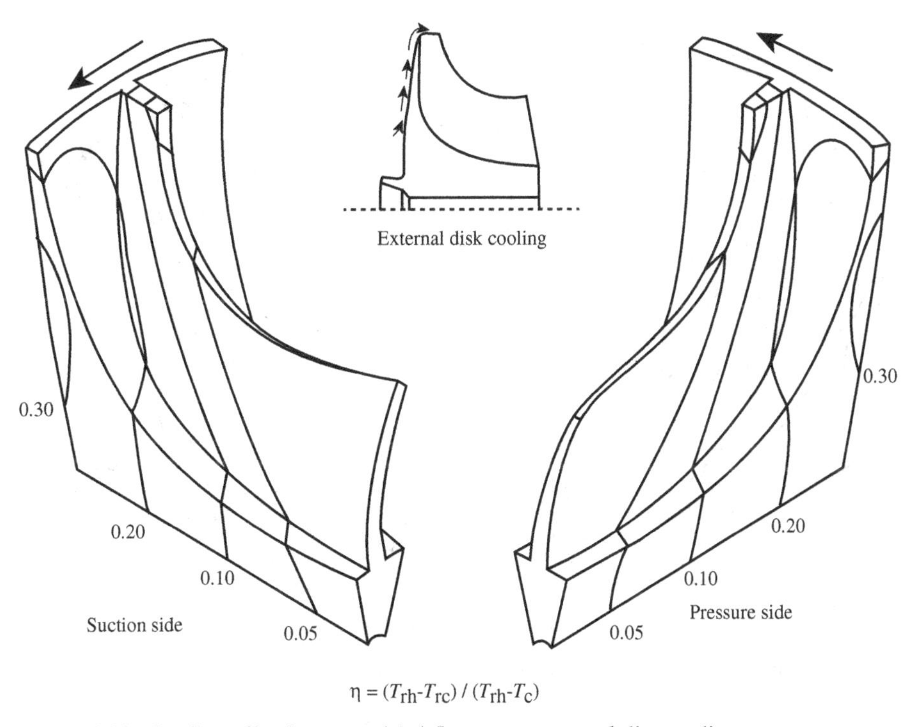

$\eta = (T_{rh}\text{-}T_{rc}) / (T_{rh}\text{-}T_c)$

Figure 8.10. Cooling effectiveness with 1.5 percent external disc cooling.

The temperature distribution corresponding to the case of internal convection cooling through a slot in the rotor blade, for the same 1.5 percent cooling percentage, is shown in Figure 8.11. This figure shows a considerable temperature reduction in the region adjacent to the coolant passage. The average value of temperature in this region of higher stress levels is reduced to approximately 1360F.

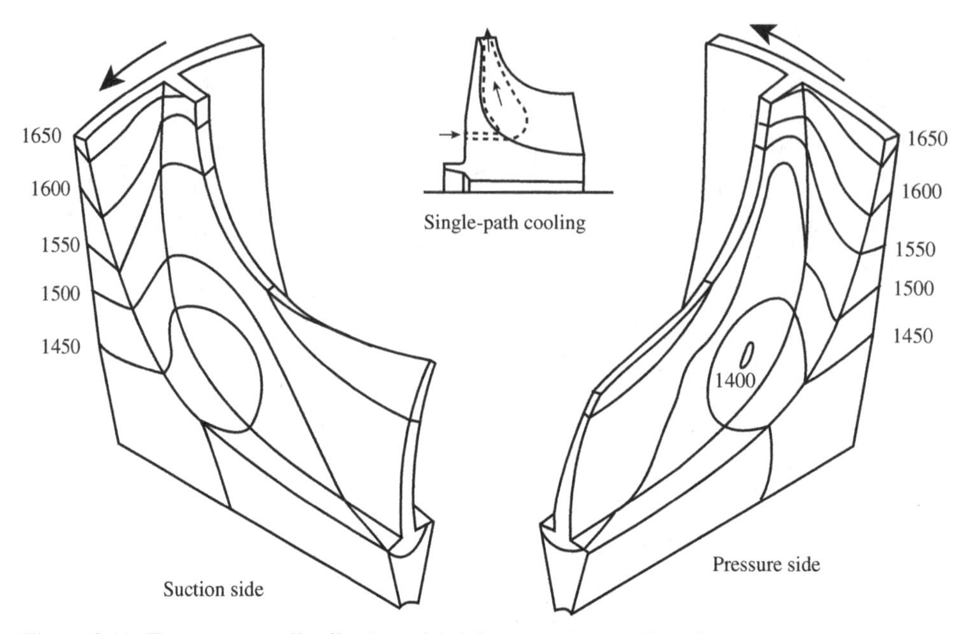

Figure 8.11. Temperature distribution with 1.5 percent internal cooling.

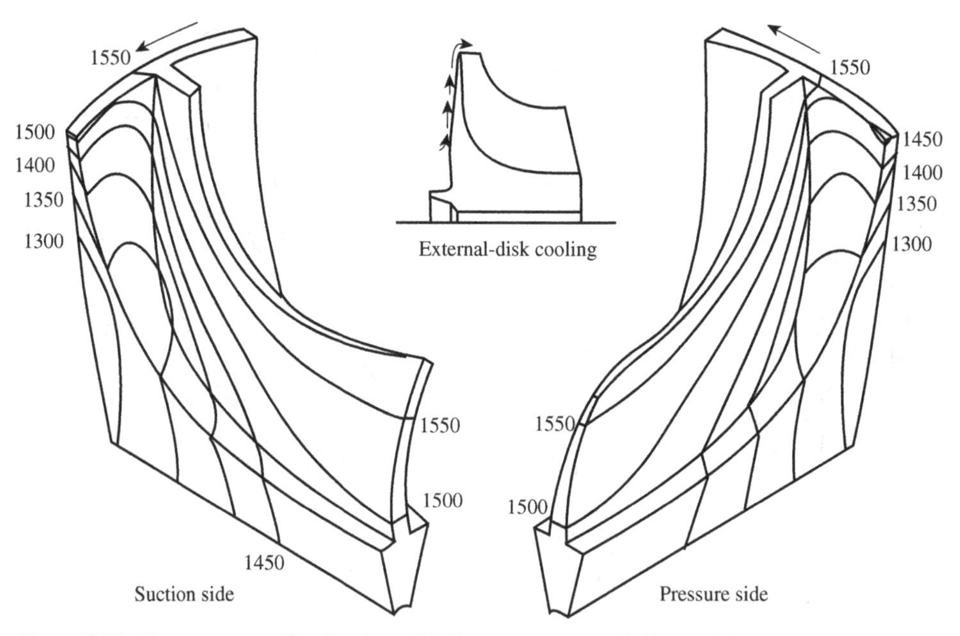

Figure 8.12. Temperature distribution with 3 percent external disc cooling.

Figure 8.12 shows the temperature distribution corresponding to a 3 percent cooling-to-primary mass-flow rate percentage. The coolant here flows radially outward over the disk backside and is injected into the primary-flow stream at the blade leading edge. The critically stressed region of the blade–hub surface interface is now as low as 1300°F, in comparison with 1350°F for a coolant/primary mass-flow ratio of 1.5 percent. Figure 8.13 shows the cooling effectiveness for the same cooling configuration.

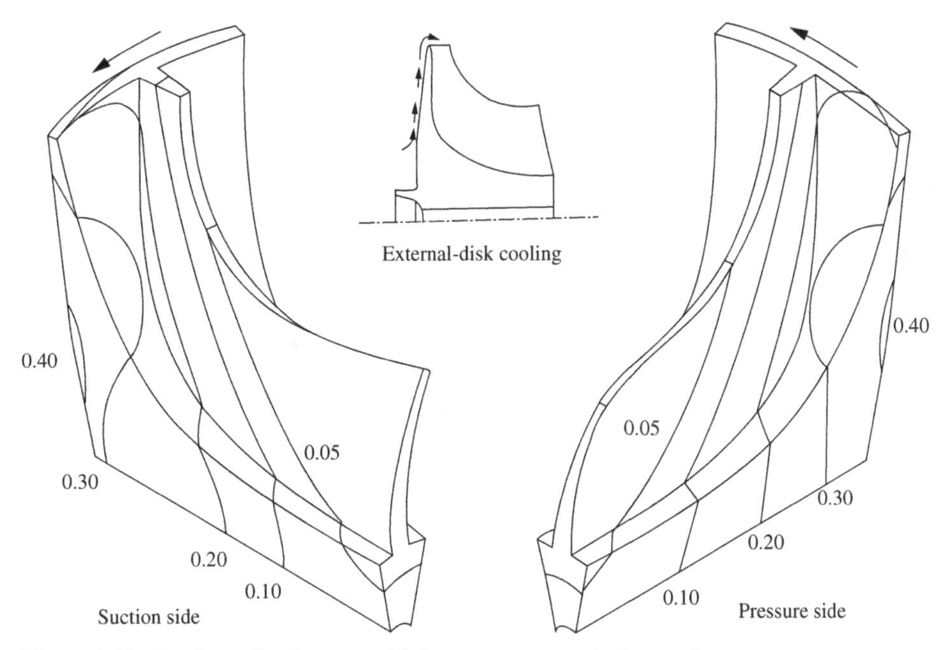

Figure 8.13. Cooling effectiveness with 3 percent external disc cooling.

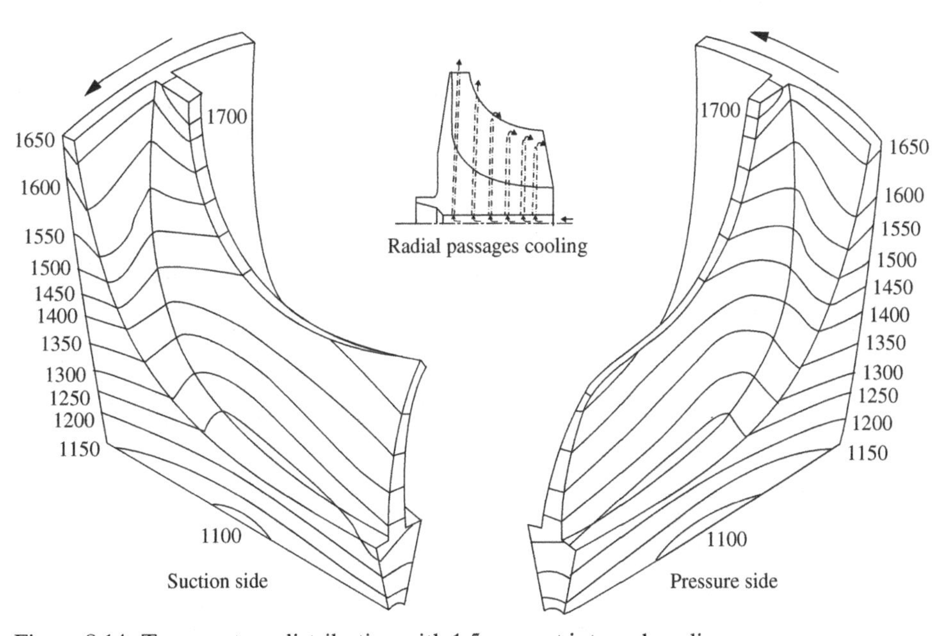

Figure 8.14. Temperature distribution with 1.5 percent internal cooling.

Figure 8.14 shows the temperature distribution associated with 1.5 percent cooling-to-primary flow percentage. The coolant here proceeds radially outward through radially drilled circular holes in the blade/hub unit. Again, the coolant is discharged over the blade suction side, where it is aerodynamically needed, as explained earlier in this section. In view of this figure, it is clear that this particular type of cooling, with only 1.5 percent cooling/primary flow ratio, is the absolute most effective, considering all the other cooling techniques investigated in this study.

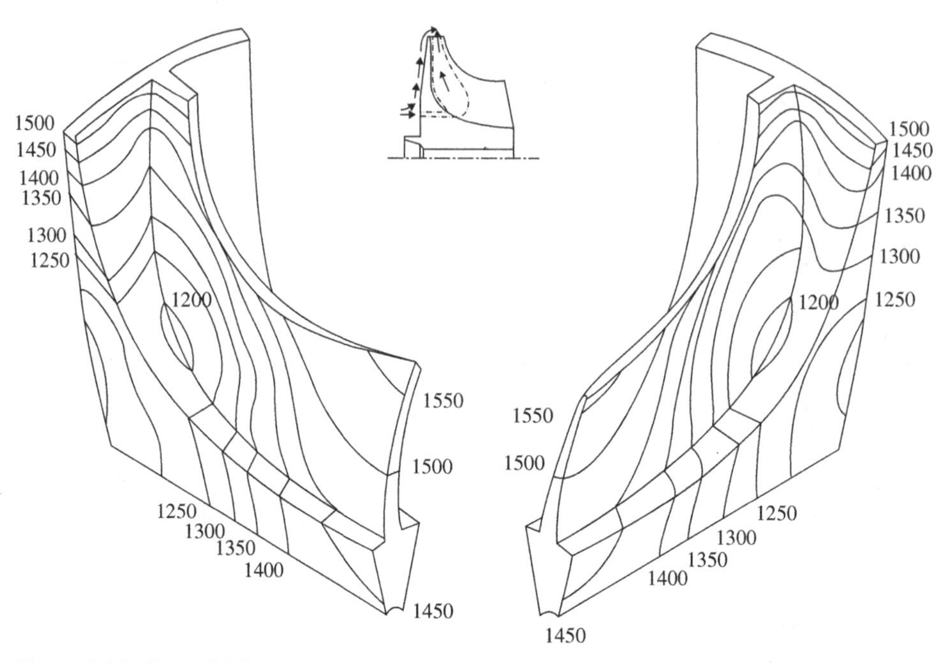

Figure 8.15. Case of 1.5 percent external cooling and 1.5 percent internal cooling.

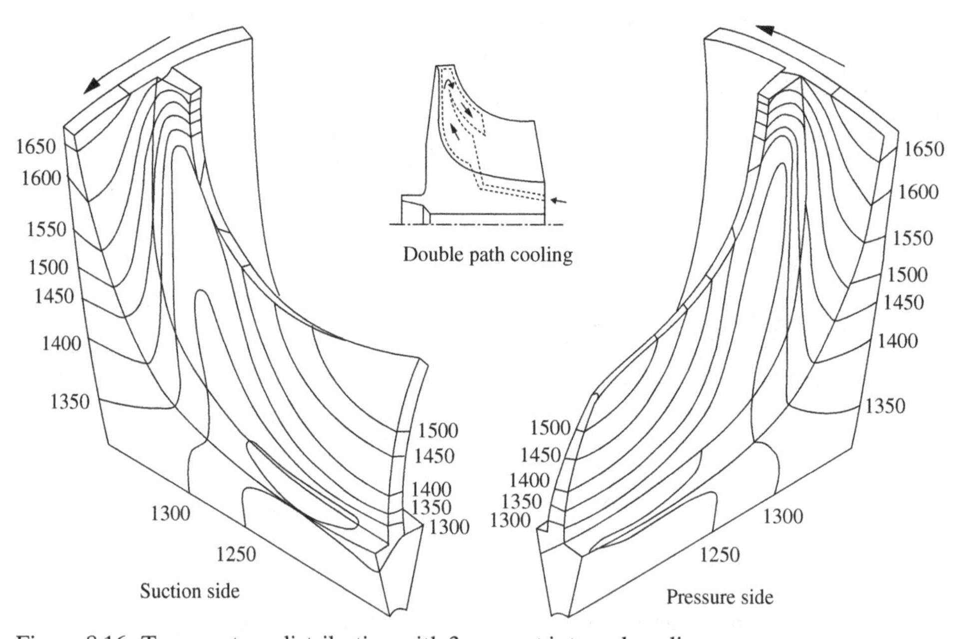

Figure 8.16. Temperature distribution with 3 percent internal cooling.

Note the overall temperature reduction and, consequently, the cooling-effectiveness elevation produced by this cooling method.

Figures 8.15 and 8.16 correspond to the cases of internal convection cooling through a slot coupled with, and then without, external cooling along the disk backside, with 1.5 percent cooling flow rate for both the internal and external cooling. The first of these two figures corresponds to the case in which the external cooling air direction is radially outward, whereas the other one corresponds to the coolant direction being radially inward. As seen in these two figures, both cases result in approximately the same cooling effectiveness distribution. On the other hand, these figures show that a radially outward flow of the coolant along the rotor-disk backside produces a lower temperature in the highly stressed region than the radially inward flow does.

Remark

Note, again, that any internal-cooling category here was represented, in the problem analysis, by a succession of heat sinks. These are properly placed at the coolant-affected elements in the investigated rotor segments.

PROBLEMS

P.1 Consider the heat-conduction problem in a slab with specified wall temperatures (Figure P.1). Write the matrix equation for computing the nodal temperatures using linear finite-elements.

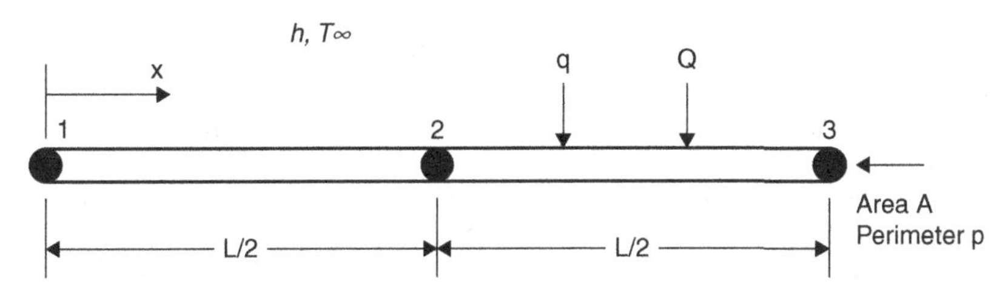

Figure P.1 Input variables for Problem P.1

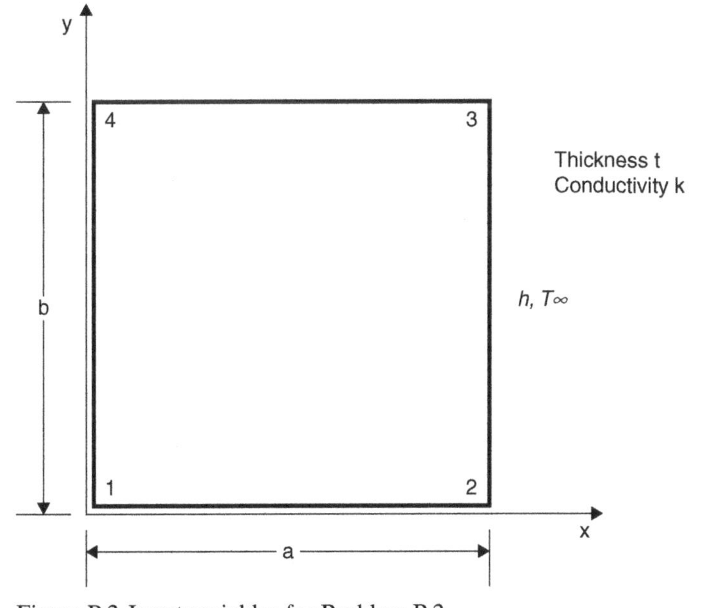

Figure P.2 Input variables for Problem P.2

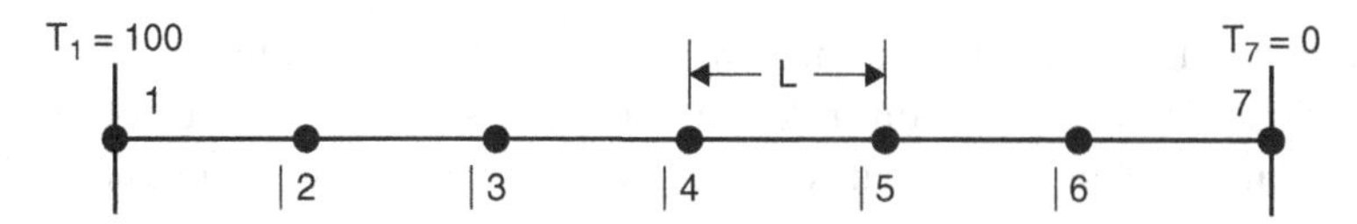

Figure P.3 Input variables for Problem P.3

P.2 Consider a three-node quadratic conduction/convection element (Figure P.2). Derive the following for that element:
   a. Conduction matrix
   b. Convection matrix
   c. Heat-load vector for internal heat generation.

P.3 Consider a rectangular two-dimensional conduction element (Figure P.3). Derive the following for that element:
   a. Conduction matrix
   b. Heat-load vector for surface heating q

REFERENCES

[1] Zienkewicz, O.C., and Cheung. Y.K., *The Finite Element Method in Structural and Continuum Mechanics*, McGraw-Hill, New York, 1967.
[2] Huebner, K.H., *The Finite Element Method for Engineers*, Wiley, New York, 1975.
[3] Myers, G.E., *Analytical Methods in Conduction Heat Transfer*, McGraw-Hill, New York, 1971.
[4] Okapuu, U., and Calvert, G.S., An Experimental Cooled Radial Turbine," *AGARD Conference Proceedings. No. 73*, 1971.
[5] Calvert, G.S., and Okapuu, U., "Design and Evaluation of a High-Temperature Radial Turbine," *USAAVLABS Technical Report 68-69*, 1969.
[6] Dunham. J., and Edwards, J.P., *Heat Transfer Calculations for Turbine Blade Design*, AGARD-CP-73, Natural Gas Turbine Establishment, Pyestok, Farnborough, Hants, United Kingdom, 1971.
[7] Kreith, F., *Principles of Heat Transfer*, International Textbook Company, Scranton, PA, 1965.

9 One-Dimensional Transient Problems

Perhaps the most representative problem in this category is the plane-wall problem. The partial differential equation here is

$$k\frac{\partial^2 T}{\partial x^2} = \rho c\frac{\partial T}{\partial t}$$

where c is the specific heat of the metal.

The boundary conditions are as follows:

$$T(0,t) = T_O$$

$$\left(\frac{\partial T}{\partial x}\right)_{x=L} = 0$$

$$T(x,0) = T_i$$

Variational Statement

The variational statement for this one-dimensional transient problem involves the minimization of a single integral over the space coordinate as shown below:

$$I = \int_{x=0}^{L} F(x,t,u,u_x,u_t)\,dx$$

This integral is to be minimized at every point in time.

Following the standard procedure in variational calculus, a trial solution of the form $\bar{u}(x,t,\epsilon) = u(x,t) + \epsilon\eta(x,t)$ can be substituted into the preceding expression to obtain [for specified values of $u(x,t)$ on the boundaries] the Euler-Lagrange equation, namely

$$\frac{\partial F}{\partial u} - \frac{\partial}{\partial x}\left(\frac{\partial F}{\partial u_x}\right) = 0$$

We could then work backwards from the given partial differential equation by comparison with the preceding equation to reduce the variational statement to

$$I = \frac{1}{2}\int_{x=0}^{L}\left[k\left(\frac{dT}{dx}\right)^2 + \rho c\frac{\partial T}{\partial t}\right]dx$$

We want to minimize this function at every instant in time while satisfying the boundary conditions and the earlier initial condition.

Finite Element Formulation

The preceding given function may be considered in two parts by giving it as

$$I = I_k + I_c$$

where

$$I_k = \frac{1}{2}\int_{x=0}^{L} k\left(\frac{\partial T}{\partial x}\right)^2 dx$$

$$I_c = \frac{1}{2}\int_{x=0}^{L} \rho c \frac{\partial T}{\partial t} dx$$

We then want to minimize this function with respect to the nodal temperatures at every instant in time. That is, we want

$$\frac{dI}{dT} = \frac{dI_k}{dT} + \frac{dI_c}{dT} = 0$$

As usual, we will subdivide the interval into E elements and consider each element individually. That is, we will write the preceding derivatives as follows:

$$\frac{dI_k}{dt} = \sum_{e=1}^{E} \frac{dI_k{}^{(e)}}{dt}$$

$$\frac{dI_c}{dt} = \sum_{e=1}^{E} \frac{dI_c{}^{(e)}}{dt}$$

We will again use the $D^{(e)}$ mapping-matrix concept to write the preceding as follows:

$$\frac{dI_k}{dt} = \sum_{e=1}^{E} D^{(e)} \frac{dI_k{}^{(e)}}{dt}$$

$$\frac{dI_c}{dt} = \sum_{e=1}^{E} D^{(e)} \frac{dI_c{}^{(e)}}{dt}$$

where
$$[D]^{(e)} = \begin{bmatrix} 0 & 0 \\ 0 & 0 \\ . & . \\ . & . \\ 1 & 0 \\ 0 & 0 \\ 0 & 0 \\ . & . \\ . & . \\ 0 & 1 \\ 0 & 0 \\ . & . \\ . & . \\ 0 & 0 \end{bmatrix}$$

where the nonzero entries correspond to the ith and jth rows, respectively.

At any instant in time, we will assume that the temperature distribution within any element (e) varies linearly between the two nodal temperatures T_i and T_j. Therefore, following the same line of attack we have used previously, we can express the elemental temperature distribution in a matrix form as follows:

$$T^{(e)} = p^T R^{(e)} T^{(e)}$$

where
$$p^T = [1 \text{ x}]$$

$$R^{(e)} = \frac{1}{x_{ij}} \begin{bmatrix} x_j & -x_i \\ -1 & 1 \end{bmatrix} T^{(e)} = \begin{Bmatrix} T_i \\ T_j \end{Bmatrix}$$

Following the same familiar procedure (as before), we find

$$\frac{dI_k{}^{(e)}}{dT^{(e)}} = K^{(e)} T^{(e)}$$

where
$$K^{(e)} = \frac{k^{(e)}}{x_{ij}} \begin{bmatrix} 1 & -1 \\ -1 & 1 \end{bmatrix}$$

We can replace the element's temperatures $T^{(e)}$ by $D^{(e)^T} T$ and then sum the contributions of each element to arrive at the following result:

$$\frac{dI_k}{dT} = \sum_{e=1}^{E} D^{(e)} K^{(e)^T} T$$

A global conduction matrix $[K]$ may now be defined. We can then rewrite the preceding expression as follows:

$$\frac{dI_k}{dT} = KT$$

and

$$D^{(e)}K^{(e)}D^{(e)^T} = \begin{bmatrix} 0 & 0 \\ 0 & 0 \\ k_{11} & k_{12} \\ 0 & 0 \\ 0 & 0 \\ k_{21} & k_{22} \\ 0 & 0 \\ 0 & 0 \end{bmatrix}$$

where the entries are in the ith and jth columns and in the ith and jth rows. Also, k_{11}, k_{12}, k_{21}, and k_{22} are now the components of $K^{(e)}$ we were given previously. This matrix is seen to be just a simplified version of the two-dimensional conduction matrix given earlier.

We are now ready to consider dI_c/dT, which will be handled just the as dI_c/dT expression given earlier. In other words,

$$I_c^{(e)} = \frac{(\rho c)^{(e)}}{2} \int_{x_1}^{x_2} \frac{\partial}{\partial t}(T^{(e)^2})\, dx$$

In writing the preceding expression, we have assumed that the element is sufficiently small that an average value of ρc for the entire element can be used. If we now recognize that the time derivative may be pulled outside the integral sign, we find that we can rewrite the same expression as follows:

$$I_c^{(e)} = \frac{(\rho c)^{(e)}}{2} \frac{d}{dT} \int_{x=x_1}^{x_2} (p^T R^{(e)} T^{(e)})^2\, dx$$

This can be differentiated with respect to $T^{(e)}$ to give

$$\frac{dI_c^{(e)}}{dT^{(e)}} = \frac{(\rho c)^{(e)}}{2} \frac{d}{dt} \int_{x_1}^{x_2} 2(p_T r^{(e)} T^{(e)})(p^T R^{(e)})^T\, dx$$

In writing this, we have assumed that ρc is independent of temperature. If it is temperature-dependent, a modification would have to be made.

The order of multiplication can be interchanged and $p^T R^{(e)}$ may be transposed to give

$$\frac{dI_c^{(e)}}{dT^{(e)}} = (\rho c)^{(e)} \int_{x_1}^{x_2} R^{(e)^T} p p^T T^{(e)}\, dx$$

From the preceding equation we observe that $R^{(e)}$ and $T^{(e)}$ are both independent of x and, consequently, may be removed from the integral to give

$$\frac{dI_c^{(e)}}{dT^{(e)}} = (\rho c)^{(e)} \frac{d}{dt} (R^{(e)^T} \int_{x_1}^{x_2} p p^T\, dx R^{(e)} T^{(e)})$$

The only term in parentheses that is a function of time is the element's temperature vector $T^{(e)}$. Note that p, p^T, $R^{(e)}$, and $R^{(e)^T}$ are all independent of time. This means that we may write the preceding expression in the following form:

$$\frac{dI_c{}^{(e)}}{dT^{(e)}} = (\rho c)^{(e)} R^{(e)^T} \int_{x_1}^{x_2} p p^T R^{(e)} T^{(e)} \, dx$$

where we have used $T^{(e)}$ to denote the time derivative of $T^{(e)}$.

The steps of integration and multiplication of these matrices are quite similar to those we have seen before. The final result is

$$\frac{dI_c{}^{(e)}}{dT^{(e)}} = C^{(e)} T^{(e)}$$

where the element capacitance matrix is given by

$$C^{(e)} = \frac{(\rho c)^{(e)} x_{ij}}{6} \begin{bmatrix} 2 & 1 \\ 1 & 2 \end{bmatrix}$$

The derivative of the elemental temperature vector may then be replaced in terms of T by using $D^{(e)^T}$ in the usual way to give

$$\frac{dI_c{}^{(e)}}{dT^{(e)}} = C^{(e)} D^{(e)^T} T$$

This can be summed over all elements to give

$$\frac{dI_c}{dT} = \sum_{e=1}^{E} D^{(e)} C^{(e)} D^{(e)^T} T$$

This can be further simplified by defining a global capacitance matrix as follows

$$C = \sum_{e=1}^{E} D^{(e)} C^{(e)} D^{(e)^T}$$

We can also write the following expression:

$$\frac{dI_c}{dT} = C\dot{T}$$

As the matrix multiplications are carried out, we find that

$$D^{(e)} C^{(e)} D^{(e)^T} = \begin{bmatrix} 0 & 0 \\ 0 & 0 \\ c_{11} & c_{12} \\ 0 & 0 \\ 0 & 0 \\ c_{21} & c_{22} \\ 0 & 0 \\ 0 & 0 \end{bmatrix}$$

Nodal coordinates		Element information						
i	x_i	(e)	i	j	k	ρ	c
1	0	1	1	2	k	ρ	c
2	$L/4$	2	2	3	k	ρ	c
3	$L/2$	3	3	4	k	ρ	c
4	$3L/4$	4	4	5	k	ρ	c
5	L						

Figure 9.1. Computer input data for plane-wall transient.

where the entries are in the matrix places explained earlier. Also, c_{11}, c_{12}, c_{21}, and c_{22} are the components of $C^{(e)}$ as given earlier.

Proceeding on, we get

$$\frac{dI}{dT} = KT + C\dot{T} = 0$$

or

$$C\dot{T} = -KT$$

This is a relationship between the nodal temperature T and the time derivative $\bar{T}$ that must be satisfied at each point in time to minimize I. It represents a system of ordinary differential equations for the nodal temperatures as functions of time. At time zero we are given the initial temperature distribution, which can be substituted into the right-hand side of the preceding matrix equation. The system of equations then can be solved for the initial temperature derivative necessary to minimize I at that instant. We would then make use of these derivatives to move ahead in time, as discussed next.

The problem formulation in the computer is very similar to problems we have discussed previously. Let us consider the plane-wall transient. If we take the spacing between nodes to be $L/4$, we will have four elements and five nodal temperatures. The computer input data will consist of nodal coordinates information as well as element information, as shown in Figure 9.1.

The computer will then consider the element (1) and compute $K^{(1)}$. This would be stored in the global matrix K as follows:

$$K = \frac{k}{\Delta x}\begin{bmatrix} 1 & -1 & 0 & 0 & 0 \\ -1 & 1 & 0 & 0 & 0 \\ 0 & 0 & 0 & 0 & 0 \\ 0 & 0 & 0 & 0 & 0 \end{bmatrix}$$

We have factored out $k/\Delta x$ for simplicity. Then the computer would move to element (2), compute $K^{(2)}$, and add it to K to give

$$K = \frac{k}{\Delta x}\begin{bmatrix} 1 & -1 & & \\ -1 & 2 & -1 & \\ & -1 & 1 & \end{bmatrix}$$

This process would be repeated until all elements have been incorporated. The final result is

$$\begin{bmatrix} 1 & -1 & & & \\ -1 & 4 & 1 & & \\ & -1 & 2-1 & & \\ & & -1 & 2 & -1 \\ & & & -1 & 1 \end{bmatrix}$$

The global capacitance matrix could be constructed in the same way. The final result is

$$C = \frac{\rho c \Delta x}{6} \begin{bmatrix} 2 & 1 & & & \\ 1 & 4 & 1 & & \\ & 1 & 4 & 1 & \\ & & 1 & 4 & 1 \\ & & & 1 & 2 \end{bmatrix}$$

In a practical case, the components below the main diagonal would not have to be stored in the computer because of the matrix symmetry.

Numerical Solutions

In practice, the system of ordinary differential equations will be tedious. There will be too many unknown nodal temperatures, and the system must be solved numerically with the aid of a computer. The techniques used to solve these problems are Euler, Crank-Nicolson, and pure implicit. These three methods will be discussed next.

Euler's Method

In the Euler's method, the solution is advanced in time by the following relationship:

$$T^{(\nu+1)} = T^{(\nu)} + \Delta t \dot{T}^{(\nu)}$$

This equation can be premultiplied by C to give

$$CT^{(\nu+1)} = CT^{(\nu)} + \Delta t C\dot{t}^{\nu}$$

or

$$CT^{(\nu+1)} = (C - \Delta t K)T^{(\nu)}$$

On dividing by $\rho c \Delta x/6$ and defining $u = (T - T_O)/(T_i - T_O)$, this equation can be rewritten as:

$$Au^{(\nu+1)} = Bu^{(\nu)}$$

where, for $E = 4$,

$$A = \begin{bmatrix} 2 & 1 & & & \\ 1 & 4 & 1 & & \\ & 1 & 4 & 1 & \\ & & & 1 & 2 \end{bmatrix}$$

and

$$B = \begin{bmatrix} 2-6p & 1+6p & & & \\ 1+6p & 4-12p & 1+6p & & \\ & 1+6p & 4-12p & 1+6p & \\ & & 1+6p & 4-12p & 1+6p \\ & & & 1+6p & 2-6p \end{bmatrix}$$

where $p = \alpha \Delta t/(\Delta x)^2$.

This is a tridiagonal system of equations to be solved for $u^{(\nu+1)}$. Here the Gaussian elimination method can be used. Advantages can be taken of the fact that the matrix $[A]$ is symmetrical to save a little on computation time and memory usage. It should be noted that in the finite element method, the Euler method gives an implicit set of equations to be solved.

Before solving the system of equations, they must be modified to be sure that the boundary condition at $x = 0$ is satisfied. The technique for doing this when $u_1 = u_O \neq 0$ is indicated. Observe that we have forced the first equation to have the solution ${u_1}^{(\nu+1)} = u_O$. We have also transferred all the unknown effects of $u_1 = u_O$ out of the matrices $[A]$ and $[B]$ into two constant column matrices on the right-hand side of the equation. It should be noted that it is not necessary to store any entry underneath the diagonal due to the matrix symmetry. Because of the normalization in the example, we are considering $u_O = 0$. Consequently, the two extra column matrices are not needed (they are identically zeros). The matrices $[A]$ and $[B]$ thus are given as follows:

$$A = \begin{bmatrix} 1 & 0 & & & \\ & 4 & 1 & & \\ & & 4 & 1 & \\ & & & 4 & 1 \\ & & & & 2 \end{bmatrix}$$

and

$$B = \begin{bmatrix} 1 & 0 & & & \\ & 4-12p & 1+6p & & \\ & & 4-12p & 1+6p & \\ & & & 4-12p & 1+6p \\ & & & & 2-6p \end{bmatrix}$$

where both matrices are symmetric. If the system of equations is solved using $p = 0.25$, the solution turns out to be unstable. In fact, the oscillations are so out of hand at $t = 1$ that the following temperature distribution is obtained:

$$u_1 = 0$$

$$u_2 = 0.533 \times 10^{13}$$

$$\cdots$$

and so on.

Clearly, this solution is unacceptable. This means that the oscillatory limits for the finite element method are severe.

Crank-Nicolson Method

Here we expect this method to be more accurate than the Euler's method. However, the Crank-Nicolson method requires no more work because we have just seen that Euler's method is also implicit in the finite element formulation.

The Crank-Nicolson method moves the solution ahead in time according to the following relationship:

$$T^{(\nu+1)} = T^{(\nu)} + \frac{\Delta t}{2}(\dot{T}^{(\nu)} + \dot{T}^{(\nu+1)})$$

This may be premultiplied by C to give

$$CT^{(\nu+1)} = CT^{(\nu)} + \frac{\Delta t}{2}(C\dot{T}^{(\nu)} + C\dot{T}^{(\nu+1)})$$

The temperature at the new time step $T^{(\nu+1)}$ can be moved to the left-hand side to give

$$\left(C + \frac{\Delta t}{2}K\right)T^{(\nu+1)} = \left(C - \frac{\Delta t}{2}K\right)T^{(\nu)}$$

On normalizing the problem, we get

$$Au^{(\nu+1)} = Bu^{(\nu)}$$

where, after incorporating the boundary condition at $x = 0$,

$$[A] = \begin{bmatrix} 1 & 0 & & & \\ & 4+6p & 1-3p & & \\ & & & & \\ & & & 4+6p & 1-3p \\ & & & & 2+3p \end{bmatrix}$$

$$[B] = \begin{bmatrix} 1 & 0 & & & \\ & 4-6p & 1+3p & & \\ & & 4-6p & 1+3p & \\ & & & 4-6p & 1-3p \\ & & & & 2-3p \end{bmatrix}$$

where both matrices are symmetric. The numerical results for this method are shown in Figure 9.2 for a p value of 1.0.

Purely Implicit Method

The purely implicit method moves the solution ahead in time according to the following relationship:

$$T^{(\nu+1)} = T^{(\nu)} + \Delta t\dot{T}^{(\nu+1)}$$

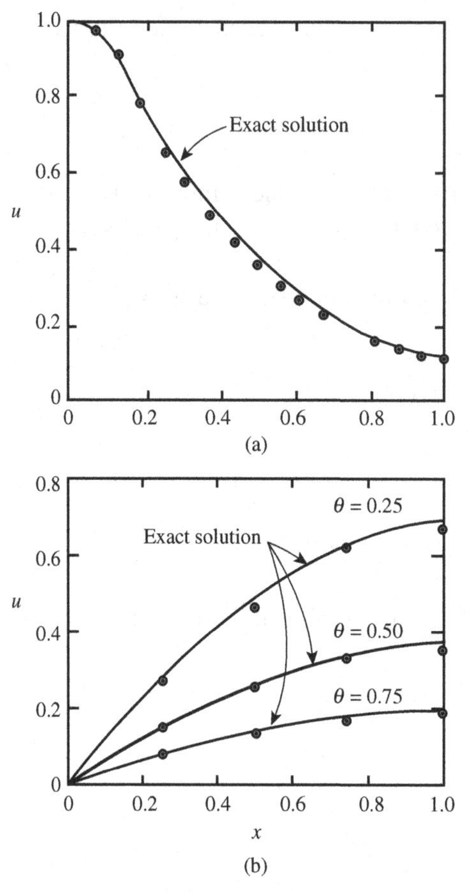

Figure 9.2. Crank-Nicolson finite element solution for $E = 4$ and $p = 1.0$.

On premultiplying this equation by C (to eliminate the time derivatives) in the usual manner and rearranging the result, the following familiar equation is obtained:

$$AT^{(\nu+1)} = BT^{(\nu)}$$

where, in this case,

$$[A] = \begin{bmatrix} 1 & 0 & & & & 4-12p & 1-6p \\ & & 4-12p & 1+6p & & & \\ & & 4+12p & 1-6p & & & \\ & & & & 2-6p & & \end{bmatrix}$$

and

$$[B] = \begin{bmatrix} 1 & 0 & & & \\ & 4 & 1 & & \\ & & 4 & 1 & \\ & & & 4 & 1 \\ & & & & 2 \end{bmatrix}$$

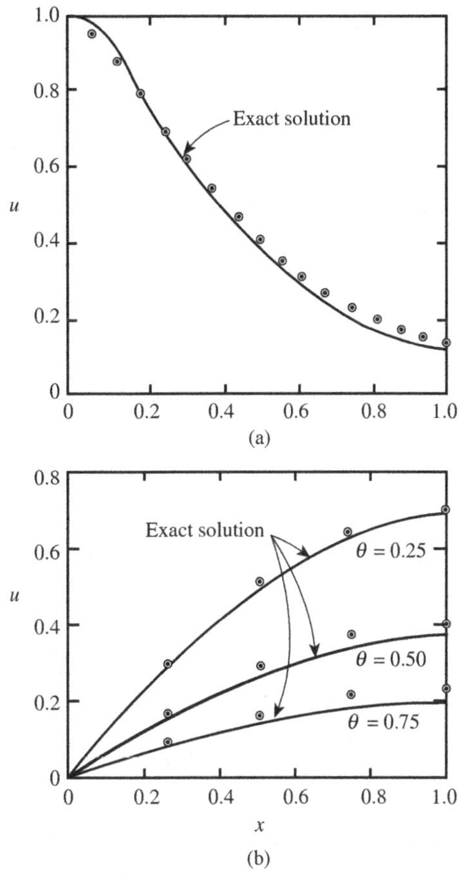

Figure 9.3. Pure implicit finite element solution for E = 4 and p = 1.0.

where both matrices are symmetric. The purely implicit solution is shown in Figure 9.3, and it looks as good as the Crank-Nicolson method solution.

REFERENCES

[1] Huebner, K.H., and Thornton, E.A., *The Finite Element Method for Engineers*, 2nd ed., Wiley, New York, 1982.

[2] Myers, G.E., *Analytical Methods in Conduction Heat Transfer*, Genium, Amsterdam, NY, 1987.

10 Fluid Mechanics Finite Element Applications

Introduction

During the past half century, engineering analysis has relied on the traditional finite-difference method to obtain computer-based solutions to difficult flow problems. The progress and success achieved in these pursuits have been, in many cases, noteworthy. Slow viscous flows, boundary layer flows, diffusion flows, and variable-property flows are just some examples of areas for which analysts have developed refined calculation procedures based on the finite-difference method.

Yet there remains a number of problems for which the finite-difference methods were proven inaccurate. Problems involving complex geometries, multiply-connected domains, and complicated boundary conditions always pose quite a challenge. Finite element methods can help in alleviating these difficulties but should not be expected to triumph in every case where the finite-difference methods have failed. Instead, the finite element methods offer easier ways to treat complex geometries requiring irregular meshes, and they provide a more consistent way of using higher-order approximations. In some cases, the finite element approach can provide an approximate solution of the same order of accuracy as the finite-difference method but at less expenses. Regardless of the method used, the accurate numerical solution of most of the viscous-flow problems requires vast amounts of computer time and data storage, and of course, problems of numerical stability and convergence can occur with either method.

Only since the early 1970s has the finite element method been recognized as an effective means for solving difficult fluid mechanics problems. Literature on the application of finite element methods to fluid mechanics is rapidly increasing, with contributions being made virtually daily.

This part of the book is devoted to the description of finite element methods in the fluid mechanics discipline. For incompressible and compressible inviscid and viscous flows, we will lay the theoretical foundation, develop the element equations, and report the most promising findings of recent research efforts. Some of the finite element example problems here serve as test cases that demonstrate feasibility. Once known solutions are accurately matched and the calculation procedures are established, the extension to problems of arbitrary geometries is obvious.

First, the reader should be reminded that Appendix B contains almost all classifications of fluid flow structures, both inviscid and viscous. Examination of this appendix may just place the reader at a position to judge where the major flow patterns differ from one another and the kinds of applications to be expected under each heading.

Inviscid Incompressible Flows

Because all fluids are viscous, an inviscid fluid is simply a hypothetical concept that simplifies the mathematics of the fluid flow problem. Inviscid fluids experience no shearing stress, and when they come into contact with a solid boundary, they slip tangentially along without resistance. Real fluids, of course, produce shear stresses, and they adhere to flow boundaries so that no slip occurs at the fluid/solid interface.

Despite these differences between viscous and inviscid fluids, there are many practical problems in fluid mechanics that can be analyzed with good accuracy when the inviscid-flow theory is invoked. Problems such as the flow around streamlined objects, and flow through, in particular, converging passages are just a couple of significant examples.

Problem Statement

For a two-dimensional irrotational (potential) flow, the velocity components u and v may be expressed in terms of the stream function $\psi(x,y)$ as follows:

$$u = \frac{\partial \psi}{\partial y}$$

$$v = -\frac{\partial \psi}{\partial x}$$

or, in terms of the velocity potential ϕ, as follows:

$$u = \frac{\partial \phi}{\partial x}$$

$$v = \frac{\partial \phi}{\partial y}$$

For a two-dimensional (two-dimensional) flow, both the stream function and the velocity potential satisfy Laplace's equation:

$$\nabla^2 \psi = 0$$

$$\nabla^2 \phi = 0$$

Dirichlet (specified magnitude of the field variable) and Neumann (specified magnitude of the normal derivative) boundary conditions apply to both the preceding equations. At a solid boundary, the velocity component perpendicular to the boundary must be the same as the boundary velocity in the normal direction. Hence we have

$$\vec{V}.\vec{n} = \vec{V}_B.\vec{n}$$

or
$$un_x + vn_y = (V_B)_x n_x + (V_B)_y n_y$$

where $\vec{V}_B$ is the boundary local velocity, with n_x, n_y being the direction cosines of the local outward unit vector perpendicular to the local boundary segment. Should the local boundary segment be stationary, then $V_B = 0$, and the preceding boundary conditions become

$$\frac{\partial \psi}{\partial s} = \frac{\partial \psi}{\partial y} n_x - \frac{\partial \psi}{\partial x} n_y = 0$$

$$\frac{\partial \phi}{\partial n} = \frac{\partial \phi}{\partial x} n_x + \frac{\partial \phi}{\partial y} n_y = 0$$

These two expressions state that the tangential derivative of the stream function along the wall is zero, whereas the normal derivative of the velocity potential is zero.

The boundary value problem for potential flow in terms of ϕ then may be stated as follows: Given a solution domain Ω, bounded by the curve $C = C_1 + C_2$, find the velocity potential such that:

$$\nabla^2 \phi = 0 \qquad \text{in } \Omega$$

$$\phi = q(x, y) \qquad \text{on } C_1$$

$$\frac{\partial \phi}{\partial n} = V_B.\vec{n} = V^* \qquad \text{on } C_2$$

This problem is equivalent to finding the function ϕ that minimizes the following integral:

$$I(\phi) = \frac{1}{2} \int_\Omega \left[\left(\frac{\partial \phi}{\partial x} \right)^2 + \left(\frac{\partial \phi}{\partial y} \right)^2 \right] d\Omega - \int_{C_1} V^* \phi ds$$

subject to the same boundary conditions. Because details of the finite element formulation of this problem were, in general terms, given earlier, our aim here is to summarize the relevant finite element equations. The nodal values of the velocity potential (sometimes called the *potential function*) $\phi(x, y)$ are computed from the system equation

$$[K]\{\phi\} = \{R\}$$

subject to the appropriate Dirichlet boundary conditions. We form the system's coefficient matrix $[K]$ from element coefficient matrices

$$k_{ij}^{(e)} = \int_\Omega \left(\frac{\partial N_i}{\partial x} \frac{\partial N_j}{\partial x} + \frac{\partial N_i}{\partial y} \frac{\partial N_j}{\partial y} \right) d\Omega$$

where $i, j = 1, 2, \ldots, r$, with r denoting the number of element's nodes. In discussing the stream function formulation, we consider the Neumann boundary condition through the element's *load* vector

$$R_i^{(e)} = \int_{S_2} V^* N_i \, ds$$

where S_2 denotes the element's boundary for which the Neumann boundary condition applies (note that in the definition of the load vector given here, there is a sign change due to a difference in the statement of the Neumann boundary condition). The formulation given earlier is for two-dimensional flows, but it readily extends to three-dimensional flows. A similar finite element formulation exists for two-dimensional flows specified in terms of the stream-function, but the stream function approach is only limited to two-dimensional flows. Because the details of evaluating the preceding integrals were discussed in earlier chapters, we will discuss, instead, the special considerations applicable to potential flows.

Velocity Potential and Stream-Function Formulations

Consider the problem of determining the flow pattern around a body of an arbitrary shape immersed in a uniform flow stream field (Figure 10.1). Because the actual solution domain for this problem is infinite in extent, it is necessary to construct some finite domain. This is done as indicated in Figure 10.2 by taking a boundary sufficiently far from the body that the flow field at every point on the boundary is known in terms of the given approaching flow. A rectangular boundary is the most convenient choice here. We will consider the solution procedure first in terms of the velocity potential ϕ and then in terms of the stream function ψ.

When formulating the problem in terms of ϕ, we note that all the boundary conditions are given in terms of $\partial\phi/\partial n$, and incorporating this Neumann boundary condition requires a special procedure. On solid stationary boundary segments, $\partial\phi/\partial n = 0$, but on the outer boundary, $\partial\phi/\partial n \neq 0$ in general. The variational statement automatically accounts for these conditions via the line integral. However, solution of the preceding equation, subject to only Neumann boundary conditions,

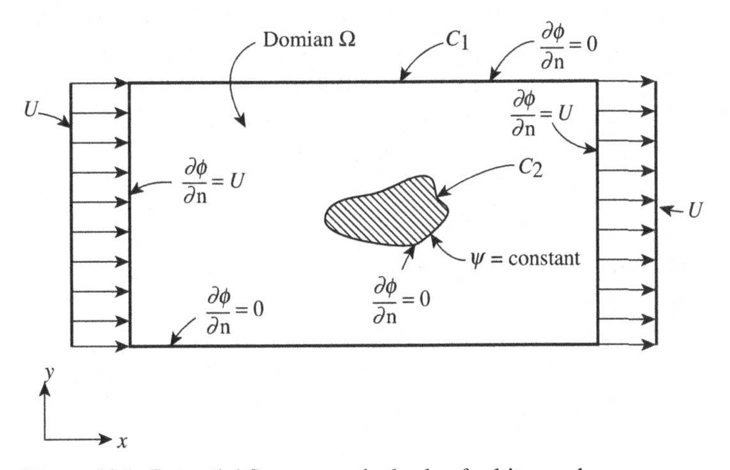

Figure 10.1. Potential flow around a body of arbitrary shape.

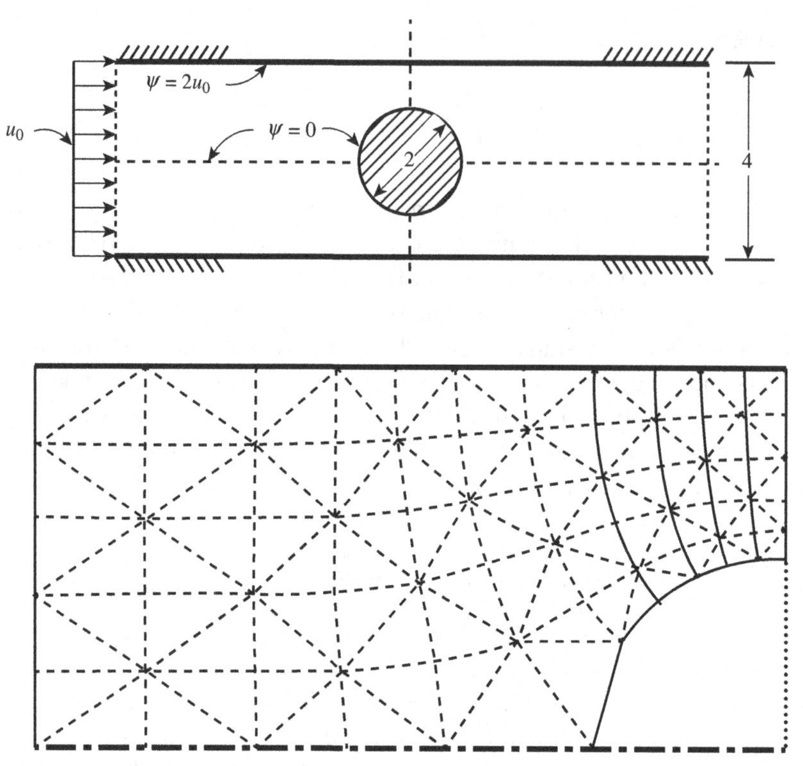

Figure 10.2. Potential flow around a cylinder between parallel plates.

is non unique. For this reason, when the solution domain is discretized and the element equations are formulated and assembled, the system matrix $[K]$ is singular. To overcome this difficulty, we arbitrarily select one node and specify the value of ϕ at it (e.g., $\phi = 0$). Essentially, we impose a Dirichlet boundary condition at one node. When incorporated by one of the means indicated in earlier chapters, this Dirichlet condition removes the singularity and allows us to proceed with the solution as usual.

Formulating the problem in terms of the stream function ψ requires a different technique. In this case, we know that ψ = constant or $\partial\psi/\partial s = 0$ on all the solid boundary segments, but the values of the constant are unknown. To proceed, we can use a superposition technique suggested by deVaries and Norrie [1]. On the outer boundary C_1, we recognize that ψ is known in terms of the inlet velocity U; hence we can write $\psi = g(x,y)$ on C_1, where $g(x,y)$ is a known function. The next step is to represent the complete solution as a sum of two separate parts, that is,

$$\psi(x,y) = \psi_1(x,y) + b\psi_2(x,y)$$

where b is a constant to be determined. The problem, then, reduces to two separate problems, described as follows:

$$\begin{aligned} \nabla^2\psi_1 &= 0 && \text{in } \Omega \\ \psi_1 &= g(x,y) && \text{on } C_1 \\ \psi_1 &= 0 && \text{on } C_2 \end{aligned}$$

$$\nabla^2\psi_2 = 0 \qquad \text{in } \Omega$$
$$\psi_2 = 0 \qquad \text{on } C_1$$
$$\psi_2 = 1 \qquad \text{on } C_2$$

These two sets of equations are easily solved using the finite element techniques as discussed in earlier chapters. Once $\psi_1(x,y)$ and $\psi_2(x,y)$ are known, we return to the original equation and find the constant b by evaluating $\psi(x,y)$ at one point inside Ω and close to the boundary C_1 where $\psi(x,y)$ is known. This gives an equation to be solved for b and completes the solution procedure.

Figure 10.2 shows an example of an internal flow studied by Martin [2], together with the discretization model [note that the domain is symmetric around two (vertical and horizontal) axes of symmetry].

Flow around Multiple Bodies by Superposition

The same superposition technique (above) can be used for either ϕ or ψ formulations when we have N arbitrarily shaped bodies immersed in a flow stream (Figure 10.3). For the ψ formulation, for example, we would write

$$\psi = \psi_1 + b_2\psi_2 + b_3\psi_3 + \cdots + b_N\psi_N$$

and then form N problems as follows:

$$\nabla^2\psi_1 = 0 \qquad \text{in } \Omega$$
$$\psi_1 = g \qquad \text{on } C_1$$
$$\psi_1 = 0 \qquad \text{on } C_2, C_3, \ldots, C_N$$
$$\nabla^2\psi_2 = 0 \qquad \text{in } \Omega$$
$$\psi_2 = 1 \qquad \text{on } C_2$$

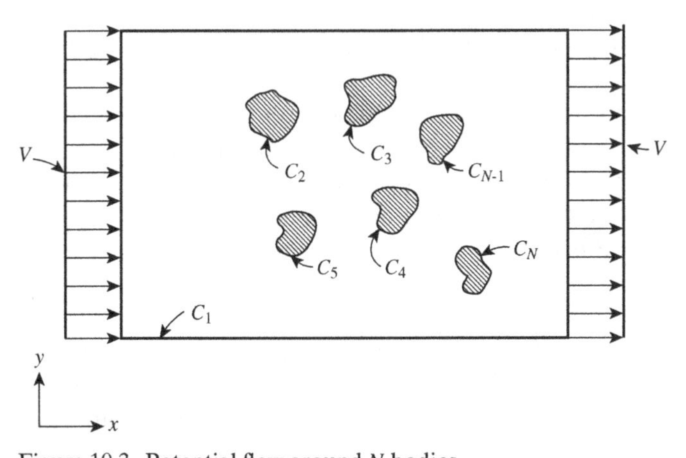

Figure 10.3. Potential flow around N bodies.

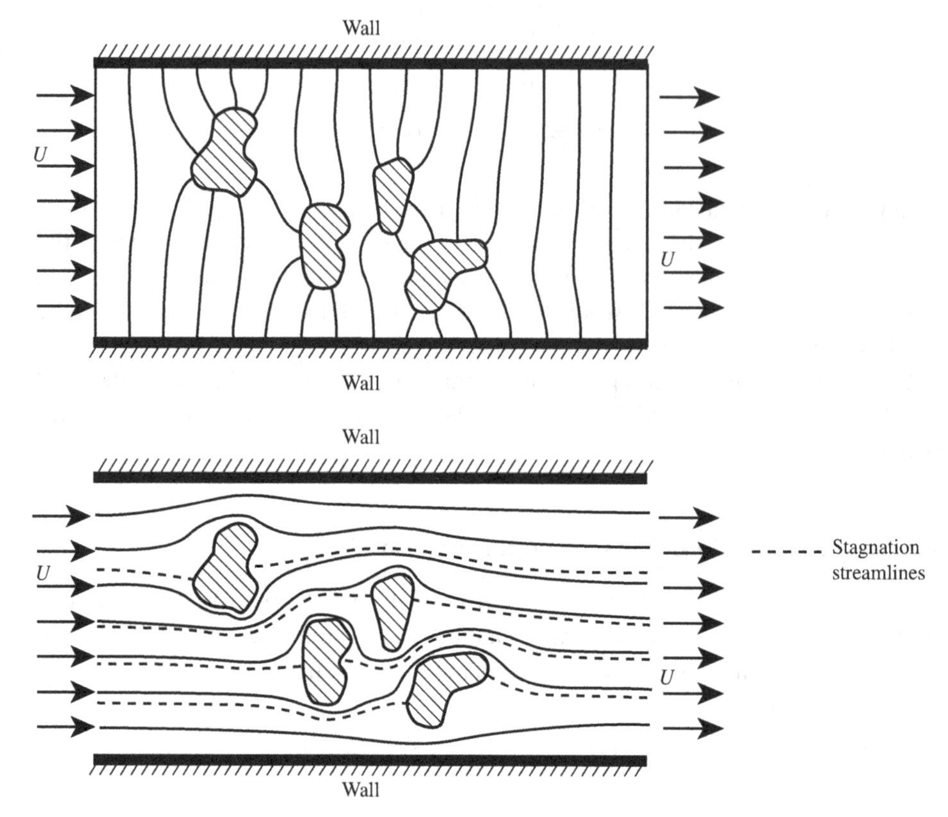

Figure 10.4. Potential flow solution for irregularly shaped bodies.

$$\begin{aligned} \psi_2 &= 0 && \text{on } C_1, C_2, \ldots, C_N \\ \nabla^2 \psi_N &= 0 && \text{in } \Omega \\ \psi_N &= 1 && \text{on } C_N \\ \psi_N &= 0 && \text{on } C_1, C_2, \ldots, C_{N-1} \end{aligned}$$

Each of the preceding sets of equations is solved by the finite element techniques discussed in earlier chapters. After ψ_i are found, the original equation is evaluated at $N-1$ points to form $N-1$ linear algebraic equations to be solved for the constants $b_2, b_3, \ldots, b_N$. Figure 10.4 shows some sample of flow patterns obtained through this method [1].

Once the stream function or velocity potential is known for a given problem, we can find the pressure distribution throughout the flow domain through the use of Bernoulli's equation (note that this equation is applicable because we are dealing with a potential and incompressible flow field). For a steady potential (or, equivalently, irrotational) and inviscid two-dimensional flow, Bernoulli's equation can be written as follows:

$$\frac{1}{2}V^2 + \eta + \frac{p}{\rho} = \text{constant}$$

where $V^2 = u^2 + v^2$, and η is the body-force potential, p is the (static) pressure, and ρ is the (static) density.

REFERENCES

[1] deVries, G., and Norrie, D. H., "The Application of the Finite Element Technique to Potential Flow Problems," *Journal of Applied Mechanics*, Vol. 38, 1971.
[2] Martin, H. C., "Finite Element Analysis of Fluid Flows," *Proceedings of the Second Conference on Matrix Methods in Structural Mechanics (AFFDL-TR-68-150)*, Wright-Patterson Air Force Base, Dayton, OH, October 1968.

11 Use of Nodeless Degrees of Freedom

Overview

Of the different problem categories in the remainder of this text, this is the simplest and, appropriately, a good starting point. A potential flow field is one where a single field variable suffices and a single flow-governing equation applies. This variable has typically been chosen as either the stream function ψ or the velocity potential ϕ. This apparent simplicity, nevertheless, may (in the larger picture) underestimate the critical role a potential-flow code often plays in a typical cascade-design setting, as well as the inherent analytical difficulty in securing a single-valued flow solution in a multiply-connected domain, with the latter being the focus of this chapter.

Beginning as early as the 1930s, several methods were devised for the problem of potential flow past a cascade of lifting bodies. Some of these methods were based on the use of conformal transformation [1–5], where one or more transformation step(s) are used in mapping the computational domain into a set of ovals or a flat-plate cascade [4, 5]. A separate category of analytical solutions [6] is based on the so-called singularity method, whereby sequences of sources and sinks and/or vortices are used to replace the airfoil itself. Next, the streamline-curvature method was established [7] as a viable approach to the airfoil–cascade-flow problem. With advent of the computer revolution came several numerical models of the problem based on the finite-difference method [9], finite element [9–12], and finite-volume [13] computational techniques.

From an analytical viewpoint, a flow passage will conceptually suffer one level of multiconnectivity at any point where two streams with two different histories are allowed to mix together. Referring to Figure 11.1, it is clear that the airfoil trailing edge is such a point. From a purely geometrical standpoint, a multiply connected domain is one where one can draw a closed-ended contour (e.g., contour C in Figure 11.1) and then fail to shrink it down to a point without having to cross the domain boundary. The two equivalent characteristics just described clearly apply to the flow domain highlighted in Figure 11.1. The problem here is that the flow field, in this category of domains, fails to be unique unless an external constraint is imposed. Essentially implying an externally prescribed circulation magnitude, such a constraint has typically taken the form of externally imposing the flow exit angle a priori. Although this alleviates the solution multi-valuedness problem, it simply

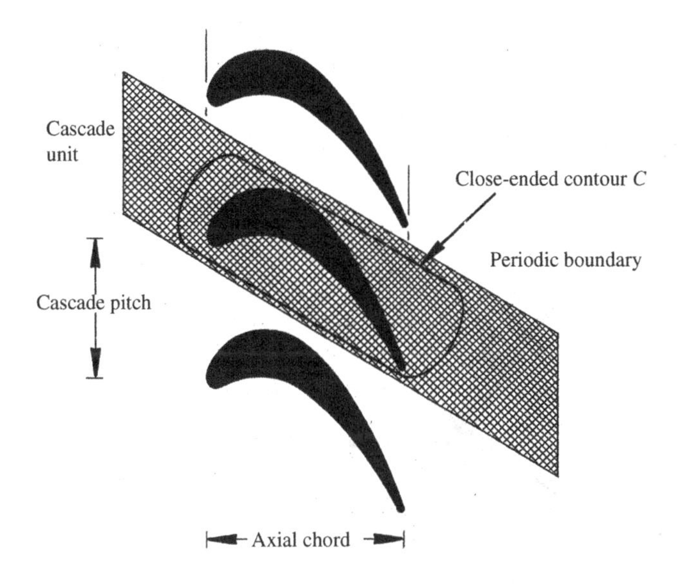

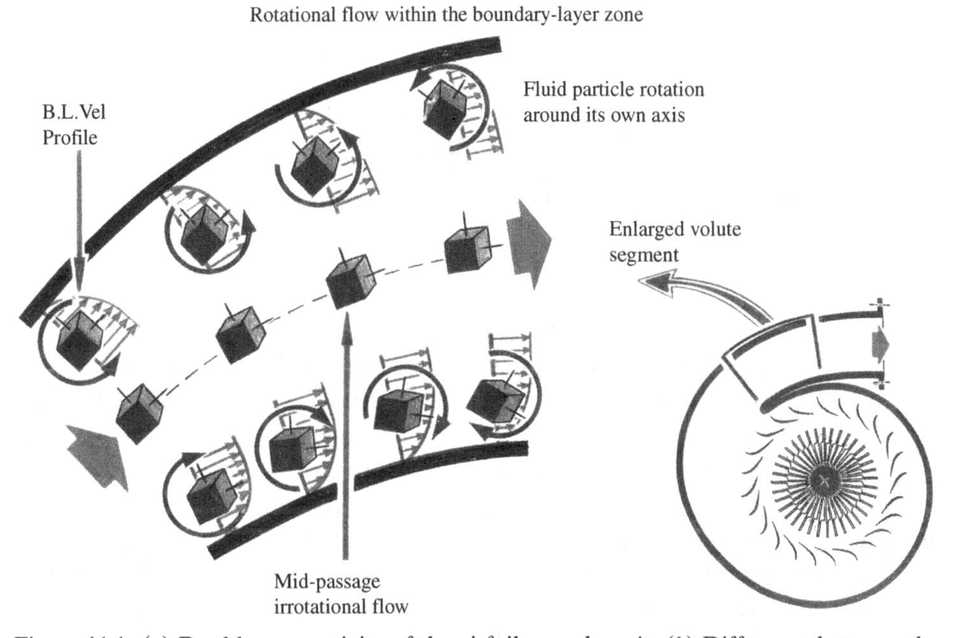

Figure 11.1. (*a*) Double connectivity of the airfoil cascade unit. (*b*) Difference between the irrotational core flow and near-wall rotational flow.

means that the outcome is essentially one of the analyst's own choosing. The present analysis (Baskharone 1979; Baskharone and Hamed 1981) is but an implementation of the simple fact that given the cascade geometry (Figure 11.2) and the flow inlet conditions, there should be a unique airfoil circulation magnitude, giving rise to a unique flow field, without resorting to any fictitious constraints on the flow behavior. Figure 11.1*b*, however, offers an interesting point. The shown geometry is that of the inlet segment to a radial-inflow turbine scroll. Examination of this important Figure reveals two distinct behaviors of the fluid particles as they traverse this

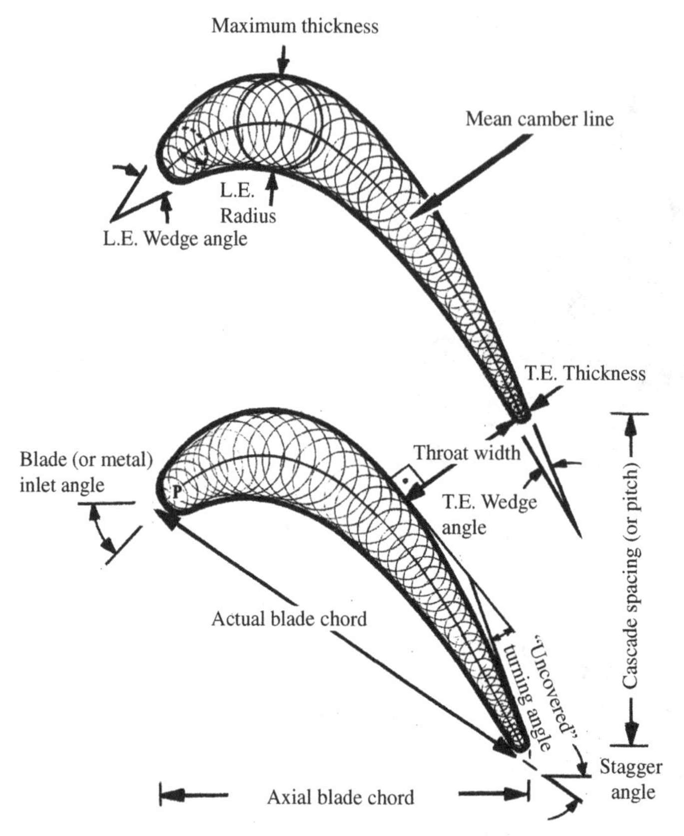

Figure 11.2. Geometrical details of a typical-turbine cascade.

passage. Midway in this passage you find fluid particles which are spinning around an external axis, unaffected by any such factors as boundary layer characteristics or vorticity. However, those particles near the passage walls are clearly affected by their presence within the boundary layers. Due to the severe velocity gradients in this thin layer, you find that one of the particle faces, away from the wall, is moving at a faster rate than the particle face near the wall. The result is a rotational motion around the particle's own axis. This is what bring the vorticity as the strongest component in this particle's motion. Of course, this kind of motion charges its own degradation to particles within this thin layer. The term "degradation" here is the substantial total pressure loss due to such undesirable motion.

Barring the current analysis, some of the existing computational models externally enforce satisfaction of the well-known Kutta condition on an iterative basis. In its original form, the Kutta condition simply states that the velocity vector in the vicinity of the airfoil trailing edge is to be both finite and continuous. The condition, by definition, applies to a theoretically sharp trailing edge such as the wedge or cusp trailing-edge configurations in Figure 11.3. Assuming full guidedness of the flow stream by the airfoil surface, a physically meaningful flow field in the wedge category is attainable only under the condition of zero velocity precisely at the trailing edge. Short of this, a fictional flow field would arise in which the velocity vector at the trailing edge has two different directions corresponding to the slopes of the

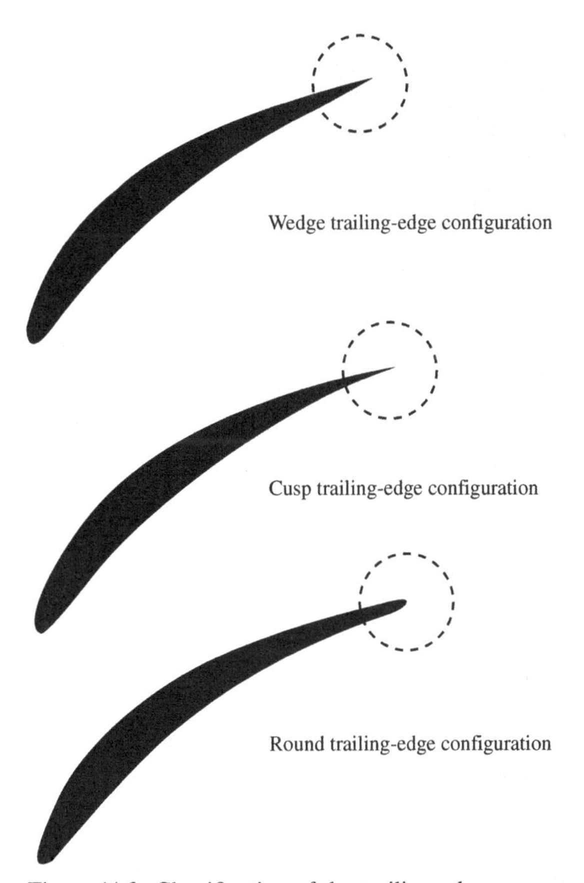

Figure 11.3. Classification of the trailing-edge geometry.

pressure- and suction-side tangents at the trailing-edge. As for the cusp trailing-edge configuration, where the two tangents coincide, a trailing-edge condition applies, whereby the suction- and pressure-side velocity vectors are equated as the trailing edge is approached. Easier said than done, implementation of this simple condition, within a computational model, is anything but easy. The airfoil aerodynamic loading in Figure 11.4 validates this very statement. The figure contains the results of applying McFarland's finite-volume code [13] to the 72 percent span section of the first-stage rotor in the high-pressure turbine of the F109 turbofan engine (a subsonic trainer designed for the U.S. Air Force). Of interest here is the fictitious flow behavior on the pressure side as the trailing edge is approached, an outcome of externally imposing a cusp trailing-edge condition at the trailing edge that is not, and can practically never be, a theoretical cusp. Needless to say, that the equality of suction- and pressure-side velocities in this figure is meaningless, for it is the result of a fictional flow behavior over the rear segment of the airfoil pressure side, as shown in the figure.

The common problem with the trailing-edge conditions (earlier) is that the trailing edge itself is assumed to be infinitely thin. In real-life applications, however, the trailing edge will always be round with a finite nonzero thickness. Moreover, there is a whole set of airfoil-geometry variables (see Figure 11.2) that collaborate with the trailing-edge thickness to give rise to a unique flow structure around the trailing

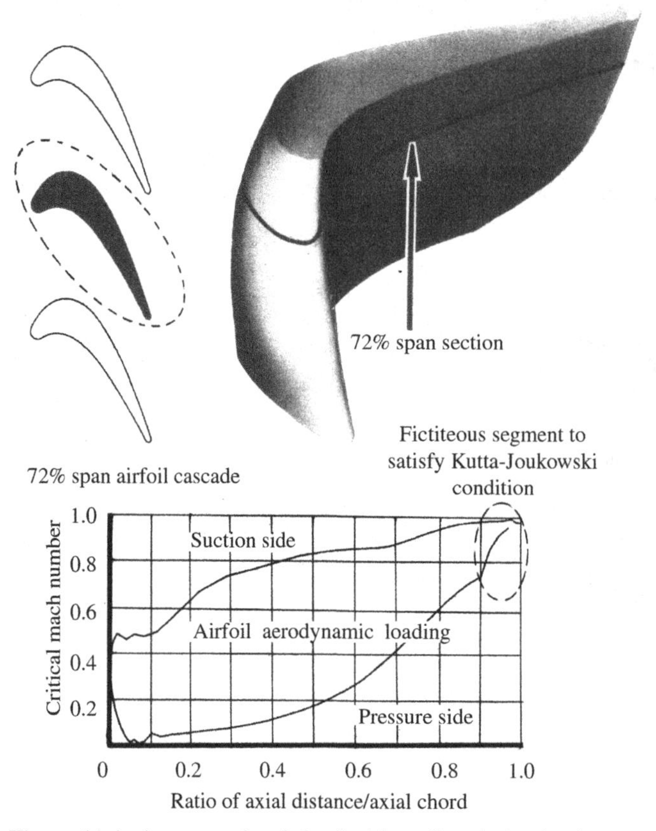

Figure 11.4. An example of the fictitious flow behavior imparted by an inaccurate trailing-edge condition.

edge. This very fact underscores the danger in allowing a largely simplified trailing-edge condition to play the only role in determining the circulation magnitude and, consequently, the entire flow field for that matter.

In this chapter, a weighted-residual-based finite element model is developed for the rectilinear cascade unit of the type shown in Figure 11.1. The term *rectilinear* here defines a cascade of blades, all existing between two parallel planes, and is famous in the area of wind-tunnel flow measurements. Uniqueness of this computational model (Baskharone and Hamed 1981) stems from the fact that the flow behavior is not externally subjected to any fictitious constraints such as a much-simplified trailing-edge condition or a rough magnitude of the exit flow angle. By including the circulation as a "nodeless" field variable (within the finite element computational model), the flow field becomes a function of only the cascade geometry and the flow inlet conditions, just as it should be. The term *nodeless* here defines the situation where a degree of freedom, namely, the circulation around the airfoil, is not assigned to any particular computational node, as is normally the case in a typical field-discretization step of a typical finite element analysis. As described later in this chapter, this particular variable is introduced in the form of a velocity potential discontinuity over a pair of infinitely close domain-splitting boundaries. Referring to Figure 11.5, these boundaries convert what is clearly a doubly connected cascade

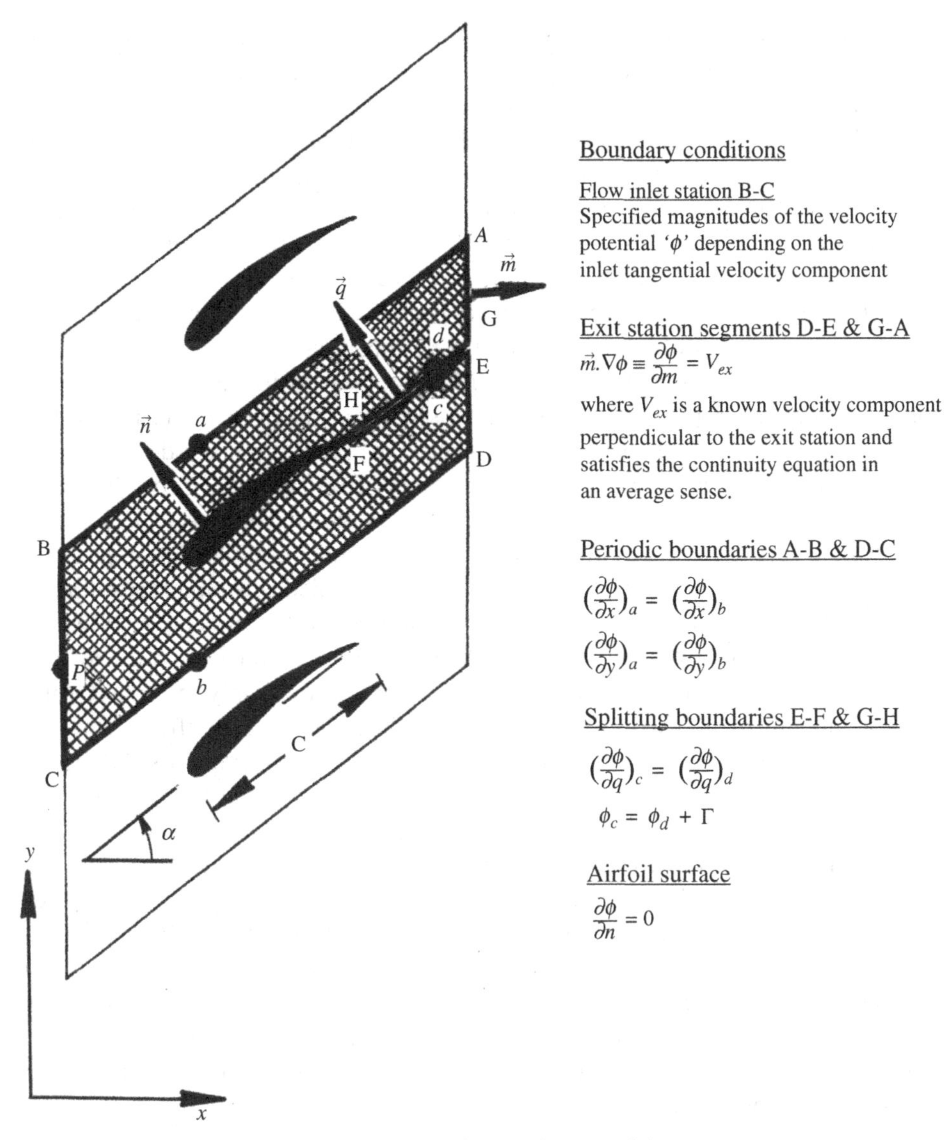

Figure 11.5. Classification of the cascade-unit boundary conditions.

unit into one that is simply connected. As the geometric problem of this domain is resolved, so is the analytical challenge of what would otherwise be a multivalued solution to the same physical problem in the absence of the airfoil circulation now being part of the problem formulation.

The previously-described rectilinear-cascade problem is one of two (earlier) is one of two applications covered in this chapter. The other concerns the potential flow field in a radial-turbine scroll. As discussed later in this chapter, this stationary turbomachinery component equally falls under the multiply-connected domain category. In the interest of producing practically usable results, the scroll problem is handled on a three-dimensional basis. Different in some details, the general approach to the two problems is conceptually the same.

It may appear that the potential flow simplification as it applies to the airfoil-cascade problem, for instance, is excessively degrading from an analytical standpoint. This is contrary to the standard approach in a typical real-life cascade design setting. The fact is that the major part of design decisions is made on the basis of aerodynamic loadings resulting from potential-flow solvers in the form of velocity or pressure distributions over the blade suction and pressure sides. To the trained designer, specific features of the flow behavior, such as excessive suction-surface velocity decline (or, equivalently, an adverse pressure gradient), are commonly called *diffusion* and are aerodynamically damaging. The reason is that such a situation invites some harmful viscosity-related deficiencies, including a rapid boundary layer buildup and the likelihood of boundary-layer separation and flow recirculation. Other equally important design-related issues, such as front versus rear airfoil loading, are also handled, perhaps exclusively, at this preliminary design phase, where the flow field is potential. In an essentially trial-and-error procedure, the designer would normally make the appropriate geometry adjustments to achieve the best possible aerodynamic loading prior to any subsequent detailed design phase.

The flow model that is offered in this chapter is applied to a typical rectilinear-cascade flow domain. This should not give the impression that it is only applicable to this type of geometric configurations. The model is centered around the idea of converting the domain into one that is simply connected by simply splitting it using a pair of infinitesimally close boundaries (extending, for instance, from the trailing edge to the flow exit station) and then patching it up through a specific (yet unknown) discontinuity in the field variable across the splitting boundaries. The magnitude of this discontinuity is numerically that of the circulation around the lifting body. Treating this discontinuity as a field variable, to be attained as part of the final solution, is where this computational model deviates from any other flow models in this area. This, as will be seen later, will eliminate the solution multivaluedness obstacle without the need to enforce any externally imposed constraint that is conceptually fictitious. The common thread, therefore, connecting the two problems under investigation in this chapter is what is referred to as a *nodeless degree of freedom*, a generic term that describes the discontinuity (or jump) in the field variable across a pair of splitting boundaries.

Flow-Governing Equations

The velocity potential ϕ is defined as follows:

$$\vec{V} = \nabla\phi \tag{11.1}$$

where $\vec{V}$ is the local velocity vector.

In a potential flow field, the flow-governing equations consist of the continuity equation and the irrotationality condition. Note that the latter has nothing to do with any flow turning within its bounding passage, for the rotationality of the fluid particle means that it spins around its own axis (due to shear stresses, for example). For a potential (i.e., irrotational) flow field, such motion ceases to exist.

The two equations governing the potential (or irrotational) flow field are the previously cited continuity and irrotationality equations. These can be expressed in terms of the velocity potential as follows:

$$\nabla.(\rho\vec{V})=0 \tag{11.2}$$

$$\nabla\times\vec{V}=0 \tag{11.3}$$

The (static) density (earlier) is obtained by applying the following isentropic flow relationship (note that a potential flow will have to be irrotational as well as isentropic):

$$\rho=\frac{p_t}{RT_t}\left[1-\frac{(\gamma-1)V^2}{2\gamma T_t}\right]^{\frac{1}{\gamma-1}} \tag{11.4}$$

Worth noting is the fact that both the total pressure p_t and total temperature T_t remain constant for the previously noted fact that a potential flow field is also isentropic. Of course, R is the gas constant. Note that we are treating the flow as *compressible*; otherwise, who cares about the density. Finally, γ is the specific-heat ratio.

The mere definition of the velocity potential ϕ automatically satisfies the irrotationality condition, for the simple fact that the term $\nabla\times(\nabla\phi)$ is identically zero, with the two remaining equations assuming the following forms:

$$\nabla.(\rho\nabla\phi)=0$$

$$\rho=\frac{p_t}{RT_t}\left[1-\frac{(\gamma-1)}{2})(\nabla\phi.\nabla\phi)2\gamma RT_t\right]^{1/(\gamma-1)}=0 \tag{11.5}$$

In the case of a low-subsonic incompressible flow, Equation (11.5) assumes the following Laplacian form:

$$\nabla^2\phi=0 \tag{11.6}$$

For a compressible flow field, however, the static density ρ, being a field variable, will cause Equation (11.5) to be nonlinear. The system's linearity, under such circumstance, can be restored by using the density distribution corresponding to the preceding iteration or an initial guess. Consistent with this approach, Equation (11.5) can now be rewritten as follows:

$$\nabla.\left[\rho^{(n)}\nabla\phi^{(n+1)}\right]=0 \tag{11.7}$$

with the density ρ^n taking the following form:

$$\rho^{(n)}=\frac{p_t}{RT_t}\left[1-\frac{(\gamma-1)}{2\gamma RT_t}(\nabla\phi^{(n)}.\nabla\phi^{(n)})\right]^{1/(\gamma-1)} \tag{11.8}$$

The superscripts (n) and $(n+1)$ signify two subsequent iterative steps.

The numerical procedure, therefore, is to substitute an initial guess for the static density (perhaps the total density magnitude everywhere in the solution domain) in Equation (11.7) and then use the newly computed magnitudes of ϕ in re-calculating the static-density magnitudes throughout the computational domain. These magnitudes are then carried over (see Figure 11.8). These magnitudes are then carried over to the next iterative step, and the process is repeated toward convergence.

Boundary Conditions

These are summarized in Figure 11.5, including those around the cascade-unit outer contour, as well as the pair of splitting boundaries.

Flow Inlet Station $(B - C)$

Assuming that the inlet segment is "sufficiently" far upstream from the airfoil leading edge, we can treat the inlet velocity vector as uniform. Of this known vector, let us refer to the tangential (upward) component by V_t. Recalling the definition of the velocity potential ϕ, we can express V_t as follows:

$$\frac{\phi_B - \phi_C}{S} = V_t \tag{11.9}$$

where S is the cascade pitch defined as the blade-to-blade spacing.

Now, setting the velocity potential ϕ to an arbitrary value (say zero) at point C in Figure 11.5, the potential magnitude at any point P on the inlet boundary segment (B – C) can be expressed as follows:

$$\phi_P = \phi_C + SV_t \tag{11.10}$$

where S is the distance between the two points.

Note that it is legitimate, as well as necessary, to set the potential magnitude at any arbitrary point (C in this case) to a fixed magnitude, as we just did. As will be explained next, we will assign a Neumann (or derivative)–type boundary condition to the flow exit station (D – A). Without a potential "datum" magnitude, therefore, the problem solution would fail to be unique.

Flow Exit Station $(D - A)$

Refering to the unit vector perpendicular to this station by $\vec{m}$, the following relationship applies:

$$\vec{m}.\nabla\phi \equiv \frac{\partial\phi}{\partial m} = (V_m)_{ex} \tag{11.11}$$

where $(V_m)_{ex}$ is the exit-velocity component normal to the exit boundary. This discharge-velocity component [which also can be called $(V_x)_{ex}$] can be obtained by applying the continuity equation at the exit station with prior knowledge of the mass-flow rate through the cascade unit. Again, the assumption of a location that is sufficiently far downstream from the airfoil trailing edge applies to the exit station, which, in turn, justifies a uniform discharge-velocity component. Shown differently in Figure 11.5, note that points E and G on this boundary segment are physically coincident. Nevertheless, these will be assigned two different magnitudes of potential, with the difference being numerically identical to the airfoil circulation magnitude Γ, as will be explained in the finite element analysis section later in this chapter.

Periodic Boundaries ($A - B$ and $D - C$)

Across these boundary segments, the flow behavior is to be identical, a feature that is referred to as the cascade *periodicity condition*, because they are parallel and one pitch S apart from one another. Using the velocity potential ϕ, the following two conditions apply at points a and b in Figure 11.5:

$$\left(\frac{\partial \phi}{\partial x}\right)_a = \left(\frac{\partial \phi}{\partial x}\right)_b \tag{11.12}$$

$$\left(\frac{\partial \phi}{\partial y}\right)_a = \left(\frac{\partial \phi}{\partial y}\right)_b \tag{11.13}$$

Domain-Splitting Boundaries ($E - F$ and $G - H$)

These two boundary segments are physically coincident, but not analytically. There is a need for them in order to convert the solution domain into a simply connected one, as explained in the introductory remarks of this chapter, should a single-valued solution be sought. Note that these two lines are arbitrary in shape and location, as long as they extend from the airfoil surface to any flow-permeable boundary segment. The chosen location in Figure 11.5 favorably affects the numerical accuracy because it is at the airfoil surface, in general, and the trailing edge, in particular, that the intensity of finite elements is (as it should be) relatively high (Figures 11.6 and 11.7), an advantage that makes it geometrically easier to refine the region in the immediate vicinity of the splitting boundaries.

Across the splitting lines in Figure 11.5, the velocity potential ϕ will experience a uniform "jump," the magnitude of which is numerically identical to the yet-unknown circulation Γ around the airfoil. In an equation form,

$$\phi_c = \phi_d + \Gamma \tag{11.14}$$

where c and d are two corresponding points, one on each boundary segment (see Figure 11.5). Applying this flow constraint along the entire splitting pair of boundaries is, in effect, guaranteeing that the velocity components, in the boundary direction, are identical at each pair of corresponding points (e.g., c and d) becaues Γ is simply a constant. As for the equality of the perpendicular velocity component, a derivative-type boundary condition is invoked as follows:

$$\left(\frac{\partial \phi}{\partial q}\right)_c = \left(\frac{\partial \phi}{\partial q}\right)_d \tag{11.15}$$

where $\vec{q}$ is the unit vector that is perpendicular to the two boundary segments as indicated in Figure 11.5.

Airfoil Suction and Pressure Surfaces

Over these solid-boundary segments, the so-called no-penetration boundary condition applies as follows:

$$\frac{\partial \phi}{\partial n} = 0 \tag{11.16}$$

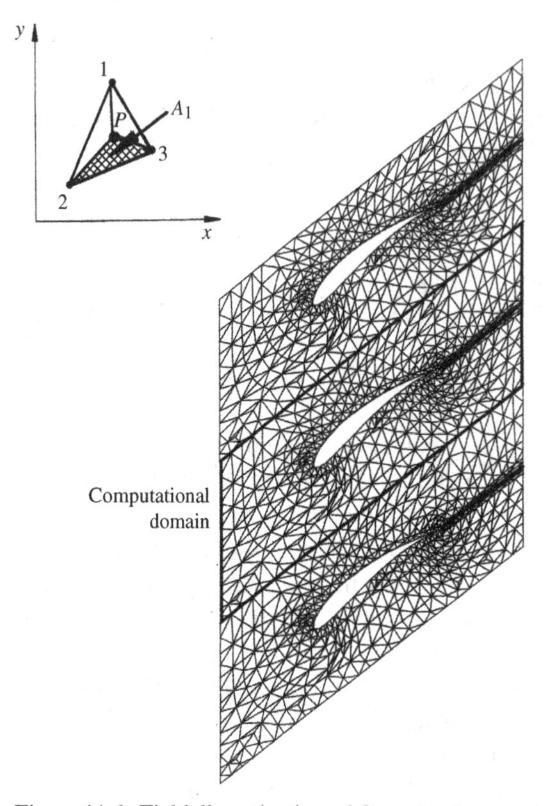

Figure 11.6. Field discretization of Gostelow's cascade-unit using a linear triangular elements.

where $\vec{n}$ is the local unit vector that is locally perpendicular to the airfoil surface. Within the finite element model, this so-called "natural" boundary condition does not require external enforcement because it is automatically satisfied at each and every boundary segment unless a different boundary condition is imposed instead, which is not the case here.

Finite Element Analysis

As indicated in Chapter 10, the first step in creating a typical finite element analysis is what is referred to as the field *discretization* process. In this process, the flow domain is replaced by an assembly of nonoverlapping subdomains termed *elements* that may vary in shape and size in light of how steep the field variable (the velocity potential ϕ in this case) is anticipated to vary in one subregion relative to another. Figures 11.6 and 11.7 show the outcome of this process, with the linear triangular element being the discretization unit in Figure 11.6 and the eight-noded curve-sided quadratic element being the case in Figure 11.7.

For simplicity, let us first consider a typical triangular element (e), as shown in Figure 11.6. Over this element, the velocity potential is interpolated as follows:

$$\phi^{(e)} = N_1\phi_1 + N_2\phi_2 + N_3\phi_3 \tag{11.17}$$

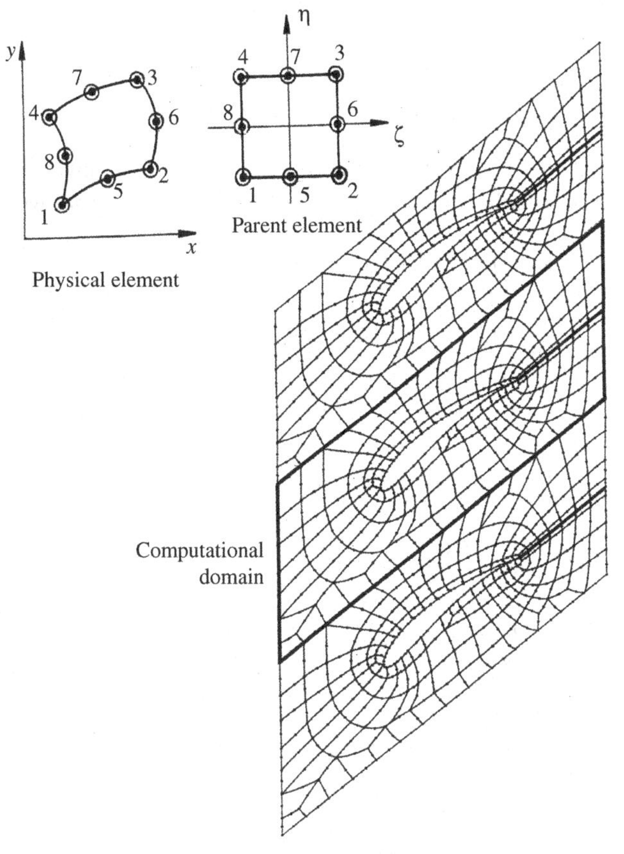

Figure 11.7. Field discretization of Gostelow's cascade unit using a curve-sided quadrilateral element.

where N_1, N_2, and N_3 are termed the *shape functions* associated with the element nodes 1, 2, and 3, respectively. These are functions of the spatial coordinates (x and y) and vary between 0.0 and 1.0. Let us, for the purpose of generality, derive the expression for N_1 that is associated with an arbitrary interior point P in Figure 11.6. This, in fact, is simply the ratio between the area of the triangle (P–2–3) and that of the entire element (1–2–3), or

$$N_1 = \frac{A_1}{A^{(e)}}$$

where

$$A_1 = (y_2 - y_3)x_P + (x_3 - x_2)y_P + (x_2y_3 - x_3y_2)$$

$$A^{(e)} = (x_2y_3 - x_3y_2) + (x_3y_1 - x_1y_3) + (x_1y_2 - x_2y_1)$$

The eight-noded curve-sided element in Figure 11.7 is one of a family that is referred to as the *serendipity* family of elements and is, by definition, a quadratic element.

We can write the interpolation function for the potential ϕ within the typical quadratic element (e) as follows:

$$\phi^{(e)} = \sum_{i=1}^{8} N_i \phi_i$$

The shape functions N_i in this case are not as simple as with the preceding triangular element. These are defined, in the so-called parent element (Figure 11.7), as follows:

$$N_1 = -\frac{(1-\zeta)(1-\eta)(1+\zeta+\eta)}{4}$$

$$N_2 = -\frac{(1+\zeta)(1-\eta)(1-\zeta+\eta)}{4}$$

$$N_3 = -\frac{(1+\zeta)(1+\eta)(1-\zeta-\eta)}{4}$$

$$N_4 = -\frac{(1-\zeta)(1+\eta)(1+\zeta-\eta)}{4}$$

$$N_5 = \frac{(1-\zeta^2)(1-\eta)}{2}$$

$$N_6 = \frac{(1+\zeta)(1-\eta^2)}{2}$$

$$N_7 = \frac{(1-\zeta^2)(1+\eta)}{2}$$

$$N_8 = \frac{(1-\zeta)(1-\eta^2)}{2}$$

where the ζ and η coordinate system defines a fictitious frame of reference to which the physical element is mapped, as described next.

The physical finite element in Figure 11.7 is first mapped into the so-called parent element, with the latter being a simple square. In doing so, and in order to achieve a square parent element, the Cartesian coordinates themselves are mapped into the local $(\zeta - \eta)$ frame of reference as follows:

$$x = \sum_{i=1}^{8} N_i(\zeta, \eta) x_i$$

$$y = \sum_{i=1}^{8} N_i(\zeta, \eta) y_i$$

Now the velocity potential itself can be interpolated as follows:

$$\phi^{(e)} = \sum_{i=1}^{N} N_i \phi_i \tag{11.18}$$

where N is the total number of nodes, including (in the case of the curve-sided element) the midside nodes associated with this quadratic element in Figure 11.7.

Galerkin's Weighted-Residual Approach

Substituting the potential-approximating relationship (11.18) into the flow-governing equation (11.5) results in an error (or residual) $E^{(e)}$ within the typical element (e) as follows:

$$E^{(e)} \doteq \nabla.(\rho^{(e)} \nabla \phi^{(e)}). \tag{11.19}$$

The element equations are derived in a manner that can be viewed as a process where

- The *weighted* residual is set to zero over the finite element in an integral sense.
- The residual is made orthogonal to a set of *weight* functions over the entire finite element.

Adopting Galerkin's weighted-residual method [15], the above-referenced weight functions are precisely the element shape functions N_i. Regardless of how the process is viewed, the finite element equation comes out to be

$$\int_{A^{(e)}} N_i E^{(e)} = 0$$

or

$$\int_{A^{(e)}} N_i[\nabla.(\rho^{(e)} \nabla \phi^{(e)})] = 0 \tag{11.20}$$

where i varies from 1 to N, the total number of nodes associated with the element (e). Now, applying Gauss's theorem, we get the following set of elemental equations:

$$\int_{A^{(e)}} \rho[\nabla N_i.\nabla N_j]\phi_j dA = \int_{L^{(e)}} N_i \frac{\partial \phi}{\partial n} dL \tag{11.21}$$

where both i and j vary from 1 to N, the total number of the element's nodes, which is 8 in this case, with $A^{(e)}$ and $L^{(e)}$ referring to the element area and perimeter, respectively, with the vector $\vec{n}$ being the local unit vector that is perpendicular to the element's boundary. The term $(\vec{n}.\nabla\phi)$ appearing in Equation 11.21, is simply equal to the velocity component perpendicular to the local boundary segment where the unit vector $\vec{n}$ is defined. The right-hand side of Equation (11.21) will appear only wherever one side of the element exists on a boundary segment where a nonzero Neumann-type boundary condition applies (e.g., over the flow-domain exit station). The differential area and line increment (dA and dL, respectively) can be expressed for a general curve-sided element (see Figure 11.7) as follows:

$$dA = dxdy = | J | \, d\zeta \, d\eta$$

$$dL = G(\eta) d\eta$$

where the length element dL is assumed to exist on a constant ζ side of the element (say $\zeta = 1$) in the local frame of reference, with $| J |$ being the Jacobian of the Cartesian-to-local-coordinate transformation, which can be expressed as follows:

$$| J | = \frac{\partial x}{\partial \zeta}\frac{\partial y}{\partial \eta} - \frac{\partial x}{\partial \eta}\frac{\partial y}{\partial \zeta}$$

The function $G(\eta)$ is defined as follows:

$$G(\eta) = \left\{ \left[\left(\frac{\partial x}{\partial \eta} \right)^2 + \left(\frac{\partial y}{\partial \eta} \right)^2 \right]^{1/2} \right\}_{\zeta=1} \tag{11.22}$$

In a compact form, the set of element equations can be rewritten as follows:

$$[k]\{\phi\} = \{f\} \tag{11.23}$$

where the $N \times N$ matrix $[k]$ and the $N \times 1$ vector $\{f\}$ are normally referred to as the element's *stiffness matrix* and *load vector*, respectively.

Assembled among all elements, the final set of finite element equations can be represented as follows:

$$[K]\{\phi\} = \{F\} \tag{11.24}$$

The so-called stiffness matrix $[K]$ is diagonally dominant but is not as bounded as one would wish. The reason is that it contains circulation-related terms, which were discussed earlier. One appropriate method of inversion, in this case, is that due to Gupta and Tanji [16], where the matrix contents are stored in two partially packed arrays, one carrying the nonzero contents of each row in the original matrix $[K]$ and the other containing the column indices where these contents actually exist in the matrix. The method is but one of several numerically efficient means of matrix inversion.

Applications

Flow Analysis in a Rectilinear-Airfoil Cascade

The preceding analysis was applied to an airfoil cascade for which the airfoil geometry and pressure distribution were obtained by Gostelow [4] through application of the Merchant and Collar transformation into a set of ovals. The generated cascade has a pitch-chord (S/C) ratio of 0.99 (see Figure 11.5), a stagger angle α of 37.5 degrees, and a flow inlet angle of 53.5 degrees.

Field Discretization Model

Two different finite element models were used to analyze Gostelow's cascade flow. These are shown in Figure 11.6, where a linear triangular element is the case, and Figure 11.7, where the quadratic eight-noded element is the discretization unit. Whereas the latter element category is known to be one of the most accurate, it is one objective of this study to weigh the accuracy enhancement against the drastic increase in computational resources this high-order element requires.

Computational Results and Accuracy Assessment

Figure 11.8 shows a comparison between the computed results and the exact solution obtained by Gostelow [4]. Under examination here is the airfoil-surface-pressure coefficient C_P, where

$$C_P = \frac{p - p_{in}}{\rho_{in} {V_{in}}^2} \tag{11.29}$$

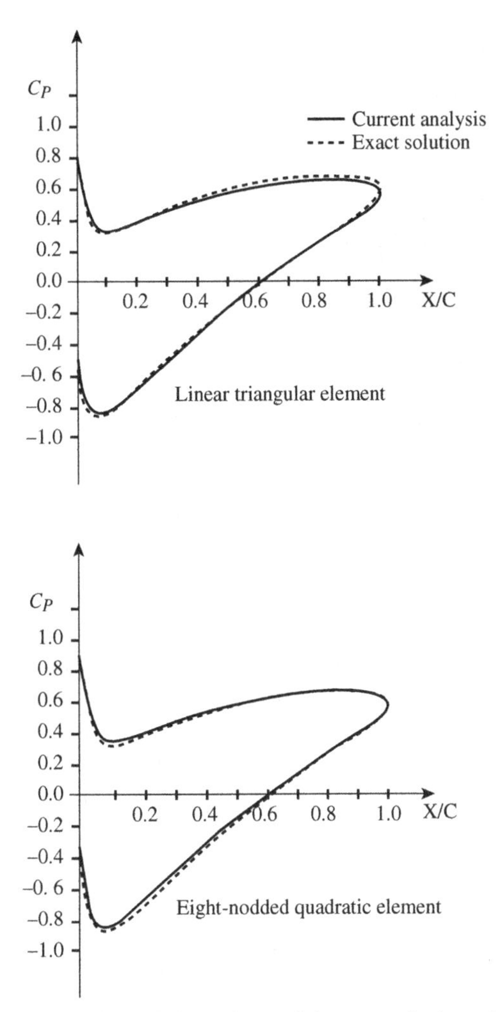

Figure 11.8. Comparison of the numerical results with Gostelow's exact solution.

with p being the (static) pressure over the airfoil surface, and the constants p_{in}, ρ_{in}, and V_{in} are the inlet magnitudes of pressure, density and velocity, respectively. In order to compute the local pressure p, the local velocity V has to be computed first:

$$V = \left[\left(\frac{\partial \phi}{\partial x} \right)^2 + \left(\frac{\partial \phi}{\partial y} \right)^2 \right]^{1/2} \tag{11.30}$$

While the finite element model in Figure 11.6 shows a generally good agreement with the exact solution, the model composed of quadratic elements (see Figure 11.7) seems to yield better agreement by comparison. Worth noting is the fact that the two models were established using a total of 691 linear triangular elements and 179 eight-noded quadratic elements, respectively. However, it was the latter model that consumed approximately 39 times the CPU seconds that the triangular-element model required.

Of particular interest here is the flow exit angle, for it is directly indicative of the circulation magnitude. Computed by Gostelow, the exact value of this angle

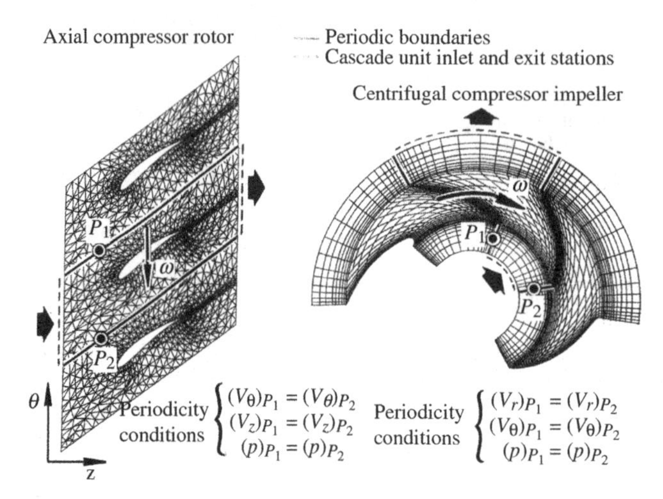

Figure 11.9. Comparison of the periodicity conditions between axial and radial turbomachinery components.

is 30.01 degrees. On the other hand, the computed magnitudes are 30.53 and 29.64 degrees for the linear and quadratic-element models, respectively. The errors, therefore, are 1.7 percent and 1.2 percent, respectively. Such low error magnitudes go toward validating the very foundation of the current analytical approach, including, in particular, the very concept of using a nodeless degree of freedom to what would, otherwise be a problem with (theoretically) an infinite number of solutions depending on the fictitious condition that is externally dictated.

Periodicity Conditions in Radial Cascades

As seen in the preceding problem, the flow periodicity conditions constitute an essential part of this and any cascade-flow problem. Figure 11.9 shows a comparison of these conditions between a rectilinear cascade and a radial-airfoil cascade, with the latter being the rotor (or impeller) of a centrifugal compressor. Note that the airfoils, in the latter case, are separated by one *angular* pitch, as contrasted to a "linear" pitch in the rectilinear-cascade problem. As the figure indicates, the periodicity conditions in both cases involve the equality of the velocity vector and, subsequently, the static pressure at the periodic-boundary computational nodes that correspond to one another. Another difference, however, lies in the type of velocity components to be equated. These, in the case of the radial-airfoil cascade, are appropriately chosen to be the radial and tangential velocity components because it is the cylindrical polar frame of reference that is uniquely suitable for this and similar types of radial-cascade problems.

Flow Investigation in a Radial-Turbine Scroll

The scroll (or distributor) passage in a radial-inflow turbine is the very first component of the turbine stage. The function of this passage (shown in Figure 11.10) is to discharge the flow continuously , and hopefully uniformly, around the circumference at a radius that is just higher than the rotor tip radius (not shown in the figure). Because of the continuous flow discharge, the scroll cross-flow sectional

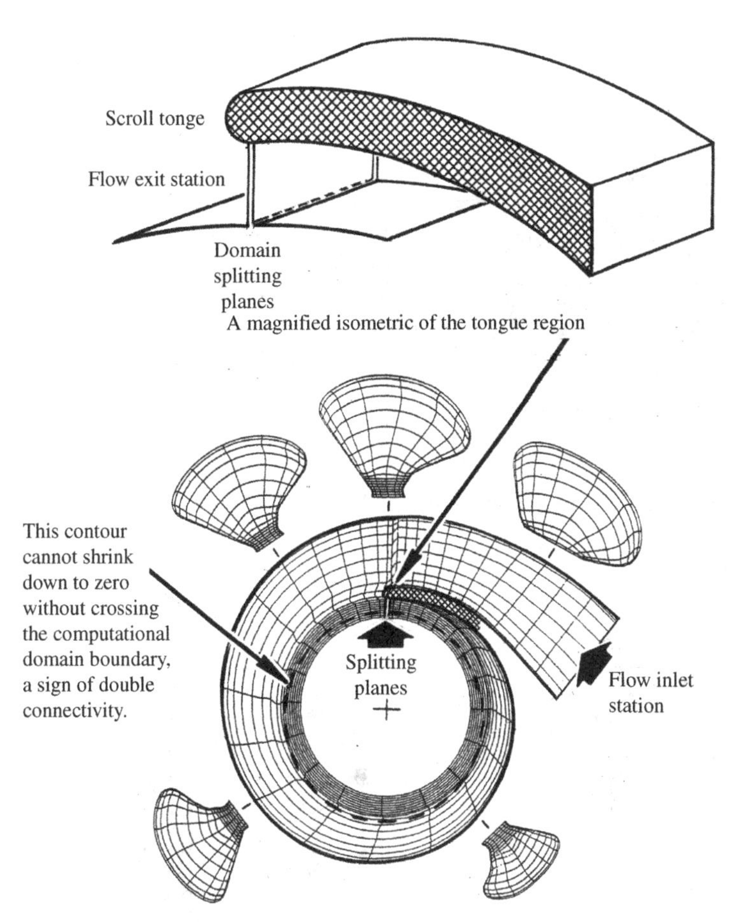

Figure 11.10. Double connectivity of an unvaned radial-inflow turbine scroll.

area is designed to gradually decline in the through-flow direction in order to maintain a nearly constant through-flow velocity. The circumferential rate at which this area declines is not only critical to the scroll-exit mass flux distribution but also to the uniformity of the exit flow angle. Unfortunately, little attention has been paid to the design optimization of this passage and, in fact, to the general radial-turbine category for that matter. Perhaps the reason behind this neglect is that this particular turbine, despite its high shaft-work capacity, has never found its way to propulsion applications because of its relative "bulkiness," large "envelope" requirement (due to the large rotor-tip radius), and heaviness.

The scroll shown in Figure 11.10 is the final outcome of a NASA-sponsored Technology Demonstration Program undertaken by Allied-Signal Aerospace Corporation (Phoenix, AZ), with the bigger objective of designing fabricating, and testing an efficient radial-turbine stage. The actual component (designed by the author) is shown in Figure 11.11 in the form of a cutaway of this passage showing the scroll's so-called tongue at the tip of which two flow streams with two different histories mix together, a definite sign of double connectivity, as indicated in Figure 11.10. Figure 11.11 shows a normally unacceptable level of surface finish, except that it is acceptable here, owing to the traditional "excuse" that the passage belongs

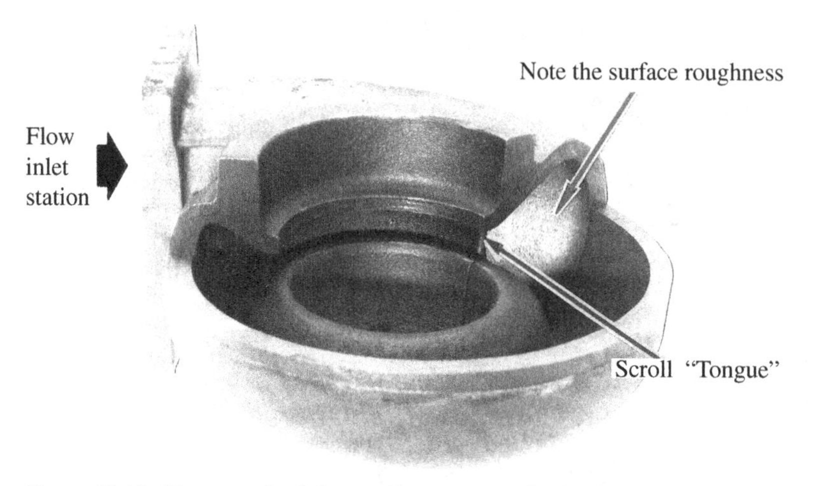

Figure 11.11. Photograph of the scroll component in the TV-71 radial-turbine stage.

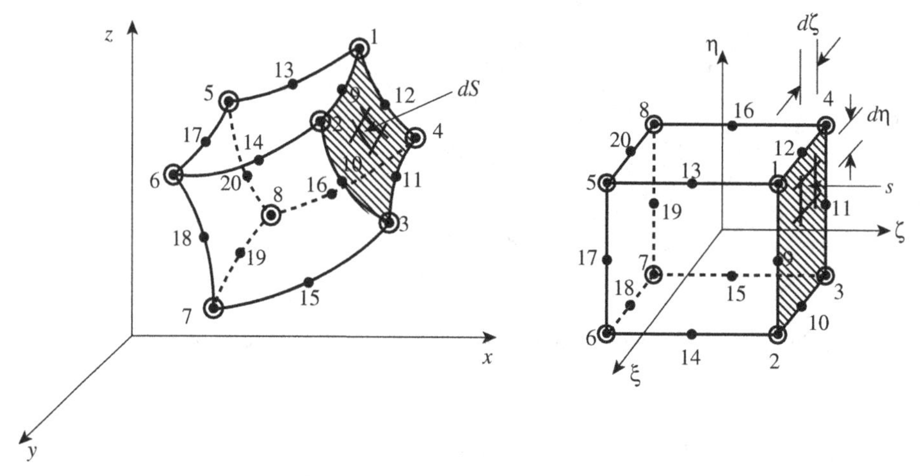

Figure 11.12. Three-dimensional curve-sided isoparametric finite element for the potential flow analysis in a radial-turbine scroll.

to a radial turbine, which is categorically unused in propulsion applications, where "clean," well-designed axial-flow turbine components are expected to be.

Finite Element Analysis

Shown in Figure 11.10 is a cutaway of the scroll passage. The grid lines in the figure arise from the field discretization process in which the three-dimensional 20-noded isoparametric element in Figure 11.12 is used as the discretization unit. The 20 shape functions associated with this element are all functions of ζ, η, and ξ and can be found in Ref. [14].

Within a typical element (e), let us interpolate the velocity potential ϕ and spatial coordinates x, y, and z using their nodal magnitudes and the element's shape (or interpolation) functions as follows:

$$\phi^{(e)} = \sum_{i=1}^{20} N_i(\zeta, \eta, \xi)\phi_i \tag{11.31}$$

$$x^{(e)} = \sum_{i=1}^{20} N_i(\zeta, \eta, \xi)x_i \tag{11.32}$$

$$y^{(e)} = \sum_{i=1}^{20} N_i(\zeta, \eta, \xi)y_i \tag{11.33}$$

$$z^{(e)} = \sum_{i=1}^{20} N_i(\zeta, \eta, \xi)z_i \tag{11.34}$$

Substituting expressions (11.31) through (11.34) into the flow-governing equation (11.7), we get a residual function. Applying Galerkin's principle in the same manner as discussed earlier in conjunction with the two-dimensional curve-sided finite element (see Figure 11.7), we arrive at the three-dimensional equivalent of Equation (11.21) as follows:

$$\int_{V^{(e)}} \rho \left[\frac{\partial N_i}{\partial x}\frac{\partial N_j}{\partial x} + \frac{\partial N_i}{\partial y}\frac{\partial N_j}{\partial y} + \frac{\partial N_i}{\partial z}\frac{\partial N_j}{\partial z} \right] dV = \int_{S^{(e)}} \rho(\vec{m}.\nabla\phi)\, dS \tag{11.35}$$

where $\vec{m}$ is the local outward unit vector that is perpendicular to the differential surface area dS on the element boundary. The differential volume dV can be expressed as follows:

$$dV = |J|\, d\zeta\, d\eta\, d\xi$$

where $|J|$ is the Jacobian of physical-to-local-coordinate transformation and can be expressed as follows:

$$|J| = \begin{vmatrix} \sum_{i=1}^{20} \frac{\partial N_i}{\partial \zeta} x_i & \sum_{i=1}^{20} \frac{\partial N_i}{\partial \zeta} y_i & \sum_{i=1}^{20} \frac{\partial N_i}{\partial \zeta} z_i \\ \sum_{i=1}^{20} \frac{\partial N_i}{\partial \eta} x_i & \sum_{i=1}^{20} \frac{\partial N_i}{\partial \eta} y_i & \sum_{i=1}^{20} \frac{\partial N_i}{\partial \eta} z_i \\ \sum_{i=1}^{20} \frac{\partial N_i}{\partial \xi} x_i & \sum_{i=1}^{20} \frac{\partial N_i}{\partial \xi} y_i & \sum_{i=1}^{20} \frac{\partial N_i}{\partial \xi} z_i \end{vmatrix}$$

As for the differential surface area over the element boundary, it should first be stated that any finite element (in the physical space) can be manipulated in such a way as to ensure that the element's face that contributes to the right-hand side of Equation (11.35) is always the $\zeta = 1$ plane in the local frame of reference (see Figure 11.12). With this condition satisfied, the expression for the differential surface area dS is as follows:

$$dS = G(\eta, \psi)\, d\eta\, d\xi$$

where the function $G(\eta,\xi)$ assumes the following form:

$$G(\eta,\xi) = \left\{ \left[\left(\frac{\partial y}{\partial \eta}\frac{\partial x}{\partial \xi} - \frac{\partial y}{\partial \xi}\frac{\partial x}{\partial \eta} \right)^2 + \left(\frac{\partial x}{\partial \xi}\frac{\partial z}{\partial \eta} - \frac{\partial x}{\partial \eta}\frac{\partial z}{\partial \xi} \right)^2 + \left(\frac{\partial x}{\partial \eta}\frac{\partial y}{\partial \xi} - \frac{\partial z}{\partial \xi}\frac{\partial y}{\partial \eta} \right)^2 \right]^{1/2} \right\}_{\zeta=1}$$

Introduction of a Velocity Potential Discontinuity

Thorough examination of the scroll and, in particular, its tongue in Figure 11.10 answers the question of whether we are dealing with a doubly connected solution domain. It is at this particular point (meaning the tip of the scroll tongue) that two flow streams on both sides of the tongue and with two different histories mix together, reminding us that the pressure- and suction-side flow streams mix together in the immediate vicinity of the airfoil trailing edge. Just as it was valid then, the criterion is still valid now that such a feature of the flow field classifies the flow domain as doubly connected. To emphasize, the flow field, under such circumstances, fails to be unique.

Applying the same logic of splitting the computational domain, a pair of infinitesimally close planes extending from the scroll tongue down to the flow exit station is employed, with the objective of converting the domain into a simply connected one. These domain-splitting planes are shown in Figure 11.10, where the velocity potential is allowed to experience an abrupt discontinuity, one that is circulation-like.

Computational Results

The preceding finite element flow analysis was applied to the TV-71 scroll (by Allied-Signal Aerospace Corporation) during the scroll design procedure with highly promising results. In fact, the computational results in Figure 11.13 were obtained within an analytical effort to optimize the scroll inlet segment. With the two (curved and straight) inlet-passage configurations under investigation, the two most important characteristics, being the exit-station circumferential distributions of mass flux and flow angle, were particularly monitored. Note that a highly nonuniform scroll-exit mass flux, in any scroll design, means that the blade-to-blade passages, as the scroll passage approaches its end, will receive vastly different mass-flow-rate magnitudes. This, by itself, creates a significant tangential imbalance with serious mechanical consequences.

Analysis of the two scroll configurations gave rise to the velocity components' nodal magnitudes and static density. Using the velocity components, the calculation of flow angle (relative to the local radial direction) at all nodes on the exit station was straightforward.

The exit-station mass-flux distribution is also shown in Figure 11.13, where the straight inlet segment seems to produce a circumferential angle distribution that is much closer to being uniform by comparison. Worth noting is the fact that the thinner tongue associated with this configuration had a positive impact on the numerical results, with the straightness of the scroll admission passage having its own favorable effect. Note that the angle deviation immediately downstream from the tongue section is understandably tolerable in this case. After all, it is this region where two streams abruptly mix together.

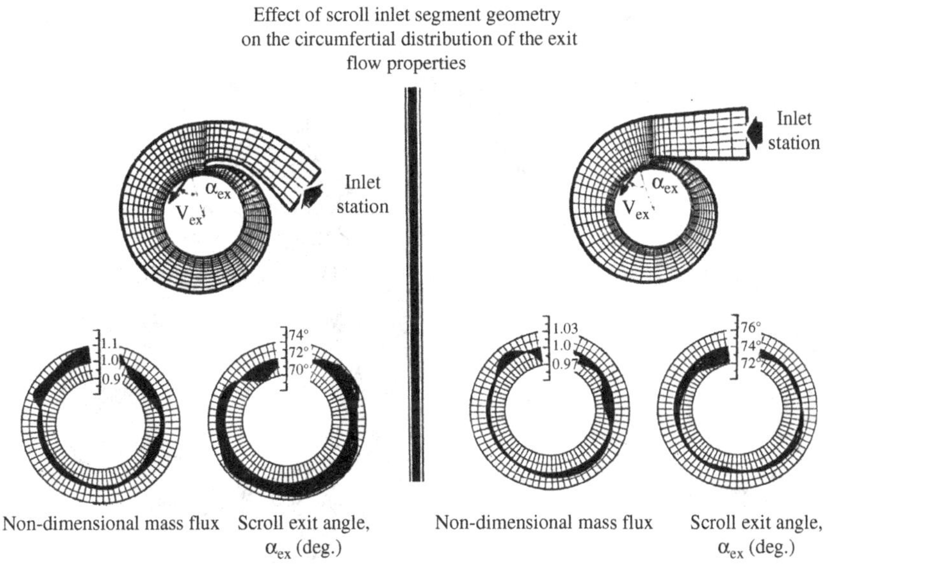

Effect of scroll inlet segment geometry on the circumfertial distribution of the exit flow properties

Inlet station

Non-dimensional mass flux

Scroll exit angle, α_{ex} (deg.)

Figure 11.13. Scroll flow analysis: circumferential distributions of mass flux and exit flow angle.

The exit-station mass-flux distribution in Figure 11.13 seems to exhibit the same characteristics of the exit-angle distribution. The mass flux, in this figure, is nondimensionalized using the average exit magnitude. The decision, which was favorable to the straight inlet segment, was influenced in the first place by what was considered to be an excellent mass-flux distribution for a scroll with no guide vanes.

It should be noted here that the scroll-exit boundary conditions did not involve the specification of any velocity component. Instead, the free-vortex boundary conditions were applied as follows:

$$\frac{dV_r}{dr} = -\frac{V_r}{r} \qquad \text{(continuity equation)}$$

$$\frac{dV_\theta}{dr} = -\frac{V_\theta}{r} \qquad \text{(angular momentum conservation)}$$

where V_r and V_θ are the radial and tangential velocity components, respectively. Note that these two boundary conditions are linear and are, as a result, introduced noniteratively in the numerical solution process. The two conditions are simply the differential forms of the following free-vortex equations:

$$rV_r = \text{constant}$$

$$rV_\theta = \text{constant}$$

Proposed Analysis Upgrades

Domains with High Degrees of Multiconnectivity

The two problems discussed thus far have commonly involved doubly connected flow domains. Required in this case is a single pair of domain-splitting boundaries (see Figures 11.5 and 11.10) to convert the computational domain into one

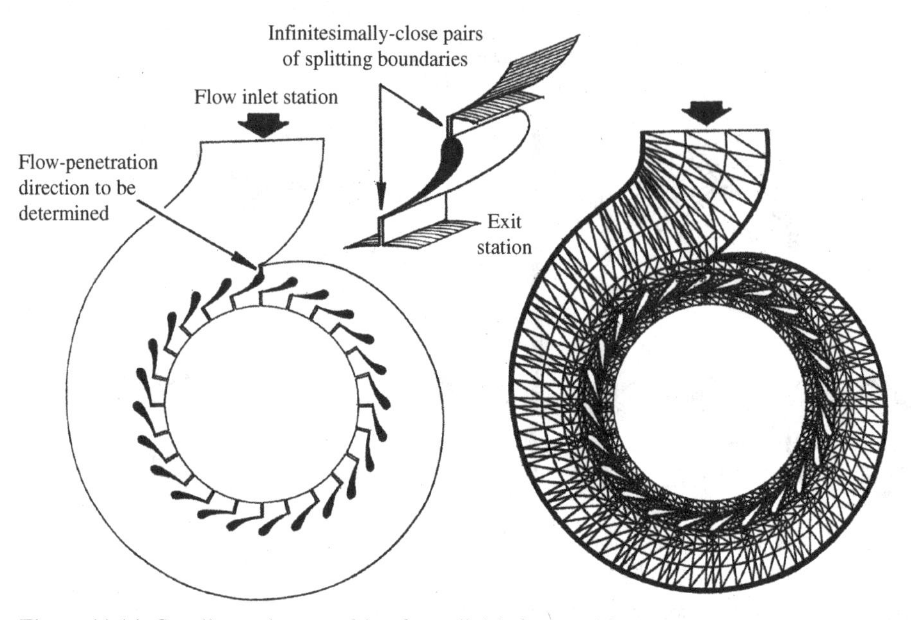

Figure 11.14. Scroll-nozzle assembly of a radial-inflow turbine.

that is simply connected. The velocity potential discontinuity across these infinitesimally close boundaries was, in each case, declared as an added (nodeless) degree of freedom in the problem's finite element formulation.

In many engineering applications, the existence of multiple lifting bodies raises the level of domain multiconnectivity. One example, in this category, is the scroll-nozzle assembly shown in Figure 11.14. Existence of the scroll passage here precludes the possibility of isolating a vane-to-vane flow passage as representative of all the radial cascade flow passages because each of these receives a different share of the total mass-flow rate. The degree of multiconnectivity here is equal to $N_v + 1$, where N_v is the total number of stator vanes. Note that the scroll passage itself is doubly connected, which is the reason why the assembled-domain multi-connectivity level is one larger than the number of vanes.

Shown in Figure 11.14 is the arrangement of splitting boundaries that, combined, reduces the computational domain to a simply connected one. Of these, a pair of boundaries extends from the scroll tongue down to the top-vane leading edge. Across each pair of domain-splitting boundaries, the potential is allowed to "jump" by a yet-unknown amount that is to be treated as a circulation or circulation-like nodeless variable in the same manner as explained earlier in this chapter.

The splitting boundaries in this case, as well as in any other problem in this category, are flow-permeable. Whereas the mass-flux direction across each of the vane-trailing-edge-attached boundaries is obvious, the flow-stream penetration of the tongue-to-vane pair of boundaries is neither known nor obvious. It is, in fact, the (static) pressure differential across this particular pair of boundaries that determines the flow direction in this region. Although one of the scroll design intents is to maintain a more or less constant exit static pressure throughout the entire passage, the pressure differential (above) will always have a finite nonzero magnitude despite

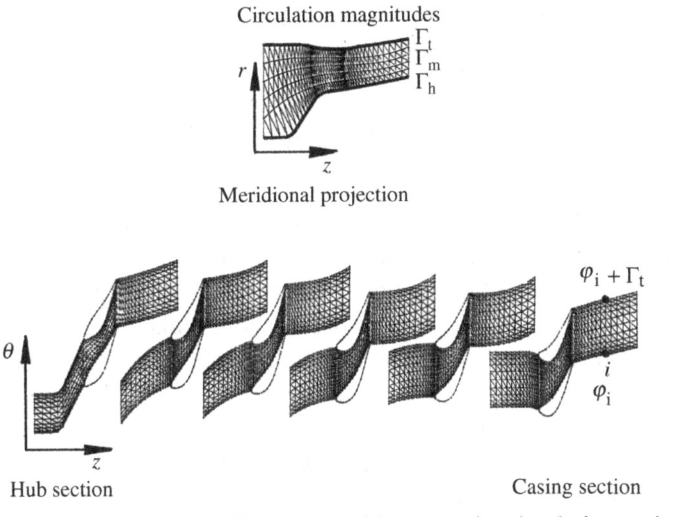

Figure 11.15. An axial-flow stator with a spanwise circulation variation.

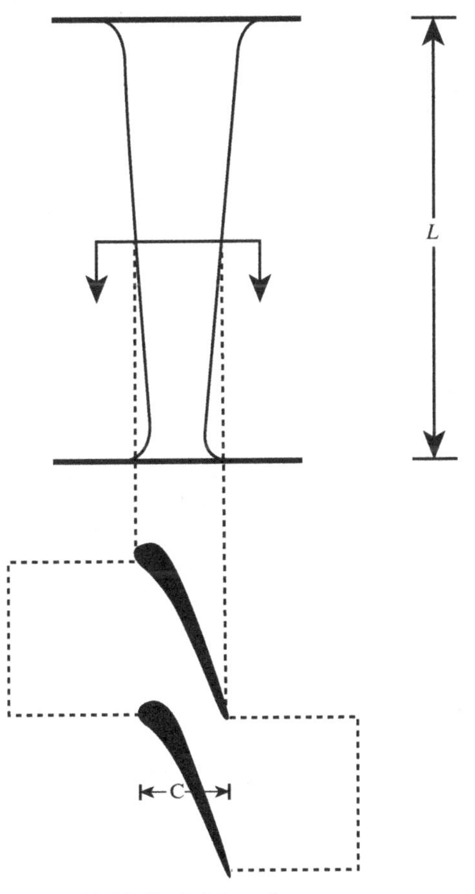

Figure 11.16. Definition of a stator-vane aspect ratio.

the fact that the nearly constant magnitudes of the flow thermophysical properties will always be indicative of how aerodynamically sound the scroll design is.

Axial-Flow Stator with a Spanwise Circulation Variation

Figure 11.15 shows an axial-flow stator in a commercial turbocharger with a spanwise variation in the blade cross section. This is an example of a turbomachinery component where the vane-section circulation would experience significant spanwise changes, as shown in the figure. With the flow-governing equations being the same as cited earlier in this chapter, the problem geometry here makes it advantageous to use the usual cylindrical frame of reference, where the boundary conditions can be imposed with relative ease.

The most practical means of effecting a hub-to-casing circulation variation is to declare a different circulation magnitude to each computational surface, as shown in the meridional view in Figure 11.15. Over each surface, the circulation is represented by a uniform velocity-potential "jump" across the pair of periodic boundaries extending from the airfoil trailing edge down to the exit station. These potential discontinuities are declared as nodeless degrees of freedom in the finite element model, in the same manner presented earlier in this chapter, with their magnitudes being part of the final numerical solution, as opposed to guessing their magnitudes at the analysis starting point.

Generally speaking, rotor blades (or stator vanes) with high aspect ratios (tall and skinny) are those where the spanwise circulation variation is most notable. The blade-vane aspect ratio (by reference to Figure 11.16) is defined as follows:

$$AR = \frac{L}{C}$$

where

- L is the blade/vane hight (or radial extension)
- C is the mean value of the blade (or vane) axial chord length

PROBLEMS

P.1 Figure 11.17 shows the arrangement of finite-elements near a flow-permeable boundary segment, where the following general boundary condition applies:

$$\frac{\partial \phi}{\partial n} = C_1 \phi + C_2$$

Considering the highlighted element in the figure, and by substituting the right-hand side of this condition in the line integral of Equation (11.21), show how this boundary condition contributes to both sides of this equation. Hint: It may be helpful to express ϕ over the element side coinciding with the boundary segment as follows:

$$\phi_{1-8-4} = \left(\sum_{j=1}^{8} N_j \phi_j \right)_{\zeta=1}$$

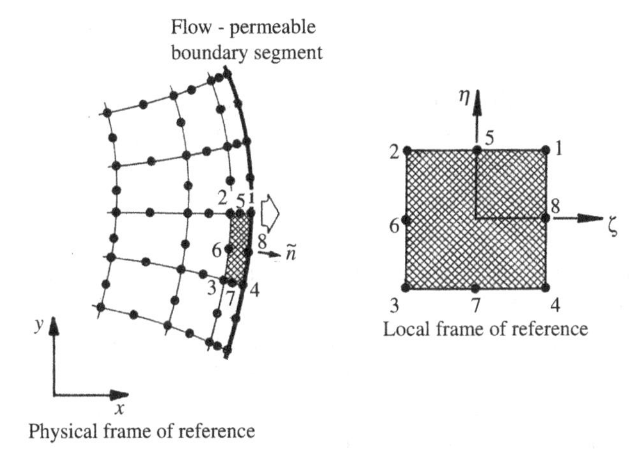

Figure 11.17. Element arrangement near a general-condition boundary segment.

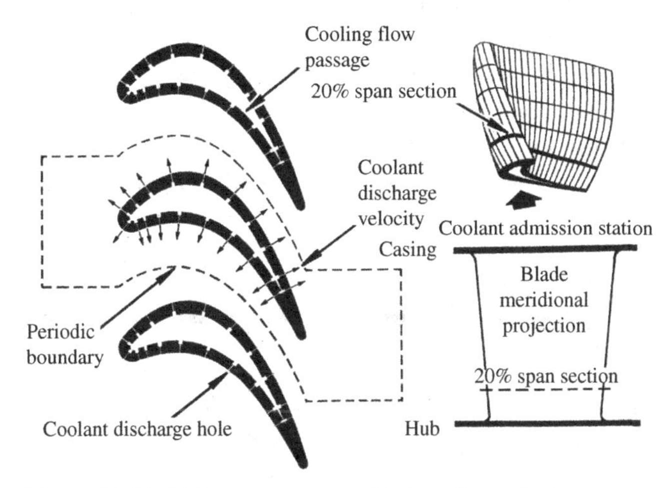

Figure 11.18. Stationary cascade of internally cooled vanes.

P.2 In many high-pressure turbine stages, and for thermal-stress considerations, one or more cooling technique(s) are applied. Figure 11.18 shows an internally cooled vane of a stator, with the core region being the passage of the coolant (normally air that is extracted at the compressor exit station), which extends from the hub surface (where the coolant is introduced) up to the casing surface (where part of it is discharged). The difference in the coolant mass-flow rate is continuously discharged through narrow holes (shown in the figure) into the primary flow stream. To simplify the problem, let us assume the following:

- The airfoil cascade is rectilinear, that is, defined in the x-y plane, instead of being the annular cascade it really is.
- The coolant discharge is perfectly uniform over the entire vane surface, which is the case where the number of discharge holes is theoretically infinite, with the discharge velocity being V_d.

Using these approximations, show how this boundary condition will contribute to the line integral in Equation (11.21).

P.3 Referring to the third Analysis Upgrade section (where the stream function was used as the field variable), consider two finite-elements sharing two corresponding segments of the downstream pair of periodic boundaries. Beginning with the ψ interpolation functions within each of these elements, derive the periodic boundary conditions briefly stated at the end of this section. Your boundary conditions should include some or all of the eight ψ values associated with each finite element.

P.4 Figure 11.19 shows the midsection of a stationary rectilinear airfoil cascade. The figure in particular highlights two finite-elements that correspond to one another, with each element sharing one side with the vane surface. The volumetric flow rate per unit length of the vane height is χ, and the discretization unit is the eight-noded curve-sided finite element previously discussed in this chapter. Referring to the pattern of numbering (in the figure), the computational nodes in the magnified view, construct a 16×16 matrix to contain the influence coefficients associated with the finite element equations at each node, as well as entries that reflect the implementation of boundary conditions over the two sides coinciding with the vane suction and pressure sides. Also construct the 16 × 1 "load" vector that is composed of the free terms in the finite element equations, as well the boundary-conditions-related entries. In these two arrays, write only the nonzero entries, knowing that the vector of unknown variables begins with the eight stream-function values associated with the upper element.

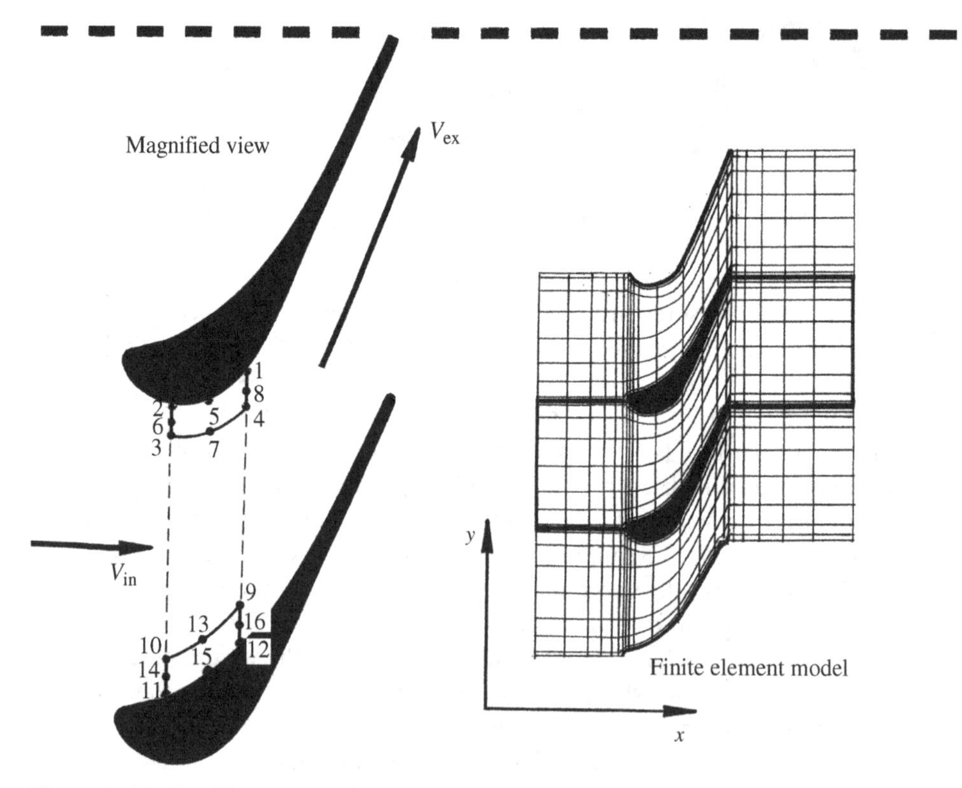

Figure 11.19. Rectilinear-stator flow analysis under a stream-function problem formulation.

These are followed by the lower-element stream-function magnitudes corresponding to the computational nodes comprising that element. Note that the flow-governing equation here is the following:

$$\int_{A^{(e)}} \left(\frac{\partial N_i}{\partial x} \frac{\partial N_j}{\partial x} + \frac{\partial N_i}{\partial y} \frac{\partial N_j}{\partial y} \right) dA = \int_{L^{(e)}} N_i \frac{\partial \phi}{\partial n} dL$$

where $\vec{n}$ is the local outward normal unit vector on the element's boundary.

P.5 Figure 11.20 shows the finite element discretization model for a stationary rectilinear cascade of turbine airfoils, where the discretization unit is a simple linear triangle. Assuming an incompressible flow field and a velocity-potential formulation, the flow-governing equation assumes the Laplacian form. Associated with the typical element (e), the finite element equivalent of the flow-governing equation can be written as follows:

$$\int_{A^{(e)}} \left(\frac{\partial N_i}{\partial x} \frac{\partial N_j}{\partial x} + \frac{\partial N_i}{\partial y} \frac{\partial N_j}{\partial y} \right) dA = \int_{L^{(e)}} N_i \left(\frac{\partial \phi}{\partial n} \right) dL$$

where $L^{(e)}$ refers to the element's perimeter, and $\vec{n}$ is the outward unit vector that is perpendicular to the element boundary. Referring to Figure 11.20, and treating the two elements a and b as one unit, we can express this equation in

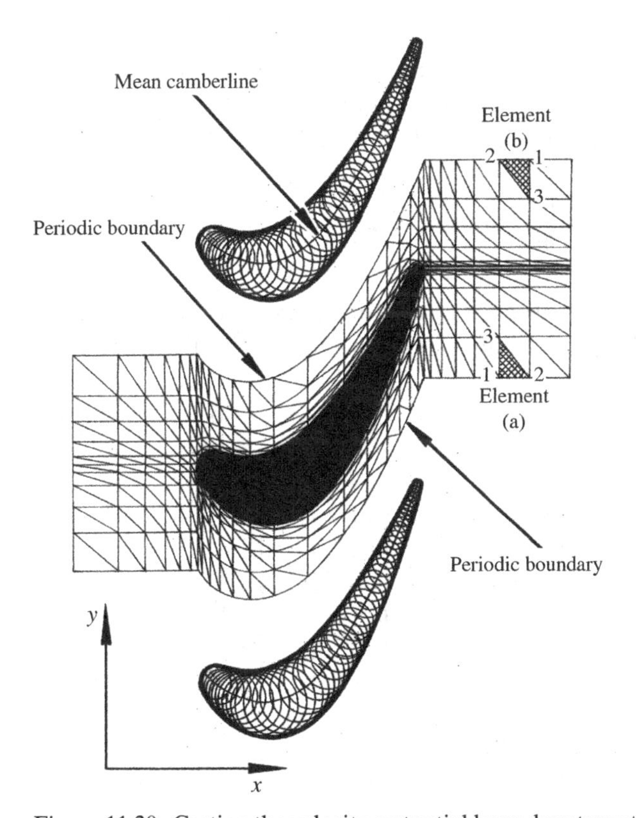

Figure 11.20. Casting the velocity-potential boundary term through implicit means.

the following matrix form:

$$\begin{bmatrix} a_{11} & a_{12} & a_{13} & & & \\ a_{21} & a_{22} & a_{23} & & & \\ a_{31} & a_{32} & a_{33} & & & \\ & & & b_{11} & b_{12} & b_{13} \\ & & & b_{21} & b_{22} & b_{23} \\ & & & b_{31} & b_{32} & b_{33} \end{bmatrix} \begin{Bmatrix} \phi_1{}^{(a)} \\ \phi_2{}^{(a)} \\ \phi_3{}^{(a)} \\ \phi_1{}^{(b)} \\ \phi_2{}^{(b)} \\ \phi_3{}^{(b)} \end{Bmatrix} = \begin{Bmatrix} q_1{}^{(a)} \\ q_2{}^{(a)} \\ q_3{}^{(a)} \\ q_1{}^{(b)} \\ q_2{}^{(b)} \\ q_3{}^{(b)} \end{Bmatrix}$$

The focus, in this problem, is on enforcing the periodicity of ϕ normal derivatives over a corresponding pair of periodic boundaries (see Figure 11.20). In physical terms, we are to equate the x and y velocity components ($\partial\phi/\partial x$ and $\partial\phi/\partial y$ in this case) at corresponding points on each pair of corresponding boundaries. Let us make it our task to ensure the equality of v (meaning $\partial\phi/\partial y$) at the two computational nodes $1^{(a)}$ and $2^{(b)}$, belonging to the finite-elements (a) and (b) in Figure 11.20, respectively. Two ways to achieve this follow.

Because the ϕ interpolation functions within the elements (a) and (b) are linear, we would expect the derivative $\partial\phi/\partial y$ to be constant over each element. In fact, the periodicity condition of concern can be enforced by removing (in the preceding matrix equation) the finite element equation associated with node $2^{(b)}$ and replacing it with the following equation:

$$\phi_1{}^{(b)} - \phi_3{}^{(b)} = \phi_3{}^{(a)} - \phi_1{}^{(a)} \tag{P.1}$$

An implicit and more accurate means of enforcing the same is to leave the $2^{(b)}$-associated equation in place while casting the boundary term in it using the $\partial\phi/\partial y$ magnitude in element (a) (which is an implicit way of doing the same thing). On an element basis, this equation will now appear as follows:

$$\int_{A^{(b)}} \left(\frac{\partial N_2}{\partial x}\frac{\partial N_2}{\partial x} + \frac{\partial N_2}{\partial y}\frac{\partial N_j}{\partial y} \right) dA = \int_{L^{(b)}} N_2 \left(\frac{\phi_3{}^{(a)} - \phi_1{}^{(a)}}{y_3{}^{(a)} - y_1{}^{(a)}} \right) dL \tag{P.2}$$

As seen, the line integral in Equation (P.2) will alter previously zero entries in the matrix equation (earlier), providing a zero contribution to the right-hand side vector.

Let us now rewrite the finite element set of equations associated with the elements (a) and (b), combined, as follows:

$$\begin{bmatrix} a_{11} & a_{12} & a_{13} & \alpha_{11} & \alpha_{12} & \alpha_{13} \\ a_{21} & a_{22} & a_{23} & \alpha_{21} & \alpha_{22} & \alpha_{23} \\ a_{31} & a_{32} & a_{33} & \alpha_{31} & \alpha_{32} & \alpha_{33} \\ \beta_{11} & \beta_{12} & \beta_{13} & b_{11} & b_{12} & b_{13} \\ \beta_{21} & \beta_{22} & \beta_{23} & b_{21} & b_{22} & b_{23} \\ \beta_{31} & \beta_{32} & \beta_{33} & b_{31} & b_{32} & b_{33} \end{bmatrix} \begin{Bmatrix} \phi_1{}^{(a)} \\ \phi_2{}^{(a)} \\ \phi_3{}^{(a)} \\ \phi_1{}^{(b)} \\ \phi_2{}^{(b)} \\ \phi_3{}^{(b)} \end{Bmatrix} = \begin{Bmatrix} q_1{}^{(a)} \\ q_2{}^{(a)} \\ q_3{}^{(a)} \\ q_1{}^{(b)} \\ q_2{}^{(b)} \\ q_3{}^{(b)} \end{Bmatrix}$$

Rewrite the matrix equation (earlier) on separately invoking expressions (P.1) and (P.2) in the finite element equation associated with the computational node $2^{(b)}$

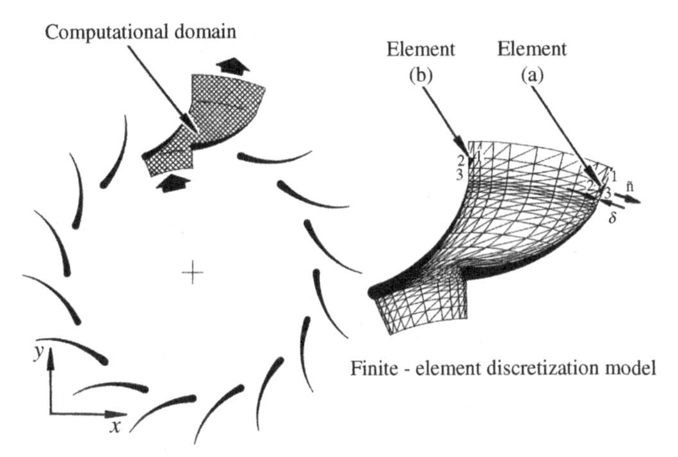

Figure 11.21. Definition of the computational domain in the diffuser component of a typical centrifugal compressor stage.

P.6 Figure 11.21 shows the vaned stator in a typical centrifugal compressor stage, the computational domain, and its finite element model, where the discretization unit is a simple linear triangular element. Darkened in this figure are two finite-elements (a) and (b), each sharing one side with a branch of a periodic boundary pair. It is across this pair of boundaries that the velocity potential ϕ experiences a uniform shift, the magnitude of which is the circulation Γ around the airfoil. The latter, as discussed earlier, is treated as a nodeless degree of freedom. Applied to the elements (a) and (b) in Figure 11.21, the following equations are particularly relevant:

$$\phi_2{}^{(b)} = \phi_1{}^{(a)} + \Gamma \tag{P.3}$$

$$\phi_3{}^{(b)} = \phi_3{}^{(a)} + \Gamma \tag{P.4}$$

$$\begin{aligned} \phi^{(b)} &= N_1{}^{(b)}\phi_1{}^{(b)} + N_2{}^{(b)}\phi_2{}^{(b)} + N_3{}^{(b)}\phi_3{}^{(b)} \\ &= N_1{}^{(b)}\phi_1{}^{(b)}\phi_1{}^{(b)} + N_2{}^{(b)}(\phi_1{}^{(a)} + \Gamma) + N_3{}^{(b)}(\phi_3{}^{(a)} + \Gamma) \\ &= N_1{}^{(b)}\phi_1{}^{(b)} + N_2{}^{(b)}\phi_1{}^{(a)} + N_3{}^{(b)}\phi_3{}^{(a)} + N_\Gamma\Gamma \end{aligned} \tag{P.5}$$

where Γ's shape function N_Γ is as follows:

$$N_\Gamma = N_2{}^{(b)} + N_3{}^{(b)} \tag{P.6}$$

Equations (P.3) and (P.4), will be assigned to computational nodes $1^{(a)}$ and $3^{(a)}$, respectively, where they satisfy the equality of the velocity component over both branches of the periodic-boundary pair. Note that both velocity components are derivatives of a linear ϕ-interpolating function and are, therefore, constant throughout each finite element.

Equality of the velocity components perpendicular to the periodic boundaries can be implicitly achieved by expressing this component over element (a) and then using the expression in evaluating the boundary integral term associated with element (b). Referring to element (a), the normal velocity component is precisely the normal ϕ derivative, which can be expressed as follows:

$$\left(\frac{\partial \phi}{\partial n}\right)^{(a)} = V_n = \left(\frac{\phi_3{}^{(a)} - \phi_2{}^{(a)}}{\delta}\right)^{(a)} \tag{P.7}$$

Combining all degrees of freedom associated with the two elements, as well as the circulation Γ, the combined set of finite element equations can be written as follows:

$$\begin{bmatrix} B_{11} & B_{12} & B_{13} & \beta_{11} & \beta_{12} & \beta_{13} & C_{17} \\ B_{21} & B_{22} & B_{23} & \beta_{21} & \beta_{22} & \beta_{23} & C_{27} \\ B_{31} & B_{32} & B_{33} & \beta_{31} & \beta_{32} & \beta_{33} & C_{37} \\ \alpha_{11} & \alpha_{12} & \alpha_{13} & A_{11} & A_{12} & A_{13} & C_{47} \\ \alpha_{21} & \alpha_{22} & \alpha_{23} & A_{21} & A_{22} & A_{23} & C_{57} \\ \alpha_{31} & \alpha_{32} & \alpha_{33} & A_{31} & A_{32} & A_{33} & C_{67} \\ C_{71} & C_{72} & C_{73} & C_{74} & C_{75} & C_{76} & C_{77} \end{bmatrix} \begin{Bmatrix} \phi_1{}^{(b)} \\ \phi_2{}^{(b)} \\ \phi_3{}^{(b)} \\ \phi_1{}^{(a)} \\ \phi_2{}^{(a)} \\ \phi_3{}^{(a)} \\ \Gamma \end{Bmatrix} = \begin{Bmatrix} q_1{}^{(b)} \\ q_2{}^{(b)} \\ q_3{}^{(b)} \\ q_1{}^{(a)} \\ q_2{}^{(a)} \\ q_3{}^{(a)} \\ q_\Gamma \end{Bmatrix} \tag{P.8}$$

In this set of equations, the submatrices $[B]$ and $[A]$ are those associated with the *uncoupled* elements (a) and (b), respectively. The submatrices $[\beta]$ and $[\alpha]$ and $[C_{i,7}, i = 1,2,\ldots,7]$ contain, in part, entries that relate the nodal magnitudes of ϕ (in both finite-elements) to one another, which is where the circulation Γ gets into the picture. Other entries represent the right-hand-side boundary integrals of the form $\int_{L^{(b)}} N_i \left(\frac{\partial \phi}{\partial n}\right)^{(b)} dL$, for example, on invoking expression (P.7) for $(\partial\phi/\partial n)^{(b)}$.

On reviewing the section entitled, "The Circulation: A Nodeless Degree of Freedom," and using (where applicable) Equations (P.3) through (P.7), prove that the final structure of (P.8) can be written as follows:

$$\begin{bmatrix} B_{11} & B_{12} & B_{13} & 0 & 0 & 0 & 0 \\ B_{21} & B_{22} & B_{23} & 0 & \beta_{22} & \beta_{23} & 0 \\ B_{31} & B_{32} & B_{33} & 0 & \beta_{32} & \beta_{33} & 0 \\ 0 & 1 & 0 & -1 & 0 & 0 & -1 \\ 0 & 0 & 1 & 0 & 0 & -1 & -1 \\ 0 & 0 & 0 & A_{31} & A_{32} & A_{33} & 0 \\ C_{71} & 0 & 0 & C_{74} & C_{75} & C_{76} & C_{77} \end{bmatrix} \begin{Bmatrix} \phi_1{}^{(b)} \\ \phi_2{}^{(b)} \\ \phi_3{}^{(b)} \\ \phi_1{}^{(a)} \\ \phi_2{}^{(a)} \\ \phi_3{}^{(a)} \\ \Gamma \end{Bmatrix} = \begin{Bmatrix} 0 \\ 0 \\ 0 \\ 0 \\ 0 \\ 0 \\ 0 \end{Bmatrix}$$

where

$$C_{73} = \int_{A^{(b)}} \left(\frac{\partial N_\Gamma}{\partial x}\frac{\partial N_3}{\partial x} + \frac{\partial N_\Gamma}{\partial y}\frac{\partial N_3}{\partial y}\right) dA$$

$$C_{74} = \int_{A^{(b)}} \left(\frac{\partial N_\Gamma}{\partial x}\frac{\partial N_1{}^{(b)}}{\partial x} + \frac{\partial N_\Gamma}{\partial y}\frac{\partial N_1{}^{(b)}}{\partial y}\right) dA$$

$$C_{75} = \int_{A^{(b)}} \left(\frac{\partial N_\Gamma}{\partial x}\frac{\partial N_1{}^{(b)}}{\partial x} + \frac{\partial N_\Gamma}{\partial y}\frac{\partial N_1{}^{(b)}}{\partial y}\right) dA - \frac{2}{\delta}\int_{L^{(b)}} N_\Gamma \, dL$$

$$C_{76} = \int_{A^{(b)}} \left(\frac{\partial N_\Gamma}{\partial x}\frac{\partial N_3{}^{(b)}}{\partial x} + \frac{\partial N_\Gamma}{\partial y}\frac{\partial N_3{}^{(b)}}{\partial y}\right) dA + \frac{2}{\delta}\int_{L^{(b)}} N_\Gamma \, dL$$

$$C_{77} = \int_{A^{(b)}} \left(\frac{\partial N_\Gamma}{\partial x}\frac{\partial N_\Gamma}{\partial x} + \frac{\partial N_\Gamma}{\partial y}\frac{\partial N_\Gamma}{\partial y}\right) dA$$

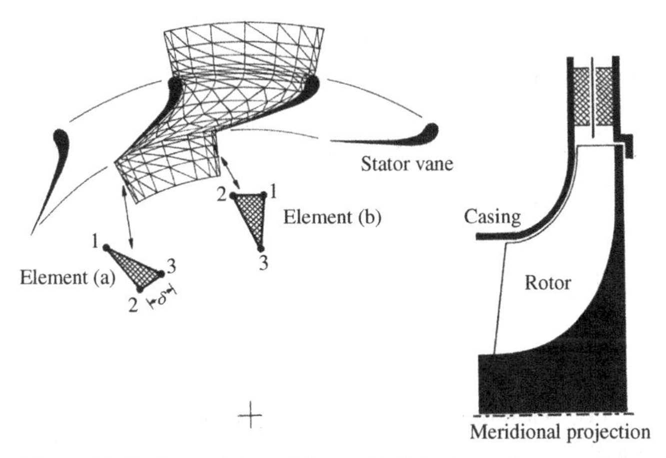

Figure 11.22. Imposition of the periodicity boundary conditions in a radial-turbine stator.

P.7 Shown in Figure 11.22 is the cascade unit in the stator of a radial-inflow turbine where two elements, coinciding with the downstream periodic boundaries, are particularly highlighted. In addition to the circulation-governing equation, contribution of these two elements [labeled (a) and (b)] can be expressed in the following matrix form:

$$\begin{bmatrix} a_{11} & a_{12} & a_{13} & \alpha_{12} & \alpha_{12} & \alpha_{13} & c_{17} \\ a_{21} & a_{22} & a_{23} & \alpha_{21} & \alpha_{22} & \alpha_{23} & c_{27} \\ a_{31} & a_{32} & a_{33} & \alpha_{31} & \alpha_{32} & \alpha_{33} & c_{37} \\ \beta_{11} & \beta_{12} & \beta_{13} & b_{11} & b_{12} & b_{13} & c_{47} \\ \beta_{21} & \beta_{22} & \beta_{23} & b_{21} & b_{22} & b_{23} & c_{57} \\ \beta_{31} & \beta_{32} & \beta_{33} & b_{31} & b_{32} & b_{33} & c_{67} \\ c_{71} & c_{72} & c_{73} & c_{74} & c_{75} & c_{76} & c_{77} \end{bmatrix} \begin{Bmatrix} \phi_1{}^{(a)} \\ \phi_2{}^{(a)} \\ \phi_3{}^{(a)} \\ \phi_1{}^{(b)} \\ \phi_2{}^{(b)} \\ \phi_3{}^{(b)} \\ \Gamma \end{Bmatrix} = \begin{Bmatrix} q_1{}^{(a)} \\ q_2{}^{(a)} \\ q_3{}^{(a)} \\ q_1{}^{(b)} \\ q_2{}^{(b)} \\ q_3{}^{(b)} \\ q_\Gamma \end{Bmatrix}$$

Let us now make the following approximation along the element (a) side $1^{(a)} - 2^{(a)}$:

$$V_n = \left(\frac{\phi_3 - \phi_2}{\delta}\right)^{(a)}$$

where n refers to the direction of the inward unit vector that is perpendicular to the left-hand-side periodic boundary. Prove that the imposition of periodic boundary conditions in both elements will alter the preceding set of equations to the following form:

$$\begin{bmatrix} 1 & 0 & 0 & -1 & 0 & 0 & -1 \\ 0 & 1 & 0 & 0 & 0 & -1 & -1 \\ 0 & 0 & 0 & \alpha_{31} & \alpha_{32} & \alpha_{33} & c_{37} \\ 0 & \beta_{12} & \beta_{13} & b_{11} & b_{12} & b_{13} & c_{47} \\ 0 & 0 & 0 & b_{21} & b_{22} & b_{23} & c_{57} \\ 0 & \beta_{32} & \beta_{33} & b_{31} & b_{32} & b_{33} & c_{67} \\ c_{71} & c_{72} & c_{73} & c_{74} & c_{75} & c_{76} & c_{77} \end{bmatrix} \begin{Bmatrix} \phi_1{}^{(a)} \\ \phi_2{}^{(a)} \\ \phi_3{}^{(a)} \\ \phi_1{}^{(b)} \\ \phi_2{}^{(b)} \\ \phi_3{}^{(b)} \\ \Gamma \end{Bmatrix} = \begin{Bmatrix} 0 \\ 0 \\ 0 \\ 0 \\ 0 \\ 0 \\ q_\Gamma \end{Bmatrix}$$

where

$$\beta_{12} = \frac{1}{\delta} \int_{(3^{(b)}-1^{(b)})} N_1^{(b)} \, dL$$

$$\beta_{13} = -\frac{1}{\delta} \int_{(3^{(b)}-1^{(b)})} N_1^{(b)} \, dL$$

$$\beta_{32} = \frac{1}{\delta} \int_{(3^{(b)}-1^{(b)})} N_3^{(b)} \, dL$$

$$\beta_{33} = -\frac{1}{\delta} \int_{(3^{(b)}-1^{(b)})} N_3^{(b)} \, dL$$

P.8 Figure 11.23 shows an isometric view of the discretized cascade unit of the vaned diffuser in a typical centrifugal compressor, with the discretization unit being the simple linear tetrahedron. Highlighted in the figure are two elements, each with a base that exists on one periodic boundary. Referring to the left-hand-side periodic boundary, the velocity component V_n perpendicular to this boundary (i.e., in the direction of the unit vector $\vec{n}$) can be expressed as follows:

$$V_n = \vec{n}.(\nabla \phi)^{(n)} = n_x \frac{\partial \phi}{\partial x} + n_y \frac{\partial \phi}{\partial y} + n_z \frac{\partial \phi}{\partial z} = \left(n_x \frac{\partial N_1{}^{(a)}}{\partial x} + n_y \frac{\partial N_1{}^{(a)}}{\partial y} + n_z \frac{\partial N_1{}^{(a)}}{\partial z} \right)$$
$$+ \left(n_x \frac{\partial N_2{}^{(a)}}{\partial x} + n_y \frac{\partial N_2{}^{(a)}}{\partial y} + n_z \frac{\partial N_2{}^{(a)}}{\partial z} \right)$$
$$+ \left(n_x \frac{\partial N_3{}^{(a)}}{\partial x} + n_y \frac{\partial N_3{}^{(a)}}{\partial y} + n_z \frac{\partial N_3{}^{(a)}}{\partial z} \right)$$
$$+ \left(n_x \frac{\partial N_4{}^{(a)}}{\partial x} + n_y \frac{\partial N_4{}^{(a)}}{\partial y} + n_z \frac{\partial N_4{}^{(a)}}{\partial z} \right)$$

which can be compacted as follows:

$$V_n = \lambda_1 \phi_1{}^{(a)} + \lambda_2 \phi_2{}^{(a)} + \lambda_3 \phi_3{}^{(a)} + \lambda_4 \phi_4{}^{(a)}$$

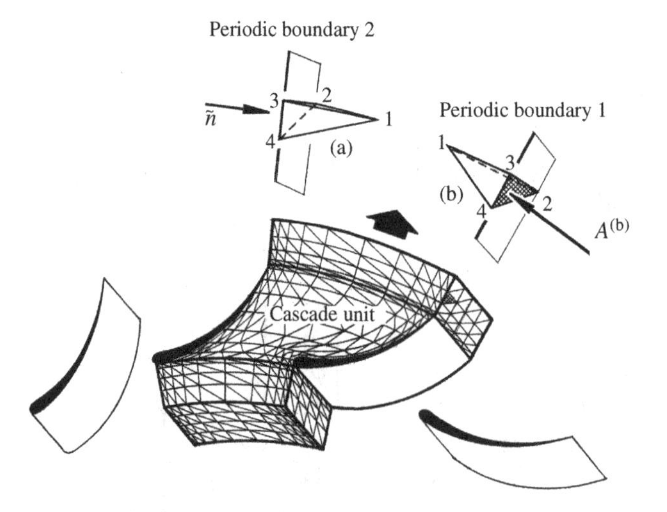

Figure 11.23. Isometric of the centrifugal-compressor stator in Problem (P.8).

Prove that the imposition of periodic conditions over the two periodic boundaries [where the equality of the normal velocity component is handled through the boundary integral term of element (*b*)] gives rise to the following set of equations for the two elements combined:

$$\begin{bmatrix} a_{11} & a_{12} & a_{13} & a_{14} & 0 & 0 & 0 & 0 & 0 \\ 0 & +1 & 0 & 0 & 0 & -1 & 0 & 0 & -1 \\ 0 & 0 & +1 & 0 & 0 & 0 & -1 & 0 & -1 \\ 0 & 0 & 0 & +1 & 0 & 0 & 0 & -1 & -1 \\ 0 & 0 & 0 & 0 & b_{11} & b_{12} & b_{13} & b_{14} & 0 \\ \beta_{21} & \beta_{22} & \beta_{23} & \beta_{24} & b_{21} & b_{22} & b_{23} & b_{24} & 0 \\ \beta_{31} & \beta_{32} & \beta_{33} & \beta_{34} & b_{31} & b_{32} & b_{33} & b_{34} & 0 \\ \beta_{41} & \beta_{42} & \beta_{43} & \beta_{44} & b_{41} & b_{42} & b_{43} & b_{44} & 0 \\ c_1 & c_2 & c_3 & c_4 & c_5 & c_6 & c_7 & c_8 & c_9 \end{bmatrix} \begin{Bmatrix} \phi_1^{(a)} \\ \phi_2^{(a)} \\ \phi_3^{(a)} \\ \phi_4^{(a)} \\ \phi_1^{(b)} \\ \phi_2^{(b)} \\ \phi_3^{(b)} \\ \phi_4^{(b)} \\ \Gamma \end{Bmatrix} = \begin{Bmatrix} 0 \\ 0 \\ 0 \\ 0 \\ 0 \\ 0 \\ 0 \\ 0 \\ 0 \end{Bmatrix}$$

Interpretation of the preceding symbols is as follows:

$$\beta_{ij} = \int_{A^{(b)}} N_i^{(b)} \lambda_j dA \quad i = 1,2,3,4 \text{ and } j = 1,2,3,4$$

$$c_i = \int_{A^{(b)}} N_\Lambda \lambda_i 1 \le i \le 4$$

$$c_i = \int_{V^{(b)}} \left(\frac{\partial N_\lambda}{\partial x}\frac{\partial N_i^{(b)}}{\partial x} + \frac{\partial N_\lambda}{\partial y}\frac{\partial N_i^{(b)}}{\partial y} + \frac{\partial N_\lambda}{\partial z}\frac{\partial N_i^{(b)}}{\partial z} \right) dV, \qquad 5 \le i \le 8$$

$$c_9 = \int V^{(b)} \left(\frac{\partial N_\lambda}{\partial x}\frac{\partial N_\lambda}{\partial x} + \frac{\partial N_\lambda}{\partial y}\frac{\partial N_\lambda}{\partial y} + \frac{\partial N_\lambda}{\partial z}\frac{\partial N_\lambda}{\partial z} \right) dV$$

where

$$N_\Gamma = N_2^{(b)} + N_3^{(b)} + N_4^{(b)}$$

REFERENCES

[1] Carter, A. D. S. and Hughes, H. P., "A Theoretical Investigation into the Effect of Profile Shape on the Performance of Airfoils in Cascade," *R&W No. 2884* Aeronautical Research Council, London, 1946.

[2] Theodorsen, T., "Theory of Wing Sections of Arbitrary Shape," *NACA Report No. 411*, National Advisary Committee for Aeronautics, Washington, DC, 1932.

[3] Goldstein, S., "Low Drag and Suction Airfoils," *Journal of Aeronautical Science*, Vol. 15, 1948, pp. 189–220.

[4] Gostelow, J. P., Potential Flow through Cascades: A Comparison Between Exact and Approximate Solutions," *CP No. 807*, Aeronautical Research Council, London, 1965.

[5] Merchant, W., and Collar, A. R., "Flow of an Ideal Fluid Past a Cascade of Blades," Part II, *R&M No. 1893*, Aeronautical Research Council, London, May 1941.

[6] Martensen, E., "The Calculation of Pressure Distribution on a Cascade of Thick Airfoils by Means of Fredholm Integral Equations of the Second Kind," *NASA TTF-702*, NASA, Washington, DC, July 1971.

[7] Bindon, J. P., and Carmichael, A. D., "Streamline Curvature Analysis of Compressible and High Mach Number Cascade Flows," *Journal of Mechanical Engineering Science*, Vol. 13, No. 5, October 1971, pp. 344–357.

[8] Katsanis, T., "Fortran Program for Calculating Transonic Velocities on a Blade-to-Blade Stream Surface of a Turbomachine," *NASA TN D-5427*, NASA, Washington, DC, September 1969.

[9] Prince, T. C., "Prediction of Transonic Inviscid Steady Flow in Cascades by Finite Element Methods," Ph.D. dissertation, University of Cincinnati, 1976.

[10] Laskaris, T. E., "Finite Element Analysis of Three-Dimensional Potential Flow in Turbomachines," *AIAA Journal*, Vol. 16, July 1978, pp. 717–722.

[11] Thompson, D. S., "Finite Element Analysis of the Flow Through a Cascade of Airfoils," *Turbo/TR 45*, Engineering Department, Cambridge University, 1973.

[12] Schmidt, G., "Inviscid Flow in Multiply Connected Regions," *Numerical Methods in Fluid Dynamics, Proceedings of the International Conference*, University of Southampton, England, September 1973, pp. 153–171.

[13] McFarland, E. R., "A Rapid Blade-to-Blade Solution for Use in Turbomachinery Design," *Journal of Gas Turbine and Power*, Vol. 106, 1984, pp. 376–382.

[14] Ergatoudis, J. G., Irons, B. M., and Zienkiewicz, O. C., "Curved, Isoparametric Quadrilateral Elements for Finite Element Analysis," *International Journal of Solids and Structures*, Vol. 14, 1968, pp. 31–42.

[15] Zienkiewicz, O. C., *The Finite Element Method in Engineering Science*, McGraw-Hill, NewYork, 1971.

[16] Gupta, S. K., and Tanji, K. K., "Computer Program for Solution of Large, Sparse, Unsymmetric System of Linear Equations," *International Journal of Numerical Methods in Engineering*, Vol. 11, No. 8, 1977, pp. 1251–1259.

[17] Baskharone, E., and Hamed, A., "A New Approach in Cascade Flow Analysis Using the Finite Element Method," *AIAA Journal*, Vol. 19, No. 1, January 1981, pp. 65–71.

[18] Baskharone, E. A., "A New Approach in Turbomachinery Flow Analysis Using the Finite Element Method," Ph.D. dissertation, University of Cincinnati, 1979.

[19] Hamed, A., and Baskharone, E., "Analysis of the Three-Dimensional Flow in a Turbine Scroll," *Journal of Fluids Engineering* Vol. 102, No. 3, 1980, pp. 297–301.

[20] Baskharone, E. A., "Optimization of the Three-Dimensional Flow Path in the Scroll-Nozzle System in a Radial Inflow Turbine," *Journal of Engineering for Gas Turbine and Power*, Vol. 106, No. 2, 1984, pp. 511–515.

12 Finite Element Analysis in Curvilinear Coordinate

In this chapter we apply Galerkin's weighted-residual finite element approach to a special category of flow problems. This is where only the through-flow and tangential momentum equations (beside the continuity equations, of course) suffice as the flow-governing equations. This problem is perhaps best represented by the so-called quasi-three-dimensional flow field in analyzing airfoil cascades. Some terms within the finite element formulation are presented and modeled as "source" terms, in analogy with a special problem category in heat conduction. Also, implicit means are used in enforcing the cascade periodicity conditions.

Introduction

In the cascade theory discipline, the basic problem is that of a three-dimensional periodic flow in the blade-to-blade hub-to-casing passage (Figure 12.1). In modeling this flow type, it is crucial to account for such real-flow effects as boundary layer separation, flow recirculation, and trailing-edge mixing losses. Existing numerical models in this area vary in complexity from the potential flow category [1–3] to that of the fully three-dimensional viscous flow field [4, 5]. Compared with the strictly two-dimensional and three-dimensional flow models, the quasi-three-dimensional approach (which is the topic in this chapter) to the cascade flow problem has been recognized as a sensible compromise in terms of both economy and precision. It is, however, the viscous flow version of the problem, under this approach, that is in need of further enhancement, particularly in the area of simulating the hub-to-casing flow interaction effects on the blade-to-blade flow field. In other words, addressing the flow field over a blade-to-blade stream filament (we will be) in an environment that totally isolates it (Figure 12.1), instead of allowing what is a natural interaction of this filament with those adjacent to it, brings nothing but accuracy degradation to the flow model. Tolerating this interaction is at the heart of the computational model in this chapter.

Existing viscous-flow models of airfoil cascades can be classified into two major categories: one of simple analyses and various levels of uncertainty and another of full-scale models with undesirably high requirements of computational resources. Of these, the first category includes strictly two-dimensional (e.g., [4])

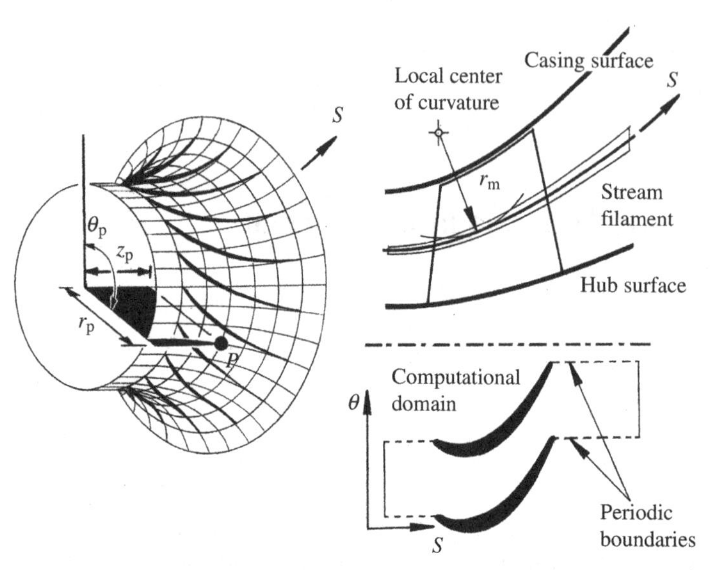

Figure 12.1. Meridional and blade-to-blade views of the computational domain in the quasi-three-dimensional flow analysis.

and quasi-three-dimensional (e.g., [5–7]) flow models. A common feature here is that ellipticity of the viscous-flow field in this case is either partially or totally ignored. A relatively recent model in this context is based on the thin-layer approximation [4–7] whereby different levels of inviscid/viscous-flow interactions are introduced in an attempt to partially preserve the elliptic nature of the flow field. Another common feature is that the flow passage is defined as bounded by two stagnation streamlines, the shape of which is part of the outcome provided by a preprocessor in the form of a commonly inviscid (often potential) flow solver. Given that the geometry of these streamlines does not reflect such real-flow features as the viscous-flow mixing pattern within the airfoil wake, resorting to this method in defining the computational domain would naturally have its own accuracy-degradation impact. It is, however, the extraction of the streamwise diffusion terms from the Navier-Stokes equations that is in the majority of these models, which is most damaging, for it unjustifyably forbids the downstream effects from influencing the flow behavior. Note that a fluid particle in a subsonic (or elliptic) flow field will "sense" any and all disturbances (or obstacles) downstream and properly adjust to them before physically reaching them. Figure 12.2 aims to illustrate this very fact because it shows the stagnation streamlines computed for the hub, mean, and casing sections of an axial-flow stator in a commercial gas-turbine engine. In each of these three sections, note that the stagnation streamline senses the existence of the blade leading edge because it gradually curls up well before the leading edge is physically reached. Under the assumption of zero streamwise diffusion component, and with the exception of exit pressure, which is externally fixed, the flow exit conditions are normally obtained by extrapolation and are generally inaccurate. Moreover, a fictitiously parabolized flow field, where the solution is obtained by marching along a midchannel "master" streamline, is one that conceptually prohibits boundary layer separation anywhere and precludes the existence of any flow-recirculation

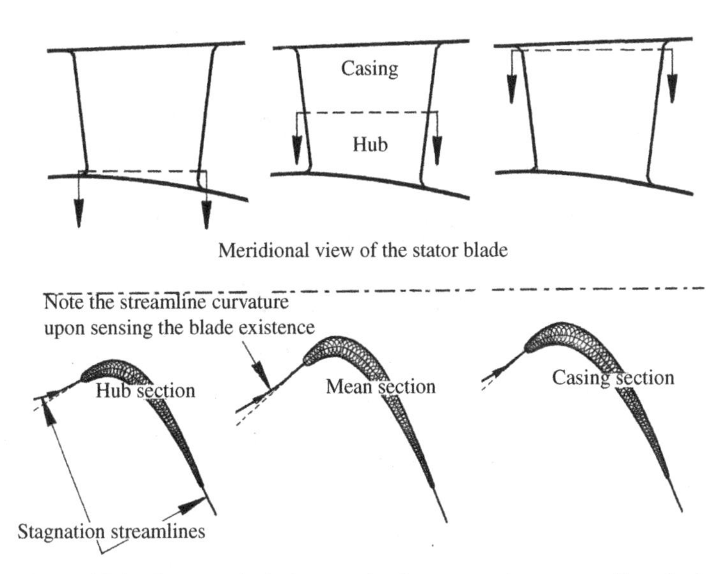

Figure 12.2. Geometrical changes in the stagnation streamlines before reaching the leading edge, a sign of an elliptic flow field.

zones when these are generally part and parcel of a typical cascade-flow field, both in turbine and, in particular, in compressor-cascade applications. Note that the flow stream in a compressor is proceeding against an adverse pressure gradient, which aggravates the boundary layer buildup and invites zones of boundary layer separation and flow recirculation.

Specific details of the current analysis advocate a perhaps bigger objective than the mere derivation of the flow-governing equations and their finite element formulation. On many occasions the analyst is forced, for practicality purposes, to isolate a flow subdomin and investigate it with much more accuracy and higher grid resolution than would otherwise be possible. The question, then, is whether to account for the rest of the larger flow domain and whether sufficient care has been paid to the mutual flow interaction between the chosen subdomain and the rest of the flow region. In addressing such issues, we should first accept the very fact that full accounting of such an interaction is perhaps unthinkable, for it requires prior knowledge of the flow structure throughout the bigger flow domain, which, if available, would make it silly to reinvestigate a smaller subdomain. However, in situations such as the present, a separate low-order analysis can act as a preprocessor for the more involved flow analysis in our isolated subregion, which is the blade-to-blade stream filament in Figure 12.3.

The present finite element analysis fits as is under the category of quasi-three-dimensional flow models but deviates in the aspects of accuracy and completeness. The idea here is to solve the blade-to-blade hub-to-casing flow field over two mutually interacting families of orthogonal stream surfaces and was first devised by Wu [8] for a rotational yet inviscid cascade flow problem. Unfortunately, existing blade-to-blade flow models (e.g., [2–4) that were aimed at extending Wu's theory to viscous flow computations have commonly ignored the viscosity-related shear stresses on the blade-to-blade stream surfaces. In the absence of these stresses, the geometrically isolated flow domain (see Figure 12.3) becomes analytically isolated

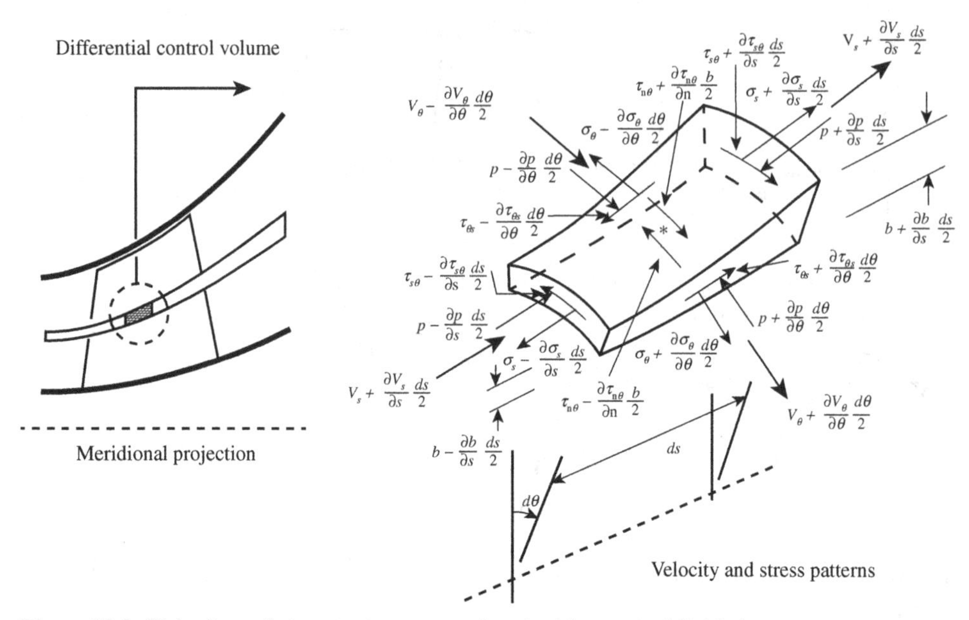

Figure 12.3. Velocity and stress patterns associated with a typical fluid element.

as well. Flow interaction effects such as these would, according to Wu's procedure, be the outcome of the meridional (hub-to-casing) flow computation phase.

The starting point in building up the current computational model is a thorough examination of the different stress patterns on the typical fluid element in Figure 12.3. The flow-governing equations are then derived in accordance with the basic conservation principles, giving rise to a set of new terms that, in the end, were proven to be of substantial influence on flow behavior. One of these newly included terms is proportional to the second derivatives of the stream tube thickness b and radius r (see Figure 12.3), as well as a key term through which the shear stress (identified by double arrowheads in Figure 12.3) over the lower and upper surfaces of the control volume are introduced in the equations of motion. Disappearance of the former term in the standard problem formulation is a result of treating b and/or r as linear functions of the meridional distance s (e.g., [5]) which is often a fictitious constraint. This is true in the sense that b depends primarily on the endwall contours, which are often designed to be nonlinear in an attempt by the designer to alleviate excessive diffusion (flow deceleration) over the blade suction side. On the other hand, ignoring the above-mentioned shear-stress components, dictated by the spanwise gradients of V_s and V_θ, implies an overly simplified hub-to-casing free vortex flow structure that is simply inconsistent with long-standing gas turbine design trends, particularly in the propulsion area. In this case, the desire to minimize the weight and maximize the propulsive efficiency often leads to low-aspect-ratio blading with spanwise stacking patterns that are virtually unrestricted, particularly in the area of turbine blading (Figure 12.4). In order to capture such important flow zones as those of separation and recirculation, the flow ellipticity (a crucial subsonic flow feature) is preserved throughout the computational domain. This, added to a new implicit means of enforcing the cascade periodicity conditions, makes it possible to obtain details of the viscous-flow mixing region (wake) of not only the

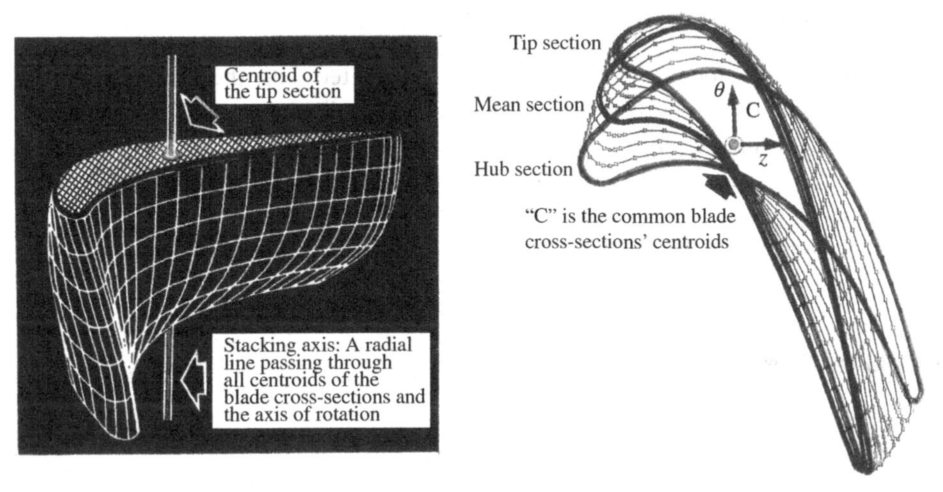

Figure 12.4. Stacking pattern of the first-stage rotor blade in the high-pressure turbine of the F109 turbofan engine.

airfoil contained in the computational domain but also of others that are analytically inactive airfoils outside the computational domain.

Analysis Guidelines and Limitations

Figure 12.1 shows different views of the computational domain under consideration. Of these, the meridional ($r - z$) view shows radius and stream-tube thickness variations that are totally unrestricted. The arbitrariness of those views qualifies the flow model to handle virtually all bladed turbomachinery components regardless of their meridional projections. The stream tube thickness variation in this figure is assumed to be known a priori. This is usually the outcome of a straightforward axisymmetric flow analysis. It is worth noting that even the simple programs frequently used in the gas turbine industry for the purpose of solving the hub-to-casing axisymmetric flow problem (e.g., [2, 9]) normally account, to a good extent and through implicit means, for different real-flow effects such as the blade-caused blockage of the flow stream and the profile (friction-related) losses, both of which are functions of the spanwise location. As a result, the gradients of velocity components produced by these programs are highly representative of the real-flow effects in the meridional plane. These are the same gradients used in the current model to compute the surface shear stresses exerted on the stream tube in Figure 12.3. Applicable assumptions in the present model include a steady, adiabatic, and laminar flow throughout the computational domain. Of these, the laminar-flow assumption (removed in all succeeding flow models) confines the model's applicability to operation modes with a sufficiently low Reynolds number (typically below 2×10^5). Despite the strict limitation implied by such an assumption, the choice was, in part, motivated by the desire to extract the inaccuracy associated with what would have realistically been a low-order turbulence closure, concentrating, instead, on the manner in which refinement of the flow equations impacts the flow behavior. Meanwhile, the two test cases chosen to validate the model meet the above-cited low-Reynolds-number limitation.

Flow-Governing Equations

Referring to the velocity and stress patterns associated with the fluid element in Figure 12.3, and denoting the local unit vector perpendicular to the stream tube by $\vec{n}$, the flow-governing equations can be cast as follows:

Continuity Equation

$$\frac{1}{b}\frac{\partial}{\partial s}(rbV_s)+\frac{\partial V_\theta}{\partial \theta}=0 \tag{12.1}$$

Through-Flow Momentum Equation

$$\begin{aligned}
&\rho\left(b+\frac{db}{2}\right)dsd\theta\left(V_s\frac{\partial V_s}{\partial s}+\frac{V_\theta}{r}\frac{\partial V_s}{\partial \theta}-\frac{{V_\theta}^2}{r}\frac{dr}{ds}\right)\\
&=\left(p+\frac{\partial p}{ds}ds\right)(b+db)(r+dr)d\theta+pbrd\theta\\
&+\left(\sigma_s+\frac{\partial \sigma_s}{\partial s}ds(b+db)(r+dr)d\theta-\sigma_s brd\theta\right)+\left(\tau_{\theta s}+\frac{\partial \tau_{\theta s}}{\partial \theta}d\theta\right)\left(b+\frac{db}{2}\right)ds\\
&-\tau_{\theta s}\left(b+\frac{db}{2}\right)ds+\left[\tau_{ns}+\frac{\partial \tau_{ns}}{\partial n}\left(b+\frac{db}{2}\right)\right]dA_o-\tau_{ns}dA_i
\end{aligned} \tag{12.2}$$

where dA_i and dA_O are the inner and outer areas of the differential control volumes in Figure 12.3, respectively. These infinitesimally small surface areas can be expressed as follows:

$$dA_i=ds\left(r+\frac{dr}{2}\right)d\theta \tag{12.3}$$

$$dA_O=\left(ds+\frac{bds}{R_m}-\frac{1}{2R_m}dbds\right)\left(r+\frac{dr}{2}b\cos\phi+\frac{db}{2}\cos\phi\right)d\theta \tag{12.4}$$

where R_m is the local stream-tube radius of curvature in the meridional projection (see Figure 12.1), and ϕ is the slope angle of the stream tube in reference to the positive axial direction. Referring to Figure 12.3, the following relationship also applies:

$$\cos\phi=\frac{dr}{ds} \tag{12.5}$$

Tangential Momentum Equation

$$\begin{aligned}
&\rho\left(b+\frac{db}{2}\right)\left(r+\frac{dr}{2}\right)dsd\theta\left(V_s\frac{\partial V_\theta}{\partial s}+\frac{V_\theta}{r}\frac{\partial V_\theta}{\partial \theta}-\frac{V_sV_\theta}{r}\frac{dr}{ds}\right)\\
&=-\left(p+\frac{\partial p}{\partial \theta}d\theta\right)\left(b+\frac{db}{2}\right)ds+p\left(b+\frac{db}{2}\right)ds+\left(\sigma_\theta+\frac{\partial \sigma_\theta}{\partial \theta}d\theta\right)\left(b+\frac{db}{2}\right)ds\\
&-\sigma_\theta\left(b+\frac{db}{2}\right)ds+\left(\tau_{s\theta}+\frac{\partial \tau_{s\theta}}{\partial s}\right)(b+db)(r+dr)\,d\theta-\tau_{s\theta}br\,d\theta\\
&\times\left[\tau_{n\theta}+\frac{\partial \tau_{n\theta}}{\partial n}\left(b+\frac{db}{2}\right)\right]dA_O-\tau_{n\theta}\,dA_i
\end{aligned} \tag{12.6}$$

where dA_i and dA_O are the same as defined earlier. On substitution and algebraic manipulation, we get the two momentum equations in the following forms:

$$\rho\left(V_s\frac{\partial V_s}{\partial s}+\frac{V_\theta}{r}\frac{\partial V_s}{\partial \theta}-\frac{V_\theta{}^2}{r}\frac{dr}{ds}\right)=-\frac{\partial p}{\partial s}-\left(\frac{1}{b}\frac{db}{ds}+\frac{1}{r}\frac{dr}{ds}\right)p+\frac{\partial \sigma_s}{\partial s}$$
$$+\left(\frac{1}{b}\frac{db}{ds}+\frac{1}{r}\frac{dr}{ds}\right)\sigma_s+\frac{1}{r}\frac{\partial \tau_{\theta s}}{\partial \theta}+\frac{\partial \tau_{ns}}{\partial n}+\frac{1}{r}\left(\frac{dr}{ds}-\frac{r}{R_m}\right)\tau_{ns} \tag{12.7}$$

$$\rho\left(V_s\frac{\partial V_\theta}{\partial s}+\frac{V_\theta}{\partial \theta}-\frac{V_sV_\theta}{r}\frac{dr}{ds}\right)=-\frac{1}{r}\frac{\partial p}{\partial \theta}+\frac{1}{r}\frac{\partial \sigma_\theta}{\partial \theta}+\frac{\partial \tau_{s\theta}}{\partial s}$$
$$+\left(\frac{1}{b}\frac{db}{ds}+\frac{1}{r}\frac{dr}{ds}\right)\tau_{s\theta}+\frac{\partial \tau_{n\theta}}{\partial n}+\frac{1}{r}\left(\frac{dr}{ds}-\frac{r}{R_m}\right)\tau_{n\theta} \tag{12.8}$$

Note that the radius of curvature R_m appearing in the last two terms of the preceding two equations would be infinite in the case of a purely axial turbomachinery component. In this case, the derivative dr/ds would vanish along the stream tube. Both these conditions are met with reasonable accuracy in axial-flow turbomachines, where the meridional flow trajectory is predominantly in the axial direction.

Assuming a Newtonian fluid, the stress-strain relationships expressed in the curvilinear frame of reference are as follows:

$$\sigma_s=2\mu\frac{\partial V_s}{\partial s} \tag{12.9}$$

$$\sigma_\theta=2\mu\left(\frac{1}{r}\frac{\partial V_\theta}{\partial \theta}+\frac{V_s}{r}\frac{dr}{ds}\right) \tag{12.10}$$

$$\tau_{s\theta}=\tau_{\theta s}=\mu\left(\frac{1}{r}\frac{\partial V_s}{\partial \theta}+\frac{\partial V_\theta}{\partial s}-\frac{V_\theta}{r}\frac{dr}{ds}\right) \tag{12.11}$$

$$\tau_{ns}=\mu\frac{\partial V_s}{\partial n} \tag{12.12}$$

$$\tau_{n\theta}=\mu\left(\frac{1}{r}\frac{\partial V_\theta}{\partial n}-\frac{V_\theta}{r^2}\right) \tag{12.13}$$

where μ is the dynamic viscosity coefficient. Substitution of the preceding relations in the two momentum equations gives rise to the final set of linearized momentum equations, which can be expressed in terms of the kinematic viscosity coefficient ν as follows:

$$\nu\nabla^2 V_s=\frac{1}{b}\frac{db}{ds}+\frac{1}{r}\frac{dr}{ds}\Bigg)p+\frac{1}{\rho}\frac{\partial p}{\partial s}+\alpha_1\frac{\partial V_s}{\partial s}+\alpha_2\frac{\partial V_s}{\partial \theta}+\alpha_3V_s+\alpha_4V_\theta+\alpha_5 \tag{12.14}$$

$$\nu\nabla^2 V_\theta=\frac{1}{\rho r}\frac{\partial p}{\partial \theta}+B_1\frac{\partial V_\theta}{\partial s}+\beta_2\frac{\partial V_\theta}{\partial \theta}+\beta_3V_\theta+\beta_4\frac{\partial V_)s}{\partial \theta}+\beta_5V_s+\beta_6 \tag{12.15}$$

With ν being the kinematic viscosity coefficient, the different coefficients in the two momentum equations are defined as follows:

$$\alpha_1=\bar{V}_s-\nu\left(\frac{1}{b}\frac{db}{ds}\right)$$

$$\alpha_2=\frac{\bar{V_\theta}}{r}$$

$$\alpha_3 = \nu \frac{d}{ds}\left(\frac{1}{b}\frac{db}{ds} + \frac{1}{r}\frac{dr}{ds}\right)$$

$$\alpha_4 = -\nu \frac{\partial^2 V_s}{\partial n^2} - \frac{\nu}{r}\left(\frac{dr}{ds} - \frac{r}{R_m}\right)\frac{\partial V_s}{\partial n}$$

$$\beta_1 = \bar{V}_s - \nu\left(\frac{1}{b}\frac{db}{ds} - \frac{1}{r}\frac{dr}{ds}\right)$$

$$\beta_2 = \frac{\bar{V}_\theta}{r}$$

$$\beta_3 = -\frac{\bar{V}_s}{r}\frac{dr}{ds} + \nu\left(\frac{1}{br}\frac{db}{ds}\frac{dr}{ds} + \frac{1}{r}\frac{d^2r}{ds^2}\right)$$

$$\beta_4 = -\frac{\nu}{r^2}\frac{dr}{ds}$$

$$\beta_5 = 0$$

$$\beta_6 = -\frac{\nu}{r}\frac{\partial^2 V_\theta}{\partial n^2} + \frac{\nu}{r^2}\frac{\partial V_\theta}{\partial n} - \frac{\nu}{r^2}\frac{\partial V_\theta}{\partial n} - \frac{\nu}{r}\left(\frac{dr}{ds} - \frac{r}{R_m}\right)\left(\frac{1}{r}\frac{\partial V_\theta}{\partial n} - \frac{V_\theta}{r^2}\right)$$

with the overbar signifying a value that is known from a previous iteration or an initial guess and ρ being the fluid density.

It is meaningful to point out that the contents of Equations (12.14) and (12.15) are hardly those in a standard quasi-three-dimensional flow model. In deriving these equations, different manipulated versions of the continuity equation (12.1) were used in an effort to compact these equations. Also used in the process was the Laplacian operator expression in the current curvilinear frame of reference as follows:

$$\nabla^2 = \frac{\partial^2}{\partial s^2} + \frac{1}{r^2}\frac{\partial^2}{\partial \theta^2} + \frac{1}{r}\frac{dr}{ds}\frac{\partial}{\partial s}$$

As seen, the free terms α_5 and β_6 are proportional to the lateral gradients of velocity components and are treated, within the finite element model, as "source" terms in analogy with the problem of conduction heat transfer.

Boundary Conditions

In an elliptic flow field, the solution is highly sensitive to the prescribed flow constraints over the different boundary segments. Of these, the periodic and/or exit boundary segments (Figure 12.5) are largely within the viscous mixing zone (blade wake), where an artificially imposed boundary condition can lead to substantial inaccuracies.

Three of the applicable boundary conditions are relatively straightforward. At the flow inlet station (see Figure 12.5), both the through-flow and tangential velocity components V_s and V_θ are specified in accordance with the cascade operating conditions. As for the blade surface, the "no-slip" boundary condition applies (note that we are now dealing with a viscous flow field), whereby the two velocity components are both set to zero, assuming the case of non-rotating-blade cascade at this point. Lastly, the pressure magnitude is set to an arbitrary value at any computational node, setting a "datum" for all other pressure magnitudes.

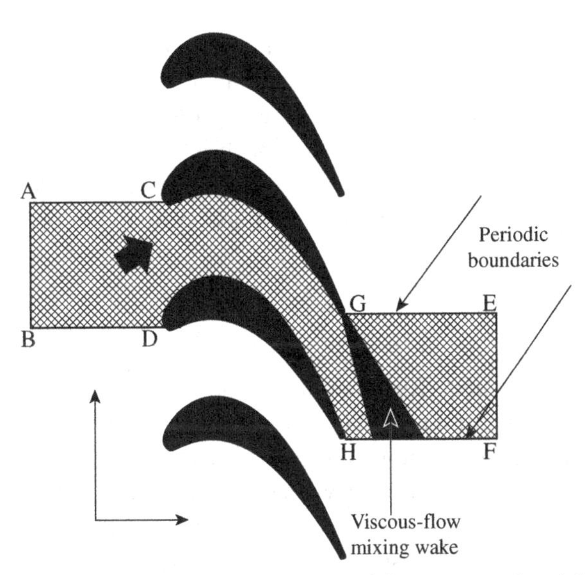

Figure 12.5. Boundary segments of the computational domain.

The flow exit boundary conditions are based on the assumption that the exit station is located sufficiently far from the blade trailing edge for a nearly complete mixing of the suction- and pressure-side flow streams to occur. In this case, the exit flow behavior is dictated by the following free-vortex-like Neumann-type boundary conditions:

$$\frac{\partial V_s}{\partial s} = -\frac{V_s}{rb}\frac{d}{ds}(rb) \tag{12.16}$$

$$\frac{\partial V_\theta}{\partial s} = -\frac{V_\theta}{r}\frac{dr}{ds} \tag{12.17}$$

where r and b are the local magnitudes of radius and stream-tube thickness, respectively. Equation (12.16) is recognized as a special form of the continuity equation (12.1), whereas Equation (12.17) is a general derivative-type version of the angular momentum conservation principle that would, with reasonable accuracy, prevail at such a far-downstream location. Note that the exit flow behavior, in the current analysis is unrestricted to artificially satisfy any uniform distribution of, for instance, a velocity component or the flow angle. Also note that the special geometric features of the solution domain make it inappropriate to specify any relatively simple exit conditions, such as zero surface tractions, because this would be incompatible with, for example, the stream-tube thickness variation as the exit station is approached.

The cascade flow periodicity in the current analysis is implicitly enforced in a manner that eliminates abrupt changes in the field variables close to the periodic boundaries (see Figure 12.5) while minimizing the core size during execution. Considering two corresponding nodes, each on one periodic boundary, the traditional approach is to delete the two finite element equations associated with one of them. These two equations are then replaced by explicit relationships, one equating the radial velocity components at both nodes and the other equating the two tangential components. In the current approach, both the finite element equations are preserved. Assigning the nodal numbers of one periodic boundary to nodes on the other periodic boundary not only reduces the size of the finite element model, but also takes care of equating the radial components at both computational nodes. As for the equality of the tangential velocity component (which is perpendicular to the periodic boundary), an expression of this velocity component from a node on one periodic boundary is used in casting the boundary term associated with the corresponding node on the other boundary.

Finite Element Formulation

In the following, a Galerkin finite element formulation of the flow-governing equations is developed, with the discretization unit being a biquadratic curve-sided element of the Lagrangian type. Figure 12.6 shows the nine-noded element that was proven [11] to be highly accurate in the current velocity-pressure (so-called primitive-variables) formulation of Navier-Stokes equations. Elemental interpolation of the flow variables, on the other hand, is established in such a manner as to ensure satisfaction of well-established compatibility requirements [12, 13]. Applied to the current problem, these requirements give rise to a subfamily of velocity-pressure interpolation combinations under which a stable and convergent solution is achievable.

Throughout a typical element (e), let the velocity components and (static) pressure be interpolated as follows:

$$V_s^{(e)} = \sum_{i=1}^{9} N_i V_{s,i} \tag{12.18}$$

$$V_\theta^{(e)} = \sum_{i=1}^{9} N_i V_{\theta,i} \tag{12.19}$$

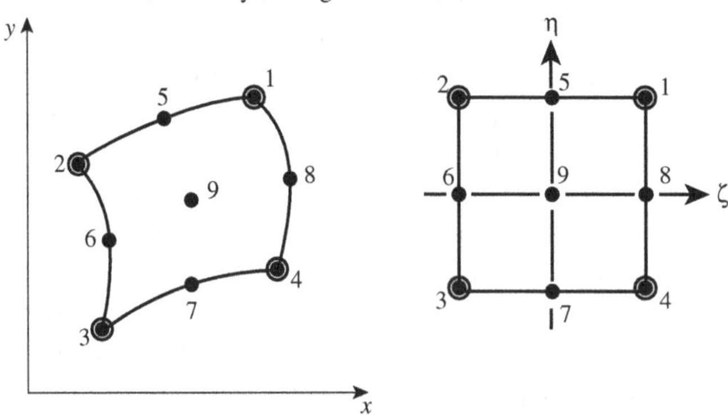

Figure 12.6. Biquadratic curve-sided Lagrangian element used in the current finite element flow model.

$$p^{(e)} = \sum_{i=1}^{4} M_k p_k \tag{12.20}$$

where N_i is a quadratic shape function of the local coordinates ζ and η that is associated with the ith corner, midside, or interior node of the element, with M_k being a linear function of ζ and η that is associated with the kth corner node of the element. As for the field variables $V_{s,i}$, $V_{\theta,i}$, and p_k, these are the nodal magnitudes of velocity components and pressure. Moreover, the physical coordinates s and θ are themselves mapped into the local frame of reference as follows:

$$s = \sum_{i=1}^{9} N_i S_i \tag{12.21}$$

$$\theta = \sum_{i=1}^{9} N_i \theta_i \tag{12.22}$$

In expressions (12.18) through (12.22), the different quadratic and linear shape functions associated with the finite element in its $(\zeta - \eta)$ local frame of reference are as follows:

$$N_1 = \frac{\zeta\eta(1-\zeta)(1-\eta)}{4}$$

$$N_2 = \frac{\zeta\eta(1+\zeta)(1+\eta)}{4}$$

$$N_3 = \frac{\zeta\eta(1+\zeta)(1+\eta)}{4}$$

$$N_4 = \frac{\zeta\eta(1+\zeta)(1+\eta)}{4}$$

$$N_5 = \frac{\zeta(1+\zeta)(1-\zeta)(1+\eta)}{2}$$

$$N_6 = \frac{\zeta(1+\zeta)(1+\eta)(1-\eta)}{2}$$

$$N_7 = \frac{\eta(1+\zeta)(1-\zeta)(1+\eta)}{2}$$

$$N_8 = \frac{\zeta(1+\zeta)(1+\eta)(1-\eta)}{2}$$

$$N_9 = (1+\zeta)(1-\zeta)(1+\eta)(1-\eta)$$

$$M_1 = \frac{1}{4}(1-\zeta)(1-\eta)$$

$$M_2 = \frac{1}{4}(1+\zeta)(1-\eta)$$

$$M_3 = \frac{1}{4}(1+\zeta)(1+\eta)$$

$$M_4 = \frac{1}{4}(1-\zeta)(1+\eta)$$

Application of Galerkin's method [12] to the flow-governing Equations (12.1), (12.14), and (12.15) for a typical finite element gives rise to the following set of equations:

Continuity Equation

$$A_{kj}V_{s,j} + B_{kj}V_{\theta,j} = 0 \tag{12.23}$$

Through-Flow Momentum Equation

$$C_{ij}V_{s,j} + D_{ij}V_{\theta j} + E_{ik}p_k = I_{sji} \tag{12.24}$$

Tangential Momentum Equation

$$F_{ij}V_{sj} + G_{ij}V_{\theta j} + H_{ik}p_k = I_{\theta i} \tag{12.25}$$

where i and j vary from 1 to 9, whereas k varies from 1 to 4.

In a more visible form, Equations (12.23) through (12.25) can be written for a typical finite element as follows:

$$\begin{bmatrix} [A] & [B] & 0 \\ [C] & [D] & [E] \\ [F] & [G] & [H] \end{bmatrix} \begin{Bmatrix} \{V_s\} \\ \{V_\theta \\ \{P\} \end{Bmatrix} = \begin{Bmatrix} 0 \\ \{I_s\} \\ \{I_\theta\} \end{Bmatrix}$$

The different arrays in the preceding matrix equation are defined as follows:

$$A_{kj} = \int_{-1}^{+1}\int_{-1}^{+1} |J|\, rM_k \left(br\frac{\partial N_i}{\partial s} + N_j \frac{d(br)}{ds} \right) d\zeta\, d\eta \tag{12.26}$$

$$B_{kj} = \int_{-1}^{+1}\int_{-1}^{+1} |J|\, brM_k \frac{\partial N_i}{\partial \theta}\, d\zeta\, d\eta \tag{12.27}$$

$$C_{ij} = \int_{-1}^{+1}\int_{-1}^{+1} |J|\, br \left[\nu \left(\frac{\partial N_i}{\partial s}\frac{\partial N_j}{\partial s} + \frac{1}{r^2}\frac{\partial N_i}{\partial \theta}\frac{\partial N_j}{\partial \theta} \right) \right.$$
$$\left. + \alpha_1 N_i \frac{\partial N_j}{\partial s} + \alpha_2 N_i \frac{\partial N_j}{\partial \theta} + \alpha_3 N_i N_j \right] d\zeta\, d\eta \tag{12.28}$$

$$D_{ij} = \int_{-1}^{+1}\int_{-1}^{+1} |J|\, br\alpha_4 N_i N_j\, d\zeta\, d\eta \tag{12.29}$$

$$E_{ik} = \int_{-1}^{+1}\int_{-1}^{+1} |J| \frac{1}{\rho} br \left[N_i \frac{\partial M_k}{\partial s} + \left(\frac{1}{b}\frac{db}{ds} + \frac{1}{r}\frac{dr}{ds} \right) N_i M_k \right] d\zeta\, d\eta \tag{12.30}$$

$$F_{ij} = \int_{-1}^{+1}\int_{-1}^{+1} |J|\, br \left(\beta_4 N_i \frac{\partial N_j}{\partial \theta} + \beta_5 N_i N_j \right) d\zeta\, d\eta \tag{12.31}$$

$$G_{ij} = \int_{-1}^{+1}\int_{-1}^{+1} |J|\, br \left[\nu \left(\frac{\partial N_i}{\partial s}\frac{\partial N_j}{\partial s} + \frac{1}{r^2}\frac{\partial N_i}{\partial \theta}\frac{\partial N_j}{\partial \theta} \right. \right.$$
$$\left. + \beta_1 N_i \frac{\partial N_j}{\partial s} + \beta_2 N_i \frac{\partial N_j}{\partial \theta} + \beta_3 N_i N_j \right] d\zeta\, d\eta \tag{12.32}$$

$$H_{ik} = \int_{-1}^{+1}\int_{-1}^{+1} |J| \frac{1}{\rho r} bN_i \frac{\partial M_k}{\partial \theta}\, d\zeta\, d\eta \tag{12.33}$$

$$I_{s,i} = \oint \nu N_i \frac{\partial V_s}{\partial n}\, dl - \int_{-1}^{+1} |\, J \,|\, brN_i \alpha_5 \, d\zeta \, d\eta \tag{12.34}$$

$$I_{\theta,i} = \oint \nu N_i \frac{\partial V_\theta}{\partial n}\, dl - \int_{-1}^{+1} |\, J \,|\, brN_i \beta_6 \, d\zeta \, d\eta \tag{12.35}$$

where $|\, J \,|$ is the Jacobian of Cartesian-to-local coordinate transformation, I_s and I_θ are closed integrals performed along the element boundary, and n is the direction of the local outward unit vector that is perpendicular to the element boundary. Referring to the Lagrangian element in Figure 12.6, and recalling that the velocity components are degrees of freedom at all nine nodes, whereas the static pressure is a declared variable only at the four corner nodes, the subscripts i and j vary between 1 and 9, whereas the subscript k varies between 1 and 4. With this in mind, the arrays $[A]$ and $[B]$ are 4×9 matrices; $[C]$, $[D]$, $[F]$, and $[G]$ are 9×9 matrices; and the arrays $[E]$ and $[H]$ are 9×4 matrices. The arrays $\{I_s\}$ and $\{I_\theta\}$, on the other hand, are 9×1 vectors. Note that the double integrals appearing in $I_{s,i}$ and $I_{\theta,i}$ result from the shear stress acting over the control-volume inner and outer surfaces (see Figure 12.3) and are modeled as "source" terms. This becomes clear by examination of the previously cited expressions of α_5 and β_6.

Noting the large number of per-element nodal variables and the need for near-wall element refinement (Figure 12.7) in order to capture the boundary-layer flow structure, the final set of assembled finite element equations is substantially large. The equations therefore were, stored in partially packed arrays in the interest of minimizing the core size and subsequently solved using the numerical procedure of Gupta and Tanji [14], which is based on the Gauss-Jordan elimination technique.

Iterative Solution Procedure

The flow solution is obtained iteratively through progressively updating the velocity components $\bar{V}_s$ and $\bar{V}_\theta$ in the convection (inertia) terms of the momentum equations in a successive-substitution fashion. The computational procedure in this case is initialized by solving the flow equations for a theoretically creeping flow, where the variables $\bar{V}_s$ and $\bar{V}_\theta$ are both set to zero in the inertia terms, leaving only the

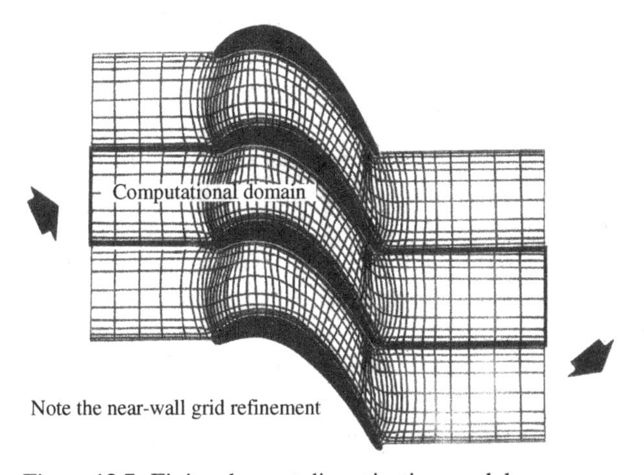

Figure 12.7. Finite element discretization model.

diffusion terms in what theoretically amounts to a Reynolds number of zero. The resulting velocity field is then used to update the finite element equations, and the process is repeated toward convergence. This inner loop is reentered every time the magnitude of the Reynolds number is elevated. Note that the nonlinearity of the momentum equations grows every time the Reynolds number is increased, and so does the number of iterations needed for convergence. In updating the velocity components, underrelaxation factors of 0.8 (for the lower magnitude of Reynolds number) and 0.6 (for the higher magnitudes) were used to ensure monotonic convergence. This means, for instance, that the pressure ${p_i}^{n+1}$ at node i and at the upcoming iteration $n+1$ is expressed as follows:

$${p_i}^{n+1} = \lambda {p_i}^{n} + (1-\lambda){p_i}^{n-1}$$

with λ being the underrelaxation factor.

The state of numerical convergence is defined to occur whenever the summation of nondimensionalized errors in the velocity components and pressure at all nodes maintains a sufficiently small value in two successive iterations, that is,

$$\sum_i \left| \frac{\phi_i^{(n)} - \phi_i^{(n-1)}}{\phi_i^{(n)}} \right|^2 \leq \epsilon$$

$$\sum_i \left| \frac{\phi_i^{(n-1)} - \phi_i^{(n-2)}}{\phi_i^{(n-1)}} \right|^2 \leq \epsilon$$

where ϕ_i refers to the nodal value of a velocity component or the pressure, n is the iterative step at convergence, and ϵ is a tolerance factor, arbitrarily set to 3 percent.

The reason for invoking three successive numerical solutions in the convergence criterion is to provide a means for distinguishing the state of actual convergence from that of numerical fluctuation. It was generally observed that the error jumped abruptly and a state of numerical oscillation occurred once the Reynolds number was elevated, particularly at the computational stages where the final magnitude was closely approached.

Figure 12.7 shows a typical discretization model for a blade cascade. Note the elemental refinement near, and downstream from, the trailing edge. This is essential to capture the blade-wake details, where steep velocity gradients prevail.

Application Examples

Among the airfoil cascades used to validate the current analysis are those of two turbine stators, with the author being the primary investigator in both cases. The first is that of the second stage in a commercial auxiliary power unit and is shown in Figure 12.8. The stream tube appearing in the figure is a strip that is bounded by two successive meridional streamlines that are the postprocessed outcome of the axisymmetric meridional flow analysis that preceded execution of the current computational model. The second example involves a low-aspect-ratio stator (Figure 12.9) that was part of a stage designed and tested for NASA–Lewis Research Center (Cleveland, OH) in the early and middle seventies within a NASA Technology Demonstration Program. Of these, the latter example is one of the most suited

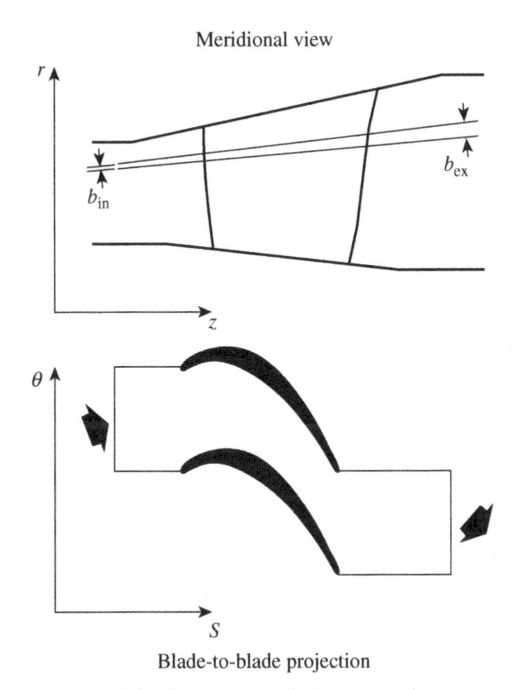

Figure 12.8. Geometry of the second-stage stator in the turbine section of a commercial auxiliary-power unit.

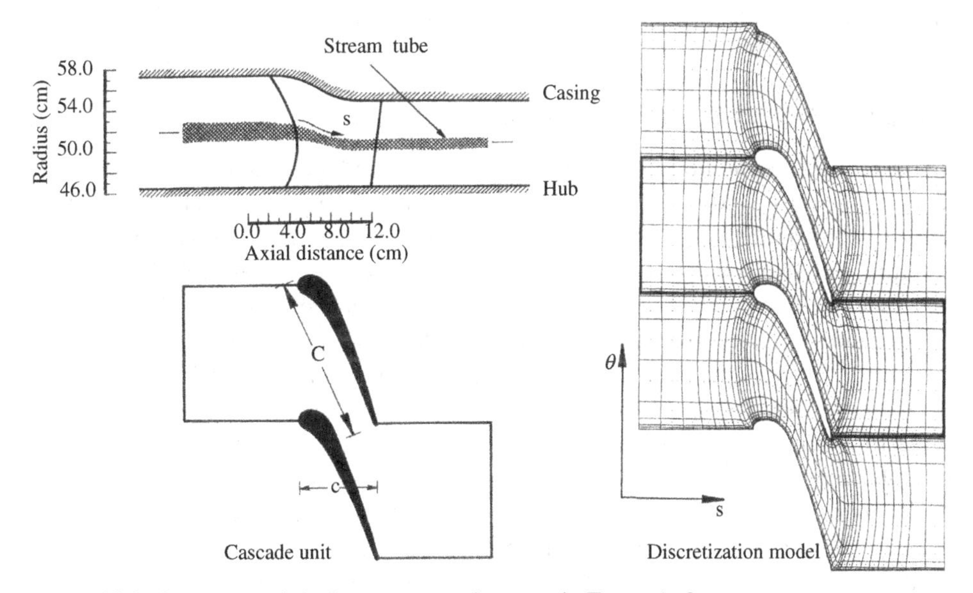

Figure 12.9. Geometry of the low-aspect-ratio stator in Example 2.

cases for the current analysis due to the closeness of the endwalls to each other. This particular feature creates a situation where the flow-field diffusion forces (as opposed to the inertia forces) are dominant, creating a fairly low-Reynolds-number flow field. In comparing the results of the second example with the experimental data, an additional and meaningful step was carried out, whereby the newly introduced source terms, carrying (among others) the shear-stress contributions to the

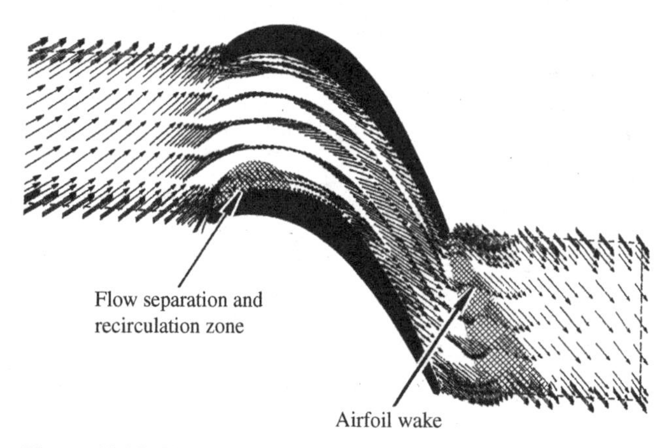

Figure 12.10. Velocity-vector plot in the computational domain in Example 1.

momentum equations, were purposely eliminated in an effort to assess the accuracy they impart to the current analysis. In both examples, special emphasis was placed on the viscous-flow mixing region (blade wake) downstream from the airfoil trailing edge. The point is emphasized that the special application of periodicity conditions, where the finite element equations associated with nodes on the periodic boundaries are unconventionally retained (with the boundary conditions implicitly enforced), makes it possible for the wakes, which begin outside the computational domain, to present themselves within the chosen domain.

Example 1: Second-Stage Stator of a Gas Turbine

Figure 12.7 shows the finite element discretization model created for this stator in the curvilinear $s-\theta$ coordinate system. Refinement of the elements near the airfoil surface in this figure is intended to accommodate not only the boundary layer flow structure, but also zones of flow separation and recirculation, particularly over the airfoil suction side. Note that zones such as these are conceptually outside the scope of flow models that are based on a parabolized flow field (e.g., Chima [15]).

The finite element model in Figure 12.7 is the outcome of a preliminary study to test the sensitivity of the computed results to both model precision (in terms of the total number of elements and the near-wall elemental refinement) and the Reynolds number magnitude. Considering a relatively coarse grid, the preliminary results made it cleat that significant details of the boundary layer development over the airfoil surface, as well as the velocity profiles within the airfoil wake, were largely sacrificed. Note that it is, at least theoretically, possible to obtain a convergent flow solution with a discretization model in which the widths of elements near the airfoil surface are larger than the boundary layer thickness. The issue, however, is that of solution reliability. As for the speed with which a convergent solution was attained, it was observed (as well as expected) that this was rather low because the nonlinearity of the flow-governing equations was elevated or, equivalently, because the Reynolds number was raised.

Figure 12.10 shows a plot of velocity vectors throughout the computational domain. Of particular interest here is the flow structure within the airfoil wake.

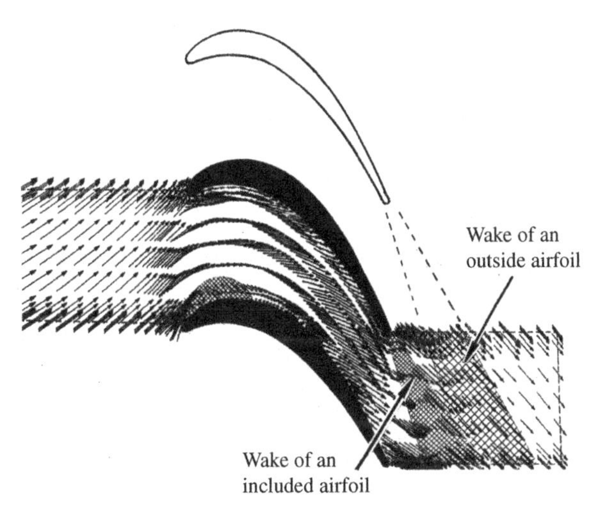

Figure 12.11. Wake of an airfoil outside the computational domain, an outcome of the solid means with which the periodicity conditions are enforced.

The figure reflects the well-known fact that the velocity deficit within the wake is largest in the immediate vicinity of the trailing edge. The wake strength, measured by this very deficit, appears to diminish downstream to the point where mixing of the suction- and pressure-side flow streams is nearly complete. Figure 12.11 is intended to illustrate the accuracy of enforcing the periodic boundary conditions in the manner explained earlier. The figure shows the downstream segment of the wake that is associated with an airfoil that is outside the computational domain.

Example 2: Low-Aspect-Ratio Turbine Stator

Figure 12.12 shows the geometric configurations of this low-aspect-ratio (simply meaning short and axially wide) stator where the endwalls are notably contoured. The figure also shows the mean section stream tube under consideration. Aside from the existence of experimental data, for comparison purposes, another motivating factor in selecting this particular stator was the significant reversed "lean" in the blade stacking pattern or, in other words, the significant spanwise tangential shifting of the blade cross sections as shown in Figure 12.12. This particular feature gives rise to substantial spanwise gradients of the velocity components, which, in turn, create two significant shear-stress components on the inner and outer surfaces of the control volume (see Figure 12.9). In fact, the choice of this particular stator created an environment where all the newly introduced terms (in the momentum equations) provided strong contributions to the finite element equations. The finite element discretization model created for this problem is shown in Figure 12.9, where, again, refinement of the finite-elements is gradually applied as the airfoil surface is approached. The same is applied to those elements in the near-wake region (immediately following the trailing edge) for a better resolution of this high-gradients subregion.

Preparation of the input data was based on the sets of cascade geometry and operating conditions provided by Waterman [16, 17]. A preprocessing step involved the meridional ($r-z$) axisymmetric flow analysis of the stator domain. The results

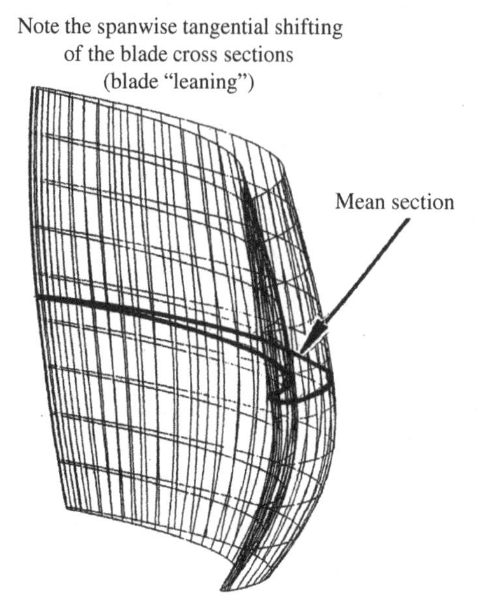

Figure 12.12. Isometric view of the stator blade in Example 2.

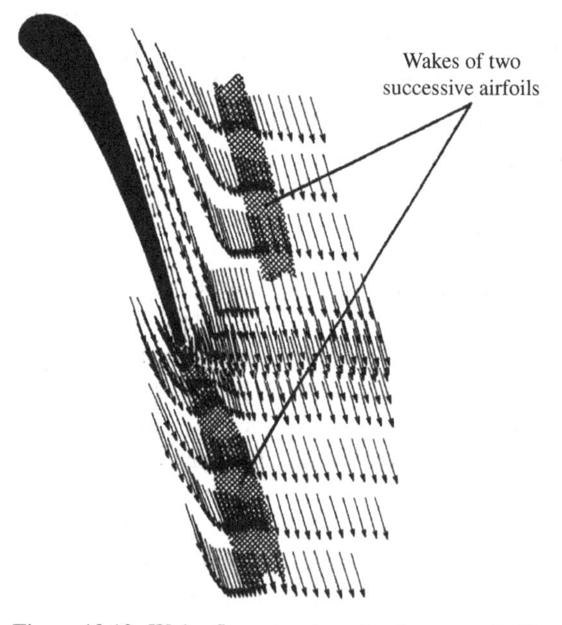

Figure 12.13. Wake-flow structure for the stator in Example 2.

of this preliminary step provided some needed data, including the streamwise stream-tube thickness variation as well as the spanwise gradients of the velocity components, which were subsequently used in computing the shear stresses over the inner and outer surfaces of the stream tube. Figure 12.13 shows a plot of the velocity vector in the blade-to-blade region emphasizing, in particular, the wake flow structure, with one resulting from a blade that is physically outside the computational domain.

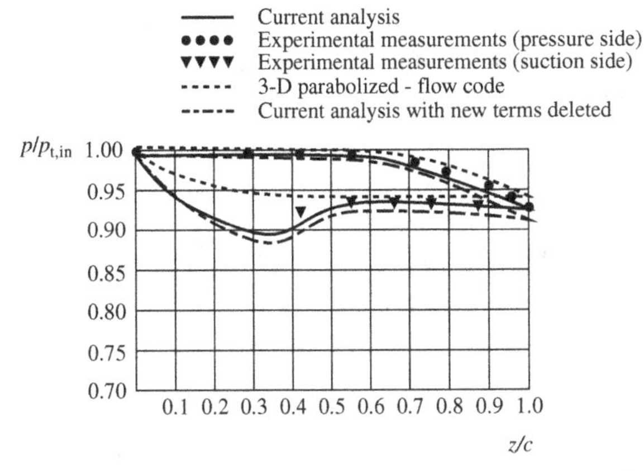

Figure 12.14. Comparison of the computed blade surface-pressure distribution with existing data.

Shown in Figure 12.14 is a comparison of the blade aerodynamic loading between the current analysis and the experimental data. The results are also compared, in the same figure, with those obtained through a three-dimensional parabolized-flow program [17]. The figure shows better agreement of the blade surface-pressure distribution obtained with the experimental data compared with the parabolized-flow results. Superimposed on the same figure are the results obtained via the current flow model, with the exception that the governing-equation newly introduced terms, which account primarily for the shear stresses over the stream-tube surfaces, were intentionally eliminated. Compared with the experimental results (with these terms retained), it is cleat that the accuracy of the current solution is significantly compromised by removing these terms.

Proposed Analysis Upgrades

Adaptation to a Rotating-Blade Cascade

The quasi-three-dimensional flow model in this chapter has so far been tailored to a stationary-airfoil cascade. In this section a rotating-cascade upgrade is proposed whereby the absolute velocity components (V_s and V_θ) are now replaced by their relative counterparts (W_s and W_θ), and the rotation-related acceleration terms (those associated with the centripetal and Coriolis acceleration components) are added to the flow-governing equations. Note that the boundary conditions will have to be correspondingly altered because they now constitute flow constraints on the fluid-particle velocity relative to the spinning blades. In doing so, and wherever needed, the following well-known kinematics relationship (between the absolute and relative velocities) is used:

$$\vec{W} = \vec{V} - \omega r \vec{e}_\theta$$

where ω is the rotational speed, r is the local radius, and $\vec{e}_\theta$ is the local unit vector in the circumferential direction. This relationship influences, in particular, the tangential-velocity-component conversion as follows:

$$W_\theta = V_\theta - \omega r$$

As for the flow-governing equations, the coefficients α_1 through α_5 and β_1 through β_6 will remain the same as derived earlier for a stationary-blade cascade, with the exception of α_4, α_5, and β_5, which are now defined as follows:

$$\alpha_4 = -\frac{W_\theta}{r}\frac{dr}{ds} - 2\omega\frac{dr}{ds}$$

$$\alpha_5 = -\nu\frac{\partial^2 W_s}{\partial n^2} - \omega^2 r\frac{dr}{ds}$$

$$\beta_5 = 2\omega\frac{dr}{ds}$$

In the preceding three expressions, note that the free term $\omega^2 r$ is the centripetal acceleration. Also note that the quantity 2ω essentially multiplies a velocity component, a product that is recognized to give rise to a Coriolis-type acceleration component. Both these accelerations are proportional to the rotor speed ω, and they present themselves once the fluid is passing through a rotating passage.

One of the nost representative rotors, where contributions of the added terms are comparatively substantial, is the centrifugal-compressor impeller in Figure 12.15. The figure shows the geometry of the stream tube midway between the endwalls. As seen in this figure, the significant streamwise radius shift magnifies the impact of both the centripetal and Coriolis acceleration components on the final flow solution.

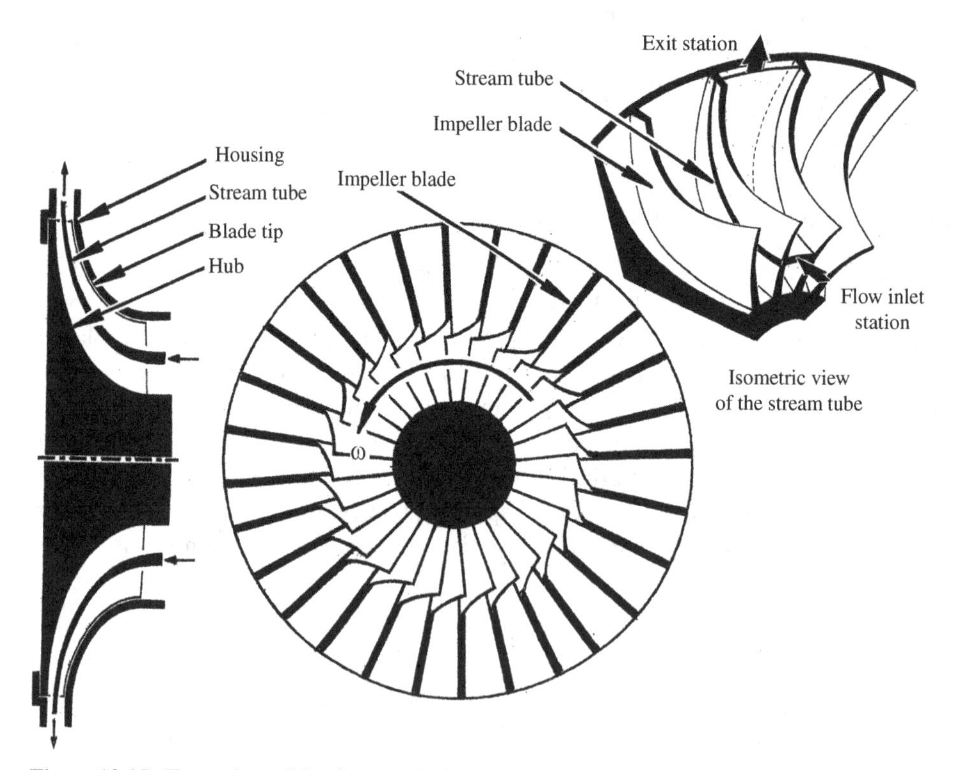

Figure 12.15. Extension of the flow analysis to a rotating-blade cascade: centrifugal compressor impeller.

Inclusion of the Flow Turbulence Aspect

The flow model in this chapter is based on laminar flow behavior, an assumption that is hardly applicable to most turbomachinery components. In fact, it was by no means an easy task to identify the two test cases (discussed earlier in this chapter) where the Reynolds number was nearly close for this condition to prevail. The fact of the matter is that gas turbine cascades will always operate at Reynolds numbers well above 5×10^5, with the exception of far off-design operation modes.

Part of Chapter 13 is dedicated to the development of a turbulence closure that is based on the Baldwin and Lomax [10] algebraic turbulence model. This vorticity-based model is enhanced by adopting the near-wall flow analysis of Benim and Zinser [18] in an effort to elevate the accuracy of the eddy viscosity calculations within critical subregions. By definition, this turbulence closure does not add any new degrees of freedom to the finite element set of influence coefficients. Details of this closure include the means of computing the eddy viscosity coefficient within the airfoil wake, and not only in the vicinity of solid boundary segments. Although this turbulence closure is formulated for a three-dimensional flow field, there should not be any major obstacle in applying it to what is essentially a two-dimensional flow model for implementation as part of the current flow analysis.

PROBLEMS

P.1 Consider the perfectly axial stator configuration in Figure 12.16, where the stream filament is one with constant radius and constant thickness. Beginning with the free diagram of the general stream filament in Figure 12.4,

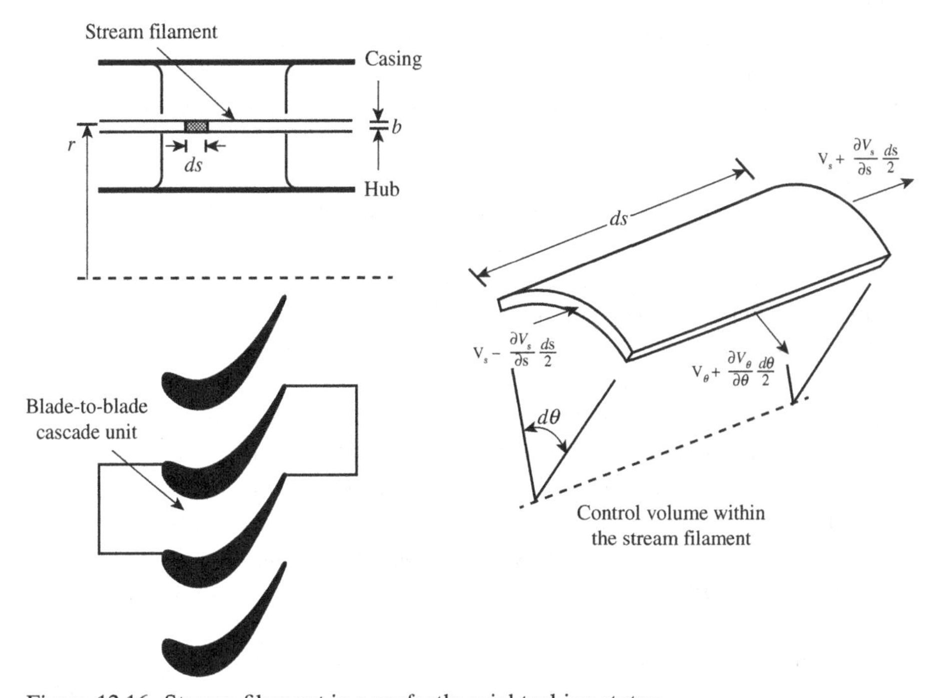

Figure 12.16. Stream filament in a perfectly axial turbine stator.

and eliminating the inapplicable stress terms in this figure, rederive the two momentum equations and their finite element counterparts.

P.2 Figure 12.17 shows a radial turbine stator with parallel endwalls. Also shown in the figure is the stream filament in the stator meridional view, as well as the stress and velocity patterns over a control volume within this filament. Noting that $ds = -dr$ in this case, rederive the two momentum equations for this problem category.

P.3 In the preceding problem, consider the case where the flow field is convection-dominated in its totality, with the diffusion (or viscosity-related) stresses being zero everywhere (i.e., a case where the Reynolds number is theoretically infinite). In this case, the flow field does *not* become automatically potential. In fact, the flow vorticity could be imparted by the flowing medium as the effect of an upstream flow component and will appear in the inlet profiles of the velocity components. Considering this special case:

a. Deduce the two (s and θ) momentum equations governing the flow field.
b. Explain how the solid-surface boundary condition (namely, the no-slip boundary condition) will differ now that the flow field is inviscid.

P.4 Figure 12.18 shows a centrifugal compressor impeller that is equipped with incomplete blades in the "exducer" segment (where the blades are farther apart from one another), with the objective being better flow guidedness as the exit station is approached. The figure also shows the meridional projection of the stream tube under focus and the manner in which the computational domain is defined. As seen in the figure, this domain is tangentially limited by two successive complete blades containing one that is incomplete. Because this subregion is rotating, it would be more convenient to cast the flow-governing equations in a rotating frame of reference, with ω (the impeller speed) being the coordinate axes' rotational frequency. By implementing the α_4, α_5, and β_5 changes indicated in the first proposed upgrade, and in terms of the relative velocity components (W_s and W_θ), rewrite the flow-governing equations [(12.1), (12.14), and (12.15), and express the boundary conditions over the solid and periodic boundary segments in Figure 12.18.

P.5 One of the most important fluid-exerted forces on a rotor-blade cascade is the thrust (or axial) force F_x. Figure 12.19, which shows a turbine rotor cascade, also shows the effect of this force direction on the tip-clearance height. Provided that the force magnitude is significant, the result could be closing down this gap or opening it up depending on the force direction. These would, respectively, cause rubbing against the casing or magnifying the problem of flow migration over the tip from the pressure to the suction side of the blade.

In a quasi-three-dimensional flow analysis, consider the stage where all velocity components and pressure are known at the computational nodes, where they are declared as degrees of freedom. Derive an expression for computing this force at the mean radius (in Figure 12.19), where the stream tube is defined, by integrating the pressure-force axial component over the blade pressure and suction sides in terms of the local unit vector ($\vec{n}$) that is perpendicular to the blade surface.

P.6 Partial admission is a design technique in the event of an excessively small mass-flow rate. In this case, a percentage of the blade-to-blade channels is totally

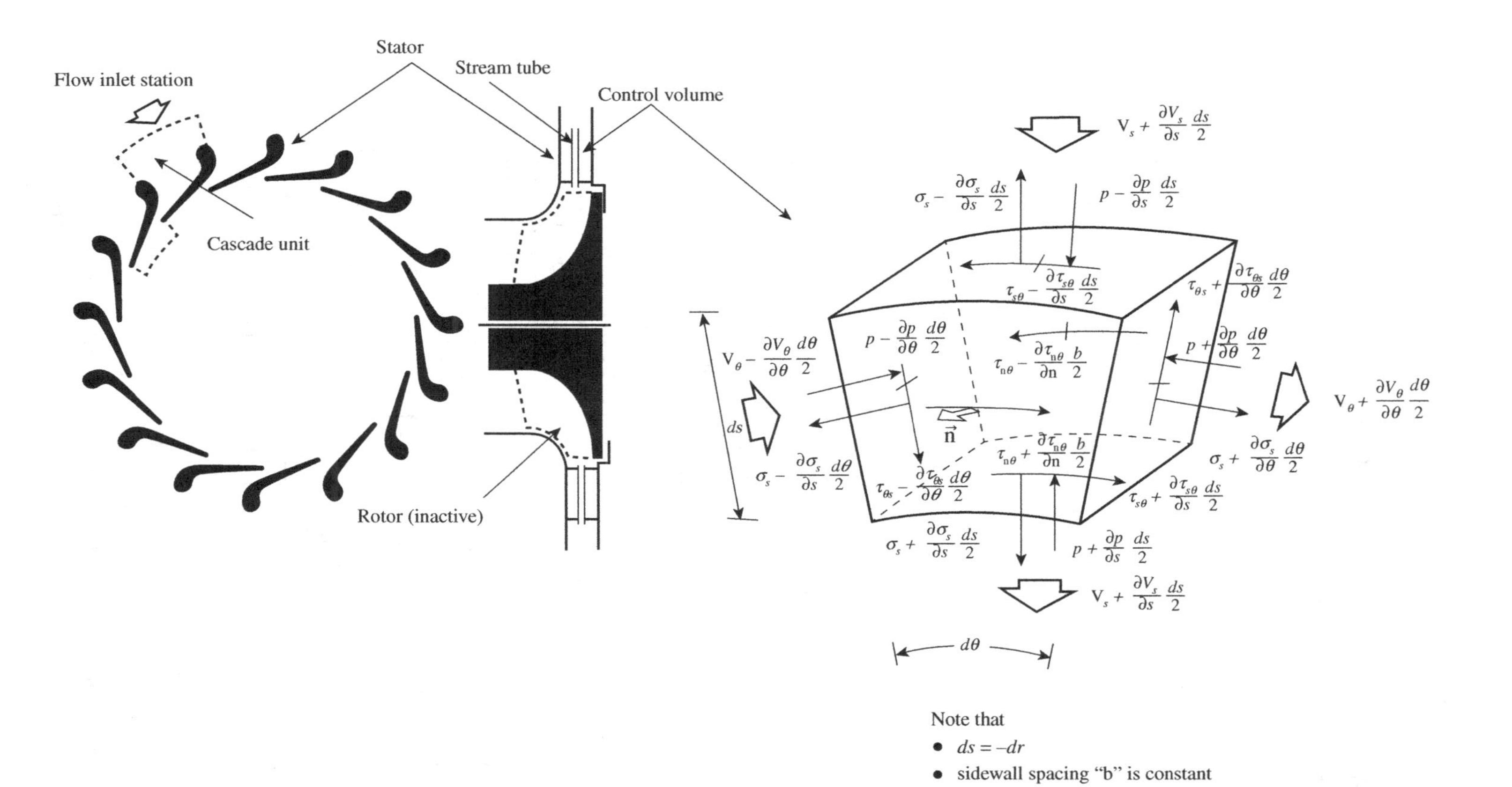

Figure 12.17. Stream-tube definition in a radial-turbine stator.

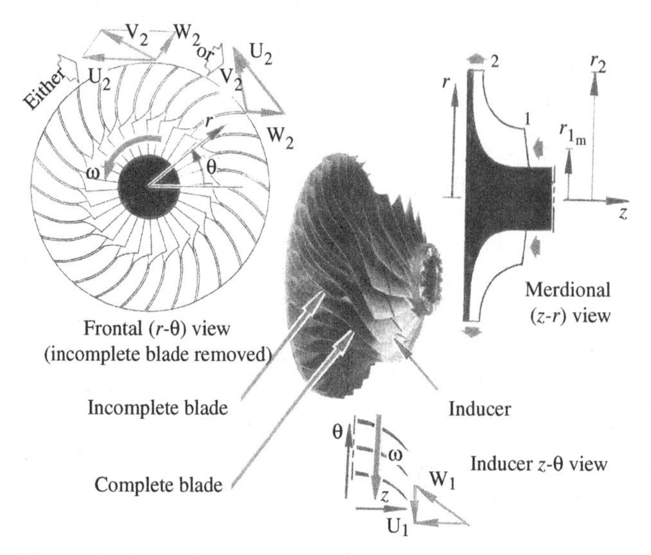

Figure 12.18. Definition of the control volume.

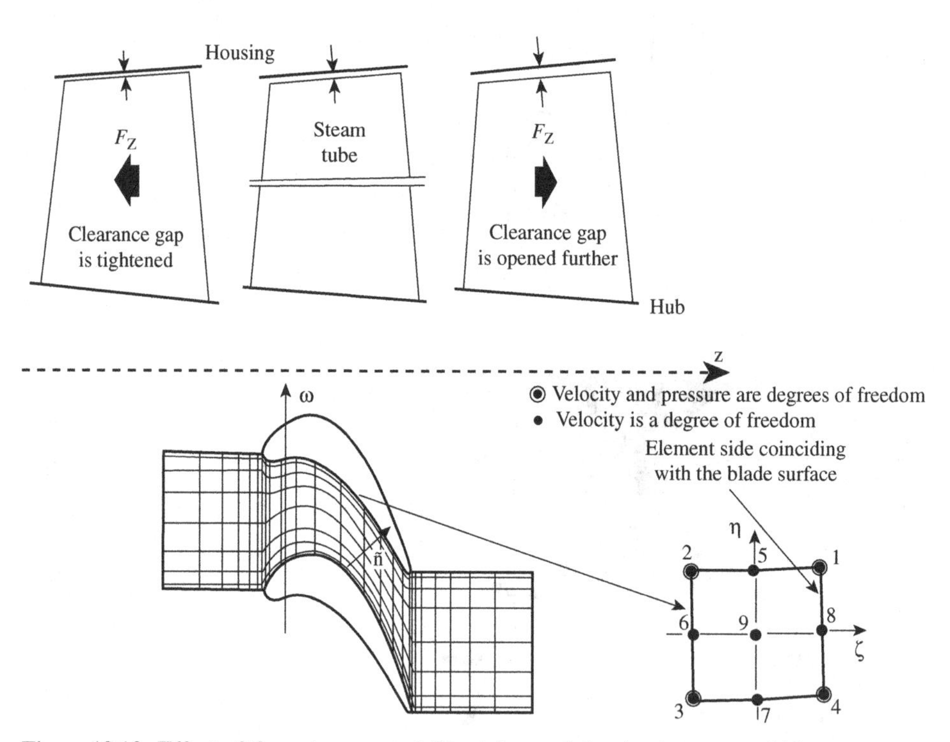

Figure 12.19. Effect of the rotor-exerted thrust force of the tip-clearance width.

blocked in such a way as to elevate the through-flow velocity component toward a larger exit flow angle than would otherwise be the case. Figure 12.20 shows the application of this technique to the stator of a radial turbine where every other blade-to-blade channel is blocked. The two pairs of periodic boundaries in this figure are $a-b$ and $c-d$ upstream of the leading edge and $e-f$ and $g-h$ downstream the trailing edge.

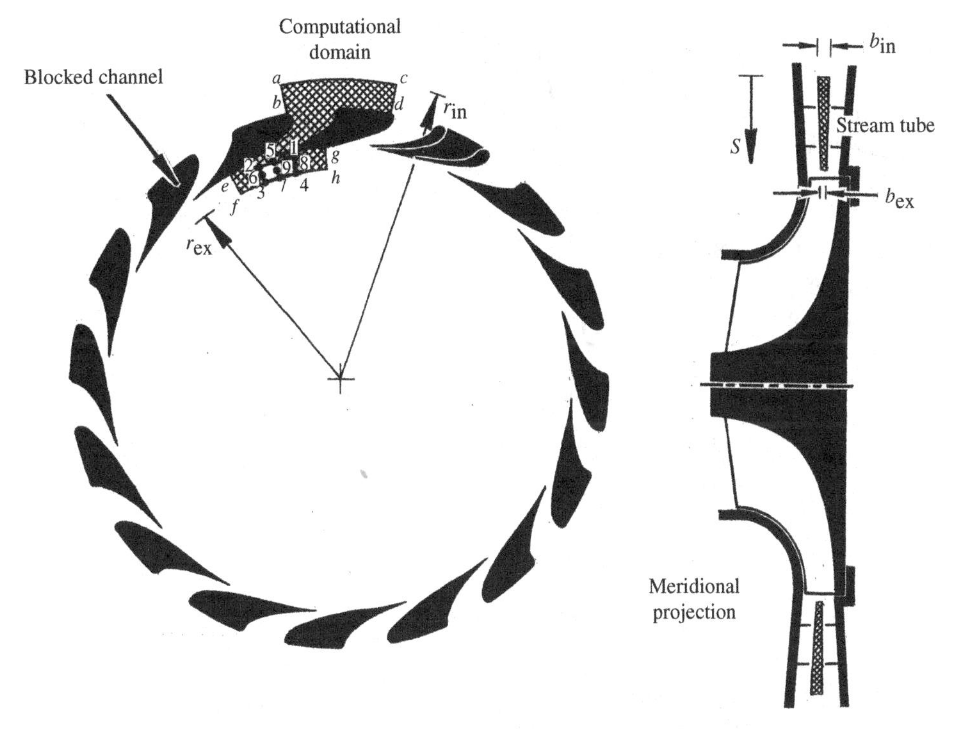

Figure 12.20. Partial admission through the stator of a radial-inflow turbines stage.

The stream tube shown in Figure 12.20 has a linear thickness, reflecting a linearly varying axial distance between the stator endwalls. This will simplify the flow-governing equations because the following two conditions are applicable:

$$b = b_{in} + \left[\frac{r - r_{in}}{r_{ex} - r_{in}}\right](b_{ex} - b_{in})$$

$$s = r_{in} - r$$

The exit boundary conditions will have to reflect the mass and angular-momentum conservation principles, in an average sense, as follows:

$$rV_r = C_1$$

$$rV_\theta = C_2$$

which can be rewritten in the following derivative form:

$$\frac{\partial V_r}{\partial n} = -\frac{V_r}{r_{ex}}$$

$$\frac{\partial V_\theta}{\partial n} = -\frac{V_\theta}{r_{ex}}$$

where $\vec{n}$ represents the direction of the outward unit vector that is perpendicular to the exit station. Referring to the problem's geometric features and the preceding simplification:

a. Derive simpler forms of the flow-governing equations [(12.1), (12.14), and (12.15)].

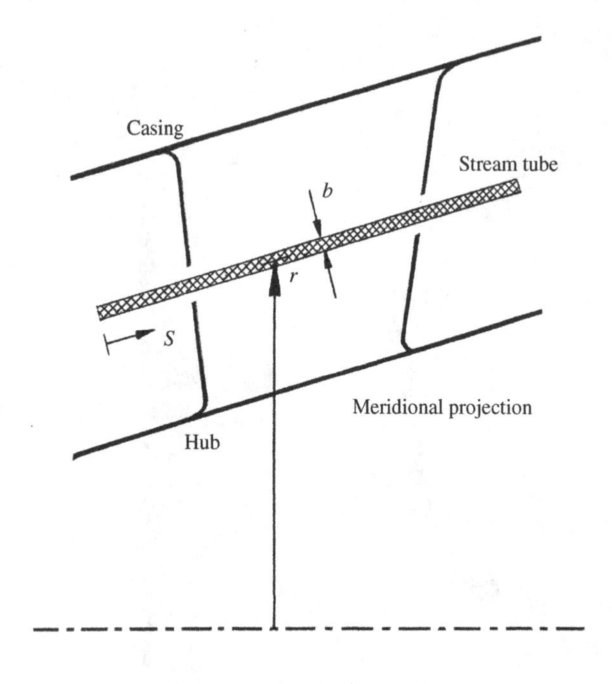

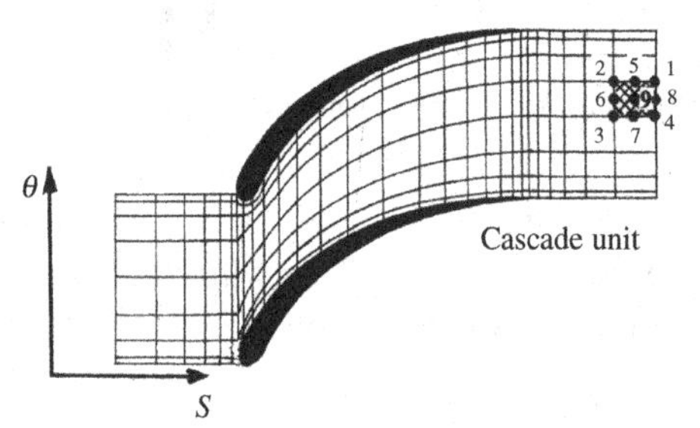

Figure 12.21. A cascade of exit guide vanes with straight and parallel endwalls.

b. Considering the nine-noded finite element shown in Figure 12.21, which shares one side with the exit station, explain the effect of implementing the exit boundary conditions on the matrix of influence coefficients.

In answering part b, note that the two exit boundary conditions do not produce free terms because the right-hand side, in each condition, contains the unknown velocity components.

REFERENCES

[1] Baskharone, E., and Hamed, A., "A New Approach in Cascade Flow Analysis Using the Finite Element Method," *AIAA Journal*, Vol. 19, 1981, pp. 65–71.

[2] McFarland, E. R., "A Rapid Blade-to-Blade Solution for Use in Turbomachinery Design," *Journal of Engineering for Gas Turbine and Power*, Vol. 106, 1984, pp. 376–382.

[3] Koya, M., and Kotake, S., "Numerical Analysis of the Fully Three-Dimensional Periodic Flows through a Turbine Stage," *ASME Paper 85-GT-57*, March 1985.

[4] Rai, M. M., "Navier-Stokes Simulation of Rotor/Stator Interaction Using Patched and Overlaid Grids," *AIAA Paper 85-1519*, 1985.

[5] Davis, R. L., Ni, R. H., and Carter, J. E., "Cascade Viscous Flow Analysis Using Navier-Stokes Equations," *AIAA Journal of Propulsion*, Vol. 3, 1987, pp. 397–405.
[6] Chima, R. V., "Explicit Multigrid Algoritm for Quasi-Three-Dimensional Viscous Flows in Turbomachinery," *AIAA Journal of Propulsion*, Vol. 3, 1987, pp. 397–405.
[7] Chima, R. V., "Inviscid and Viscous Flows in Cascades with an Explicit Multiple-Grid Algorithm," *AIAA Journal*, Vol. 23, 1985, pp. 1556–1563.
[8] Wu, C. H., "A General Theory of Three-Dimensional Flow in Subsonic and Supersonic Turbomachines of the Axial-, Radial-, and Mixed-Flow Types," *NACA TN 2604*, January 1952.
[9] Katsanis, T., "Fortran Program for Calculating Transonic Velocities in a Blade-to-Blade Stream Surface of a Turbomachine," *NASA TN-D 5427*, September 1969.
[10] Baldwin, B. S., and Lomax, H., "Thin Layer Approximation: Algebraic Model for Separated Turbulent Flows," *AIAA Paper No. 78-257*, January 1978.
[11] Huyakorn, P. S., Taylor, C., Lee, R. L., and Gresho, P. M., "A Comparison of Various Mixed-Interpolation Finite Elements in the Velocity-Pressure Formulation of the Navier-Stokes Equations," *Journal of Computational Methods in Fluids*, Vol. 6, 1978, pp. 25–35.
[12] Carey, G. F., and Oden, J. T., *Finite Elements: Fluid Mechanics*, (The Texas Finite Element Series), Vol. VI, Prentice-Hall, Englewood Cliffs, NJ, 1986.
[13] Baker, A. J., *Finite Element Computational Fluid Mechanics*, Hemisphere, New York, 1983.
[14] Gupta, S. K., and Tanji, K. K., "Computer Program for Solution of Large, Sparse, Unsymmetric Systems of Linear Equations," *International Journal of Numerical Methods in Engineering*, Vol. 11, 1977, pp. 1251–1259.
[15] Baskharone, E. A., and McArthur, D. R., "A Comprehensive Analysis of the Viscous Incompressible Flow in Quasi-Three-Dimensional Aerofoil Cascades," *International Journal for Numerical Methods in Fluids*, Vol. 11, 1990, pp. 227–245.
[16] Chima, R. V., "Development of an Explicit Multigrid Algorithm for Quasi-Three-Dimensional Viscous Flows in Turbomachinery," *AIAA Paper No. 86-0032*, 1986.
[17] Waterman, W. F., and Tall, W. A., "Measurement and Prediction of 3D Viscous Flows in Low-Aspect-Ratio Turbine Nozzle," *ASME Paper 76-GT-73*, 1976.
[18] Waterman, W. F., "Low-Aspect-Ratio Turbine Technology: Final Report for Phase I," *AiResearch Report 75-21701 (2)*, August 1975.
[19] Benim, A. C., and Zinser, W., "Investigation into the Finite Element Analysis of Confined Turbulent Flows Using a $\kappa - \epsilon$ Model of Turbulence," *Computer Methods in Applied Mechanics and Engineering*, Vol. 51, 1985.

13 Finite Element Modeling of Flow in Annular Axisymmetric Passages

Introduction

As an illustrative problem in this area, Figure 13.1 shows the exhaust diffuser, designed by the author, for an existing commercial turboprop engine. Such a diffuser is a critical component of any gas turbine engine in both propulsion and power-system applications. Exhaust diffusers, in general, are of primary impact on the total-to-static efficiency of the turbine/exhaust-diffuser system assembly because they raise the static pressure up to the ambient magnitude, making use of existing kinetic energy in the flow stream. Note that the exit pressure value of the last turbine stage is normally less than the ambient pressure. Exit diffusing passages are also common in such power systems as auxiliary power units, which are frequently used in aircraft and various ground applications. In all cases, the demands for maximum static pressure recovery and minimum total pressure loss across the diffuser are often difficult design tasks. This is particularly true for the engine off-design operation due to the excessive amount of turbine exit swirl angle, which may lead to flow separation, normally over the diffuser inner wall.

A literature survey reveals that a considerable amount of experimental research has been focused on annular-diffuser flows [12, 15–17]. Although several crucial design aspects have, in these and other publications, been addressed, most of this experimental work has been centered predominantly around straight-wall diffusers with modest area ratios. Nevertheless, the empirical relationships formulated in these studies and the charts produced in the process have long been appreciated and used extensively in the gas turbine industry.

In the area of diffuser flow analysis, emphasis historically has been placed on the radial [13], planar [3], and conical [7, 14] diffuser configurations, with notably fewer models of the annular diffuser type [12]. These models have progressed from the simple one-dimensional approach to fictitiously parabolized flows that are governed by the boundary layer integral equations [3, 12]. To date, the most detailed methods are those by Hah [7] and Lai and So [14]. While the accuracy of Hah's conical diffuser model is primarily a result of maintaining the flow ellipticity throughout the computational domain, the major contribution of the study by Lai and So is in the area of the adverse pressure-gradient effect on the flow structure in the near-wall zone of planar and conical diffusers. Models with such a level of accuracy have not

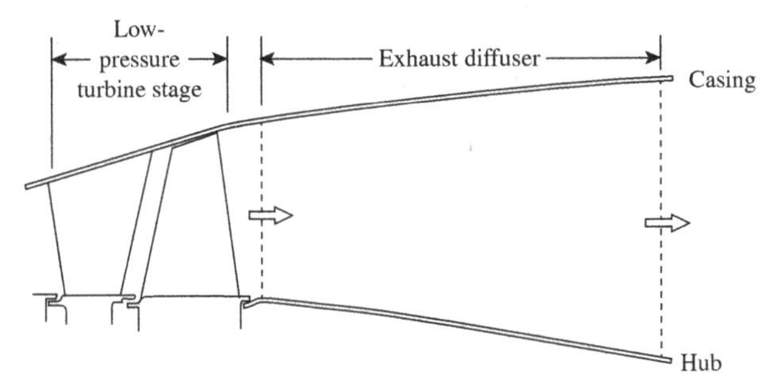

Figure 13.1. Annular exhaust diffuser of a power-system engine.

yet been devised for annular diffuser flows. The finite element formulation of the current problem alleviates the long-recognized drawbacks of using the conventional Galerkin's method in analyzing highly convective flow fields. These often appear in the form of "wiggles" in the pressure field and fictitious oscillations in the streamwise development of the through-flow velocity. Invalidity of the numerical solution, as a result, was reported by many authors [8, 11]. However, it was realized during an early phase of the current study that uniform upwinding of all terms in the flow equations of motion, in the manner devised by Heinrich and Zienkiewicz [8], would have an adverse effect on the solution accuracy as a result of improperly modeling the equations's diffusion terms. The numerical discrepancy in this case was conclusively eliminated by applying different weight functions to the different terms of these equations in the process of deriving the finite element equations. This, in fact, is consistent with the approach devised by Hughes [11], with the exception that the latter is inapplicable to the biquadratic finite element configuration, which is the discretization unit in the current analysis. Special attention in creating the finite element model was paid to the findings of Gresho and Lee [6] suggesting that unless the grid size in the high-gradient subregion is sufficiently small, a smooth flow solution that results from the upwinding scheme may be grossly inaccurate.

Analysis

Flow-Governing Equations

Figure 13.1 shows the meridional view of a typical exhaust diffuser of the type often used in gas turbine engines, particularly, in the power-system area. The flow in this passage is treated as generally swirling and turbulent. However, with the exception of far-off-design operation modes, which are not under consideration here, the diffuser inlet Mach number is assumed to be sufficiently small (0.448 at the design point of the turbine stage in Figure 13.1) to yield a practically incompressible flow at this station and therefore throughout the entire diffuser. Furthermore, the flow is assumed axisymmetric, adiabatic, and steady. Under these conditions, the momentum and mass-conservation equations can be cast in the cylindrical frame of reference (progressing from the r-momentum to the θ-momentum to the z-momentum equations and then the continuity equation) as follows:

$$V_r \frac{\partial V_r}{\partial r} + V_z \frac{\partial V_r}{\partial z} - \frac{{V_\theta}^2}{r} = -\frac{1}{\rho}\frac{\partial p}{\partial r} + \nabla \cdot \left(\nu_{eff} \nabla V_r + \frac{\partial \nu_t}{\partial r}\frac{\partial V_r}{\partial r} - \frac{\nu_{eff}}{r^2} V_r + \frac{\partial \nu_t}{\partial z}\frac{\partial V_z}{\partial r} \right) \tag{13.1}$$

$$V_r \frac{\partial V_\theta}{\partial r} + V_z \frac{\partial V_\theta}{\partial z} + \frac{V_r V_\theta}{r} = \nabla(\nu_{eff} \nabla V_\theta) - \left[\frac{1}{r}\frac{\partial \nu_t}{\partial r} + \frac{\nu_{eff}}{r^2}\right] V_\theta \tag{13.2}$$

$$V_r \frac{\partial V_z}{\partial r} + V_z \frac{\partial V_z}{\partial z} = -\frac{1}{\rho}\frac{\partial p}{\partial z} + \nabla\cdot(\nu_{eff} \nabla V_z) + \frac{\partial \nu_t}{\partial z}\frac{\partial V_z}{\partial z} + \frac{\partial \nu_t}{\partial r}\frac{\partial V_r}{\partial z} \tag{13.3}$$

$$\frac{\partial V_r}{\partial r} + \frac{\partial V_z}{\partial z} + \frac{V_r}{r} = 0 \tag{13.4}$$

where

- V_r, V_θ, and V_z are the r, θ, and z velocity components.
- p is the static pressure.
- ρ is the flow (static) density.
- ν_t and ν_{eff} are the eddy and effective kinematic viscosity coefficients, respectively (defined next).

Turbulence Closure

The algebraic eddy-viscosity model by Baldwin and Lomax [2] is extended to the current axisymmetric flow analysis. In this case, the flow variables used by the model, including the vorticity and wall shear stress, are now based on the meridional (i.e., the r and z) as well as the tangential velocity components. According to this model, and in general, the effective kinematic viscosity coefficient ν_{eff} in equations (13.1) through (13.3) is viewed as composed of two (molecular and eddy) components as follows:

$$\nu_{eff} = \nu_l + \nu_t \tag{13.5}$$

In calculating the eddy component ν_t, the procedure assumes the presence of two (inner and outer), layers. In the inner layer, the Prandtl–Van Driest formulation yields the following expression

$$\nu_{ti} = l^2 \mid \omega \mid \tag{13.6}$$

where the subscript i refers to the inner layer. The mixing length l and the vorticity magnitude $\mid \omega \mid$ in expression (13.6) are as follows:

$$l = ky\left[1 - exp\left(\frac{-y^+}{A^+}\right)\right], y^+ = \frac{\sqrt{\rho_w \tau_w}}{\mu_w} y \tag{13.7}$$

$$\omega = \left[\left(\frac{V_\theta}{r} + \frac{\partial V_\theta}{\partial r}\right)^2 + \left(\frac{\partial V_\theta}{\partial z}\right)^2 + \left(\frac{\partial V_r}{\partial z} - \frac{\partial V_z}{\partial r}\right)^2\right]^{1/2} \tag{13.8}$$

where

- y is the distance normal to the nearest wall.
- A^+ is the sublayer thickness.
- τ_w is the wall shear stress

The model switches from the Van Driest formulation to that of the outer region at the smallest value of y for which the inner and outer values of the eddy kinematic viscosity coefficients are equal. The formulation for the outer layer is given by

$$\nu_{to} = KC_{cp}F_{max}y_{max}F_{KLEB}$$

where
$$F_{max} = y_{max} \mid \omega \mid \left[1 - exp\left(\frac{-y^+}{A^+}\right)\right]$$

$$F_{KLEB} = \left[1 + 5.5\left(C_{KLEB}\frac{y}{y_{max}}\right)^6\right]^{-1}$$

with y_{max} referring to the value of y at which F_{max} occurs. The various constants in the Baldwin-Lomax model are as follows:

- $A^+ = 26.$
- $k = 0.4.$
- $K = 0.0168.$
- $C_{cp} = 1.6.$
- $C_{KLEB} = 0.3.$

Computation of the wall shear stress τ_w is based on the near-wall zone treatment proposed by Benim and Zinser [1]. The assumption here is that the universal law of the wall at any wall location is extendible to an interior computational node that is closest to the wall at this location. Referring to the distance of the interior node from the wall by y_{min}, the following recursive expression for the wall shear stress is obtained:

$$\tau_w = \frac{\nu_l \rho V_{min}}{y_{min}} \qquad \text{for } y^+{}_{min} < 11.6 \qquad (13.10a)$$

$$\tau_w = \frac{kC_D{}^{1/4}}{\rho} V_{min} k_{min}^{1/2} ln\left(EC_D{}^{1/4} y_{min} k_{min}{}^{1/2}\right) \qquad \text{for } y_{min}{}^+ \geq 11.6 \qquad (13.10b)$$

where

- $k_{min} = \dfrac{\tau_w}{\rho C_D{}^{1/2}}.$
- $C_D = 0.09.$
- $k = 0.4.$
- $E = 9.0.$

with V_{min} referring to the magnitude of velocity, with the tangential component taken into account at the interior node.

The remainder of the Benim and Zinser wall model was implemented in executing the test cases in the current study. This is pertinent to use of the wall functions to calculate the slip velocity along the edge of the computational domain that, in this case, is located at a small predetermined distance away from the wall. Along this distance, the dimensionless velocity V^+ assumes the following form:

$$V^+ = y^+ \qquad \text{for } y^+ < 11.6 \qquad (13.11a)$$

$$V^+ = \frac{1}{k} ln\left(Ey^+\right) \qquad \text{for } y^+ \geq 11.6 \qquad (13.11b)$$

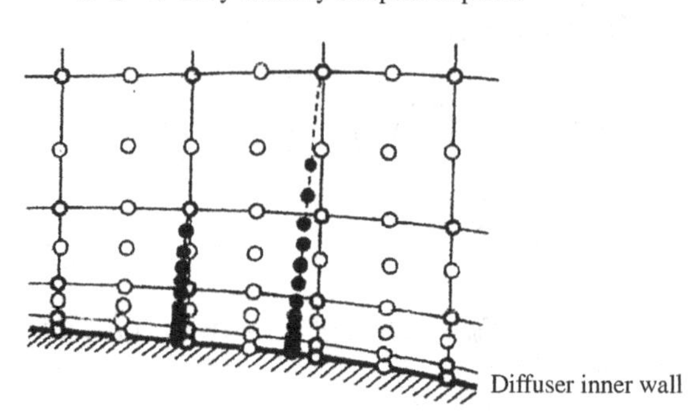

Figure 13.2. Near-wall mesh refinement for accurate eddy viscosity calculations.

Numerical implementation of the turbulence closure, including the near-wall zone analysis, is achieved using an array of points that is different from the primary set of the computational nodes. Figure 13.2 shows an enlarged segment of the computational domain near a solid wall in which the primary nodes in the finite element discretization model are identified by hollow circles, whereas the points used in the computations leading to the eddy viscosity at two selected nodes are solid circles. The objective here was twofold: to estimate the cut-off location between the inner and outer layers with sufficient accuracy and to also capture the steep gradients of the flow variables near the solid wall. A substantial enhancement of the solution was, as a result, observed during an early phase of numerical experimentation. This, for the most part, was due to the excellent accuracy with which the wall shear stress was computed, for it was generally possible to compute the stress using an interior point where the dimensionless wall coordinate y^+ was acceptably small. Worth noting is the fact that accurate evaluation of the wall shear stress would affect both the computed value of eddy viscosity and the wall slip velocity because the stress simultaneously appears in the turbulence closure and the universal law of the wall.

Boundary Conditions

Preserving the flow ellipticity throughout the computational domain requires the specification of appropriate boundary conditions over the entire boundary. Of these, the conditions imposed at the solid boundary segments, through the use of wall functions, were discussed earlier. At the flow inlet station, the entrance boundary layer thickness is assumed to be 20 percent of the inlet annulus height. Outside the boundary layer, constant core values of the inlet velocity components are specified. As for the flow exit conditions, zero surface tractions are imposed as follows:

$$\frac{\partial V_z}{\partial z} - \frac{p}{2\rho\nu_{eff}} = 0$$

$$\frac{\partial V_r}{\partial z} + \frac{\partial V_z}{\partial r} = 0$$

$$\frac{\partial V_\theta}{\partial z} = 0$$

Out of the various exit boundary conditions tested during the course of this study the preceding conditions were found to be the least fictitious and most tolerant of nonuniform exit flow patterns. This observation was previously emphasized by Gresho and Lee [6] in conjunction with their investigation of the confined and unconfined flow over a step, where the flow recirculation downstream from the step extended to and included the exit station. This type of complication was to be anticipated in the current problem should the adverse pressure gradient and excessive swirl cause flow separation anywhere on the diffuser walls. In a situation such as this, the zero-surface-traction boundary conditions would, according to Gresho and Lee, permit the flow to depart and reenter the computational domain at different segments of the exit boundary.

Finite Element Formulation

A special version of the Petrov-Galerkin weighted residual method is used to derive the finite element form of the flow-governing equations. The current approach ensures upwinding of the convection terms in the momentum equations while preserving the elliptic nature of the diffusion terms. This, for a simple orthogonal grid, would be equivalent to backward-differencing the convection terms and central-differencing the diffusion terms in the conventional finite-difference analyses of inertia-dominated flows. Successful implementation of this strategy within a finite element context was achieved by Hughes [11] for only simple (linear and bilinear) finite element configurations by modifying the integration algorithm in the process of deriving the element equations. This effectively eliminated, in the final solution, the presence of wiggles in the streamwise pressure variation, which are typically associated with the conventional Galerkin's weighted residual approach when applied to high-Reynolds-number flows.

The characteristic features of the finite element discretization model are shown in Figure 13.3. The discretization unit, which is a nine-noded curve-sided finite element of the Lagrangian type [18], is shown separately in the same figure in both the local and physical frames of reference. Within a typical element (e), let the spatial coordinates be interpolated as follows:

$$z^{(e)} = \sum_{i=1}^{9} N_i(\zeta,\eta) z_i$$

$$r^{(e)} = \sum_{i=1}^{9} N_i(\zeta,\eta) r_i$$

where the N_i's are quadratic shape functions associated with the element corner, midside, and interior nodes. Next, the flow variables are interpolated throughout the element in a similar fashion. Guided by the compatibility conditions cited in Chapter 12, the velocity components and pressure can be interpolated as follows:

$$V_z^{(e)} = \sum_{i=1}^{9} N_i(\zeta,\eta) V_{z,i}$$

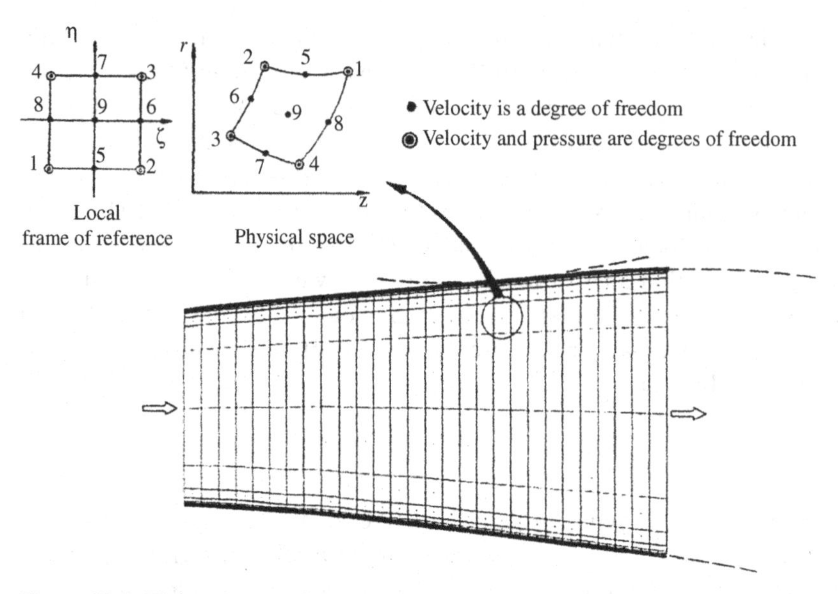

Figure 13.3. Finite element discretization model.

$$V_r^{(e)} = \sum_{i=1}^{9} N_i(\zeta,\eta) V_{r,i}$$

$$V_\theta^{(e)} = \sum_{i=1}^{9} N_i(\zeta,\eta) V_{\theta,i}$$

$$p^{(e)} = \sum_{i=1}^{4} M_k(\zeta,\eta) p_k$$

where the M_k's are the linear shape functions associated with only the element's corner nodes. According to the weighted residual method, the error functions produced by Equations (13.1) through (13.4) as a result of substituting the interpolation expressions are then made orthogonal to a special set of weight functions over the finite element subdomain. In constructing the latter set of functions, the so-called consistency criterion of Hood and Taylor [10] is implemented, whereby the element's shape functions M_k are used in conjunction with the continuity equation. On the other hand, quadratic functions, which include the element's shape functions N_i and a set of derived functions W_i (which included upwinding terms to prevent violent pressure oscillations throughout the solution domain) were used in conjunction with the momentum equations in such a way as to ensure full upwinding of the convection terms. Of these, the functions W_i were previously defined by Heinrich and Ziewnkiewicz [8] in terms of the shape functions and some upwinding constants that depend on the element's geometry and local velocity direction.

Having defined the shape functions, derivation of the finite element equivalent to Equations (13.1) through (13.4) is straightforward. The process requires linearization of these equations using known values of the velocity components and eddy viscosity (from a previous iteration or an initial guess) and use of the Gauss divergence theorem. The final form of these equations for the typical element (e) is as follows:

$$\left[\int\int_{A^{(e)}}\left\{\bar{\nu}_{eff}\left(\frac{\partial N_i}{\partial r}\frac{\partial N_j}{\partial r}+\frac{\partial N_i}{\partial z}\frac{\partial N_j}{\partial z}\right)+W_i\left(\bar{V}_r\frac{\partial N_j}{\partial r}+\bar{V}_z\frac{\partial N_j}{\partial z}-N_i\frac{\partial \bar{\nu}_t}{\partial r}\frac{\partial N_j}{\partial r}+\frac{\bar{\nu}_{eff}}{r^2}N_iN_j\right\}rdA\right]V_{r,j}$$

$$-\left[\int\int_{A^{(e)}}\frac{\bar{V}_\theta}{r}N_iN_jrdAV_{\theta,j}-\left[\int\int_{A^{(e)}}N_i\frac{\partial\bar{\nu}_t}{\partial z}\frac{\partial N_j}{\partial r}rdA\right]V_{z,j}+\left[\int\int_{A^{(e)}}\frac{1}{\rho}N_i\frac{\partial M_k}{\partial r}rdA\right]p_k\right.$$

$$=\oint_{L^{(e)}}\bar{\nu}_{eff}rN_i(\vec{n}.\nabla V_r)\,dL \tag{13.13}$$

$$\left[\int\int_{A^{(e)}}\left\{\bar{\nu}_{eff}\left(\frac{\partial N_i}{\partial r}\frac{\partial N_j}{\partial r}+\frac{\partial N_i}{\partial z}\frac{\partial N_j}{\partial z}\right)+W_i\left(\bar{V}_r\frac{\partial N_j}{\partial r}+\bar{V}_z\frac{\partial N_j}{\partial z}+\left(\frac{1}{r}\frac{\partial\bar{\nu}_t}{\partial r}+\frac{\bar{\nu}_{eff}}{r^2}+\frac{\bar{V}_r}{r}\right)N_iN_j\right\}rdA\right]V_{\theta,j}$$

$$=\oint_{L^{(e)}}\nu_{eff}rN_i(\vec{n}.\nabla V_\theta)\,dL \tag{13.14}$$

$$-\left[\int\int_{A^{(e)}}N_i\frac{\partial\bar{\nu}_t}{\partial r}\frac{\partial N_j}{\partial z}rdA\right]V_{r,j}+\left[\int\int_{A^{(e)}}\left\{\bar{\nu}_{eff}\left(\frac{\partial N_i}{\partial r}\frac{\partial N_j}{\partial r}+\frac{\partial N_i}{\partial z}\frac{\partial N_j}{\partial z}\right)+W_i\left(\bar{V}_r\frac{\partial N_j}{\partial r}+\bar{V}_z\frac{\partial N_j}{\partial z}\right)\right.\right.$$

$$\left.\left.+N_i\frac{\bar{\nu}_t}{\partial z}\frac{\partial N_j}{\partial z}\right\}rdA\right]V_{z,j}+\left[\int\int_{A^{(e)}}\frac{1}{\rho}N_i\frac{\partial M_k}{\partial r}rdA\right]p_k=\oint_{L^{(e)}}\nu_{eff}rN_i(\vec{n}.\nabla V_z)\,dL \tag{13.15}$$

$$\left[\int\int_{A^{(e)}}\left(M_k\frac{\partial N_j}{\partial r}+\frac{1}{r}M_kN_j\right)rdA\right]V_{r,j}+\left[\int\int_{A^{(e)}}M_k\frac{\partial N_j}{\partial z}rdA\right]V_{z,j}=0 \tag{13.16}$$

In these equations, the subscripts i and j vary from 1 to 9, whereas k varies from 1 to 4. Also, the symbol $(\bar{\ })$, in these equation, signifies a value that is known from a previous iteration or an initial guess. The global set of equations is achieved by assembling Equations (13.13) through (13.16) among all elements for the current iterative step, and the result is a system of linear algebraic equations in the flow nodal variables.

Method of Numerical Solution

In an attempt to reduce the consumption of computer resources, the θ-momentum equation (13.14) was uncoupled and separately solved as part of an iterative procedure that is basically similar to that by Baskharone and Hensel [4]. Note that the term containing $V_{\theta,j}$, in Equation (13.13) should now belong to the right-hand side of the equation, where it is treated as a "source" term and is progressively updated during the iterative solution procedure. The set of finite element equations in each computational step is assembled and simultaneously solved using the frontal method of Hood [9].

In progressing from one iteration to the next, an underrelaxation factor is applied to the nodal values of flow variables to ensure monotonic convergence. The optimal value of this factor in the current study was found, during an earlier numerical experimentation process, to be 0.7 for all cases considered.

Numerical Results

The current computational model was applied to a representative exhaust diffuser of a commercial turboprop engine (the TPE 331-14 turboprop engine in Figure 13.3). This is one of the so-called clean diffusers in light of its low total-pressure-loss and

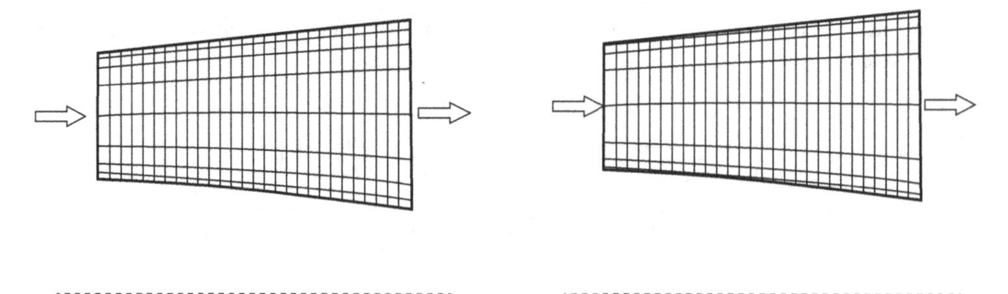

Figure 13.4. Two near-wall mesh-refinement configurations.

high static-pressure-recovery ratio. The diffuser is of the contoured-wall type, as shown in Figure 13.3, with an area ratio of 1.61, a length-to-inlet height ratio of 3.70, and a design-point Reynolds number (based on the average inlet velocity and the diffuser length) of 5.4×10^5. Performance of this diffuser at design and off-design operation modes was simulated in a cold rig by varying the diffuser inlet swirl angle. The test established the diffuser recovery characteristics over a wide range of operation, part of which is investigated in this study.

Grid Dependency of the Flow Field

Sensitivity of the numerical solution to the size of the finite element model in terms of the field resolution it provides was investigated first. In all cases considered, the number of cross-flow stations in the meridional view between the inlet and exit stations (Figure 13.4) was fixed at 29, whereas the number of grid lines between the endwalls, referred to by N_r, was varied from 5 to 11. The latter family of grid lines was constructed with varying increments in the radial direction. To better quantify this variation, the growth of increments from either one of the diffuser endwalls to the mean line at any axial location was made to be that of the geometric sequence type with a common ratio λ, where $1.5 \leq \lambda \leq 3.5$. Illustrated in Figure 13.5 are the geometric effects of varying λ from 1.75 to 2.25 for the case where N_r is fixed at 9. Note the drastic change in the width of the wall elements as a result.

Shown in Figure 13.5 is the error, based on the experimental data, of the computed recovery coefficient C_p at the diffuser design point for a matrix of finite element grids. The various combinations of N_r and λ in this figure represent finite element-model configurations that, to varied levels, ensure proper field resolution. Examination of the figure reveals that the error in the predicted recovery coefficient for $N_r = 7$ and $\lambda = 2.25$ is about 5.5 percent. This, plus the asymptotic nature of the error curve near this point, led to the choice of this combination for the remainder of the numerical investigation. The choice was, to a lesser degree, influenced by the desire to minimize the core size and CPU time consumption that would naturally be associated with larger wall-to-wall station counts.

Diffuser Flow Field and Off-Design Performance

A primary objective of this study was to apply the flow model to the diffuser off-design flow field. The intention here was to continue the process of verifying

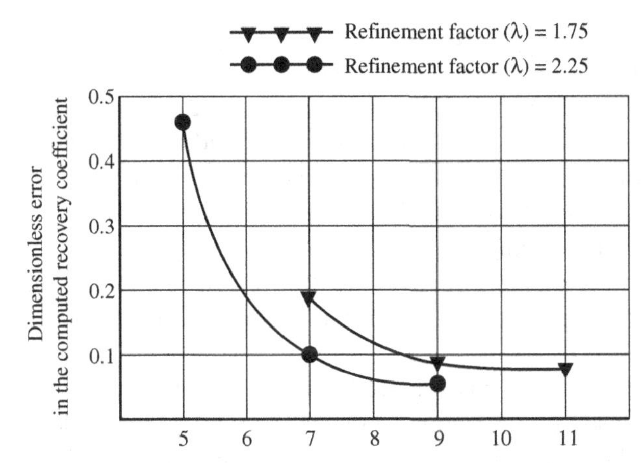

Figure 13.5. Wall-to-wall station-count effect on the solution error.

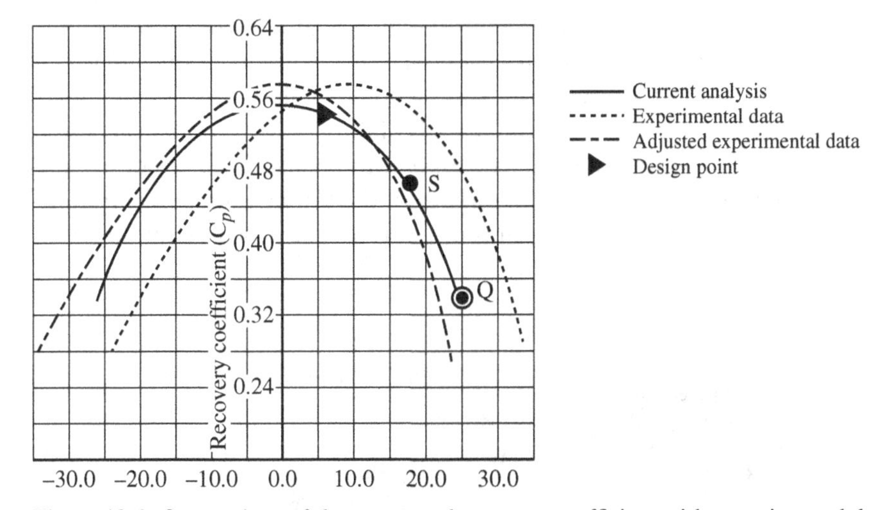

Figure 13.6. Comparison of the computed recovery coefficient with experimental data.

the computational model under some potentially challenging flow conditions and to appraise the diffuser tolerance to off-design turbine operation as an important design criterion. Because the flow compressibility, as dictated by the turbine exit Mach number, was unchangingly negligible over the entire operation range, transition to off-design diffuser operation was simulated by incrementing the average inlet swirl angle while maintaining its characteristic profile, as well as that of the through-flow velocity. Figure 13.6 shows a comparison between the computed and measured recovery coefficient C_p over a range of off-design diffuser operation. The nonsymmetric shape of the experimental data in this figure is suspected to be caused primarily by such factors as distortion in the inlet velocity profiles and the presence of exhaust struts in the actual diffuser passage. In fact, it is perhaps more meaningful, yet not sufficiently rigorous, to compare the numerical results to an "adjusted" experimental curve whereby the actual curve is shifted in such a way as to associate the point of maximum recovery with a diffuser inlet swirl angle of zero, as shown in Figure 13.6. Altering the experimental data in this manner is merely intended to establish common grounds for a qualitative comparison with the numerical results because the latter are the outcome of a flow field in which the above-mentioned complications are neglected. With this in mind, it is apparent that the overall agreement between the two sets of data in Figure 13.6 is clearly good and, to a sufficient extent of certainty, validates the various details of the computational model. Worth noting in Figure 13.6 is also the fact that the boundary layer separation over the hub surface begins to exist at an inlet swirl angle of 18 degrees (point S in Figure 13.6). The flow recirculation zone was, at first, confined to the region just upstream from the exit station but began to grow further in size as the inlet swirl angle was elevated.

It is apparent that the fairly wide range of decent pressure recovery by the diffuser (see Figure 13.6) is caused by the manner in which the diffuser boundary is contoured, particularly in the near-exit casing segment (see Figure 13.3). The concave shape of this segment is believed to retard the flow separation over the casing and, perhaps appreciably, over the hub as the exit station is approached. This favorable effect becomes more valuable at off-design operation modes where excessive inlet swirl angles create the tendency for boundary layer separation with which reattachment is, in view of the adverse pressure gradient, unlikely.

Figures 13.7 and 13.8 reveal significant details of the diffuser flow field at design and off-design operation modes. These correspond to inlet swirl angles of 6.3 and 25 degrees, respectively. First, the radial profiles of the through-flow velocity component are shown in Figure 13.7, where it is evident that the design-point flow field

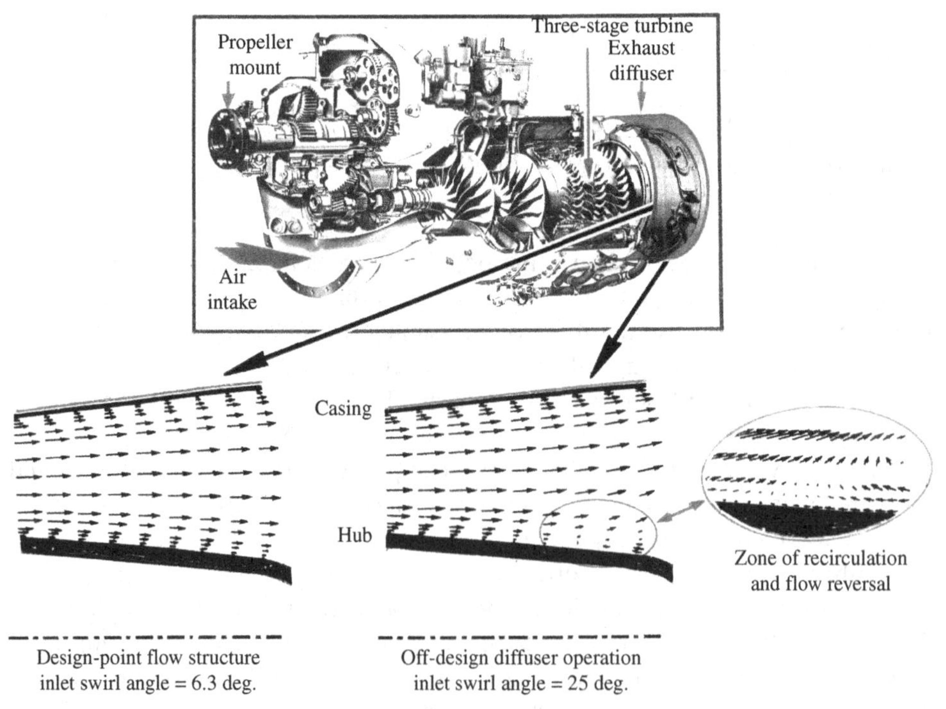

Figure 13.7. Effect of the inlet swirl angle on the exhaust-diffuser performance in a turboprop engine.

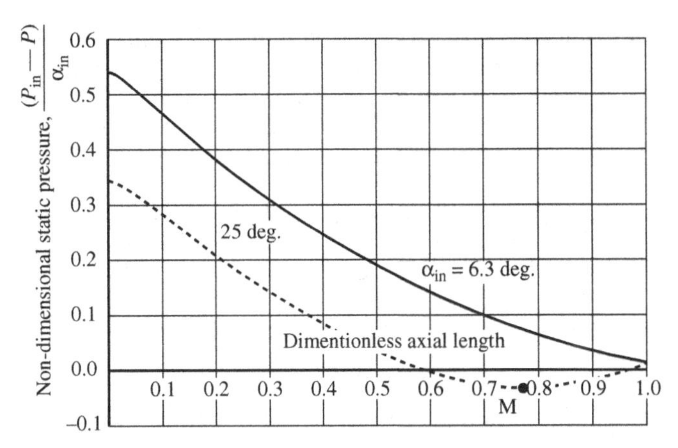

Figure 13.8. Static pressure distribution over the hub for two inlet swirl-angle magnitudes.

is separation-free. The figure also shows that boundary-layer separation indeed prevails over the hub surface at $\alpha_{in} = 25$ degrees (point Q in Figure 13.6) as a result of the substantial elongation of the flow trajectory across the diffuser. Due to the many tangential trips, in this case, the fluid particles have to go over the endwalls in progressing from one axial location to another. Note that the large centrifugal force acting on the fluid particles, as a result of the large tangential velocity component in this case, provides an additional source for flow separation because it creates a radially outward shift of the fluid particles. The impact of this flow pattern on the radially averaged static pressure variation in the axial direction is illustrated in Figure 13.8. Examination of this figure reveals that the blockage associated with the flow recirculation subdomain at $\alpha_{in} = 25$ degrees is, in an average sense, causing a local through-flow acceleration in the diffuser segment in the same point M in the same figure, where the local average pressure is less than the exit magnitude. The decline in static pressure across this segment is also a result of a significant total pressure loss in the recirculation zone.

REFERENCES

[1] Benim, A. C., and Zinser, W., "Investigation into the Finite Element Analysis of Confined Turbulent Flows Using a $\kappa - \epsilon$ Model of Turbulence," *Computer Methods in Applied Mechanics and Engineering*, Vol. 51, 1985, pp. 507–523.

[2] Baldwin, B. S., and Lomax, H., "Thin Layer Approximation and Algebraic Model for Separated Turbulent Flows," *AIAA Paper No. 78-257*, 1978.

[3] Bardina, J., Lyrio, A., Kline, S. J., Feziger, J. H., and Johnston, J. P., "A Prediction Method for Planar Diffuser Flows," *ASME Journal of Fluids Engineering*, Vol. 103, 1981, pp. 315–321.

[4] Baskharone, E. A., and Hensel, S. J., "A New Model for Leakage Prediction in Shrouded-Impeller Turbopumps," *Journal of Fluids Engineering* Vol. 111, 1989, pp. 118–123.

[5] Carey, G. F., and Oden, T. J., *Finite Elements: Fluid Mechanics* (The Texas Finite Element Series), Vol. IV, Prentice-Hall, Engelwood Cliffs, NJ, 1986.

[6] Gresho, P. M., and Lee, R. L., "Don't Suppress the Wiggles: They're Telling You Something," *Computers & Fluids*, Vol. 9, No. 2, 1981, pp. 223–253.

[7] Hah, C., "Calculation of Various Diffuser Flows with Inlet Swirl and Inlet Distortion Effects," *AIAA Journal*, Vol. 21, No. 8, 1983, pp. 1127–1133.

[8] Heinrich, J. C., and Zienkiewicz, O. C., "Quadratic Finite Element Schemes for Two-Dimensional Convective-Transport Problems," *International Journal of Numerical Methods in Engineering*, Vol. 11, 1977, pp. 1831–1844.
[9] Hood, P., "Frontal Solution Program for Unsymmetric Matrices," *International of Journal of Numerical Methods in Engineering*, Vol. 10, No. 2, 1976, pp. 379–399.
[10] Hood, P., and Taylor, C., "Navier-Stokes Equations Using Mixed Interpolation," in *Proceedings of the International Symposium on Finite Element Methods in Flow Problems*, University of Wales, Swansea, United Kingdom, 1974.
[11] Hughes, T. J. R., "A simple Scheme for Developing Upwind Finite Elements," *International Journal of Numerical Methods in Engineering*, Vol. 12, 1978, pp. 1359–1365.
[12] Japikse, D., and Pampreen, R., "Annular Diffuser Performance for an Automotive Gas Turbine," *Journal of Engineering for Power*, Vol. 101, 1979, pp. 358–372.
[13] Johnston, J. P., and Dean, R. C., "Losses in Vaneless Diffusers of Centrifugal Compressors and Pumps," *Journal of Engineering for Power*, Vol. 99, 1966, pp. 49–62.
[14] Lai, Y. G., and So, R. M. C., "Calculation of Planar Conical Diffuser Flows," *AIAA Journal*, Vol. 27, No. 5, 1989, pp. 542–548.
[15] Sovran, G., and Klomp, E. D., "Experimentally Determined Optimum Geometries for Rectilinear Diffusers with Rectangular, Conical, or Annular Cross Section," in *Fluid Mechanics of Internal Flow*, Elsevier, Amsterdam, 1967.
[16] Stevens, S. J., and Markland, E., "The Effect of the Inlet Conditions on the Performance of Two Annular Diffusers," *ASME Paper No. 68-WA/FE-38*, 1968.
[17] Takehira, A., Tanska, M., Kawashima, T., and Hanabusa, H., "An Experimental Study of the Annular Diffuser in Axial Flow Compressors and Turbines," in *Proceedings of the Tokyo Joint Gas Turbine Congress*, 1977, pp. 319–328.
[18] Zienkiewicz, O. C., *The Finite Element Method in Engineering Science*, McGraw-Hill, New York, 1971.

14 Extracting the Finite Element Domain from a Larger Flow System

Introduction

Problems in real life do not regularly come in the form of a "given" computational domain in which to solve the flow-governing equations. The fact of the matter is that we are given a large flow domain, and we are interested in finding the flow behavior over just a subregion of it. In separating our subdomain, we should be careful to add the effect of the remainder of the bigger system to our subdomain of interest. Perhaps one of the most illustrative examples of such a situation is the seal segment in the secondary (or leakage) flow passage in the the pump stage shown in Figure 14.1. Note that our focus is on this secondary passage and, in particular, the seal part of it.

Leakage flow in the shroud-to-housing gap of centrifugal pumps has significant performance and rotor-integrity consequences. First, it is the leakage flow rate, as determined by the through-flow velocity component, that is typically a major source of the stage losses. The swirl velocity component, on the other hand, is perhaps the single most predominant destabilizing contributor to the impeller rotor-dynamic behavior [3]. Control of the through-flow velocity in the clearance gap is often achieved through use of a tight-clearance seal. Suppression of the flow swirl, however, requires a careful design of the leakage passage and/or the use of such devices as the so-called swirl brakes (e.g., [7, 13]) or straightening grooves/ribs in the inner housing surface (e.g., [12]). Unfortunately, an efficient leakage-control device, such as the labyrinth seal, may itself trigger the instability problem of fluid-induced vibration [8]. In the current study, two seal configurations, comprising part of the shroud-to-housing passage in a centrifugal pump, are analyzed for comparison. In both cases, the impeller geometry and the pump operating conditions are identical.

The current computational tool is an expanded version of the finite element model that was previously proposed by Baskharone and Hensel [1] in which the mere idea of including primary-flow segments in the computational domain definition was introduced. This was a means of avoiding the need for what would otherwise be unrealistic boundary conditions at the prinary-secondary flow interface. The model, initially based on a laminar-flow assumption, has since been upgraded by recognizing such aspects as the turbulence closure (Figure 14.2) and inertia domination of the flow field.

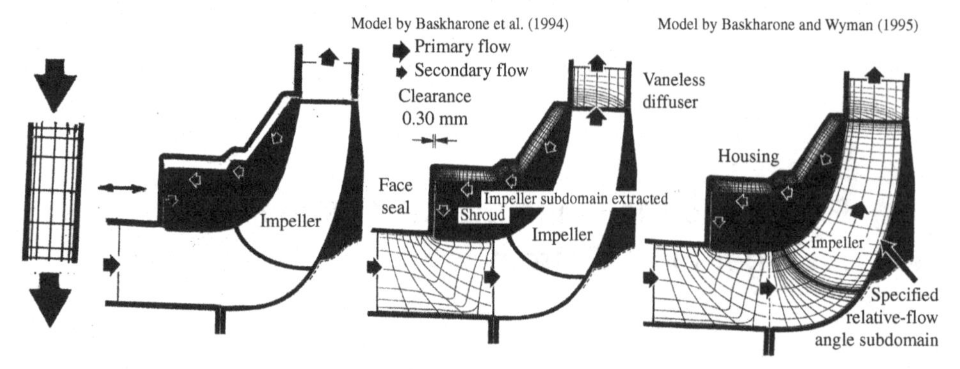

Figure 14.1. Three options for defining the seal-containing secondary passage.

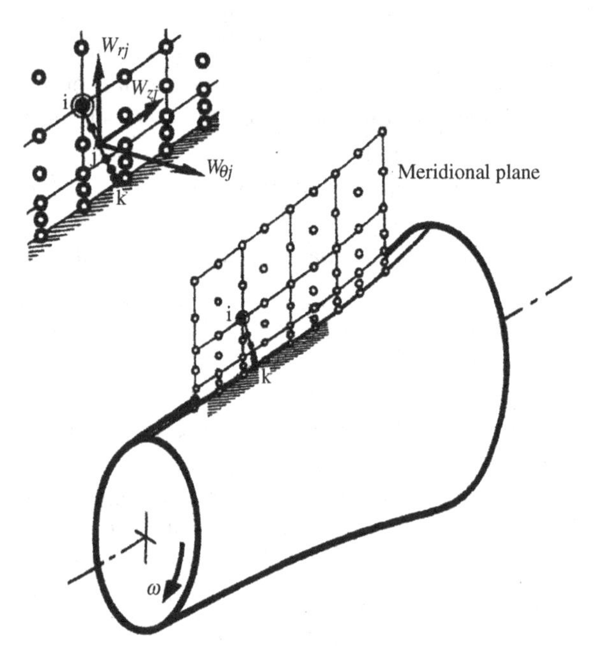

Figure 14.2. Near-wall treatment in calculating the eddy-viscosity coefficient.

The outcome of the current study is tightly linked to the rotordynamic stability analysis of shrouded pump impellers. This is true in the sense that the numerical results can essentially be used as the "zeroth-order" flow field in an existing perturbation model [2] for computing the stiffness, damping, and inertia coefficient of the fluid/shroud interaction forces as the impeller axis undergoes a "whirling" motion around the housing centerline, which is the natural, most common excitation of the pump impeller. Accuracy of these rotordynamic coefficients was reported by these two authors to be a strong function of the centered-impeller flow solution, which is under investigation here.

Analysis

Selection Options of the Computational Domain

Figure 14.1 shows three alternatives for selecting the solution domain, with the first being just the isolated seal (called a *face seal* in this configuration) [5], the leakage passage combined with two segments of the primary flow domain, and, lastly, the combined primary/secondary flow passages. Selection of the first configuration brings nothing but trouble in the inlet and exit boundary conditions. To explain, the seal inlet and exit stations connect the seal to two primary flow subdomains featuring irregular and recirculatory flow behavior on the primary-flow side. With this in mind, it is perhaps impossible to attain any sensible boundary conditions for our isolated seal, our major subregion of interest. The third configuration in Figure 14.1 embraces the impeller region itself. The finite element model, in this case, would be large in terms of the added degrees of freedom. A simplification that has to do with including the impeller in this case was devised and, in the end, implemented. We will focus our attention on the middle configuration in this figure. This includes two primary-flow segments, which renders the domain to a double-entry, double-departure flow region. Also shown in Figure 14.1 are the major features of the finite element model in which a biquadratic curve-sided element (Figure 14.3) is used as the discretization unit.

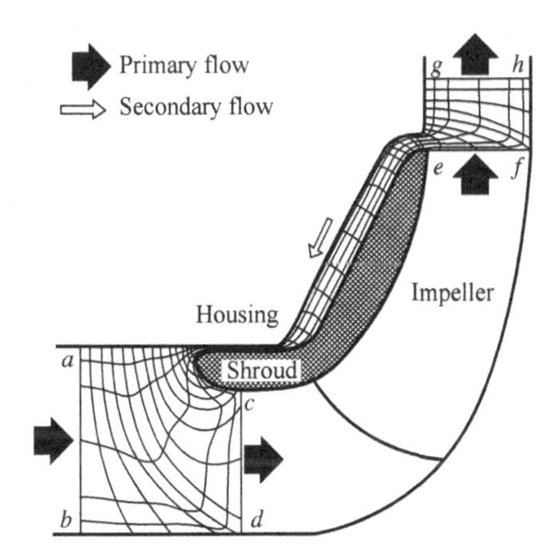

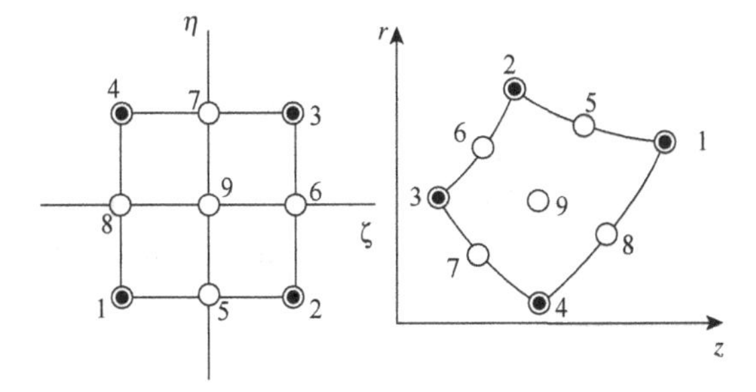

Figure 14.3. Discretization-model comparison with the clearance thickness shown.

Definition of the computational domain in the manner just described is hardly traditional. Inclusion of two primary segments in this domain is aimed at facilitating, to a reasonable level of accuracy, the primary/secondary flow interaction effects at both ends of the secondary passage. Existing computational models, by comparison, treat the secondary passage as totally isolated from the primary-flow passage segments [6, 9]. As seen in Figure 14.3 (which shows a wear-ring seal configuration instead), inclusion of the primary-flow passages (labeled *abdc* and *efhg*) clearly alleviates the need to specify what would otherwise be grossly simplified, actually inaccurate boundary conditions in the two primary/secondary-flow interaction locations at both ends of the secondary-flow passage.

It might appear, at first, that the current flow problem is solvable using one of the existing commercial flow programs. However, the very nature of the computational domain as a multiple-departure flow region and some of the corresponding boundary conditions (discussed later in this chapter) are too difficult for such programs.

Flow-Governing Equations

The momentum and mass-conservation laws governing the swirling axisymmetric flow of the incompressible flow in Figure 14.1 (the middle-domain configuration) can be expressed as follows:

$$V_r\frac{\partial V_r}{\partial r}+V_z\frac{\partial V_r}{\partial z}-\frac{{v_\theta}^2}{r}=-\frac{1}{rho}\frac{\partial p}{\partial r}+\nabla.(\nu_{eff}\nabla V_r)+\frac{\partial \nu_t}{\partial r}\frac{\partial V_r}{\partial r}-\frac{\nu_{eff}}{r^2}V_r+\frac{\partial \nu_t}{\partial z}\frac{\partial V_z}{\partial r} \tag{14.1}$$

$$V_r\frac{\partial V_\theta}{\partial r}+V_z\frac{\partial V_\theta}{\partial z}+\frac{V_rV_\theta}{r}=\nabla.(\nu_{eff}\nabla V_\theta)-\left[\frac{1}{r}\frac{\partial \nu_t}{\partial r}+\frac{\nu_{eff}}{r^2}\right]V_\theta \tag{14.2}$$

$$V_r\frac{\partial V_z}{\partial r}+V_z\frac{\partial V_z}{\partial z}=-\frac{1}{\rho}\frac{\partial p}{\partial z}+\nabla.(\nu_{eff}\nabla V_z)+\frac{\partial \nu_t}{\partial z}\frac{\partial V_z}{\partial z}+\frac{\partial \nu_t}{\partial r}\frac{\partial V_r}{\partial z} \tag{14.3}$$

$$\frac{\partial V_r}{\partial r}+\frac{\partial V_z}{\partial z}+\frac{V_r}{r}=0 \tag{14.4}$$

where

- V_r, V_θ, and V_z are the r, θ, and z velocity components.
- p is the static pressure.
- ρ is the fluid density.
- ν_t and ν_{eff} are the eddy and effective kinematic (laminar + eddy) viscosity coefficients, respectively.

The variables ν_t and ν_{eff} are obtained from the algebraic vorticity-based turbulence closure explained in Chapter 13 for the case of an annular exhaust diffuser. The difference, nevertheless, is a result of the impeller rotation ω, in which the velocity we are now speaking of is the relative velocity $\vec{W}$ (see Figure 14.2), which is related to the absolute velocity $\vec{V}$ through the well-known kinematical relationship:

$$\vec{W}=\vec{V}-\omega r\vec{e}_\theta$$

where ω is the impeller speed in radians per second.

Boundary Conditions

Referring to the flow-permeable boundary segments in Figure 14.1, the boundary conditions are as follows:

Stage Inlet Station

This is the boundary segment (*a-b*) in Figure 14.3, which is located "sufficiently" far upstream from the impeller. Fully developed flow is assumed at this location, giving rise to the following boundary condition:

$$\frac{\partial V_r}{\partial z} = \frac{\partial V_\theta}{\partial z} = \frac{\partial V_z}{\partial z} = 0$$

In addition, the stage-inlet static pressure is, arbitrarily, specified at the node midway between the endwalls on this station.

Impeller Inlet and Exit Stations

These are labeled (*c-d*) and (*e-f*) in Figure 14.3. Fixed profiles of the velocity components corresponding to the stage operating conditions are imposed over these boundary segments. Note that the the operating conditions here involve the primary (impeller) passage and have nothing to do with the secondary-flow-passage mass-flow rate or flow behavior.

Stage Exit Station

The flow behavior at this station (designated *g-h* in Figure 14.3) is viewed as predominantly confined to satisfying the mass and angular momentum conservation equations in a global sense. In their derivative form, these can be expressed as follows:

$$\frac{\partial V_r}{\partial r} = -\frac{V_r}{r}$$

$$\frac{\partial V_\theta}{\partial r} = -\frac{V_\theta}{r}$$

These two boundary conditions are linear and are, therefore, introduced non-iteratively in the numerical solution process. Moreover, a zero-normal (or radial) derivative of V_z is imposed over this station, and the stage exit static pressure is specified at the computational node midway between the endwalls on this station.

Solid Boundary Segments

As for the solid boundary segments in Figure 14.3, namely, those of the housing and shroud, as well as the hub surface segments *b-d* and *f-h*, the no-slip boundary condition applies as follows:

$$V_r = 0$$

$$V_z = 0$$

$$V_\theta = C$$

where C is equal to Ωr and zero for rotating and nonrotating boundary segments, respectively.

Finite Element Formulation

A special version of the Petrov-Galerkin weighted residual method is used to derive the finite element version of the flow-governing equations. This problem formulation and the turbulence closure were both discussed in Chapter 13 and will not be repeated here. What we will focus on is the resulting set of the finite element equations, which, by reference to Equations (14.1) through (14.4), are

$$\int_{A^{(e)}}\int\left\{\bar{\nu}_{eff}\left(\frac{\partial N_i}{\partial r}\frac{\partial N_j}{\partial r}+\frac{\partial N_i}{\partial z}\frac{\partial N_j}{\partial z}\right)+W_i\left(\bar{V}_r\frac{\partial N_j}{\partial r}+\bar{V}_z\frac{\partial N_j}{\partial z}\right)\right.$$
$$\left.-N_i\frac{\partial\bar{\nu}_t}{\partial r}\frac{\partial N_j}{\partial r}+\frac{\bar{\nu}_{eff}}{r^2}N_iN_j\right\}r\,dA\Bigg]V_{r,j}-\left[\int_{A^{(e)}}\int\frac{\bar{V}_\theta}{r}N_iN_jr\,dA\right]V_{\theta,j}$$
$$-\left[\int_{A^{(e)}}N_i\frac{\bar{\nu}_t}{\partial z}\frac{\partial N_j}{\partial r}r\,dA\right]V_{z,j}+\left[\int_{A^{(e)}}\frac{1}{\rho}N_i\frac{\partial M_k}{\partial r}r\,dA\right]p_k=\oint_{L^{(e)}}\bar{\nu}_{eff}N_i(\vec{n}.\nabla V_r)\,dL \tag{14.5}$$

$$\left[\int_{A^{(e)}}\int\left\{\bar{\nu}_{eff}\left(\frac{\partial N_i}{\partial r}\frac{\partial N_j}{\partial r}\frac{\partial N_i}{\partial z}\frac{\partial N_j}{\partial z}\right)+W_i(\bar{V}_r\frac{\partial N_j}{\partial r}+\bar{V}_z\frac{\partial N_j}{\partial z}\right.\right.$$
$$\left.\left.+\left(\frac{1}{r}\frac{\partial\bar{\nu}_t}{\partial r}+\frac{\bar{\nu}_{eff}}{r^2}+\frac{\bar{V}_r}{r}\right)N_iN_j\right\}r\,dA\right]V_{\theta,j}=\oint_{L^{(e)}}\nu_{eff}N_i(\vec{n}.\nabla V_\theta)dL \tag{14.6}$$

$$-\left[\int_{A^{(e)}}N_i\frac{\partial\bar{\nu}_t}{\partial r}\frac{\partial N_j}{\partial z}r\,dA\right]V_{r,j}+\left[\int_{A^{(e)}}\int\left\{\bar{\nu}_{eff}\left(\frac{\partial N_i}{\partial r}\frac{\partial N_j}{\partial r}+\frac{\partial N_i}{\partial z}\frac{\partial N_j}{\partial z}\right)\right.\right.$$
$$\left.\left.+W_i\left(\bar{V}_r\frac{\partial N_j}{\partial r}+\bar{V}_z\frac{\partial N_j}{\partial z}\right)+N_i\frac{\partial\bar{\nu}_t}{\partial z}\frac{\partial N_j}{\partial z}\right\}rdA\right]V_{z,j}+\left[\int_{A^{(e)}}\int\frac{1}{\rho}N_i\frac{\partial M_k}{\partial r}rdA\right]p_k$$
$$=\oint_{L^{(e)}}\nu_{eff}rN_i(\vec{n}.\nabla V_z)\,dL \tag{14.7}$$

$$\left[\int_{A^{(e)}}\left(M_k\frac{\partial N_j}{\partial r}+\frac{1}{r}M_kN_j\right)r\,dA\right]V_{r,j}+\left[\int_{A^{(e)}}\int M_k\frac{\partial N_j}{\partial z}r\,dA\right]V_{z,j}=0 \tag{14.8}$$

Again, symbols and superscripts in the preceding set of equations are defined the same as they were in Chapter 13.

Numerical Results

Two secondary-flow passage configurations corresponding to the same impeller geometry and operating conditions were chosen for comparison. These are shown in Figure 14.4 and feature a conventional wear ring and a face seal as part of the secondary-flow passage. The pump, which was the focus of rotordynamic study testing by Sulzer Bros. Corp. [4] has the following characteristics and design-point operating conditions:

- Impeller tip radius = 17.5 cm.
- Impeller speed = 2,000 rpm.
- Working medium is water at 30°C.

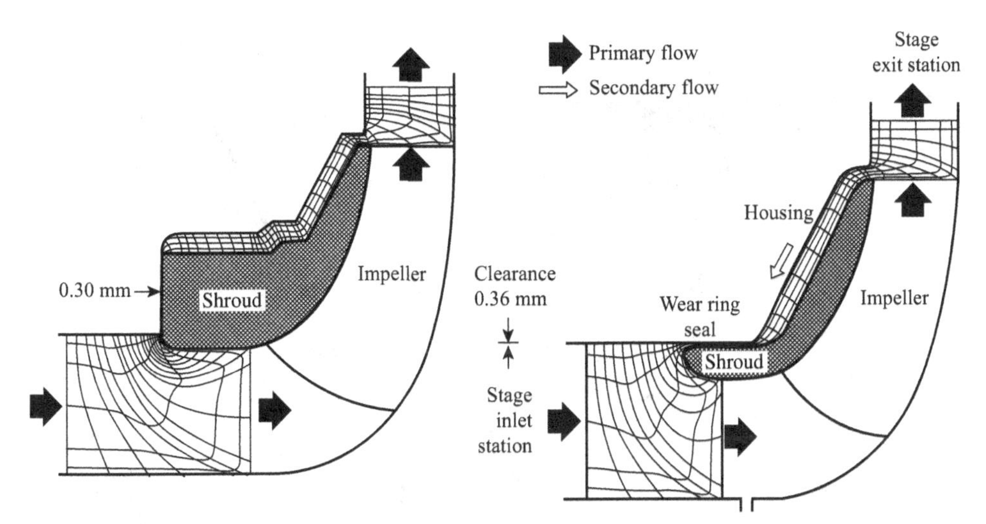

Figure 14.4. Comparison of the discretization models of the seal configurations.

- Volumetric flow rate = 130 L/s.
- Total head = 68.0 m.
- Reynolds number (based on the tip radius and speed) is 8.02×10^6.

In creating the finite element models in Figure 14.4, a total of thirteen computational nodes were placed on each cross-flow grid line in the seal region. These nodes were closely spaced near the walls in anticipation of large velocity gradients there. In an early numerical experimentation phase, this finite element grid was proven to provide good flow resolution and rule out any significant grid dependency of the computed flow field.

Figure 14.5 shows a plot of the computed meridional velocity component for the conventional wear-ring seal configuration. This component, together with the corresponding swirl velocity at the secondary-passage inlet station (Figure 14.6) constitute the desired set of parameters that are of interest to us, parameters that correspond to this seal configuration. Examination of Figure 14.5 reveals that the shroud-to-housing flow is experiencing a pronounced recirculatory motion in the secondary passage segment leading to the wear-ring seal. This is a result of the tendency of the fluid particles adjacent to the shroud to migrate radially outward due to the centrifugal force caused by the shroud rotation, on the one hand, and the tendency of those particles near the (stationary) housing to proceed radially inward as a result of the static pressure differential across the passage, on the other, with the highest pressure magnitude obviously being at the secondary-passage inlet station.

Plots of the meridional velocity component for the face-seal configuration and the secondary-passage inlet and exit swirl velocity component are shown in Figures 14.7 through 14.9, respectively. Again, the meridional velocity vectors in Figure 14.7 indicate a strong recirculatory motion in virtually all segments of the secondary passage. Reasoning of this flow behavior was discussed earlier in the wear-ring seal configuration case. However, a unique and perhaps peculiar flow structure is seen to exist in the horizontal segment of the leakage passage leading to the face seal in Figure 14.7, where radial shifting of the flow trajectories and vortex

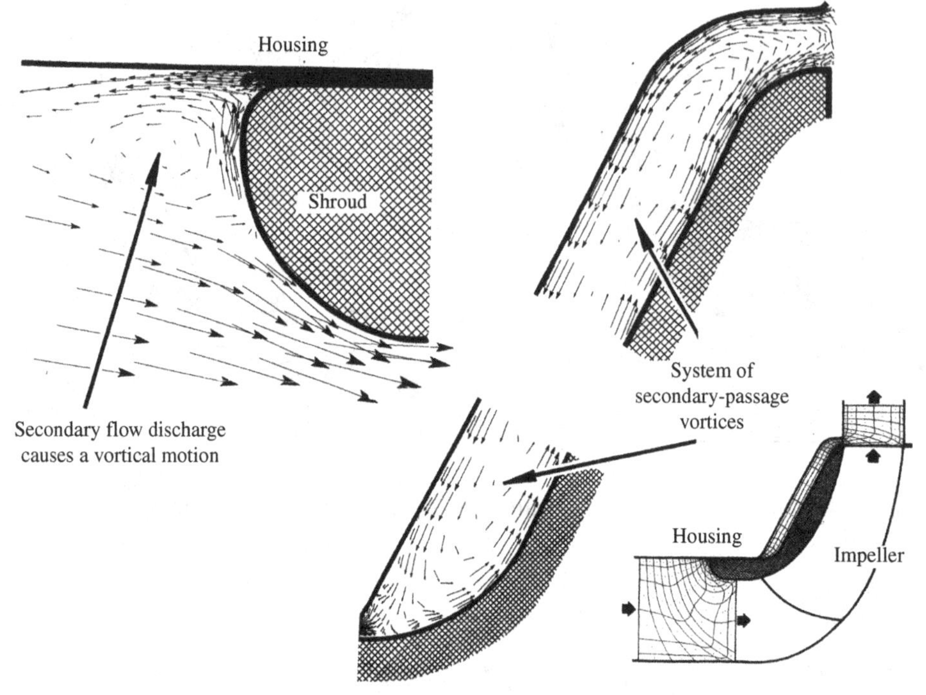

Figure 14.5. Plot of the meridional velocity in key segments for the wear-ring seal.

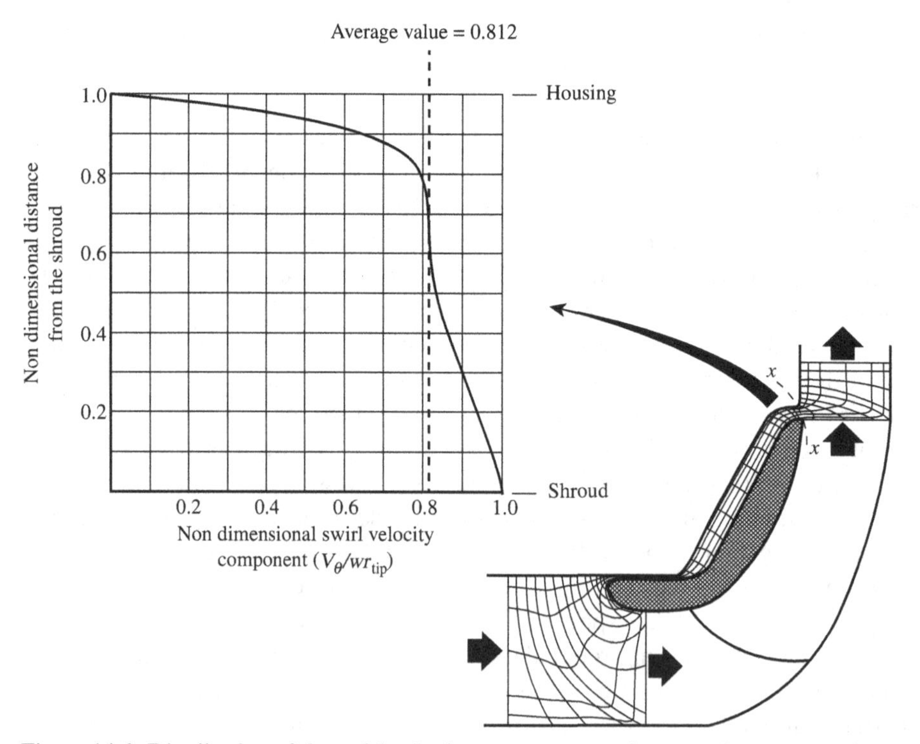

Figure 14.6. Distribution of the swirl-velocity component at the secondary-passage inlet.

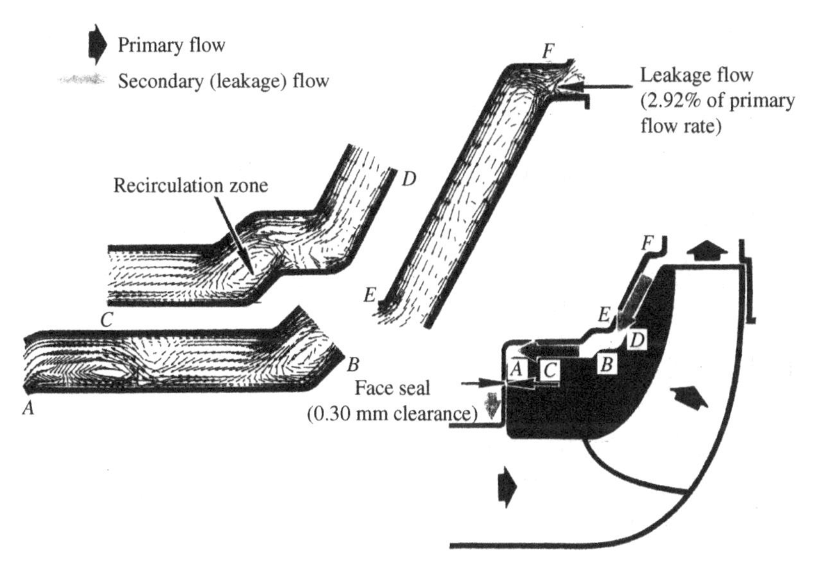

Figure 14.7. Meridional-velocity-vector plot for the face-seal pump configuration.

breakdown take place. Checking the computed pressure distribution throughout the secondary passage, it is apparent, however, that the magnitude of the pressure gradient in this low-radius region exceeds that of the centrifugal force. This seems to weaken the flow recirculation in this leakage-passage segment, confining it to the shroud side of the leakage passage.

There is a rather modest amount of experimental data to validate the computed flow field in Figures 14.7 through 14.9 [11]. First, it was indicated by Sulzer Bros. Corporation that the average swirl velocity component at the leakage-passage inlet station was measured for the face-seal pump configuration under the above-mentioned operating conditions to be approximately 0.5 of the impeller tip speed [6]. With the current finite element model, the average value of this velocity component was computed to be 0.526 of the tip speed. Note that the shroud-to-housing swirl velocity profile at the secondary-passage inlet station that gave rise to the average value is not specified a priori but is rather part of the finite element flow solution. This advantage, as indicated earlier, is a result of including primary-flow segments in the computational domain definition (see Figure 14.4). Further experimental observations that are consistent with the computed flow field in this study concern the recirculatory pattern of the meridional flow (see Figure 14.8) and were qualitatively reported by Guelich et al. [10].

Performance assessment of the two secondary-flow passages (see Figure 14.4), in light of the numerical results, involves their effectiveness as leakage suppresants and swirl dissipators. In order to determine the leakage control capacity of each passage, the mass flux was integrated at the passage inlet station. The results indicated that the leakage-flow rates in the wear-ring and face-seal pump configurations were 0.0011 and 0.0038 m^3/s, respectively. These represent 0.85 and 2.92 percent of the primary mass-flow rate, respectively, and illustrate the relative superiority of the wear-ring seal configuration as a leakage-control device over the face-seal alternative. This, in part, is due to the highly favorable streamwise

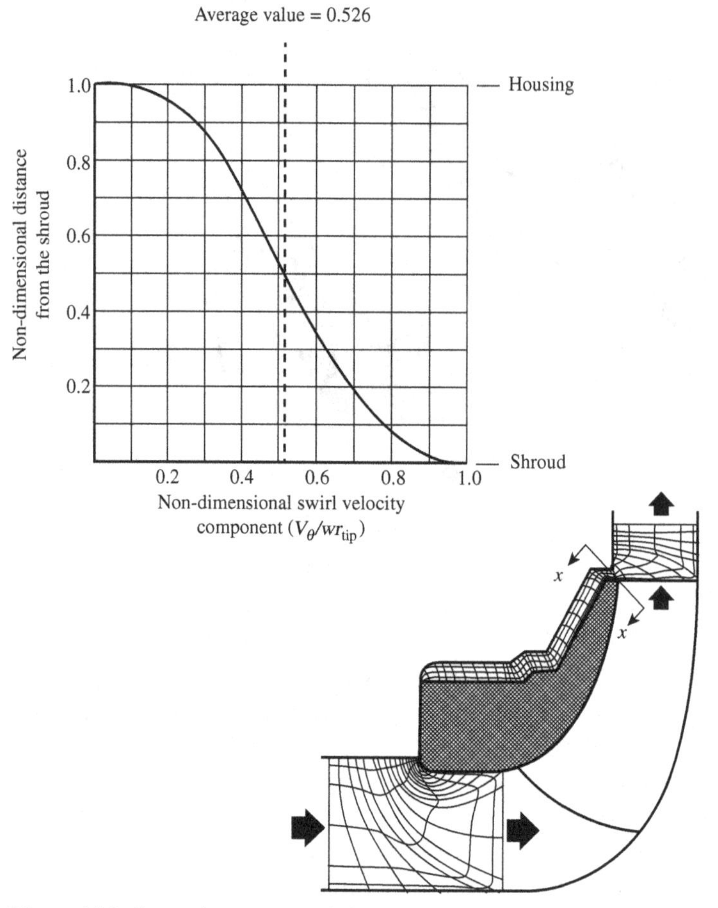

Figure 14.8. Secondary-passage inlet swirl-velocity-profile.

pressure gradient across the face seal as a result of the substantial radius decline between the seal inlet and exit stations, which, in turn, lessens the boundary-layer buildup over the solid walls and eliminates the likelihood of any flow separation. On the other hand, the tendency of the fluid particles to migrate radially outward near the constant-radius shroud surface, in the case of the wear-ring seal added to the complex flow structure in the passage leading to the seal and created an environment for an early flow separation over the shroud surface. Despite the rapid flow reattachment, in this case, the recirculation region following the separation point has the effect of enhancing the seal frictional resistance and therefore the sealing effectiveness of the wear-ring seal configuration by comparison. Another contributing factor behind the wear-ring seal effectiveness as a leakage-control device is the existence of a substantial recirculation zone at the seal exit station (see Figure 14.5) with practically no comparable seal-exit flow behavior in the case of the face-seal configuration (see Figure 14.7). As for the swirl velocity dissipation across the secondary passage, examination of Figures 14.6 and 14.8 reveals that the "kink" in the secondary passage of the face-seal pump configuration is causing a sudden and measurable reduction in the swirl velocity in the axial-passage segment with no equivalent swirl velocity reduction mechanism

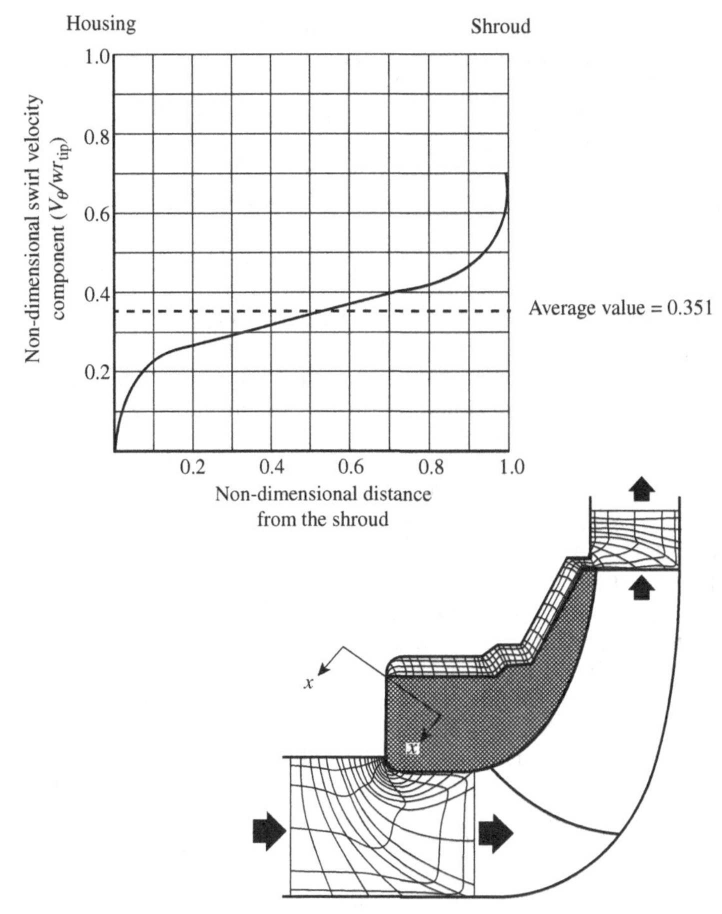

Figure 14.9. Swirl-velocity profile at a station just upstream of the face seal.

in the wear-ring seal pump configuration. More importantly, the average swirl velocity component at the secondary-passage inlet station (designated *X*-*X* in Figure 14.4) was found to be as low as 0.526 of the impeller tip speed for the face-seal configuration as opposed to 0.812 for its wear-ring counterpart. Worth noting is the fact that the choice of station *X*-*X* in Figure 14.8 is consistent with that of Childs [6] as a meaningful parameter in interpreting the rotordynamic behavior of the same pump configurations in Figure 14.8. Although these swirl characteristics of the face-seal configuration are not quantitatively convertible into a rotordynamic stability-related factor, it is established [7] that swirl suppression at the secondary-passage inlet station is among the most effective tools for shrouded-impeller rotordynamic stability enhancement. This, in view of the current results, would imply that the face-seal pump configuration provides a less destabilizing effect. The conclusion here is consistent with the experimental findings of Bolleter et al. [4].

REFERENCES

[1] Baskharone, E. A., and Hensel, S. J., "A New Model for Leakage Prediction in Shrouded-Impeller Turbopumps," *Journal of Fluids Engineering*, Vol. 113, No. 3, 1989, pp. 118–123.

[2] Baskharone, E. A., and Hensel, S. J., "A Finite Element Perturbation Approach to Fluid/Rotor Interaction in Turbomachinery Elements. 1: Theory," *Journal of Fluids Engineering*, Vol. 113, No. 3, 1991, pp. 353–361.
[3] Baskharone, E. A., and Hensel, S. J., "A Finite Element Perturbation Approach to Fluid/Rotor Interaction in Turbomachinery Elements. 2: Applications," *Journal of Fluids Engineering*, Vol. 113, No. 3, 1991, pp. 362–367.
[4] Bolleter, U., Leibundgut, E., and Sturchler, R., "Hydraulic Interaction and Excitation Forces of High Head Pump Impellers," presented at the Third Joint ASCE/ASME Mechanics Conference, University of California, La Jola, CA, 1989.
[5] Carey, G. F., and Oden, T. J., *Finite Elements: Fluid Mechanics*, (The Texas Finite Element Series), Vol. IV, Prentice-Hall, Englewood Cliffs, NJ, 1986.
[6] Childs, D. W., "Fluid-Structure Interaction Forces at Pump Impeller-Shroud Surfaces for Rotordynamic Calculations," *Journal of Vibration, Stress and Reliability in Design*, Vol. 111, 1989, pp. 216–225.
[7] Childs, D. W., Baskharone, E., and Ramsey,C., "Test Results for Rotordynamic Coefficients of the SSME HPOTP Turbine Interstage Seal with Two Swirl Brakes," *Journal of Tribology*, Vol. 113, No. 3, 1991.
[8] Childs, D., and Elrod, D., "Annular Honeycomb Seals: Test Results for Leakage and Rotordynamic Coefficients, Comparison to Labyrinth and Smooth Configurations," *NASA Conference Publication 3026: Rotordynamic Instability Problems in High-Performance Turbomachinery*, 1988, pp. 143–159.
[9] Daily, W., and Nece, R. E., "Chamber Dimension Effects on Induced Flow and Frictional Resistance of Enclosed Rotating Disks," *Journal of Basic Engineering*, Vol. 82, No. 1, 1980, pp. 217–231.
[10] Guilich, J., Florjanclcic, D., and Pace, S.E. "Influence of Flow between Impeller and Casing on Part-Load Performance of Centrifugal Pumps," *Rotating Machinery Dynamics, ASME Publication DE*, Vol. 2, 1989, pp. 227–235.
[11] Morrison, G. L., Johnson, M. C., and Tatterson, G. B., "3-D Laser Anemometer Measurements in an Annular Seal," *ASME Paper No. 88-GT-64*, 1988.
[12] Ohashi, H., Sakurai, A., and Nishima, J., "Influence of Impeller and Diffuser Geometries on the Lateral Fluid Forces of a Whirling Centrifugal Impeller," *NASA CP-3026*, 1988, pp. 285–306.
[13] Baskharone, E. A., "Swirl Brake Effect on the Rotordynamic Stability of a Shrouded Pump Impeller," *Journal of Turbomachinery*, Vol. 121, No. 1, January 1999.

15 Finite Element Application to Unsteady Flow Problems

Introduction

In this chapter we combine the flow-field time dependency with a fully three-dimensional solution domain. The result is a large-size computational model requiring a great deal of computer resources. In view of how involved the problem is, several CPU time-saving techniques are devised and implemented.

Example

The relative motion between the stator and rotor subdomains within an axial (Figure 15.1) or centrifugal turbomachinery stage creates an unsteady-flow field that is periodic in time. Limiting the discussion to the axial turbine stage case, the stator-cascade wake pattern around the circumference (see Figure 15.1) not only will shape the rotor flow behavior but also will expose its blades to a pattern of oscillating pressure that may very well lead to premature fatigue failure. In fact, the close proximity of the two (stator and rotor) cascades (Figure 15.2) in a predominantly subsonic flow field places the stator vanes in the same fluctuating-stress environment, but with lesser amplitude by comparison.

The small stator/rotor axial-gap length within a typical turbomachinery stage is a double-edge sword. On the one hand, the smaller the gap the less is the total pressure loss within it. This loss is a natural outcome of the boundary layer growth over the endwalls, which, together with the profile losses, constitutes a significant part of the stage losses. However, a small gap length magnifies the cyclic fluctuations within the rotor subdomain as a result of wake cutting, upstream vortex shedding, and potential flow interaction between the stationary and rotating blade rows. Contributions to stage efficiency, aerodynamic and aeroelastic stability, and the noise control all can be understood through an acceptably accurate resolution of the basic stator/rotor flow interaction problem. Unfortunately, progress toward a complete and reliable computational modeling in this area has been hampered by the complexity of the problem and what is understandably a tremendous amount of required computational resources.

Outlined in this chapter is an elliptic three-dimensional viscous-flow model of the stator/rotor interaction within an axial-flow turbomachinery stage. The model,

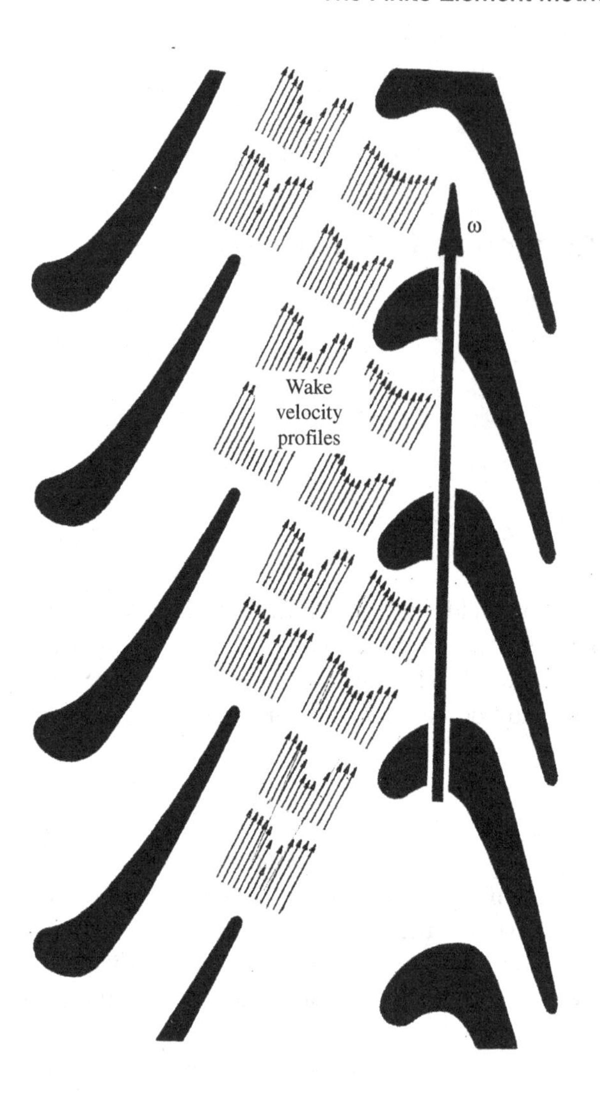

Figure 15.1. Wake pattern in the stator/rotor interface region.

however, does not accomodate the pressure-side-to-suction-side flow migration over the tip of the rotor blade. In other words, the rotor blades are necessarily of the shrouded type. In order to impart the cascade periodicity conditions, a simplifying assumption of equal stator- and rotor-blade counts is made. Rather than combining the flow fields in the stator- and rotor-blade passages, the rotor flow problem is solved independently at each time step. The uncoupled flow fields, however, are linked together through the transmission of exit conditions from the stator subdomain to that of the rotor. This method is chosen for its simplicity and in light of the established fact that the pressure fluctuations on the upstream blade row, as a result of the potential flow interaction, are significantly smaller than those caused by the flow-viscosity effects. Ellipticity of the equations of motion is enforced through retention of the streamwise diffusion terms, as opposed to the parabolized flow-solution strategy, where such terms are simply ignored. Significant as it has been, the pioneering work by Rai [2, 3] ignored the presence of these terms in a streamwise

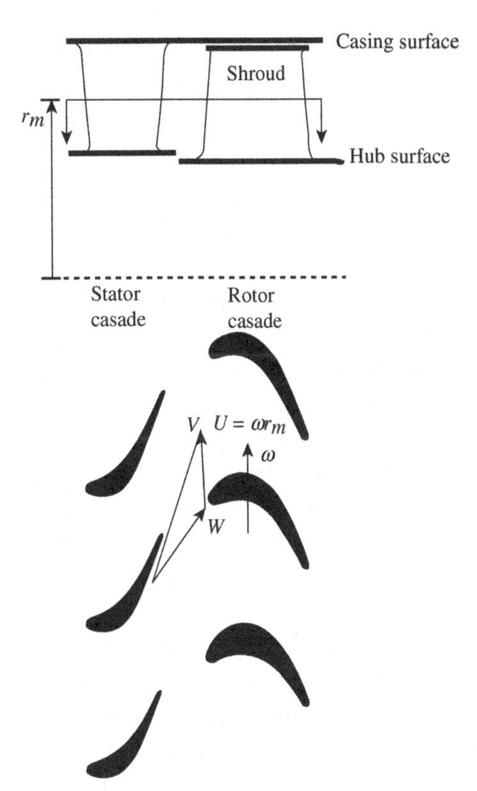

Figure 15.2. Geometrical configurations of an axial-flow turbine stage.

marching-in-time type of solution. Whereas this simplification substantially reduces the core size and CPU time consumption, it does not accommodate any flow separation or flow reversal that may be the case physically over segments of the blade suction side, whereas it eliminates the natural sensitivity of the fluid particles to downstream disturbances or abrupt geometric changes. Missing in Rai's work also were the centripetal and Coriolis acceleration components, which are functions of the rotor speed. In their study, Koya and Kotake [4] recognized these components, but their computational model was based on an entirely inviscid-flow assumption. Ignoring the impact of the viscous-flow mixing within the stator-vane wake in such a flow model is equivalent to voiding out the strongest contributor of the previously explained cyclic stresses in this problem category. Special emphasis in the current analysis is placed on the accuracy of transmitting the stator exit flow conditions to the rotor subdomain for the reason that the stator/rotor interface is typically close to or may be even within the near wake of the stator blades (see Figure 15.1), a subregion where the velocity gradients are much higher than those almost anywhere else.

Figure 15.2 shows, in part, the velocity triangle in the stator/rotor-interface gap. The absolute velocity V in this figure represents the stator exit velocity vector, whereas W stands for the rotor inlet relative velocity vector. The solid-body velocity vector of the rotor at the mean radius r_m is ωr_m, where ω is the rotor speed (in radians/second) and is totally in the tangential direction. Perhaps the most important velocity relationship this triangle spells out is

$$\vec{V} = \vec{W} + \omega r_m \vec{e}_\theta$$

This relationship will be repeatedly used in the process of stator/rotor exchange of boundary conditions. In particular, the following tangential (or θ) component of this equation has its own importance:

$$V_\theta = W_\theta + \omega r_m$$

Flow-Governing Equations

For the purpose of generality, these equations are cast for the rotor blade-to-blade hub-to-tip cascade unit. The stator flow equations can, in this case, be deduced by replacing the relative velocity components with those of the absolute velocity and dropping both the centripetal and Coriolis acceleration components. Derivation of the governing equations is conveniently done in a rotating cylindrical frame of reference, with the rotational frequency being identical to the rotor speed. In the interest of less computational resources, the stage flow is handled as incompressible, and the flow turbulence is modeled via the algebraic vorticity-based turbulence closure presented in Chapter 13, with the exception that it is now being used within a three-dimensional flow environment. Whereas the first assumption places a ceiling on the stator absolute and rotor relative Mach numbers, the choice of the simple turbulence closure is justifiable on the basis that zones of flow separation and circulation would be reasonably mild and confined to the immediate vicinity of the blade suction side, in particular, considering the overall favorable pressure gradient that is categorically associated with turbine cascades. Use of such means as one- or two-equation turbulence closures, on the other hand, would significantly increase the number of degrees of freedom to, perhaps, a practically unmanageable size of the numerical model. Note that treating the time-dependent flow field as elliptic, which is the case here, requires repetitive inversion of a substantially large matrix of "influence" coefficients that, in the absence of the previously cited simplifications, would be exceedingly costly if not practically unthinkable. To gain an appreciation of this fact, it probably suffices to state that one sweep of the rotor cascade unit past its stator counterpart consumed approximately two days of CPU time on a major IBM mainframe.

Referring to the rotor speed by ω, the conservation of mass and linear momentum equations can be expressed as follows:

Continuity Equation

$$\frac{1}{r}\frac{\partial}{\partial r}(rW_r) + \frac{1}{r}\frac{\partial W_\theta}{\partial \theta} + \frac{\partial W_z}{\partial z} = 0 \tag{15.1}$$

Radial Momentum Equation

$$\begin{aligned}
&\frac{\partial W_r}{\partial t} + W_r\frac{\partial W_r}{\partial r} + \frac{W_\theta}{r}\frac{\partial W_r}{\partial \theta} - \frac{W_\theta{}^2}{r} + W_z\frac{\partial W_r}{\partial z} - 2\omega W_\theta - \omega^2 r \\
&\quad = -\frac{\partial p}{\partial r} + 2\frac{\partial}{\partial r}\left(\nu_e\frac{\partial W_r}{\partial r}\right) + \frac{1}{r}\frac{\partial}{\partial \theta}\left[\nu_e\left(\frac{W_r}{\partial \theta} + \frac{\partial W_\theta}{\partial r} - \frac{W_\theta}{r}\right)\right] \\
&\quad + \frac{\partial}{\partial z}\left[\nu_e\left(\frac{\partial W_r}{\partial z} + \frac{\partial W_z}{\partial r}\right)\right] + \frac{2\nu_e}{r}\left(\frac{\partial W_r}{\partial r} - \frac{1}{r}\frac{\partial W_\theta}{\partial \theta} - \frac{W_r}{r}\right)
\end{aligned} \tag{15.2}$$

Tangential Momentum Equation

$$\frac{\partial W_\theta}{\partial t}+W_r\frac{\partial W_\theta}{\partial r}+\frac{W_\theta}{r}\frac{\partial W_\theta}{\partial \theta}+\frac{W_rW_\theta}{r}+W_z\frac{\partial W_\theta}{\partial z}+2\omega W_r$$
$$=-\frac{1}{\rho r}\frac{\partial p}{\partial \theta}+\frac{2}{r}\frac{\partial}{\partial \theta}\left(\frac{\nu_e}{r}\frac{\partial W_\theta}{\partial \theta}+\nu_e\frac{W_r}{r}\right)$$
$$+\frac{\partial}{\partial r}\left[\nu_e\left(\frac{1}{r}\frac{\partial W_r}{\partial \theta}+\frac{\partial W_\theta}{\partial r}-\frac{W_\theta}{r}\right)\right]+\frac{\partial}{\partial z}\left[\nu_e\left(\frac{1}{r}\frac{\partial W_z}{\partial \theta}+\frac{\partial W_\theta}{\partial z}\right)\right]$$
$$+\frac{2\nu_e}{r}\left(\frac{1}{r}\frac{W_r}{\partial \theta}+\frac{\partial W_\theta}{\partial r}-\frac{W_\theta}{r}\right) \tag{15.3}$$

Axial Momentum Equation

$$\frac{\partial W_z}{\partial t}+W_r\frac{\partial W_z}{\partial r}+\frac{W_\theta}{r}\frac{\partial W_z}{\partial \theta}+W_z\frac{\partial W_z}{\partial z}$$
$$=-\frac{\partial p}{\partial z}+2\frac{\partial}{\partial z}\left(\nu_e\frac{\partial W_z}{\partial z}\right)+\frac{1}{r}\frac{\partial}{\partial r}\left[\nu_e r\left(\frac{\partial W_r}{\partial z}+\frac{\partial W_z}{\partial r}\right)\right]$$
$$+\frac{1}{r}\frac{\partial}{\partial \theta}\left[\nu_e\left(\frac{1}{r}\frac{\partial W_z}{\partial \theta}+\frac{\partial W_\theta}{\partial z}\right)\right] \tag{15.4}$$

where W_r, W_θ, and W_z are the relative velocity components, ω is the rotor speed (in radians/second), p is the static pressure, and ν_e is the effective (laminar + eddy) viscosity coefficient. The eddy component of this parameter was presented and discussed in Chapter 13 under the heading "Turbulence Closure."

Note that the last two terms on the left-hand side of the radial momentum equation represent the Coriolis and centripetal acceleration components, respectively.

The convection (or inertia) terms on the left-hand side of Equations (15.2) through (15.4) are highly nonlinear in a typical turbomachinery stage under its normal operation modes. A converged solution, with this being the case, requires a careful iterative procedure at each individual time level during the time-marching process. For fear of numerical oscillations during the iterative procedure, a damping (underrelaxation) factor is used as a new iteration is encountered, whereby the "linearizing" array of velocity components is composed not only of the lastly obtained velocity magnitudes but also of those corresponding to the previous iteration.

Boundary Conditions

Because of the elliptic nature of the equations of motion, appropriate boundary conditions have to be specified at each and every segment of the boundary enclosing the computational domain. Because the flow field is time-dependent, a set of initial conditions also must be specified. Shown in Figure 15.2 are the different boundary segments for a typical axial-flow turbine stage. The first category in this classification is that of solid boundary segments, where the no-slip boundary condition is imposed. This is simply achieved by setting the absolute velocity components in the case of a stator passage and the relative velocity components in the case of a rotor passage to zero at all computational nodes on the solid boundary segments. Included among these segments is the blade suction and pressure surfaces, as well as the endwalls of which the rotor shroud (see Figure 15.2) is a member.

The second boundary segment is the flow inlet station. For a stator, the boundary condition is assumed to be a steady, uniform inlet velocity over the entire inlet plane, except for nodes that also exist on the hub or casing surfaces, where the no-slip boundary condition is applied instead.

The rotor inlet boundary conditions, on the other hand, are time-dependent as a result of the stator/rotor relative motion. To account for this motion, the numerical results from the stator flow solution at the stator/rotor interface plane are transmitted as inlet conditions for the rotor flow passage. In doing so, the stator/rotor relative flow motion (i.e., ωr) is first extracted from the absolute velocity vector (specifically its tangential component), giving rise to the relative-velocity-based rotor inlet flow conditions as follows:

$$W_r = V_r$$

$$W_\theta = V_\theta - \omega r$$

$$W_z = V_z$$

In transmitting the velocity components to the rotor subdomain, attention was particularly paid to the wake velocity distribution. With the rotor subdomain spinning past the stator cascade unit, there is virtually no guarantee that the rotor inlet computational nodes will coincide with the stator nodes at any given time level. In addressing this issue, a third-order polynomial was established at every time level for interpolating each absolute velocity component. The numerical outcome of this method was repeatedly checked with favorable conclusions every time.

The third boundary type is the flow exit station. In an effort to guarantee a somewhat uniform exit velocity, this station is intentionally located as far downstream as possible. At such location, mixing of the pressure- and suction-side flow streams is assumed to be complete or nearly complete, giving rise to exit-velocity profiles that are as tangentially uniform as practically possible. This, for the axial-flow stage, creates a situation where the so-called natural boundary conditions apply, meaning zero gradients of the velocity components in the direction perpendicular to the exit station. The natural boundary conditions just described are among the easiest flow constraints to impose, for they are automatically satisfied in the finite element equivalent of the governing equations, at any node on a boundary segment unless a different boundary condition is externally imposed, which is not the case here. An additional exit boundary condition concerns the static pressure, whereby an arbitrary pressure magnitude is assigned to an arbitrary exit-station node, creating a datum for referencing the pressure throughout the computational domain. In the current analysis, the same algebraic, vorticity-based turbulence closure (Figure 15.3) is again used. This may very well be the only suitable closure in light of the problem size (based on how many finite elements would be required) in such a three-dimensional flow-analysis application.

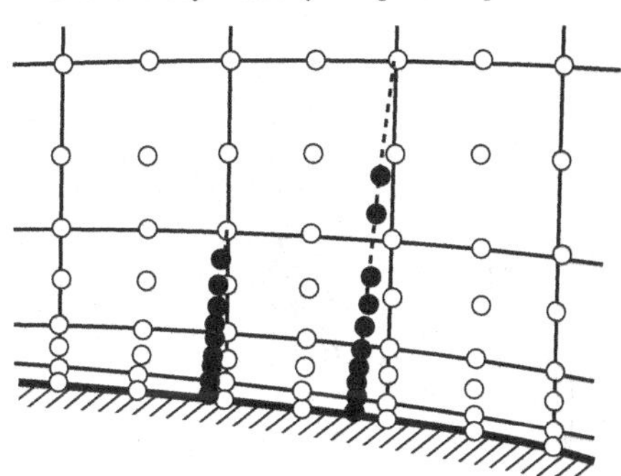

Figure 15.3. Calculation of the eddy viscosity in the near-wall subregion.

Finite Element Formulation

The current computational procedure is based on the weighted residual finite element formulation, which yields an equivalent set of linear algebraic equations in the field-variable nodal magnitudes on discretizing the computational domain. The latter process was achieved using the twenty-noded curve-sided isoparametric element in Figure 15.4 as the discretization unit. Within a typical element, a mixed-order interpolation of the velocity components and static pressure is assumed [7]. The pressure field throughout the element is interpolated using linear shape functions in terms of the nodal pressure magnitudes at the element's eight corner nodes. The velocity components, on the other hand, are interpolated using the quadratic shape functions associated with all twenty corner and midside nodes. Referring to Figure 15.4, the spatial coordinates themselves undergo a transformation from the physical to the local frame of reference. It is the quadratic shape functions that are used once again in this coordinate-mapping process.

Throughout a typical element (e), let the velocity components, pressure, and spatial coordinates be interpolated as follows:

$$W_r{}^{(e)} = \sum_{i=1}^{20} N_i(\zeta,\eta,\xi)W_{r,i}(t)$$

$$W_\theta{}^{(e)} = \sum_{i=1}^{20} N_i(\zeta,\eta,\xi)W_{\theta,i}(t)$$

$$W_z{}^{(e)} = \sum_{i=1}^{20} N_i(\zeta,\eta,\xi)W_{z,i}(t)$$

$$p^{(e)} = \sum_{k=1}^{8} M_k(\zeta,\eta,\xi)p_k(t)$$

$$r^{(e)} = \sum_{i=1}^{20} N_i(\zeta,\eta,\xi)r_i$$

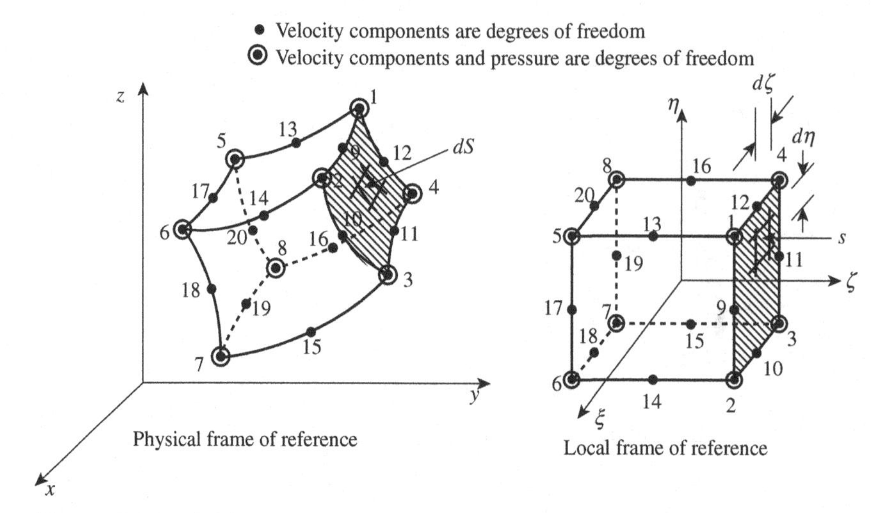

Figure 15.4. Definition of the quadrilateral 20-noded isoparametric element.

$$\theta^{(e)} = \sum_{i=1}^{20} N_i(\zeta,\eta,\xi)\theta_i$$

$$z^{(e)} = \sum_{i=1}^{20} N_i(\zeta,\eta,\xi)z_i$$

where N_i and M_k are quadratic and linear shape functions associated with the isoparametric finite element, respectively, whereas ζ, η, and ξ are coordinates in the local frame of reference (see Figure 15.4). In the preceding expressions, note that the field variables, namely, the relative-velocity components and pressure, are treated as functions of time.

Substitution of the preceding expressions in the flow-governing Equations (15.1) through (15.4) for the typical element (e) yields four different residual functions that, according to Galerkin's principle, are made orthogonal to a set of the element's shape functions throughout the entire finite element. Applying the error-consistency criterion of Hood and Taylor [8], the element's linear and quadratic shape functions are used in conjunction with the continuity and momentum equations, respectively. Applied to the residual function of each equation, the previously cited orthogonality conditions will lead to the elemental set of finite element equations that, on application of Green's theorem, can be expressed as follows:

$$\int_{-1}^{+1}\int_{-1}^{+1}\int_{-1}^{+1} A_{k,j} \mid J \mid d\zeta d\eta d\xi = 0 \tag{15.12}$$

$$\int_{-1}^{+1}\int_{-1}^{+1} [B_{i,j} + C_{k,i}] \mid J \mid d\zeta d\eta d\xi$$
$$= \int_{-1}^{+1}\int_{-1}^{+1}\int_{-1}^{+1}\int_{-1}^{+1} \mid J \mid \rho N_i r\omega^2 d\zeta d\eta d\xi + \int_{S^{(e)}} \left\{\frac{\mu_e}{Re} N_i(\vec{n}.\nabla W_r)\right\} dS \tag{15.13}$$

$$\int_{-1}^{+1}\int_{-1}^{+1}\int_{-1}^{+1}[D_{i,j}+E_{k,i}]\,|J|\;d\zeta d\eta d\xi=\int_{S^{(e)}}\left[\frac{\mu_e}{Re}N_i(\vec{n}.\nabla W_\theta)\right]\,dS \tag{15.14}$$

$$\int_{-1}^{+1}\int_{-1}^{+1}\int_{-1}^{+1}[F_{i,j}+G_{k,i}]\,|J|\;d\zeta d\eta d\xi=\int_{S^{(e)}}\left[\frac{\mu_e}{Re}N_i(\vec{n}.\nabla W_z)\right]\,dS \tag{15.15}$$

The submatrices $[A]$, $[B]$, $[C]$, $[D]$, $[E]$, $[F]$, and $[G]$ are defined as follows:

$$A_{k,j}=M_k\left[\left(r\frac{\partial N_j}{\partial r}W_{r,j}+N_jW_{r,j}\right)+\frac{\partial N_i}{\partial\theta}W_{\theta,j}+r\frac{\partial N_j}{\partial z}W_{z,j}\right.$$

$$\begin{aligned}B_{i,j}=\;&\rho rN_iN_j\frac{\partial W_{r,j}}{\partial t}+\rho r\bar{W}_{r,j}N_i\frac{\partial N_i}{\partial r}W_{r,j}+\rho\bar{W}_{\theta,j}N_i\frac{\partial N_i}{\partial\theta}W_{r,j}-\rho\bar{W}_{\theta,j}N_iN_jW_{\theta,j}\\&+\rho r\bar{W}_{z,j}N_i\frac{\partial N_j}{\partial z}W_{r,j}-2\rho r\omega N_iN_jW_{\theta,j}+\frac{\mu_e r}{Re}\frac{\partial N_i}{\partial r}\frac{\partial N_j}{\partial\theta}W_{r,j}+\frac{\mu_e}{rRe}\frac{\partial N_i}{\partial\theta}\frac{\partial N_j}{\partial\theta}W_{r,j}\\&+\frac{\mu_e r}{Re}\frac{\partial N_i}{\partial z}\frac{\partial N_j}{\partial z}W_{r,j}+\frac{\mu_e}{rRe}N_iN_jW_{r,j}+\frac{2\mu_e}{rRe}N_i\frac{\partial N_j}{\partial\theta}W_{\theta,j}\end{aligned}$$

$$C_{k,i}=rM_k\frac{\partial N_i}{\partial r}p_k$$

$$\begin{aligned}D_{i,j}=\;&\rho rN_iN_j\frac{\partial W_{\theta,j}}{\partial t}+\rho r\bar{W}_{r,j}N_i\frac{\partial N_j}{\partial r}W_{\theta,j}+\rho\bar{W}_{\theta,j}N_i\frac{\partial N_j}{\partial\theta}W_{\theta,j}+\rho\bar{W}_{r,j}N_iN_jW_{\theta,j}\\&+\rho r\bar{W}_{z,j}\frac{\partial N_i}{\partial z}W_{\theta,j}+2\rho\omega N_iN_jW_{r,j}+\frac{\mu_e r}{Re}\frac{\partial N_i}{\partial r}\frac{\partial N_j}{\partial r}W_{\theta,j}+\frac{\mu_e}{rRe}\frac{\partial N_i}{\partial\theta}\\&+\frac{\partial N_j}{\partial\theta}W_{\theta,j}+\frac{\mu_e r}{Re}\frac{\partial N_i}{\partial z}\frac{\partial N_j}{\partial z}W_{\theta,j}+\frac{\mu_e}{rRe}N_iN_jW_{\theta,j}+\frac{2\mu_e}{rRe}N_i\frac{\partial N_j}{\partial\theta}W_{r,j}\end{aligned}$$

$$E_{k,i}=M_k\frac{\partial N_i}{\partial\theta}p_k$$

$$\begin{aligned}F_{i,j}=\;&\rho rN_iN_j\frac{\partial W_{z,j}}{\partial t}+\rho r\bar{W}_{r,j}N_i\frac{\partial N_j}{\partial r}W_{z,j}+\rho\bar{W}_{\theta,j}N_i\frac{\partial N_j}{\partial\theta}W_{z,j}+\rho r\bar{W}_{z,j}N_i\frac{\partial N_j}{\partial z}W_{z,j}\\&+\frac{\mu_e r}{Re}\frac{\partial N_i}{\partial r}\frac{\partial N_j}{\partial r}W_{z,j}+\frac{\mu_e}{rRe}\frac{\partial N_i}{\partial\theta}\frac{\partial N_j}{\partial\theta}W_{z,j}+\frac{\mu_e r}{Re}\frac{\partial N_i}{\partial z}\frac{\partial N_j}{\partial z}W_{z,j}\end{aligned}$$

$$G_{k,i}=rM_k\frac{\partial N_i}{\partial z}p_k$$

where $i=1,2,\ldots,20$; $j=1,2,\ldots,20$; and $k=1,2,\ldots,8$, with $\vec{n}$ being the local outward unit vector normal to the element boundary, $|J|$ the physical-to-local coordinate transformation Jacobian, and the overbar identifying a velocity component that is obtained from a previous iteration or an initial guess for the purpose of linearizing the convection (or inertia) terms in momentum Equations (15.2) through (15.4). With the Reynolds number (Re) being based on the molecular viscosity coefficient (μ_l), the variable μ_e is defined as follows:

$$\mu_e=\frac{\mu_{eff}}{\mu_l}$$

with μ_{eff} being the result of multiplying the effective kinematic viscosity ν_{eff} (from the turbulence closure explained earlier) times the fluid density (ρ).

Note that the centripetal acceleration term [on the right-hand side of the r-momentum Equation (15.13)] is treated as a sourcelike term, giving its contribution only to the free terms in this equation, just as the surface-integral boundary-produced term does. The scalar product in the surface integral term can be expressed as follows:

$$\vec{n} \cdot \nabla \phi = \frac{\partial \phi}{\partial n}$$

where ϕ stands for W_r, W_θ, or W_z in the momentum Equations (15.13) through (15.15). With this in mind, the surface integrals appearing on the right-hand side of Equations (15.13), (15.14), and (15.15) can, respectively, be expressed as follows:

$$I_r = \int_{\xi=-1}^{+1} \int_{\eta=-1}^{+1} \frac{\mu_e}{Re} N_i \frac{\partial W_r}{\partial n} G(\eta,\xi)\, d\eta d\xi \tag{15.13a}$$

$$I_\theta = \int_{\xi=-1}^{+1} \int_{\eta=-1}^{+1} \frac{\mu_e}{Re} N_i \frac{\partial W_\theta}{\partial n} G(\eta,\xi)\, d\eta d\xi \tag{15.14a}$$

$$I_z = \int_{\xi=-1}^{+1} \int_{\eta=-1}^{+1} \frac{\mu_e}{Re} N_i \frac{\partial W_z}{\partial n} G(\eta,\xi)\, d\eta d\xi \tag{15.15a}$$

where the function $G(\eta,\xi)$ is defined as follows:

$$G(\eta,\xi) = \left\{ \left[\left(\frac{\partial y}{\partial \eta}\frac{\partial x}{\partial \xi} - \frac{\partial y}{\partial \xi}\frac{\partial x}{\partial \eta} \right)^2 + \left(\frac{\partial x}{\partial \xi}\frac{\partial z}{\partial \eta} - \frac{\partial x}{\partial \eta}\frac{\partial z}{\partial \xi} \right)^2 + \left(\frac{\partial x}{\partial \eta}\frac{\partial y}{\partial \xi} - \frac{\partial x}{\partial \xi}\frac{\partial y}{\partial \eta} \right)^2 \right]^{\frac{1}{2}} \right\}_{\zeta=1}$$

In assembling the finite element equations, note that each twenty-noded element will yield a total of sixty-eight equations in sixty-eight unknowns, namely, r-, θ- and z-velocity components per each of the twenty corner and midside nodes and one pressure magnitude per each of the eight corner node.

Time-Integration Algorithm

There exists a wide variety of techniques to integrate the governing equations in time, each with its own stability criterion and accuracy. In their assembled form, the final system of finite element equations can be written as follows:

$$[C]\{\Phi\} + [K]\{\phi\} = \{R\} \tag{15.16}$$

where $[C]$ and $[K]$ are matrices of "influence" coefficients for a given iteration, $\{\phi\}$ is the "global" vector of unknowns, $\{\Phi\}$ is the vector of time derivatives, and $\{R\}$ is the "load" vector. In general, the time-integration formulation [7] renders Equation (15.16) to

$$\left((\Theta)[K] + \frac{1}{\Delta t}[C] \right) \{\Phi\}_{n+1} = \left(\frac{1}{\Delta t}[C] - (1-\Theta)[K] \right) \{\Phi\}_n + (1-\Theta)\{R\}_n + \Theta\{R\}_{n+1} \tag{15.17}$$

where

- $\Theta = 0.0$ yields an explicit algorithm.
- $\Theta = 0.50$ yields the Crank-Nicolson algorithm.
- $\Theta = 0.66$ yields the Galerkin's algorithm.
- $\Theta = 1.00$ yields a fully implicit algorithm.

The subscripts n and $n+1$ signify values that correspond to the previous and current time levels, respectively. Implicit methods result in a fully nonlinear system of equations at each time level. An iterative solution is needed in this case to guarantee that the nonlinear terms are converged before proceeding to the next time level. Although implicit time integration is computationally more tedious per time step compared with explicit techniques, the capability to proceed with comparatively larger time steps with implicit techniques can offset the increase in computational time per step and therefore is preferable. The solution, however, may, become meaningless if too large a time step is taken owing to the large truncation error in this case.

Maintaining the same finite element formulation, an initial numerical experimentation study was separately conducted in the hope of finding the optimal magnitude of Θ in this problem category. It was concluded that for magnitudes of Θ which are less than unity, the pressure values continued to fluctuate despite a well-behaved velocity field. In fact, these fluctuations were serious enough to eventually cause instability of the solution procedure. In reasoning this phenomenon, it was felt that this was a result of the weak linkage of pressure to the time dependency of the flow field because no time derivative of the pressure appears in the momentum equations. For this reason, the fully implicit time-integration approach was naturally the choice in this study. It was necessary, meanwhile, to secure a converged solution at any given time level before incrementing to the next time level. This was achieved by continually updating the linearized convection terms in the momentum equations until the sum of all nondimensionalized errors in the nodal velocity magnitudes reach a realistically small tolerance value (set to be 3 percent), at which point the next time level was to be considered.

Numerical Procedure

The volume integrals in the finite element equations [(15.12) through (15.15)] are evaluated using Richardson's extrapolation quadrature [9]. This quadrature was previously tested with a favorable outcome by Baskharone and Hamed [10]. As for the final set of finite element equations, the matrix of influence coefficients were significantly large, considering the domain's three-dimensionality, together with the high order of the discretization unit. Moreover, this matrix is unavoidably unsymmetric due, in part, to the implementation of the cascade periodicity conditions. In the matrix inversion process, the technique developed by Gupta and Tanji [11] was used in the same manner as explained in Chapter 14.

Computational Results

The results presented in this section were obtained for the turbine-stage geometry shown in Figure 15.2. This is the same stage used by Rai in his analysis [3, 12], and is a modified version of the stage previously used by Dring et al. [13] in their experimental study. While maintaining the same rotor-cascade solidity ratio, Rai changed the rotor-cascade pitch (mean-radius blade-to-blade spacing) to match that of the stator cascade. As a result, the rotor-blade axial chord and the cascade spacing (or pitch) were both increased by 25 percent. These geometric changes were required to

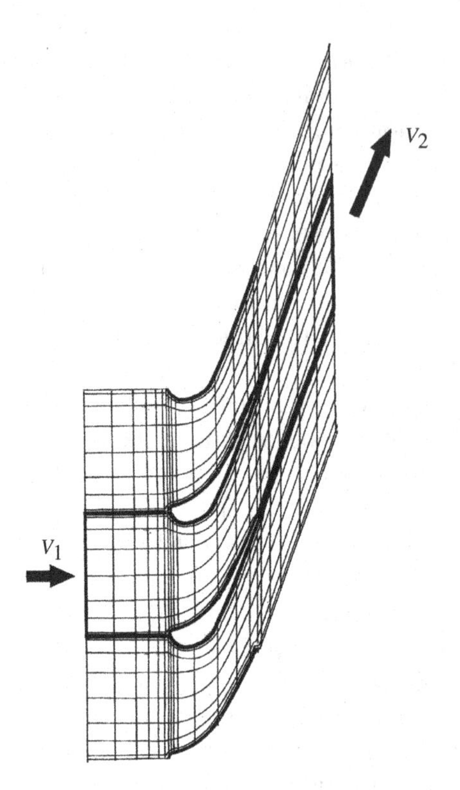

Figure 15.5. Midspan section of the stator-blade cascade with its discretization model.

enable implementation of the rotor-cascade periodicity conditions discussed earlier in this chapter. Although these changes affect the time dependency of the rotor flow field, preserving the rotor-cascade solidity ratio (axial-chord/cascade-pitch ratio) as is the case here has the effect of minimizing any discrepancy in the overall stage performance.

Figures 15.5 and 15.6 show the finite element discretization model for the mean-radius section of both the stator and rotor cascades, respectively. The rotor-cascade flow-inlet section in the latter figure is located at the position where the stator/rotor-interface surface is deemed to exist. Note that this location is really arbitrary as long as the interface surface is somewhere between the stator-vane trailing edge and the rotor-blade leading edge. The only difference would be in the location (in the stator subdomain) where the flow variables are passed to the rotor as inlet conditions. It is preferable, however, to deviate (at least slightly) from the stator trailing-edge surface. Of course, one would prefer this location to be further downstream, away from the stator-vane wake region, for obvious reasons. However, it is the rather small stator/rotor axial-gap length that has the final say in this matter.

Figure 15.7 shows a velocity-vector plot at the stage midspan section, as obtained from the stator-domain flow solution, with the wake-related velocity deficit clearly visible. In preparing the rotor inlet-velocity profile at different time levels (with the rotor blades continuing to change their tangential positions), we are at liberty to repeat the stator exit-velocity profile in the circumferential direction as many time as needed to complete the rotor inlet-velocity profiles. The repetition process here is nothing but a direct interpretation of the cascade-periodicity phrase.

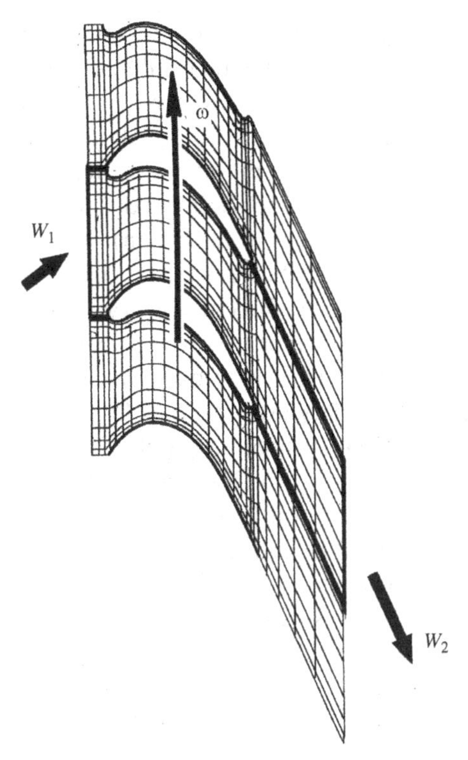

Figure 15.6. Midspan section of the rotor-blade cascade with its discretization model.

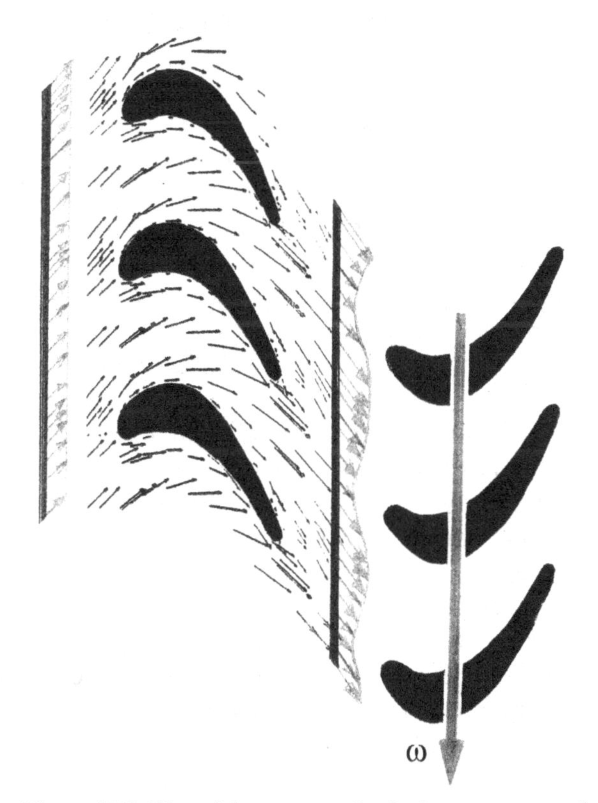

Figure 15.7. Plot of the computed velocity vector over the stator midspan section.

Figure 15.8 shows the static pressure distribution over the stator-vane surface at the hub, midspan, and casing sections, along with the experimental results of Dring et al. as well as Rai's analytical results. Examination of this figure reveals that all three sections exhibit similar aerodynamic loading characteristics, particularly over the pressure side, where a significant region of low momentum follows the leading edge, extending to almost 50 percent of the axial chord. The suction side, on the other hand, experiences a notable amount of diffusion (flow deceleration) over its rear segment, that of the casing section, occupies of the casing section, occupies roughly 66 percent of the axial chord. This feature has the effect of aggravating the boundary layer buildup and would be responsible for local or massive flow separation and recirculation. In this particular case, preliminary symptoms of such *lossy* flow behavior were apparent in examining the boundary-layer profiles but never actually materialized. It is important to point out that such an erratic flow behavior would have been detected by the current model but not by Rai's model for the reason that the latter is based on a parabolized flow pattern. To ensure monotonic conversion of the nonlinear convection terms in the flow equations of motion, an underrelaxation factor λ was used, meaning that the magnitude of any velocity component V at any node i in the current iterative step (n) is expressed in terms of the same but at the two previous steps $(n-1)$ and $(n-2)$ as follows:

$$V_i^{(n)} = \lambda V_i^{(n-1)} + (1-\lambda)V_i^{(n-2)}$$

A near-optimal λ magnitude of 0.7 was determined in the preliminary numerical experimentation study and was used subsequently in executing the iterative procedure.

The initial guess used in the rotor-flow solution process was the steady-state solution associated with a fixed set of inlet conditions taken from the stator solution at the stator/rotor-interface surface. Four sweeps of the inlet boundary conditions were performed thereafter. The first sweep consisted of ten time steps with three intermediate iterations at each time level to update the nonlinear convection terms in the flow equations of motion. The second sweep consisted of ten time steps with five intermediate iterations at each time level. The last two sweeps were performed with fifteen time steps and five intermediate iterations at each time level. In the process of updating the convection terms, the same underrelaxation factor was used for the intermediate iterations.

Figure 15.9 shows the time-averaged rotor-blade surface-pressure distribution corresponding to the final blade sweep at the hub, mean, and tip sections. These are compared with the experimental results of Dring et al. and the numerical results of Rai. The plots display similar loading characteristics to those of the stator vane, including a nearly constant pressure region over the front segment of the pressure side, followed by a rapid-acceleration segment down to the trailing edge. Agreement of the current results with the experimental data is clearly good, except for the blade-tip section, which is understandable because the rotor tips were assumed to be totally shrouded [14], which is not the case in the actual hardware.

In assessing the computed aerodynamic loading of the tip section, it is important to realize that the "indirect" (pressure-to-suction-side) leakage over the blade tip in an unshrouded rotor would, in part, tend to elevate the suction-side static pressure

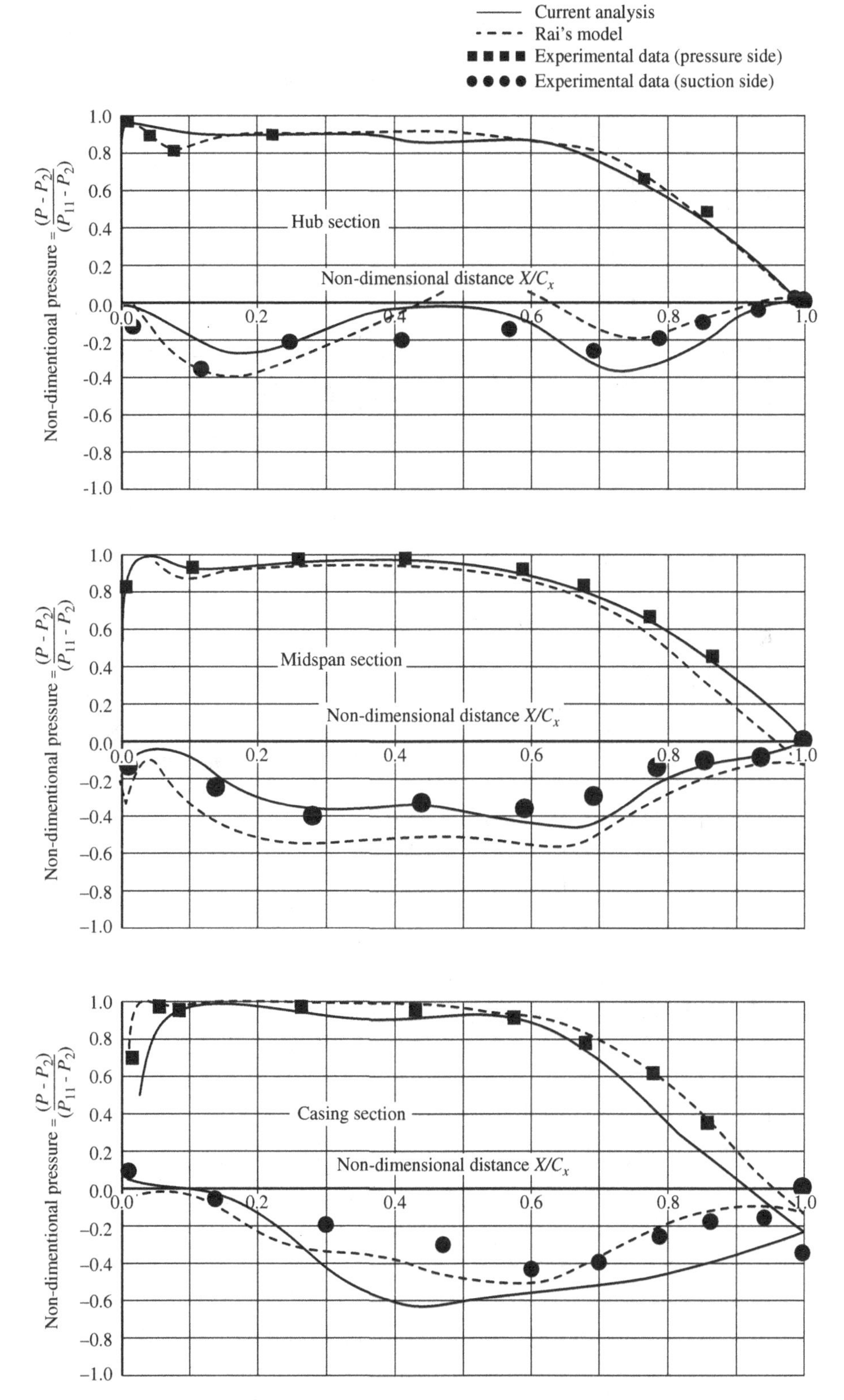

Figure 15.8. Aerodynamic loading of the stator blade at three radial locations.

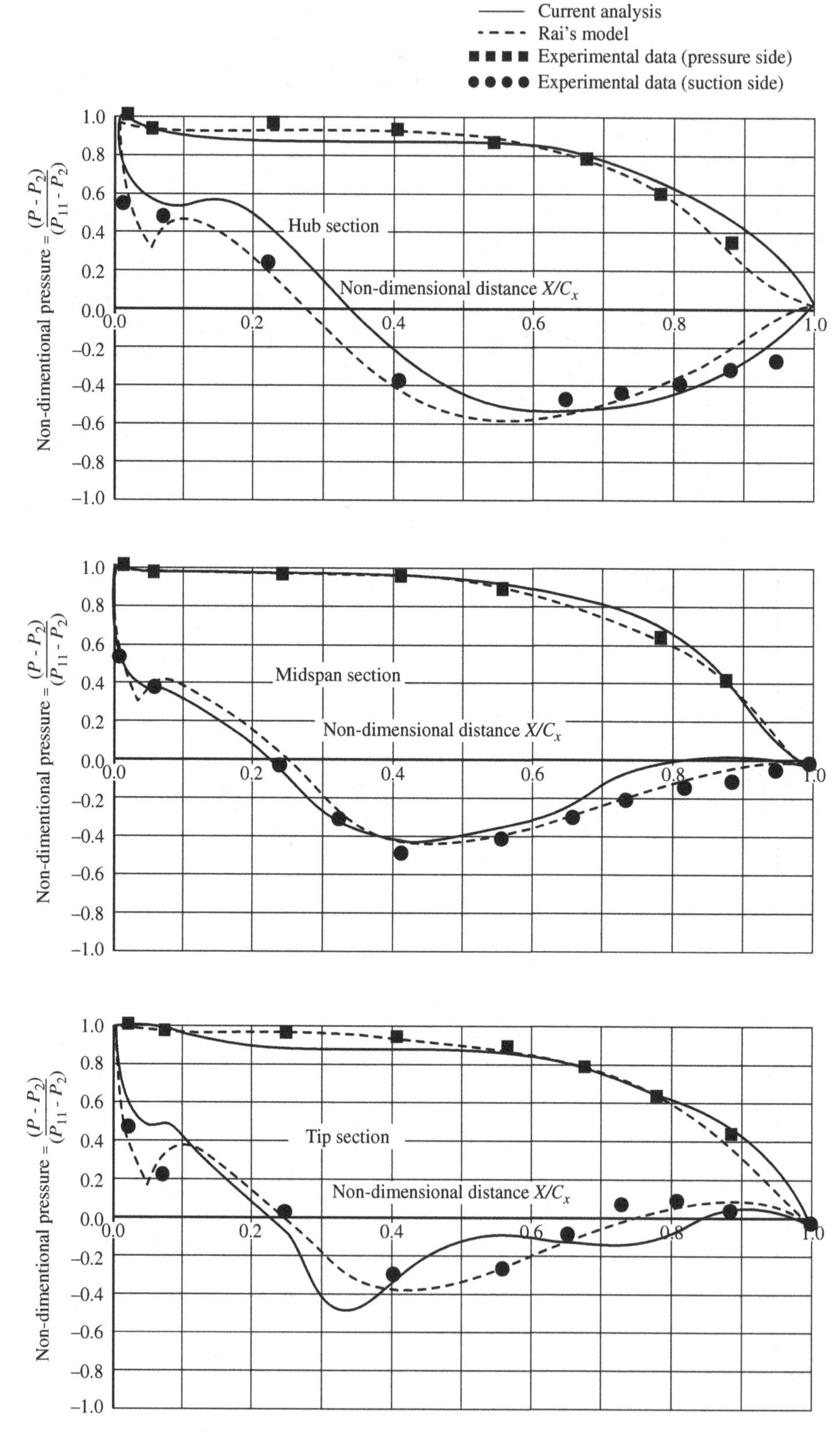

Figure 15.9. Time-averaged aerodynamic loading of the rotor blade.

because it performs, in effect, a balancing act between the pressure- and suction-side static pressures. Recalling that the rotor blades in this study were assumed to be shrouded (see Figure 15.2), such an effect naturally would be nonexistent by comparison, where the pressure magnitudes over the two sides of the blade are notably closer in the experimental results. Also note that the direct (leading-to-trailing-edge) flow leakage within the tip clearance would give rise to a smaller magnitude of circulation (or lift force) at the tip section when compared with the situation where such clearance is ignored altogether. The tip-section aerodynamic loading in Figure 15.9 confirms this fact because the enclosed area between the computed pressure curves, corresponding to the tip section, is greater than what the experimental measurements seem to suggest. The reader is reminded that rigorous, or even adequate, inclusion of the tip-clearance gap in the computational domain would require a substantial element refinement not only within this gap but also in the outer region leading to it. Realizing the complexity in terms of numerical size and tremendous computational resources of the current elliptic-flow model, such an upgrade may indeed be impractical.

Shown in Figure 15.10 is the distribution of the rotor-blade surface-pressure fluctuation amplitude for the hub midspan and tip sections. The pressure-fluctuation amplitude $\bar{p}$, in this figure is defined as the local difference between the maximum and minimum pressure magnitudes during the final blade sweep divided by the overall maximum of the local pressure difference during the same sweep, with the latter occurring at or close to the leading edge, as seen in the figure. The pressure fluctuation in Figure 15.10 seems to decrease sharply within a narrow strip of the blade just downstream from the leading edge. The fluctuation amplitude continues to decrease away from the leading edge, but at a notably slower rate in comparison with that over the suction side. This is consistent with the fact that little acceleration or deceleration of the flow occurs in this region. As the flow is accelerated over the pressure-side rear segment, the pressure fluctuation again decreases. The reason is that the viscosity effects become much less pronounced under the highly favorable (negative) pressure gradient over this segment.

The numerical results of Rai [12] and the experimental findings of Dring et al. [13] both confirm the rapid dissipation of the pressure-fluctuation as the flow accelerates around the leading edge. Both sets of results also show a subsequent increase in the pressure fluctuation amplitude (although the experimental results indicate a much narrower region where this occurs) over the forward portion of the suction surface following the abrupt decrease in the immediate vicinity of the leading edge. One possible physical explanation for such an increase in amplitude has to do with the buildup of high-turbulence-wake fluid on the suction surface as a result of what is commonly termed the *negative-jet* effect. This would also explain why this same phenomenon does not occur on the blade pressure surface. It is believed that the field resolution, in terms of the element intensity in this region, was sufficient to capture such a fine feature of flow behavior.

The experimental results for the pressure surface in Figure 15.10 cover only the front half of the axial chord. These results show that the pressure-fluctuation amplitude practically levels off over the front segment of this surface. This is generally in agreement with the current results over the same segment, with the exception that the computed amplitudes are generally higher by comparison. The numerical

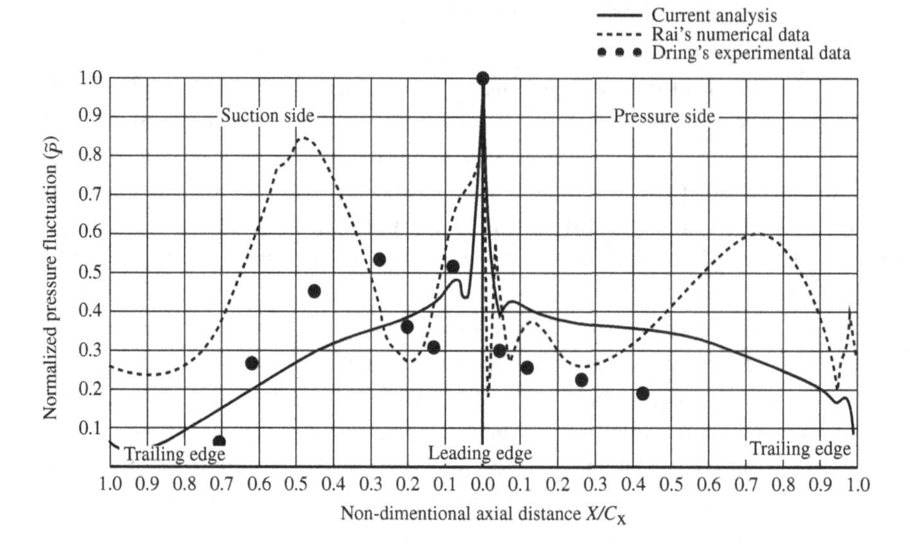

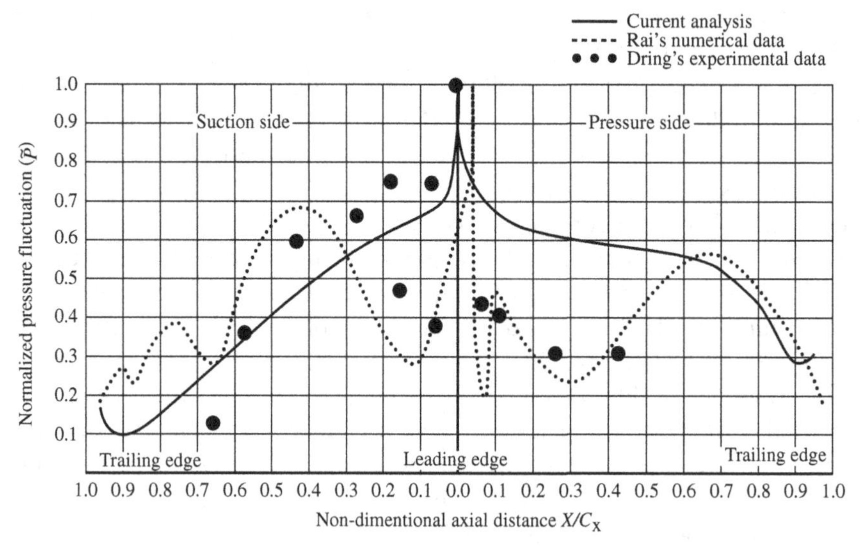

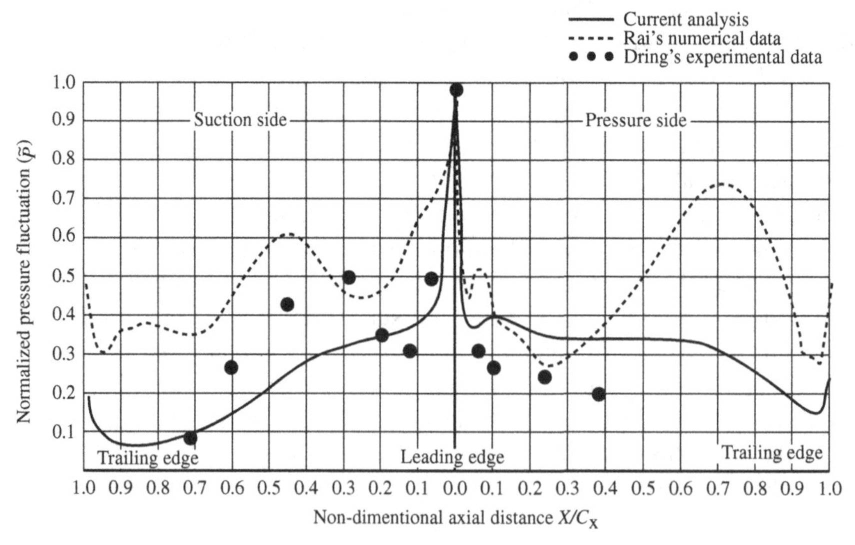

Figure 15.10. (*a*) Rotor-blade pressure-fluctuation comparison with experimental data (hub). (*b*) Rotor-blade pressure-fluctuation comparison with experimental data (midspan). (*c*) Rotor-blade pressure-fluctuation comparison with experimental data (casing).

results of Rai, however, are characterized by amplitude irregularities within the same region.

Proposed Analysis Upgrades

Bidirectional Transfer of Boundary Conditions

In laying out the computational model thus far, the simplification was introduced that the rotor blades are spinning in a stator exit subregion, details of which are obtained in a preprocessor-like solution of the time-independent stator flow. In reality, however, both the stator and rotor subdomains do "speak" to one another in the sense of exchanging boundary conditions. For instance, the stator flow behavior will very much be influenced by the presence of the rotor blades and their rotational speed, an obvious fact considering the close proximity of the two cascades to each other. Equally significant is the influence of the upstream stator vanes on the flow behavior within the rotor subdomain because the latter proceeds from one tangential location to another. In analytical terms, the ellipticity of the entire stage flow, being totally or predominantly subsonic, enables the fluid particles within the stator subdomain to "sense" the presence of the rotor blades and make the necessary trajectory adjustments well before physically reaching the blades. It is the same ellipticity feature that makes it possible for a rotor-subdomain fluid particle to track its trajectory back upstream to its origin in the stator subdomain, allowing this to influence its own aerodynamic behavior. Note that the rotor-subdomain flow field will be subsonic should the relative critical Mach number be less than unity everywhere in this subdomain regardless of the magnitudes of the absolute critical Mach number, where

$$M_{cr} = \frac{V}{\sqrt{\left(\frac{2\gamma}{\gamma+1}\right) RT_t}}$$

and

$$M_{cr,r} = \frac{W}{\sqrt{\left(\frac{2\gamma}{\gamma+1}\right) RT_{t,r}}}$$

where V and W are the local magnitudes of the absolute and relative velocities, respectively, γ is the specific-heat ratio, and R is the gas constant.

Whereas the flow adiabaticity fixes the value of the total temperature T_t throughout the stator computational domain, the total relative temperature $T_{t,r}$ is a function of both the absolute and relative velocities within the rotor subdomain as follows:

$$T_{t,r} = T_t + \frac{\gamma-1}{2\gamma R}\left(W^2 - V^2\right)$$

Note that $T_{t,r}$ will remain constant across the rotor if it is of the axial-flow type (see Figure 15.2) and the flow is adiabatic.

Starting Point

Figure 15.11 shows the succession of events in the first complete sweep of the stator-cascade pitch starting with the initial flow problem (at $t = 0$), where the rotor-cascade unit is lined up with the stator vane-to-vane unit. As the rotor unit slides upward, the stator/rotor interface will continually grow (as shown in the figure), creating two segments **A** and **B** of the stator and rotor units, respectively, that are outside the common boundary segment. Sliding of the rotor unit past the stator (see Figure 15.11) will continue to lengthen the stator/rotor interface, with the common boundary segment correspondingly shrinking, up to the point of full-sweep completion.

In the following, we focus on the boundary conditions associated with the interface line segments in Figure 15.11, which belong to the middle stator and rotor units. It should be emphasized that the interface-line segments, not simultaneously common to the stator- and rotor-cascade units (in the first stator-unit sweep), will always carry boundary conditions that are lagging in time.

Two-Way Stator/Rotor Exchange of Boundary Conditions

In this section, the first and subsequent sweeps of the stator unit by the rotor blade-to-blade passage are analyzed. In doing so, we are tracking down the sequence of events undergone by the middle stator and rotor units in Figure 15.12, which are identified by thick, tangentially bounding lines. The challenge here is in finding the appropriate boundary conditions to be specified at the stator boundary segment labeled **A2** in this figure (as an example) now that it is no longer in contact with the rotor unit. Note that there is generally an interface-line segment during this particular stator-unit sweep, which is common to both subdomains with no discontinuities in the flow variables. The exception here concerns the end of this sweep (the last column in Figure 15.12). It is at this point that a new stator-unit sweep begins to take place, with the rotor unit lined up with the stator in a manner that is geometrically identical to that at $t = 0$ in Figure 15.12. The difference, however, is that the stator/rotor-interface line begins to be one of the stator-to-rotor discontinuities in

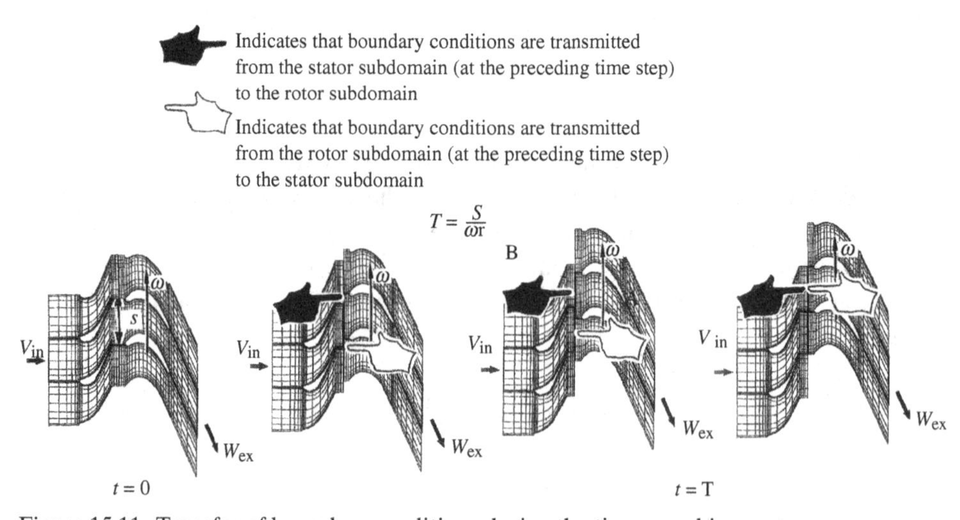

Figure 15.11. Transfer of boundary conditions during the time-marching process.

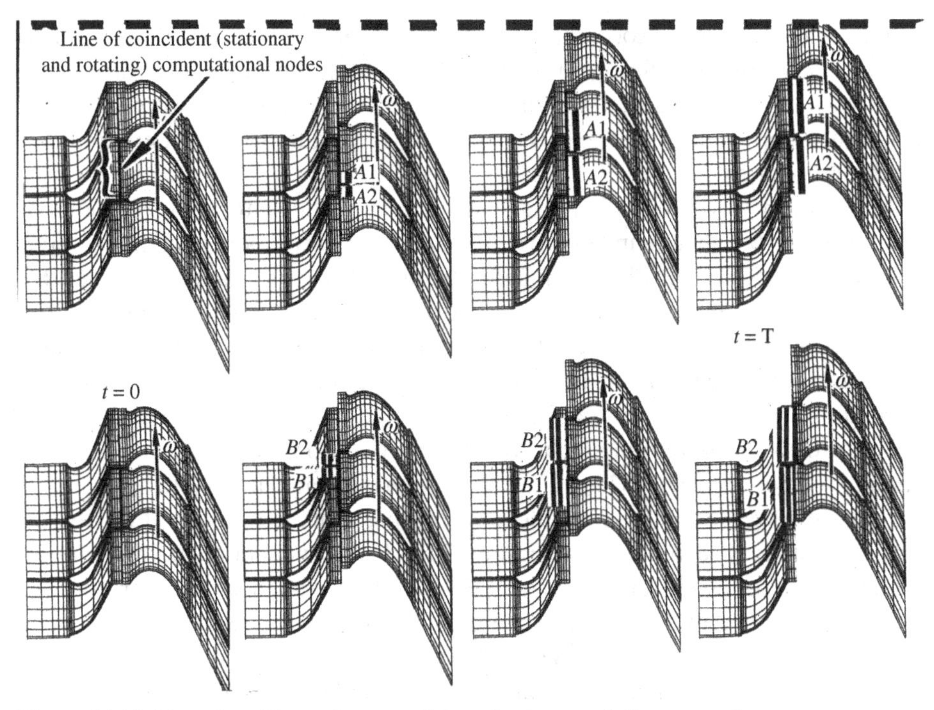

Figure 15.12. Transfer of boundary conditions during one full stator-unit sweep.

all the flow variables. In fact, convergence of the entire computational procedure is defined to occur at that point in time where the flow variables on both sides of the stator/rotor interface are compatible. The term *compatible* here is more accurate than the term *equal* in the sense that the velocity components on the left-hand side of the interface surface are defined in a stationary frame of reference, whereas those on the right are cast in rotating coordinate axes. This point was raised and resolved earlier in this chapter.

The starting point of the time-marching procedure is represented by the domain configuration at $t = 0$ in Figure 15.12, which comprises the initial condition in this time-dependent flow problem. The flow field at this point is solved in the stator and then rotor subdomains as a whole, with no discontinuities in the flow variables across the interface surface. As indicated at the top of Figure 15.12, it is helpful to use two coincident sets of computational nodes at the interface surface, one belonging to the stator and other to the rotor subdomains.

In this case, a set of additional equations will have to be added to the system, where the relationships between the stator-assigned and rotor-belonging flow variables are cast as follows:

$$W_z = V_z$$

$$W_r = V_r$$

$$W_\theta = V_\theta - \omega r$$

$$p_{rotor} = p_{stator}$$

where r is the local radial coordinate. This step eliminates the need to isolate and separately solve the flow field and relative flow field in the stator and rotor units, respectively. In fact, the same type of approach can be employed (within this first stator-unit sweep) at all interface-surface segments in Figure 15.12 that are common to both subdomains.

As the rotor blade-to-blade passage slides against the stator unit, segments of the interface surface belonging to the stator-unit boundary labeled **A2** in the upper row of Figure 15.12 and others belonging to the rotor-unit boundary labeled **B2** in the lower row will be left out of the interface segment that is common to the two cascade units being monitored. Required, therefore, are appropriate boundary conditions over these segments in order to complete the definition of the two flow subproblems in the stator and rotor units.

The process of determining the boundary conditions to be specified at the boundary segments **A2** and **B2** in Figure 15.12 is an intermediate step to be executed before progressing to the next time level. Focusing, for the purpose of clarity, on the stator boundary segment **A2** in Figure 15.12, the rotor subdomain flow solution (corresponding to the current time level) is separately examined. Of the rotor-unit flow variables (namely, the velocity components and static pressure), those associated with the segment labeled **A1** in Figure 15.12 are isolated. These are then transferred to stator boundary segment **A2** as a boundary condition in preparation for the stator-unit flow solution as the shift to the next time level is made. Noting that the rotor-unit velocity components are relative to the rotating frame of reference, and before imposing these on the stator boundary segment **A2**, the tangential velocity component along the rotor segment **A1** needs to be adjusted as follows:

$$(V_\theta)_{A2} = (W_\theta)_{A1} + \omega r$$

The rest of the flow variables are directly transmitted as follows:

$$(V_z)_{A2} = (W_z)_{A1}$$
$$(V_r)_{A2} = (W_r)_{A1}$$
$$(p)_{A2} = (p)_{A1}$$

The process of finding and imposing boundary conditions on the rotor boundary segments labeled **B2** is obtained through an approach that is similar to the previously described procedure. The difference, however, in the conversion of the absolute to the relative tangential velocity component is as follows:

$$(W_\theta)_{B2} = (V_\theta)_{B1} - \omega r$$

together with the rest of boundary-condition transfer conditions, namely,

$$(W_z)_{B2} = (V_z)_{B1}$$
$$(W_r)_{B2} = (V_r)_{B1}$$
$$(p)_{B2} = (p)_{B1}$$

Once the stator-unit sweep is completed (last column in Figure 15.12), the stator/rotor-interface surface becomes one of stator-to-rotor discontinuities in all the flow variables at the current time level. The whole process is then reexecuted for the next sweep, and so on.

The state of convergence is declared only at the end of a complete sweep of the stator-cascade unit, that is, at the point in time where the stator and rotor units are at the tangential position shown in the last column of Figure 15.12. The state of convergence is achieved should the stator and rotor variables at the interface boundary be sufficiently close to those at the end of the previous sweep (subject to the always-existing ωr difference between the stator- and rotor-side tangential velocity components).

Continuity of the Variables' Normal Derivatives through Implicit Means

The preceding three-dimensional rotor/stator unsteady-flow interaction analyses share one feature, namely, the transfer of only flow variables across the interface boundary between the two subdomains. To date, the mere idea of ensuring the continuity of the variables' normal derivatives across the interface boundary has not been part of the topic's literature.

Aside from the intuitive impression that the normal derivatives should physically suffer no discontinuities across any arbitrarily located interface surface, there are variables where such discontinuities would violate some solidly established balancing rules. Take, for instance, the static temperature T as a primary variable in the conjugate-flow/heat transfer analysis problem. Assuming the gas thermal conductivity coefficient k to be constant in the immediate vicinity of the interface surface, the temperature normal derivative would then represent the flux of heat energy, a variable that will have to be continuous across this surface from a heat-balance viewpoint.

Methodology

The strategy in this section is to achieve continuity of the velocity components' normal derivatives through implicit means. Instead of removing a finite element equation at a rotor-side node and replacing it with a specific stator-imparted magnitude, the finite element equation is left in place. In our case, the stator-obtained normal derivatives are introduced in the appropriate surface-integral part of the corresponding rotor equation. This process will affect the appropriate surface-integral term in one of three expressions (15.13a), (15.14a), or (15.15a) depending on which finite element equation (15.13), (15.14), or (15.15) is under consideration.

In the following, the point is made that once the stator flow variables are obtained at a given instant in time (during the time-marching computational procedure), the computed magnitudes of the velocity components, close to and existing on the interface surface, can be used to compute the derivatives of these components perpendicular to the interface (i.e., in the direction of the unit vector $\vec{n}$ in Figure 15.13). These normal derivatives' magnitudes then can be used to evaluate the surface-integral part of the rotor-related finite element equations of the types represented by Equations (15.13), (15.14), and (15.15). This same procedure can be reversed, with the rotor giving the stator those derivatives to be used in evaluating the stator-related boundary conditions' integrals.

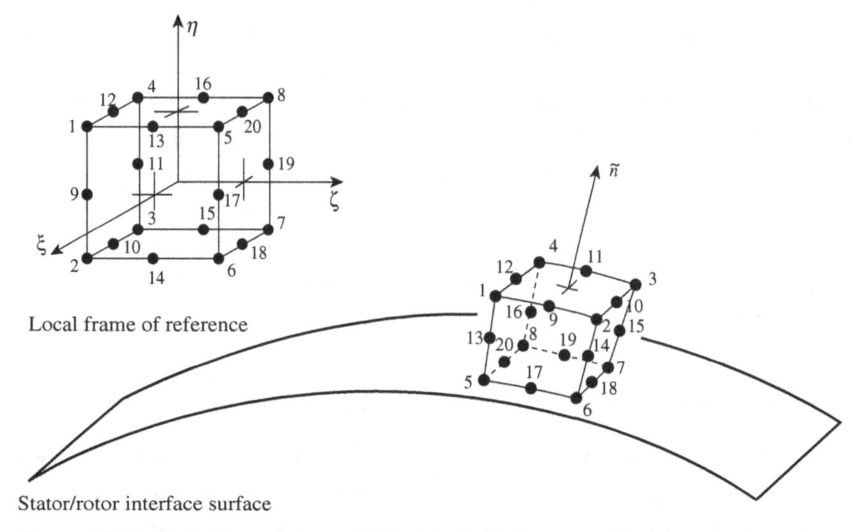

Figure 15.13. Definition of the variables' derivative perpendicular to the interface.

Analysis

Referring to Figure 15.13, let P represent any of the flow components of, for instance, velocity, and let $\vec{n}$ refer to the local unit vector that is perpendicular to the rotor/stator-interface surface. Note that Figure 15.13 shows the rotor/stator-interface surface in a radial (or centrifugal) turbomachinery stage otherwise, the interface surface would be a plane, a much easier scenario to deal with. Derivatives of the variable P in the $\vec{n}$ direction can be expressed as follows:

$$\frac{\partial P}{\partial n} = \vec{n}.\nabla P$$

The differential operator ∇ can be expressed in the cylindrical frame of reference as follows:

$$\nabla = \frac{\partial}{\partial r}\vec{e}_r + \frac{1}{r}\frac{\partial}{\partial \theta}\vec{e}_\theta + \frac{\partial}{\partial z}\vec{e}_z$$

Through simple substitution, we get

$$\frac{\partial P}{\partial n} = (\vec{n}.\vec{e}_r)\,\frac{\partial P}{\partial r} + \frac{1}{r}\,(\vec{n}.\vec{e}_\theta)\,\frac{\partial P}{\partial \theta} + (\vec{n}.\vec{e}_z)\,\frac{\partial P}{\partial z} \tag{15.18}$$

Referring to the angles between the unit vector $\vec{n}$ and the coordinate axes r, θ, and z by α_r, α_θ, and α_z, respectively, expression (15.18) can be rewritten as follows:

$$\frac{\partial P}{\partial n} = (\cos\alpha_r)\,\frac{\partial P}{\partial r} + \frac{1}{r}(\cos\alpha_\theta)\,\frac{\partial P}{\partial \theta} + (\cos\alpha_z)\,\frac{\partial P}{\partial z} \tag{15.19}$$

In the problem configuration of Figure 15.13, the normal unit vector $\vec{n}$ is totally in the radial direction, giving rise to the following simpler expression:

$$\frac{\partial P}{\partial n} \equiv \frac{\partial P}{\partial r} \tag{15.20}$$

Because the idea in this section is to equate normal derivatives of the flow variables, these derivatives should first be defined at the interface-surface computational nodes, just as the variable P itself is. Furthermore, just as P is declared, as a degree of freedom in both the rotor and stator subdomains, so must the new variable $\partial P/\partial n$. First, let us be concerned with the stator side of the interface surface, a choice that is depicted in Figure 15.13. With the intention of expressing the variable P in terms of the nodal magnitudes (recalling the difference between interpolating the velocity components and static pressure), it is probably appropriate to break up the general variable P into the three velocity components and pressure. Recalling that the velocity components are interpolated within each element using the quadratic shape functions N_i, we can express the velocity components' derivatives as follows:

$$\frac{\partial W_r}{\partial n} = \frac{\partial W_r}{\partial r} = \sum_{i=1}^{20}\left(\frac{\partial N_i}{\partial r}\right) W_{r,i} \tag{15.21}$$

$$\frac{\partial W_\theta}{\partial n} = \frac{\partial W_\theta}{\partial r} = \sum_{i=1}^{20}\left(\frac{\partial N_i}{\partial r}\right) W_{\theta,i} \tag{15.22}$$

$$\frac{\partial W_z}{\partial n} = \frac{\partial W_z}{\partial r} = \sum_{i=1}^{20}\left(\frac{\partial N_i}{\partial r}\right) W_{z,i} \tag{15.23}$$

where the shape functions N_i are defined in the element local frame of reference that is shown in Figure 15.4. The radial derivatives of the shape functions in Equations (15.21) through (15.23) should also be expressed in the local frame of reference. To this end, let us begin with the following general expression for the physical-to-local conversion of spatial derivatives:

$$\begin{Bmatrix} \dfrac{\partial}{\partial r} \\ \dfrac{\partial}{\partial \theta} \\ \dfrac{\partial}{\partial z} \end{Bmatrix} = \begin{bmatrix} \sum_{i=1}^{20} \dfrac{\partial N_i}{\partial \zeta} r_i & \sum_{i=1}^{20} \dfrac{\partial N_i}{\partial \zeta} \theta_i & \sum_{i=1}^{20} \dfrac{\partial N_i}{\partial \zeta} z_i \\ \sum_{i=1}^{20} \dfrac{\partial N_i}{\partial \eta} r_i & \sum_{i=1}^{20} \dfrac{\partial N_i}{\partial \eta} \theta_i & \sum_{i=1}^{20} \dfrac{\partial N_i}{\partial \eta} z_i \\ \sum_{i=1}^{20} \dfrac{\partial N_i}{\partial \xi} r_i & \sum_{i=1}^{20} \dfrac{\partial N_i}{\partial \xi} \theta_i & \sum_{i=1}^{20} \dfrac{\partial N_i}{\partial \xi} z_i \end{bmatrix}^{-1} \begin{Bmatrix} \dfrac{\partial}{\partial \zeta} \\ \dfrac{\partial}{\partial \eta} \\ \dfrac{\partial}{\partial \xi} \end{Bmatrix}$$

On inverting the square matrix, the following general expression for $\partial N_i/\partial r$ can be obtained:

$$\begin{aligned} \frac{\partial N_i}{\partial r} = &\left\{\left(\sum_{i=1}^{20}\frac{\partial N_i}{\partial \zeta} r_i\right)\left(\sum_{i=1}^{20}\frac{\partial N_i}{\partial \eta}\theta_i\right)\left(\sum_{i=1}^{20}\frac{\partial N_i}{\partial \xi} z_i\right) - \left(\sum_{i=1}^{20}\frac{\partial N_i}{\partial \eta} z_i\right)\left[\left(\sum_{i=1}^{20}\frac{\partial N_i}{\partial \xi}\theta_i\right)\right]\right\} \\ &\times \frac{\partial N_i}{\partial \zeta}\left\{\left(\sum_{i=1}^{20}\frac{\partial N_i}{\partial \zeta}\theta_i\right)\left[\left(\sum_{i=1}^{20}\frac{\partial N_i}{\partial \eta} r_i\right)\left[\left(\sum_{i=1}^{20}\frac{\partial N_i}{\partial \xi} z_i\right) - \left(\sum_{i=1}^{20}\frac{\partial N_i}{\partial \eta} z_i\right)\left(\sum_{i=1}^{20}\frac{\partial N_i}{\partial \xi} r_i\right)\right]\right\} \\ &\times \frac{\partial N_i}{\partial \eta}\left\{\left(\sum_{i=1}^{20}\frac{\partial N_i}{\partial \zeta} z_i\right)\left[\left(\sum_{i=1}^{20}\frac{\partial N_i}{\partial \eta} r_i\right)\left(\sum_{i=1}^{20}\frac{\partial N_i}{\partial \xi}\theta_i\right) - \left(\sum_{i=1}^{20}\frac{\partial N_i}{\partial \eta}\theta_i\right)\left(\sum_{i=1}^{20}\frac{\partial N_i}{\partial \xi} r_i\right)\right]\right\}\frac{\partial N_i}{\partial \xi} \end{aligned}$$

The preceding general expression for $\partial N_i/\partial r$ is applicable, in particular, to the computational nodes 5, 6, 7, 8, 17, 18, 19, and 20 (see Figure 15.3): all existing on the rotor/stator-interface surface), by substituting the appropriate local coordinates of each node on the right-hand side of this expression as follows:

To obtain $\left(\dfrac{\partial N_i}{\partial r}\right)_5$ set $\zeta = +1, \eta = +1$, and $\xi = +1$

To obtain $\left(\dfrac{\partial N_i}{\partial r}\right)_6$ set $\zeta = +1, \eta = -1$, and $\xi = +1$

To obtain $\left(\dfrac{\partial N_i}{\partial r}\right)_7$ set $\zeta = +1, \eta = -1$, and $\xi = -1$

To obtain $\left(\dfrac{\partial N_i}{\partial r}\right)_8$ set $\zeta = +1, \eta = +1$, and $\xi = -1$

To obtain $\left(\dfrac{\partial N_i}{\partial r}\right)_{17}$ set $\zeta = +1, \eta = 0$, and $\xi = +1$

To obtain $\left(\dfrac{\partial N_i}{\partial r}\right)_{18}$ set $\zeta = +1, \eta = -1$, and $\xi = 0$

To obtain $\left(\dfrac{\partial N_i}{\partial r}\right)_{19}$ set $\zeta = +1, \eta = 0$, and $\xi = -1$

To obtain $\left(\dfrac{\partial N_i}{\partial r}\right)_{20}$ set $\zeta = +1, \eta = +1$, and $\xi = 0$

The conclusion, then, is that the magnitude of $\partial N_i/\partial r$ is known at all corner and mid-side nodes that the finite element in Figure 15.13 shares with the interface surface. Referring to Equations (15.21), (15.22), and (15.23), it follows that the magnitudes of velocity components' nodal derivatives will also be known once the components' total magnitudes are known. In other words, once a time level has been encountered and the stator flow field is obtained, the velocity components' magnitudes at all twenty nodes of the element become known and so will be the derivatives $\partial W_r/\partial n$, $\partial W_\theta/\partial n$, and $\partial W_z/\partial n$ at, in particular, the eight corner and midside nodes existing on the interface surface. Taking the simple arithmetic average (for instance) of these normal derivatives, we obtain the derivatives we need to evaluate the surface integrals defined by Equations (15.13a), (15.14a), and (15.15a). This simply means that we are passing information (velocity components' normal derivatives) from the stator subdomain across the interface to update the boundary integrals in the elements' equations in the rotor subdomain.

PROBLEMS

P.1 Figure 15.14 shows two rectilinear (i.e., plane) cascades of compressor airfoils, one of which is sliding relative to the other at a linear velocity U. Guided by the contents of the surface integrals in Equations (15.13), (15.14), and (15.15), the two finite element equations (representing the x and y momentum equations) at sliding node 5 (on the right-hand side of the interface line) will contain the following free terms:

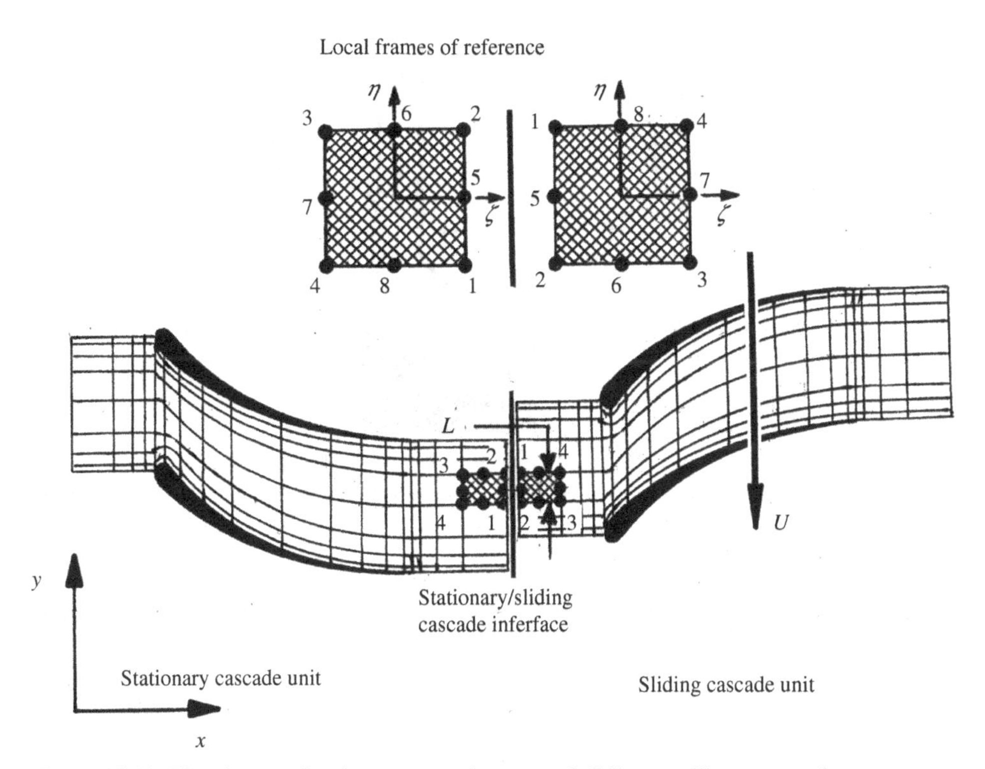

Figure 15.14. Flow interaction between stationary and sliding rectilinear cascades.

$$I_x = \int_L \frac{\mu_e}{Re} N_5 \frac{\partial W_x}{\partial x}\, dL$$

$$I_y = \int_L \frac{\mu_e}{Re} N_5 \frac{\partial W_y}{\partial x}\, dL$$

where $W_x = V_x$ and $W_y = V_y - U$, with V and W referring to the absolute and relative velocities, respectively, in the sliding cascade unit. Considering the special position of the sliding cascade unit in the figure, and assuming that the nodal magnitudes of V_x and V_y are known throughout the stationary cascade unit:

a. Express the normal derivatives $\partial V_x/\partial x$, and $\partial V_y/\partial x$ over the stationary-element side coinciding with the interface line in terms of V_x and V_y nodal magnitudes associated with this stationary element.
b. To ensure continuity of the velocity components' normal derivatives (across the interface), use the two normal derivatives obtained in item (a) to find expressions for $\partial W_x/\partial x$ and $\partial W_y/\partial x$ in the line-integral terms I_x and I_y. Write down the final expressions of these two boundary integrals.

P.2 Figure 15.15 shows a centrifugal compressor stage where a low subsonic flow prevails throughout the stage, justifying an incompressible flow field. The rotor is assumed to provide a free-vortex flow structure at the interface surface (with a radius of r_{int}), and the radial-velocity component is assumed to satisfy the

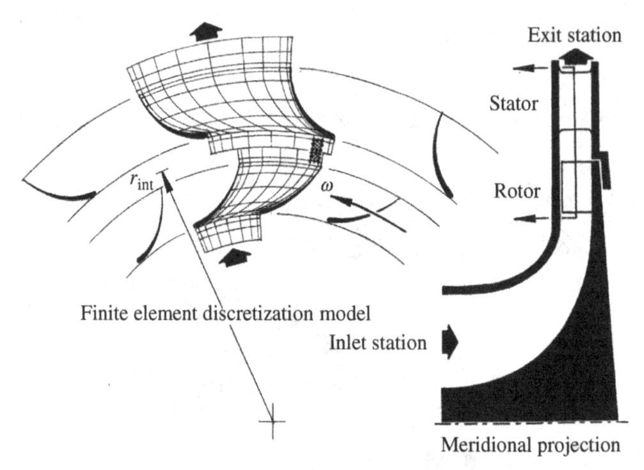

Figure 15.15. Boundary-conditions' exchange between the rotor and stator subdomains within a centrifugal compressor stage.

continuity equation in an average sense. These two conditions can be written at the interface surface as follows:

$$rV_\theta = C_1$$
$$rV_r = C_2$$

where C_1 and C_2 are constants. In a derivative form, these constraints can be written as follows:

$$\frac{\partial V_\theta}{\partial r} = -\frac{V_\theta}{r}$$
$$\frac{\partial V_r}{\partial r} = -\frac{V_r}{r}$$

which can be rewritten in terms of relative-velocity components as follows:

$$\frac{\partial V_\theta}{\partial r} = -\left(\frac{\partial W_\theta}{\partial r} + \omega\right)$$
$$\frac{\partial V_\theta}{\partial r} = -\frac{W_r}{r}$$

where ω is the rotor speed. Referring to the rotor-side cross-hatched element, consider the following velocity-component interpolating functions:

$$W_\theta = \sum_{i=1}^{8} N_i W_{\theta,i}$$
$$W_r = \sum_{i=1}^{8} N_i W_{r,i}$$

With similar interpolation functions over the stator-side cross-hatched element, the boundary-integral term at midside node k on the rotor/stator interface can

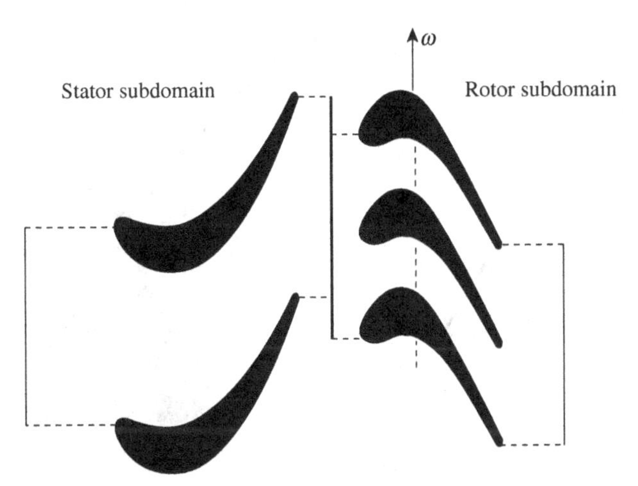

Figure 15.16. An axial turbine with a rotor-blade count that is double that of the stator stator number of vanes.

be expressed as follows:

$$I_\theta = \int_L N_k \frac{\mu_e}{Re}\frac{\partial V_\theta}{\partial r} dL$$

$$I_r = \int_L N_k \frac{\mu_e}{Re}\frac{\partial V_r}{\partial r} dL$$

Show how these boundary integrals will appear once the rotor-provided boundary conditions are passed to the stator-side element.

P.3 Figure 15.16 shows an axial-flow turbine stage where the number of rotor blades is double that of the stator vanes. Show, by reference to the tangentially limiting boundary segments (dotted in the figure), how this particular feature can be exploited to simplify the boundary conditions over these segments.

P.4 A special type of thermal science problem is that known as the *conjugate-fluid flow/heat transfer problem*. Figure 15.17 shows one of this problem category to which this approach is particularly suited. The turbine rotor-blade-cascade in this figure is internally cooled, with the cooling passages extending from the blade *platform*, (where the coolant is admitted), up to the blade-tip section (where the coolant is discharged). In a problem such as this, the adiabatic-flow condition is clearly inapplicable. On the fluid side, a new flow-governing equation representing the energy-conservation principle is added to the system. This can be expressed in the following vector form:

$$k_f \nabla^2 T = \rho c_p(\vec{V}.\nabla)T - (\vec{V}.\nabla)p \tag{A}$$

where

- k_f is the fluid thermal conductivity coefficient.
- T is the static temperature.

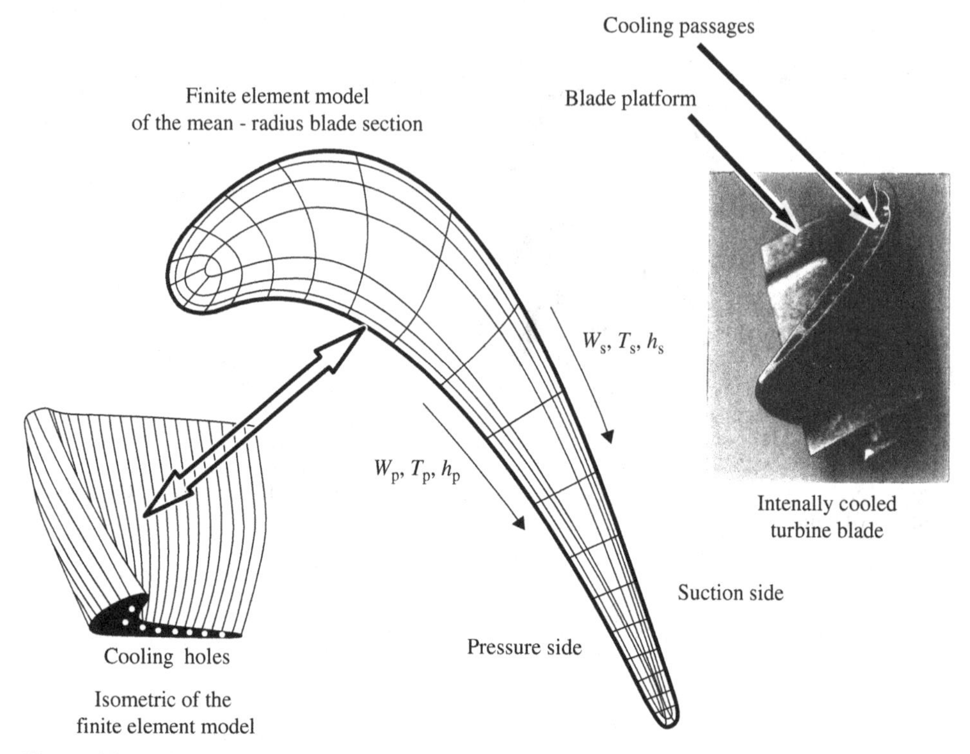

Figure 15.17. Conjugate fluid flow/heat transfer problem in a cooled rotor-subdomain blade cascade.

- ρ is the static density.
- p is the static pressure.

And the viscosity-related dissipation function is ignored.

Let us now focus on the conjugate heat-conduction problem in the blade itself. Figure 15.17 shows one of many different three-dimensional finite element models where the blade-surface computational nodes are shared with the fluid-side finite element model, linking the two temperature fields to one another. To simplify this heat-conduction problem, let us now view the blade as a heat sink that is continually loosing heat energy to the cooling flow stream. The strength of this sink, per unit volume, can be approximated as follows:

$$\psi = \frac{\dot{m}_c c_{p,c}(T_{c,ex} - T_{c,in})}{V_{blade}}$$

where

- $\dot{m}_c$ is the coolant mass-flow rate.
- $c_{p,c}$ is the coolant specific heat.
- $T_{c,in}$ is the coolant inlet temperature.
- $T_{c,ex}$ is the coolant discharge temperature
- V_{blade} is the blade volume.

Assuming a constant magnitude (k_b) of the blade thermal conductivity, the heat-conduction process is governed by the following equation:

$$k_b \nabla^2 T = 0 \tag{B}$$

Derive the finite element equivalent of equations (A) and (B).

P.5 Figure 15.18 shows the midspan section of an axial-flow turbine stage where the flow field is modeled in the $z-\theta$ frame of reference. Combining the two finite-elements (a) and (b), belonging to the stator and rotor subdomain, and referring to finite element equations (15.12), (15.114), and (15.15), with the assumption of negligible V_r and W_r magnitudes, the combined set of finite element equations associated with these two elements can be written as follows:

$$\begin{bmatrix} [A] & [B] & [C] & [\alpha] & 0 & 0 \\ [D] & [E] & [F] & 0 & [\beta] & 0 \\ [G] & [H] & 0 & 0 & 0 & 0 \\ [\gamma] & 0 & 0 & [a] & [b] & [c] \\ 0 & [\delta] & 0 & [d] & [e] & [f] \\ 0 & 0 & 0 & [g] & [h] & 0 \end{bmatrix} \begin{Bmatrix} \{V_z\} \\ \{V_\theta\} \\ \{P\} \\ \{W_z\}\,\{W_\theta\} \\ \{P\} \end{Bmatrix} = \begin{Bmatrix} \{Q_z\} \\ \{Q_\theta\} \\ \{0\} \\ \{q_z\} \\ \{q_\theta\} \\ \{0\} \end{Bmatrix}$$

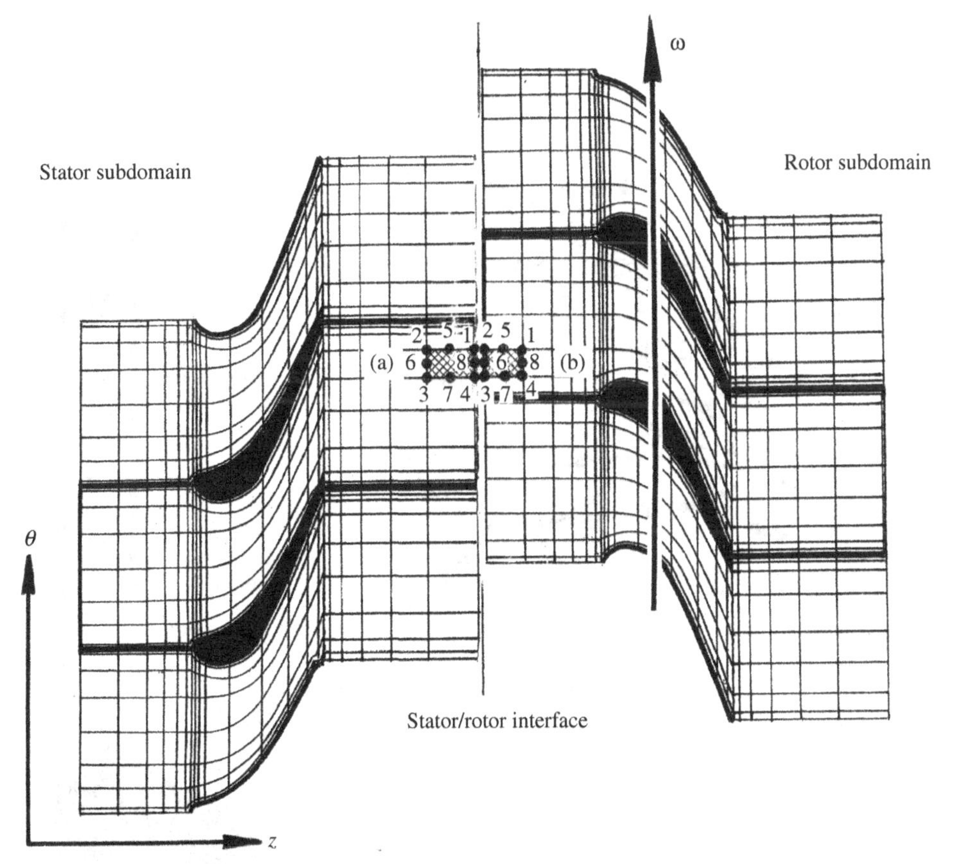

Figure 15.18. Transmission of the axial-velocity derivative from the stator-to-rotor subdomains.

Prove that the nonzero entries of the submatrix $[\gamma]$ are as follows:

$$\gamma_{2,j} = \int_{L^{(b)}} \frac{\mu_e}{Re} N_2{}^{(b)} \frac{\partial N_j{}^{(a)}}{\partial x}\, dL \qquad j = 1,2,3,\ldots,8$$

$$\gamma_{3,j} = \int_{L^{(b)}} \frac{\mu_e}{Re} N_3{}^{(b)} \frac{\partial N_j{}^{(a)}}{\partial x}\, dL \qquad j = 1,2,3,\ldots,8$$

$$\gamma_{6,j} = \int_{L^{(b)}} \frac{\mu_e}{Re} N_6{}^{(b)} \frac{\partial N_j{}^{(a)}}{\partial x}\, dL \qquad j = 1,2,3,\ldots,8$$

P.6 Figure 15.19 shows a finite element discretization model of the combined rotor and stator subdomains within a centrifugal compressor stage in a quasi-three-dimensional analysis in the stage flow field. Despite the inappropriateness of the choice, the discretization unit in this figure is a linear triangular element with equal-order interpolation of the pressure and velocity components, that is,

$$W_s{}^{(a)} = \sum_{i=1}^{3} N_i{}^{(a)} W_{s,i}$$

$$V_s{}^{(b)} = \sum_{i=1}^{3} N_i{}^{(b)} V_{s,i}$$

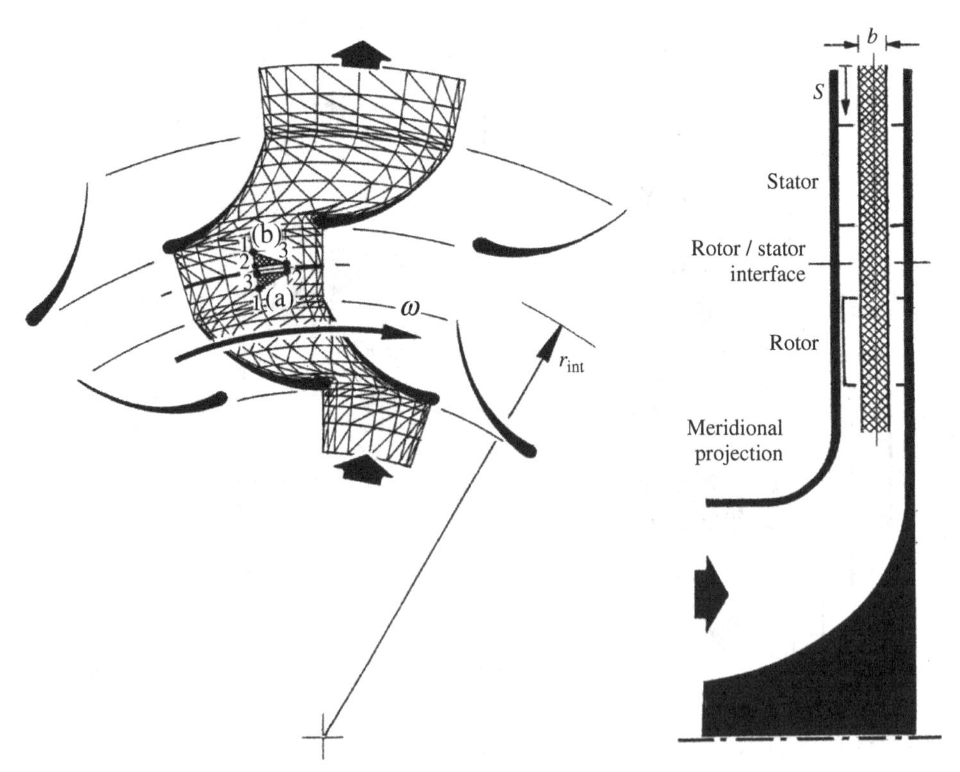

Figure 15.19. Quasi-three-dimensional analysis of the impeller-diffuser interaction.

$$W_\theta{}^{(a)} = \sum_{i=1}^{3} N_i{}^{(a)} W_{\theta,i}$$

$$V_\theta{}^{(b)} = \sum_{i=1}^{3} N_i{}^{(b)} V_{\theta,i}$$

$$p^{(a)} = \sum_{i=1}^{3} N_i{}^{(a)} p_i$$

$$p^{(b)} = \sum_{i=1}^{3} N_i{}^{(b)} p_i$$

In addition, the cross-hatched elements (a) and (b) have identical heights, and are used to illustrate the implementation of the rotor/stator compatibility conditions. Part of these conditions is particularly enforced in this problem as follows:

- Equate V_θ on both sides of the interface line.
- Cast the boundary-integral term containing $\partial V_\theta / \partial s$, on the rotor side in terms of the W_θ nodal values associated with the finite element (a).
- Equate p and $\partial p / \partial s$ at the nodes $3^{(a)}$ and $2^{(b)}$, respectively.

Under these constraints, the combined sets of finite element equations associated with the elements (a) and (b) can be expressed as follows:

$$\begin{bmatrix} [A] & [B] & [C] & [X] & [0] & [0] \\ [D] & [E] & [F] & [0] & [Y] & [0] \\ [G] & [H] & [I] & [0] & [0] & [Z] \\ [x] & [0] & [0] & [a] & [b] & [c] \\ [0] & [y] & [0] & [d] & [e] & [f] \\ [0] & [0] & [z] & [h] & [h] & [i] \end{bmatrix} \begin{Bmatrix} \{W_s{}^{(a)}\} \\ \{W_\theta{}^{(a)}\} \\ p^{(a)} \\ \{V_s{}^{(b)}\} \\ \{V_\theta{}^{(b)}\} \\ p^{(b)} \end{Bmatrix} = \begin{Bmatrix} \{Q_s\} \\ \{Q_\theta\} \\ \{0\} \\ \{q_s\} \\ \{q_\theta\} \\ \{0\} \end{Bmatrix}$$

Prove that implementation of the previously listed compatibility conditions will affect the entries of some submatrices as follows:

$$[D] = \begin{bmatrix} D_{11} & D_{12} & D_{13}0 & 0 & 0 \\ 0 & 0 & 0 & & \end{bmatrix}$$

$$[E] = \begin{bmatrix} E_{11} & E_{12} & E_{13} \\ 0 & +1 & 0 \\ 0 & 0 & +1 \end{bmatrix}$$

$$[F] = \begin{bmatrix} F_{11} & F_{12} & F_{13} \\ 0 & 0 & 0 \\ 0 & 0 & 0 \end{bmatrix}$$

$$[Y] = \begin{bmatrix} 0 & 0 & 0 \\ 0 & 0 & -1 \\ 0 & -1 & 0 \end{bmatrix}$$

$$\{Q_\theta\} = \begin{Bmatrix} Q_{s,i} \\ \omega r_{int} \\ \omega r_{int} \end{Bmatrix}$$

$$[G] = \begin{bmatrix} G_{11} & G_{12} & G_{13} \\ 0 & 0 & 0 \\ G_{31} & G_{32} & G_{33} \end{bmatrix}$$

$$[H] = \begin{bmatrix} H_{11} & H_{12} & H_{13} \\ 0 & 0 & 0 \\ H_{31} & H_{32} & H_{33} \end{bmatrix}$$

$$[I] = \begin{bmatrix} 0 & 0 & 0 \\ 0 & 0 & 0 \\ 0 & 0 & +1 \end{bmatrix}$$

$$[Z] = \begin{bmatrix} 0 & 0 & 0 \\ 0 & -1 & 0 \\ 0 & 0 & 0 \end{bmatrix}$$

$$[y] = \frac{\mu_e}{Re} \begin{bmatrix} 0 & 0 & 0 \\ \oint_{L^{(b)}} N_2{}^{(a)} \dfrac{\partial N_i{}^{(a)}}{\partial s}\, dL & \oint_{L^{(b)}} N_2{}^{(b)} \dfrac{\partial N_2{}^{(a)}}{\partial s} & \oint_{L^{(b)}} N_2{}^{(b)} \dfrac{\partial N_3{}^{(a)}}{\partial s}\, dL \\ \oint_{L^{(b)}} N_3{}^{(b)} \dfrac{\partial N_i{}^{(a)}}{\partial s}\, dL & \oint_{L^{(b)}} N_3{}^{(b)} \dfrac{\partial N_3{}^{(a)}}{\partial s} & \oint_{L^{(b)}} N_3{}^{(b)} \dfrac{\partial N_3{}^{(b)}}{\partial s}\, dL \end{bmatrix}$$

$$[d] = \begin{bmatrix} d_{11} & d_{12} & d_{13} \\ d_{21} & d_{22} & d_{23} \\ d_{31} & d_{32} & d_{33} \end{bmatrix}$$

$$[e] = \begin{bmatrix} e_{11} & e_{12} & e_{13} \\ e_{21} & e_{22} & e_{23} \\ e_{31} & e_{32} & e_{33} \end{bmatrix}$$

$$[f] = \begin{bmatrix} f_{11} & f_{12} & f_{13} \\ f_{21} & f_{22} & f_{23} \\ f_{31} & f_{32} & f_{33} \end{bmatrix}$$

$$[z] = \begin{bmatrix} 0 & 0 & 0 \\ -1 & 0 & +1 \\ 0 & 0 & 0 \end{bmatrix}$$

$$[g] = \begin{bmatrix} g_{11} & g_{12} & g_{13} \\ 0 & 0 & 0 \\ g_{31} & g_{32} & g_{33} \end{bmatrix}$$

$$[h] = \begin{bmatrix} h_{11} & h_{12} & h_{13} \\ 0 & 0 & 0 \\ h_{31} & h_{32} & h_{33} \end{bmatrix}$$

$$[i] = \begin{bmatrix} 0 & 0 & 0 \\ -1 & +1 & 0 \\ 0 & 0 & 0 \end{bmatrix}$$

REFERENCES

[1] Mikolaiczak, A. A., "The Practical Importance of Unsteady Flow," in *Unsteady Phenomena in Turbomachinery, AGARD CP 177*, September 1975.

[2] Rai, M. M., "Unsteady Three-Dimensional Navier-Stokes Simulation of Turbine Rotor-Stator Interaction," *AIAA Paper No. 87-2058*, San Diego, CA., 1987.

[3] Rai, M. M, "Three-Dimensional Navier-Stokes Simulation of Turbine Rotor-Stator Interaction. I: Methodology," *Journal of Propulsion and Power*, Vol. 5, No. 3, May–June 1989, pp. 305–311.

[4] Koya, M., and Kotake, S., "Numerical Analysis of Fully Three-Dimensional Periodic Flows through a Turbine Stage," *ASME Paper No. 85-GT-57*, Houston, TX, 1985.

[5] Baldwin, B.S., and Lomax, H., "Thin Layer Approximation and Algebraic Model for Separated Turbulent Flows," *AIAA Paper No. 78-257*, January 1978.

[6] Benim, A. C., and Zinser, W., "Investigation into the Finite Element Analysis of Confined Turbulent Flows Using $\kappa - \epsilon$ Model of Turbulence," *Computer Methods in Applied Mechanics and Engineering*, Vol. 51, 1985, pp. 507–523.

[7] Carey, G. F., and Oden, J. T., *Finite Elements: Fluid Mechanics* (Texas Finite Element Series). Prentice-Hall, Englewood Cliffs, NJ, 1986.

[8] Hood, P., and Taylor, C., "Navier-Stokes Equations Using Mixed Interpolation," in *Proceedings of the International Symposium on Finite Element Methods in Flow Problems*, University of Wales, Swansea, UK, January 1974.

[9] Stroud, A. H., *Approximate Calculation of Multiple Integrals*, Prentice-Hall, Englewood Cliffs, NJ, 1971.

[10] Baskharone, E., and Hamed, A., "A New Approach in Cascade Flow Analysis Using the Finite Element Method," *AIAA Journal*, Vol. 19, No. 1, January 1981, pp. 65–71.

[11] Gupta, S. K., and Tanji, K. K., "Computer Program for Solution of Large, Sparse, Unsymmetric Systems of Linear Equations," *International Journal for Numerical Methods in Engineering*, Vol. 11, 1977, pp. 1251–1259.

[12] Rai, M. M., "Three-Dimensional Navier-Stokes Simulation of Turbine Rotor-Stator Interaction. II: Results," *Journal of Propulsion and Power*, Vol. 5, No. 3, May–June 1989, pp. 312–319.

[13] Dring, R. P., Joslyn, H. D., Hardin, L. W., and Wagner, J. H., "Turbine Rotor-Stator Interaction," *Journal of Engineering for Power*, Vol. 104, October 1982, pp. 729–742

[14] Baskharone, E. A., *Principles of Turbomachinery in Air-Breathing Engines*, Cambridge University Press, New York, 2006.

16 Finite Element–Based Perturbation Approach to Unsteady Flow Problems

Overview

Since its inception, the finite element method has followed the tracks of other well-established computational techniques, particularly the finite-difference method. The fluid mechanics applications remained limited to mostly steady-state flow applications in two- or three-dimensional domains, with one or more complicating real-flow effects (e.g., compressibility, turbulence, etc.) being part of the computational model. Differing in complexity and accuracy, the finite-elements themselves have been taken as fixed-geometry subdomains, and this is the critical point where the following analysis categorically differs.

In many real-life applications, the flow domain itself undergoes small time-dependent changes (or distortions) that, in most cases, are periodic. The problem of wing flutter is an example of such a situation in the external aerodynamics discipline. The problem under focus here involves the vibration of a fluid-encompassed rotor in a confined-flow type of arrangement (Figure 16.1) and is known to have a major impact on the system's rotordynamic integrity.

Originally handled via finite-difference techniques, the solution strategy was to repeatedly solve the entire physical problem by marching in time while slightly altering the flow-domain geometry at each time level. Tedious as it was, this approach is hardly economical, nor is it based on prior knowledge of what controls the time increment and, often, what fluid/structure features are to be monitored. Prohibiting solid advances in addressing the problem, many believe, is that it was historically handled in either a fluid-dynamics or a mechanical-vibrations type of approach but not a combination of the two mentalities. On the one hand, fluid dynamicists have progressively sophisticated their computational modeling of what is the closest to a theoretically creeping flow through a tiny rotor-to-housing clearance gap, including complex driven-cavity subregions, which would be associated with a labyrinth seal (a reliable leakage-control device) putting up with an understandably enormous usage of computational resources. Vibration analysts, on the other hand, settled for much outdated and excessively simplified *bulk-flow* models in which crucial details of the flow field were simply ignored.

Research efforts in this rotordynamics area began to intensify in the mid-1980s, prompted by frequent mechanical failure in an early design of the liquid-oxygen

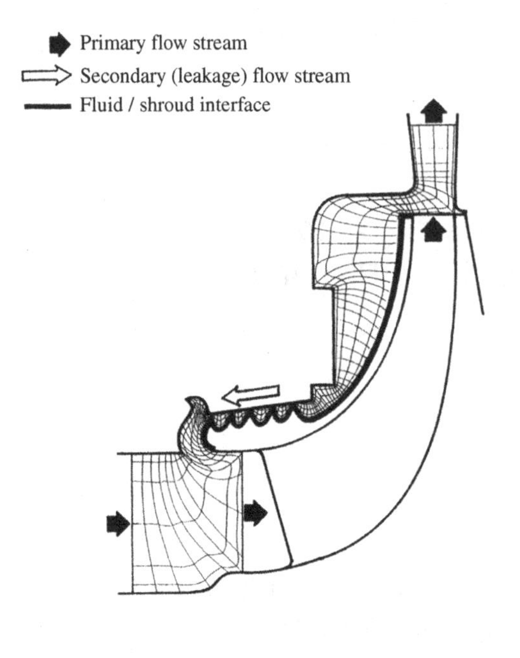

Figure 16.1. Primary and secondary flow streams in a shrouded pump impeller.

turbopump of the space shuttle main engine. A unique and potentially disastrous motion component of the shrouded impeller, in this case, was blamed as the source of the problem. Referred to as "whirl", this is an eccentric mode of operation in which the rotor axis is slightly displaced and begins to spin around the housing centerline at a finite frequency. This motion invites a set of fluid-exerted reaction forces that can either be restoring or further aggravating.

It doesn't take a great deal of rotor eccentricity (or excitation) to initiate the rotor deviation from its normal mode of operation or to eventually break the shaft for that matter. This led to the conclusion that a perturbation approach, devised by the author, is perhaps the ideal framework for handling the problem. Responding to a virtual rotor displacement, in this case, the objective is to determine the rate at which the fluid reaction forces are developed within the rotor-to-housing clearance gap (see Figure 16.1). In a postprocessing procedure, these forces are used to compute the so-called rotordynamic (force) coefficients of the rotor-fluid interaction mechanism. Details of this procedure illustrate the manner in which these forces are analyzed as the source of the stiffness, damping, and added-mass (inertia) effects of the fluid-rotor interaction system. In the end, the rotordynamic coefficients (by their signs and relative magnitudes) determine the source of instability, if any.

Foundation of the Finite Element–Based Perturbation Approach

The basic idea behind the current computational model is in the treatment of the rotor eccentricity as a virtual displacement. Once it occurs, this eccentricity is viewed as causing infinitesimally small deformations with varied magnitudes in each finite element within the rotor-to-housing gap, as illustrated for a simple cylindrical rotor

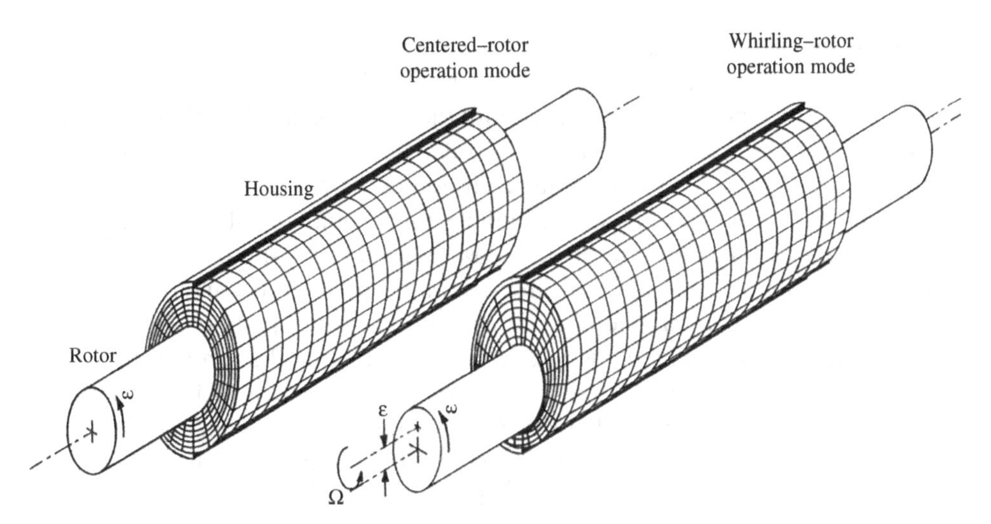

Figure 16.2. Deformations of the finite element nodes under a cylindrical rotor whirl.

in Figure 16.2. A perturbation analysis by the author [1,2] is then performed to find the incremental changes in the flow variables, particularly those contributing to the rotor fluid-exerted forces.

The computational process is initiated by securing the centered-rotor axisymmetric flow field in the rotor-to-housing gap. A transitional step is then executed, where a three-dimensional flow domain is created by repeating the finite-elements in the gap around the circumference as many times as deemed necessary and building three-dimensional elements in the process. The vector of flow-variable magnitudes associated with what is now a three-dimensional finite element assembly is created by similarly repeating the axisymmetric flow solution around the circumference. The process just described is not as silly as it may appear, for it is needed in the next step, where the rotor whirling-motion effects are brought into the picture and the difference between the two sets of results is sought. The process just described, however, doesn't exactly produce the zeroth-order flow field, for the rotor whirl frequency has to be part of it, as explained next.

In developing the distorted-domain finite element equations, it is postulated that the set of computational nodes on the rotor surface will undergo the same rotor-axis eccentricity, with those on the housing surface remaining in place. The rest of the nodes will meanwhile experience varied displacement magnitudes, causing different distortions to all finite-elements in the computational domain. While this picture may be physically comprehensible, it falls short of eliminating the unsteadiness of the flow field in the presence of the rotor whirling motion. In fact, the only simple way to remove this time dependency, as explained later in this chapter, is to cast the flow-governing equations in a rotating-translating frame of reference that is attached to the rotor and is whirling with it. Viewed from this reference frame, it is the housing inner surface that inherits the rotor whirling motion, causing the elemental distortions. Note that while the rotor eccentricity here is treated as infinitesimally small, the whirl frequency, which is the other component of this off-center operation mode, can be as large as the rotor speed (synchronous whirl) or even larger in the direction of the rotational speed (forward whirl) or opposite to it (negative whirl). Details of the zeroth-order flow equations (presented later in this chapter) reveal that this particular variable will have to be built in and that the perturbation analysis will not begin until the rotor eccentricity takes place. The point is also emphasized that the zeroth-order flow solution depends on neither the coordinate-axes conversion nor the inclusion of the whirl frequency in the flow equations of motion. This naturally simplifies the computational procedure in the sense of readily having the zeroth-order flow solution as a result of a much simpler axisymmetric-flow analysis, the very first step of the entire procedure.

The term *zeroth-order* refers to a flow field that is computationally one level higher than that implied by the term *centered-rotor* despite the fact that they both describe the same centered-rotor flow field. Beside including the whirl frequency in the former, another difference here is that the flow governing equations are now cast in a three-dimensional Cartesian frame of reference in order to ensure consistency with the perturbed flow field and the equations describing it. After all, the process of finding the differential changes in the flow variables, as a result of the rotor off-center operation, is where the two analytical models should correspond to one another.

Securing a zeroth-order flow field, an unorthodox finite element perturbation analysis (discussed in a separate section) is applied, with the outcome being the incremental changes in the fluid-induced rotor-surface forces. In a post-processing procedure, these forces are finally cast in terms of "rotordynamic" coefficients that quantify the stiffness, damping, and added-mass (inertia) interference with the fluid/rotor interaction effects. Stability related, these coefficients collectively address the likelihood of resonance in the fluid/rotor system and whether a premature mechanical failure can occur in the absence of design changes.

The rotor whirling motion discussed thus far is but one of two off-normal modes of operation covered in this chapter. Referring to Figure 16.2, note that the rotor eccentricity in this case is a lateral displacement that is uniform along the axis of rotation. The resulting rotor excitation is appropriately referred to as *cylindrical whirl*. It is not accurate to confine the fluid response to a force, nor is it correct (in the larger picture) to define the rotordynamic coefficients only on that basis. The fact is that these only define a subset of the overall set of rotordynamic matrices. What is missing, in the bigger picture, is another type of excitation, termed *conical whirl*, that can be equally threatening from a rotordynamics standpoint. This topic is separately covered later in this chapter, where the fluid response is modeled as a pure moment that may also be restoring or aggravating to what is now an angular excitation of the rotor axis. In the last section of this chapter, the two excitation modes are superimposed, and the fluid response is expanded. In fact, the stiffness, damping, and added-mass (inertia) coefficients then will be analyzed on the basis that the effects of the two combined excitations are interrelated in the sense that a purely cylindrical whirl may indeed give rise to fluid-induced moments and vice versa.

Definition of the Force-Related Rotordynamic Coefficients

In the following, a simple force-balance statement of the rotor under its whirling mode of operation is used to introduce the rotordynamic coefficients we are seeking [3]. These are the force-related direct and cross-coupled stiffness, damping, and

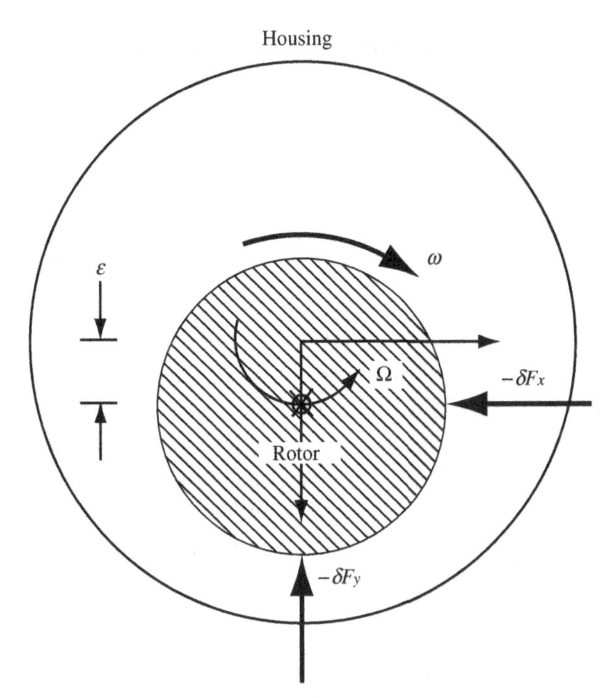

Figure 16.3. Rotor-axis rotation around the housing centerline.

inertia coefficients arising from the fluid-exerted forces (δF_x and δF_y in Figure 16.3) on the rotor surface as the latter enters its off-center operation mode. Referring to Figure 16.3, the rotor-axis locus can be described as follows:

$$x = \epsilon \sin(\Omega t), \qquad y = \epsilon \cos(\Omega t) \tag{16.1}$$

where ϵ is the rotor-axis eccentricity relative to the housing centerline, Ω is the whirl (or precession) frequency, and t refers to time. The fluid-rotor interaction under such a disturbance can be cast in terms of stiffness, damping, and inertia coefficients in the following linearized form:

$$\left\{ \begin{array}{c} -\delta F_x \\ -\delta F_y \end{array} \right\} = \left[\begin{array}{cc} K & -k \\ k & K \end{array} \right] \left\{ \begin{array}{c} x \\ y \end{array} \right\} + \left[\begin{array}{cc} C & -c \\ c & C \end{array} \right] \left\{ \begin{array}{c} \dot{x} \\ \dot{y} \end{array} \right\} + \left[\begin{array}{cc} M & -m \\ m & M \end{array} \right] \left\{ \begin{array}{c} \ddot{x} \\ \ddot{y} \end{array} \right\}$$

where δF_x and δF_y are the incremental changes in the rotor-applied force components due to the rotor disturbance, K and k are the direct and cross-coupled stiffness coefficients of the fluid-structure system, C and c are the damping coefficients, and M and m are the added-mass (inertia) coefficients.

With no lack of generality, consider the position of the rotor axis in Figure 16.3. The location, linear velocity, and acceleration at this position are

$$x = 0 \qquad y = \epsilon \tag{16.3}$$

$$\dot{x} = \Omega\epsilon \qquad \dot{y} = 0 \tag{16.4}$$

$$\ddot{x} = 0 \qquad \ddot{y} = -\Omega^2\epsilon \tag{16.5}$$

Expansion of the matrix Equation (16.2) and substitution of Equations (16.3) through (16.5) yield the following two equations:

$$\frac{\partial F_x}{\partial \epsilon} = \lim_{\epsilon \to 0} \left(\frac{\delta F_x}{\epsilon} \right) = k - \Omega C - \Omega^2 m \tag{16.6}$$

$$\frac{\partial F_y}{\partial \epsilon} = \lim_{\epsilon \to 0} \left(\frac{\delta F_y}{\epsilon} \right) = -K - \Omega c + \Omega^2 M \tag{16.7}$$

As suggested by Equations (16.6) and (16.7), determination of the stiffness, damping, and inertia coefficients requires computation of $\partial F_x/\partial \epsilon$ and $\partial F_y/\partial \epsilon$ corresponding to at least three different values of the whirl frequency Ω. Interpolation of these two derivatives as quadratic functions of Ω using curve-fitting techniques will then lead to the values of rotordynamic coefficients K, k, C, and so on simply by equating the different factors in these functions to those on the right-hand side of Equations (16.6) and (16.7). Because the resulting rotordynamic coefficients depend on the rotor speed ω, this analysis must be repeated at all operating speeds of interest.

Computational Development: Analysis of the Centered-Rotor Flow Field

Flow-Governing Equations

Figure 16.1 shows the rotor-to-housing flow domain of interest. To use the correct terminology, the rotor here is a *shrouded pump impeller*, the flow passage is rightfully termed the *secondary* (or *leakage*) *flow passage*, with the *primary passage* being that where shaft work is transferred to the fluid through the spinning impeller blades. Although the primary flow passage in this figure is extracted from the picture, the computational domain is so constructed to include some critical segments of the primary passage so that the need is not there to specify boundary conditions at locations where the two flow streams coexist [4].

In the centered-impeller operation mode, the shroud-to-housing flow is swirling, axisymmetric incompressible, and turbulent [5]. A linearized version of the mass and momentum conservation equations can then be cast in a cylindrical frame of reference as follows:

Continuity Equation

$$\frac{\partial V_z}{\partial z} + \frac{\partial V_r}{\partial r} + \frac{V_r}{r} = 0 \tag{16.8}$$

Axial Momentum Equation

$$\bar{V}_z \frac{\partial V_z}{\partial z} + \bar{V}_r \frac{\partial V_z}{\partial r} = -\frac{1}{\rho} \frac{\partial p}{\partial z} + 2 \frac{\partial}{\partial z} \left(\nu_e \frac{\partial V_z}{\partial z} \right) + \frac{1}{r} \frac{\partial}{\partial r} \left(r \nu_e \frac{\partial V_z}{\partial r} \right) + \frac{1}{r} \frac{\partial}{\partial r} \left(r \nu_e \frac{\partial V_r}{\partial z} \right) \tag{16.9}$$

Radial Momentum Equation

$$\bar{V}_z \frac{\partial V_r}{\partial z} + \bar{V}_r \frac{\partial V_r}{\partial r} - \bar{V}_\theta \frac{V_\theta}{r} = -\frac{1}{\rho}\frac{\partial p}{\partial r} + \frac{\partial}{\partial z}\left(\nu_e \frac{\partial V_r}{\partial z}\right) + \frac{2}{r}\frac{\partial}{\partial r}\left(r\nu_e \frac{\partial V_r}{\partial r}\right) + \frac{\partial}{\partial z}\left(\nu_e \frac{\partial V_z}{\partial r}\right) - 2\nu_e \frac{V_r}{r^2} \quad (16.10)$$

Tangential Momentum Equation

$$\bar{V}_z \frac{\partial V_\theta}{\partial z} + \bar{V}_r \frac{\partial V_\theta}{\partial r} - \bar{V}_r\left(\frac{V_\theta}{r}\right) = \frac{\partial}{\partial z}\left(\nu_e \frac{\partial V_\theta}{\partial z}\right) + \frac{1}{r}\frac{\partial}{\partial r}\left(r\nu_e \frac{\partial V_\theta}{\partial r}\right) - \frac{V_\theta}{r}\frac{\partial \nu_e}{\partial r} - \nu_e\left(\frac{V_\theta}{r^2}\right) \quad (16.11)$$

where

- V_z, V_r, and V_θ are the components of the absolute velocity vector,
- p is the static pressure,
- ρ is the static density,
- ν_e is the effective (molecular plus eddy) kinematic viscosity coefficient (discussed in Chapter 13), and
- $\bar{V}_z$, $\bar{V}_r$, and $\bar{V}_\theta$ are velocity components from a previous iteration or an initial guess.

Boundary Conditions

With the flow ellipticity retained throughout the computational domain, it becomes necessary to prescribe appropriate flow constraints over all segments of the domain boundary. This includes the flow exit station, where a boundary condition is not needed under many existing parabolized-flow computational models (not the situation here) and where the flow solution is obtained simply by marching along a "master" streamline down to the exit station.

The boundary conditions in this section concern a test case of a simple annular seal (see Figure 16.6), in its centered-rotor operation mode. These are classified by the boundary segment to which they belong as follows:

Flow Inlet Station

The through-flow and tangential velocity profiles at this station are assumed known. Each of these profiles can be constructed using given mean values of the inlet velocity components by assuming reasonable estimates of the boundary layer thickness and velocity profile at this station. The inlet velocity profiles must also account for the entry losses associated with the annular seal, a variable that depends on the means with which the rotor-to-housing clearance gap is naturally connected to the rest of the flow domain, where the seal exists.

Flow Exit Station

Fully developed flow is assumed at this station. As a result, the following boundary condition applies:

$$\frac{\partial V_z}{\partial z} = \frac{\partial V_r}{\partial z} = \frac{\partial V_\theta}{\partial z} = 0$$

The fully developed flow assumption here is justifiable on the basis that the length-width ratio of the annular gap in this case is substantially high. An additional boundary condition involving an arbitrary fixed value of the exit pressure is also imposed, setting a datum for the pressure magnitudes throughout the computational domain.

Solid Boundary Segments

Included in this category are the rotor and housing surfaces. Over these, the no-slip boundary condition is enforced as follows:

$V_z = V_r = 0$ over the rotor and housing surfaces

$V_\theta = 0$ over the housing surface

$V_\theta = \omega r$ over the rotor surface

where ω is the rotor speed.

Introduction of the Upwinding Technique

Adoption of the conventional Galerkin's method in deriving the finite element equivalent of the flow-governing equations was, perhaps expectedly, unsuccessful. The dominant inertia effects within the turbulent flow in this case demanded a special consideration. Surprisingly, the "wiggles" appearing in the streamwise pressure distribution were not entirely eliminated even when full upwinding of the equations of motion in the manner devised by Heinrich and Zienkiewicz [8] was implemented. The situation was further worsened by the large aspect ratio of the rotor-to-housing gap (defined as the seal-length/clearance-width ratio). The latter unavoidably led to finite-elements with similarly high aspect ratios. This had an adverse numerical effect, especially near the rotor and housing surfaces, where the need for high resolution of the flow structure gave rise to excessively narrow elements there (see Figure 16.6).

Corrective measures included deviation from the full upwinding strategy and setting a realistic limit on the element aspect ratio in the near-wall zone. First, in implementing the weighted-residual method [9], terms in the error functions resulting from the three momentum equations were weighted differently in the process of deriving the finite element equations. In this case, the special set of weight functions proposed by Heinrich and Zienkiewicz was used in conjunction with only the convection terms, whereas the element shape functions were used with all other terms. This approach is consistent with the well-known practice in the area of finite-difference modeling where backward differencing is used exclusively in modeling the convection terms. When applied to the weighted-residual method, this concept gives rise to what is categorically known as the *Petrov-Galerkin approach.*

The upwinding method proposed by Heinrich and Zienkiewicz [8] created a rather complex set of weight functions. A cubic polynomial was added to the element shape functions, with coefficients that are calculated at all the finite element computational nodes using an algebraic relationship that is highly dependent on the local velocity vector. These coefficients were continually updated at the end of each iterative step where new velocity vectors were obtained.

Finite Element Formulation

As shown in Figure 16.4, the flow domain is replaced by an assembly of non-overlapping finite-elements, each being a nine-nodded quadratic element of the Lagrangian type. Again, the turbulence model proposed by Baldwin and Lomax is used. Figure 16.5 shows how the eddy viscosity values are computed, particularly in a subregion that is adjacent to the rotor surface. As far as the elemental degrees of freedom are concerned, all three velocity components are declared as variables at all nine nodes, whereas the pressure is unknown only at the four corner nodes. Within a typical element (e), let the velocity components, pressure, and spatial coordinates be interpolated as follows:

$$V_z^{(e)} = \sum_{i=1}^{9} N_i(\zeta,\eta) V_{z,i}$$

$$V_r^{(e)} = \sum_{i=1}^{9} N_i(\zeta,\eta) V_{r,i}$$

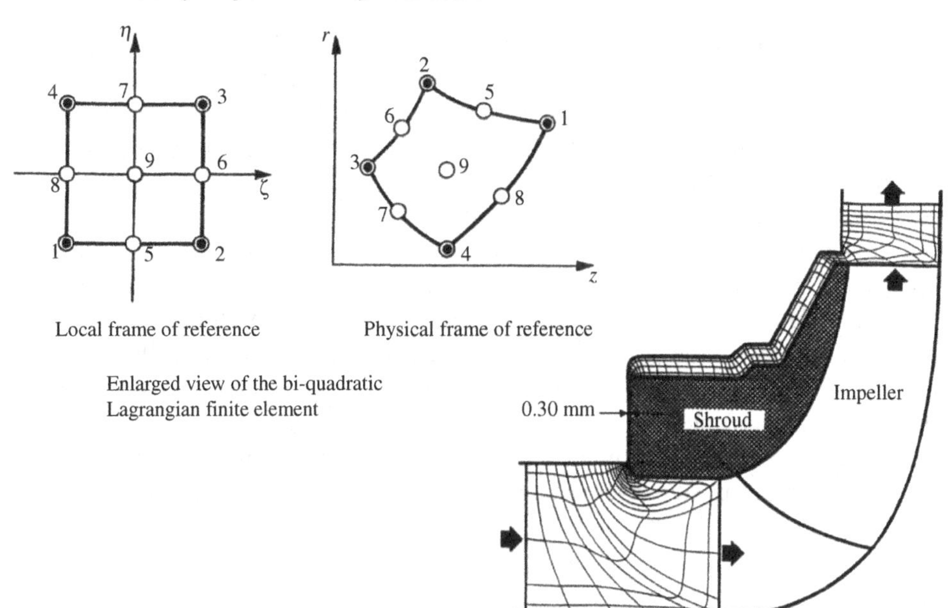

Figure 16.4. Finite element discretization model for the centered-rotor flow analysis.

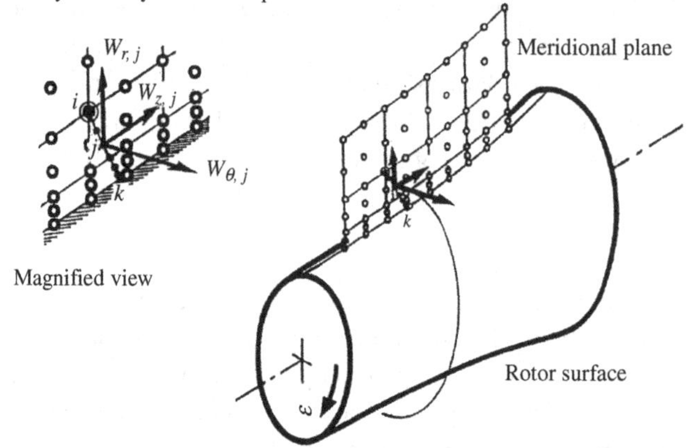

Figure 16.5. Eddy-viscosity calculation arrays in the near-wall zone.

$$V_\theta{}^{(e)} = \sum_{i=1}^{9} N_i(\zeta,\eta) V_{\theta,i}$$

$$p^{(e)} = \sum_{k=1}^{4} M_k(\zeta,\eta) p_i$$

$$z^{(e)} = \sum_{i=1}^{9} N_i(\zeta,\eta) z_i$$

$$r^{(e)} = \sum_{i=1}^{9} N_i(\zeta,\eta) r_i$$

The element's quadratic and linear shape functions N_i and M_k, respectively, were defined earlier in Chapter 12, with ζ and η being the spatial coordinates in the local frame of reference (see Figure 16.4). Substitution of the preceding interpolating functions in the flow-governing Equations (16.8) through (16.11) yields a residual (or error) function for each equation. These are required to be orthogonal to a set of weight functions throughout the entire element, a statement that (in effect) is equivalent to stating that the weighted residuals associated with each equation are set to zero in an integral sense. The weight functions used in conjunction with continuity Equation 16.8 are the linear shape functions M_k. As for the momentum equations, the quadratic shape functions N_i are assigned to all terms, with the exception of the convection (or inertia) terms of these equations. As discussed earlier, these terms are assigned a group of weight functions W_i that are combinations of the shape functions and the upwinding parameters defined by Heinrich and Zienkiewicz [8]. Execution of this step gives rise to the following set of elemental equations:

$$\mathbf{A_{kj}} V_{z,j} + \mathbf{B_{kj}} V_{r,j} = 0 \qquad (16.16)$$

$$\mathbf{C_{ij}} V_{z,j} + \mathbf{D_{ij}} V_{r,j} + \mathbf{R_{ik}} p_k = \mathbf{I_{z,i}} \qquad (16.17)$$

$$\mathbf{E_{ij}}V_{z,j} + \mathbf{F_{ij}}V_{r,j} + \mathbf{G_{ij}}V_{\theta,j} + \mathbf{S_{ik}}p_k = \mathbf{I_{r,i}} \tag{16.18}$$

$$\mathbf{H_{ij}}V_{\theta,j} = \mathbf{I_{\theta,i}} \tag{11.19}$$

The submatrices $\mathbf{A_{kj}}$, $\mathbf{B_{kj}}$, $\mathbf{C_{ij}}$, and so on and the subvectors $\mathbf{I_{z,i}}$, $\mathbf{I_{r,i}}$, and $\mathbf{I_{\theta,i}}$ are defined as follows:

$$\mathbf{A_{kj}} = \int_{A^{(e)}} \left\{ M_k \frac{\partial N_j}{\partial z} \right\} r\, dA$$

$$\mathbf{B_{kj}} = \int_{A^{(e)}} \left\{ (M_k \frac{\partial N_j}{\partial r} + \frac{1}{r} M_k N_j) \right\} r\, dA$$

$$\mathbf{C_{ij}} = \int_{A^{(e)}} \left\{ \bar{\nu}_e \left(\frac{\partial N_i}{\partial z}\frac{\partial N_j}{\partial z} + \frac{\partial N_i}{\partial r}\frac{\partial N_j}{\partial r} \right) + \bar{\nu}_e \bar{V}_z W_i \frac{\partial N_j}{\partial z} + \bar{\nu}_e \bar{V}_r W_i \frac{\partial N_j}{\partial r} \right.$$
$$\left. + 2\bar{\nu}_e N_i \frac{\partial \bar{\nu}_t}{\partial z}\frac{\partial N_j}{\partial z} + \bar{\nu}_e N_i \frac{\partial \bar{\nu}_t}{\partial r}\frac{\partial N_j}{\partial r} \right\} r\, dA$$

$$\mathbf{D_{ij}} = \int_{A^{(e)}} \left\{ N_i \frac{\partial \bar{\nu}_t}{\partial r}\frac{\partial N_j}{\partial z} \right\} r\, dA$$

$$\mathbf{R_{ik}} = \int_{A^{(e)}} \left\{ \frac{1}{\rho} N_i \frac{\partial M_k}{\partial z} \right\} r\, dA$$

$$\mathbf{E_{ij}} = \int_{A^{(e)}} \left\{ \bar{\nu}_e \left(\frac{\partial N_i}{\partial z}\frac{\partial N_j}{\partial z} + \frac{\partial N_i}{\partial r}\frac{\partial N_j}{\partial r} \right) + \frac{2\bar{\nu}_e}{r^2} N_i N_j + \bar{V}_z W_i \frac{\partial N_j}{\partial z} + N_i \frac{\partial \bar{\nu}_t}{\partial z}\frac{\partial N_j}{\partial z} \right.$$
$$\left. + 2N_i \frac{\partial \bar{\nu}_t}{\partial r}\frac{\partial N_j}{\partial r} + \bar{V}_r W_i \frac{\partial N_j}{\partial r} \right\} r\, dA$$

$$\mathbf{F_{ij}} = \int_{A^{(e)}} \left\{ N_i \frac{\partial \bar{\nu}_t}{\partial z}\frac{\partial N_j}{\partial r} \right\} r\, dA$$

$$\mathbf{G_{ij}} = \int_{A^{(e)}} \left\{ \frac{\bar{V}_\theta}{r} N_i N_j \right\} r\, dA$$

$$\mathbf{S_{ik}} = \int_{A^{(e)}} \left\{ \frac{1}{\rho} N_i \frac{\partial M_k}{\partial z} \right\} r\, dA$$

$$\mathbf{H_{ij}} = \int_{A^{(e)}} \left\{ \bar{\nu}_e \left(\frac{\partial N_i}{\partial z}\frac{\partial N_j}{\partial z} + \frac{\partial N_i}{\partial r}\frac{\partial N_j}{\partial r} \right) + \frac{\bar{\nu}_e}{r^2} N_i N_j + \bar{V}_z W_i \frac{\partial N_j}{\partial z} + \bar{V}_r W_i \frac{\partial N_j}{\partial r} + N_i \frac{\partial \bar{\nu}_t}{\partial z}\frac{\partial N_j}{\partial z} \right.$$
$$\left. + \frac{1}{r} N_i \frac{\partial \bar{\nu}_t}{\partial r}\frac{\partial N_j}{\partial r} - \frac{1}{r} N_i N_j \frac{\partial \bar{\nu}_t}{\partial r} + \frac{1}{r} \bar{V}_r N_i N_j \right\} r\, dA$$

$$\mathbf{I_z} = \oint_{L^{(e)}} \bar{\nu}_e r N_i \frac{\partial V_z}{\partial \mathbf{n}}\, dL$$

$$\mathbf{I_r} = \oint_{L^{(e)}} \bar{\nu}_e r N_i \frac{\partial V_r}{\partial \mathbf{n}}\, dL$$

$$\mathbf{I_\theta} = \oint_{L^{(e)}} \bar{\nu}_e N_i \frac{\partial V_\theta}{\partial \mathbf{n}}\, dL$$

where

- $\mathbf{n}$ denotes the outward unit vector normal to the local boundary segment
- $dA = dz\, dr = \mid J \mid d\zeta\; d\eta$, with $\mid J \mid$ referring to the physical (z&r) to local (ζ&η) coordinate transformation Jacobian (see Figure 16.4),

- $i = 1, 2, 3, \ldots, 9$,
- $j = 1, 2, 3, \ldots, 9$, and
- $k = 1, 2, 3, 4$.

The overbar in these expressions identifies a variable that is obtained from the previous iteration or an initial guess. Note that derivatives of the effective (molecular plus eddy) kinematic viscosity coefficient ν_e reduce to the derivatives of the eddy viscosity coefficient ν_t because the fluid molecular viscosity coefficient ν_l is constant.

Equations (16.16) through (16.19) represent the finite element equivalent of the flow-governing Equations (16.8) through (16.11) and can be rewritten in the following matrix form:

$$\begin{bmatrix} [\mathbf{A}] & [\mathbf{B}] & [\mathbf{0}] & [\mathbf{0}] \\ [\mathbf{C}] & [\mathbf{D}] & [\mathbf{0}] & [\mathbf{R}] \\ [\mathbf{E}] & [\mathbf{F}] & [\mathbf{G}] & [\mathbf{S}] \\ [\mathbf{H}] & [\mathbf{0}] & [\mathbf{0}] & [\mathbf{0}] \end{bmatrix} \begin{Bmatrix} \{\mathbf{V_z}\} \\ \{\mathbf{V_r}\} \\ \{\mathbf{V_\theta}\} \\ \{\mathbf{P}\} \end{Bmatrix} = \begin{Bmatrix} \{\mathbf{0}\} \\ \{\mathbf{I_z}\} \\ \{\mathbf{I_r}\} \\ \{\mathbf{I_\theta}\} \end{Bmatrix}$$

which can be further compacted as follows:

$$[k]\{\psi\} = \{f\}$$

where the matrix of influence coefficients $[k]$ is commonly referred to as the elemental *stiffness* matrix and the vector of free terms $\{f\}$ is termed the *load* vector. The vector of unknowns $\{\psi\}$ is composed of all degrees of freedom associated with the elemental nodes.

Once all elemental contributions are assembled and the boundary conditions imposed, the result is a system of linear algebraic equations at each iterative step that can be written in the following compact form:

$$[K]\{\Psi\} = \{F\} \tag{16.20}$$

where the "global" arrays $[K]$ and $\{F\}$ are the assembled versions of their elemental counterparts. The global vector of unknowns $\{\Psi\}$ is consistently composed of all nodal values of velocity components and pressure, that is,

$$\{\Psi\} = \begin{Bmatrix} \{V_z\} \\ \{V_r\} \\ \{V_\theta\} \\ \{P\} \end{Bmatrix}$$

The subvectors $\{V_z\}$, $\{V_r\}$, and $\{V_\theta\}$ contain all nodal values of the axial, radial, and tangential velocity components, respectively, and the subvector $\{P\}$ is that of nodal pressures. Solution of the system of equations (16.20) requires an iterative procedure whereby the variables $\bar{V}_z$, $\bar{V}_r$, $\bar{V}_\theta$, and $\bar{\nu}_e$ in these equations are progressively updated toward convergence.

Method of Numerical Solution

In an effort to reduce the core size and CPU time requirements, the tangential momentum Equation (16.19) is uncoupled and solved separately as part of each iterative step. The entire procedure is initiated by assuming a theoretically creeping (zero Reynolds number) flow field, where all inertia terms are extracted, with the outcome treated as an initial guess.

In progressing from one iteration to the next, an underrelaxation factor was used to ensure monotonic convergence. It is recommended that the value of this damping factor should be large enough to avoid numerical oscillations or slow convergence and small enough to avoid numerical divergence. In this particular example, convergence was achieved on execution of roughly forty iterations using an underrelaxation factor of 0.7.

Details of the algebraic equation solver are similar to those of the frontal method proposed by Hood [10]. This method is based on a partial-pivoting Gauss elimination technique, whereby only a submatrix of the coefficient matrix $[K]$ in Equation (16.20) is stored in the core at any given time during execution.

Assessment of the Centered-Rotor Flow Field

Consideration of the annular seal problem is aimed at validating the preceding centered-rotor axisymmetric flow model. Under focus, in particular, are such aspects as the type of finite element selected, the calculation of eddy viscosity through the turbulence closure, and the effectiveness of the upwinding method in analyzing the convection terms. Evaluation of the finite element solution is made in light of the flow measurements by Morrison et al. [11] for the same annular seal.

The annular seal under investigation has a gap-width/length ratio of 3.4 percent. Development of the through-flow and tangential velocity profiles along this seal were measured under a high rotor-surface/inlet-velocity ratio of 4.2 and a Reynolds number (based on the inlet velocity and clearance width) of 13,280. The finite element discretization model (Figure 16.6) is created with nine rotor-to-housing grid lines and 21 inlet-to-exit stations.

Figure 16.7 shows the set of measured through-flow and tangential velocity profiles at nondimensional axial length ratios of 0.25, 0.50, and 0.75, along with the computed profiles at these locations. As seen in the figure, the computed profiles seem to depict the experimental velocity profiles with reasonable accuracy. Note that the peak points of the through-flow velocity, particularly in the computed profiles, correspond to radii that are higher than the seal mean radius. This feature is due to the radial shifting of the fluid particles (or meridional streamlines) as a result of the high circumferential velocity near the rotor surface due to the high rotational speed of the rotor. Development of the computed tangential velocity along the seal in Figure 16.7 also seems to be in good agreement with the flow measurements.

Computational Development: Building the Zeroth-Order Flow Model

In this section, the task is to use the results obtained in the preceding section (namely, the axisymmetric-flow solution) in building a flow model that would be consistent with that associated with the displaced-rotor operation mode not only in the set of flow-governing equations but also in the frame of reference. It is true that the centered-rotor domain geometry and flow solution in the preceding section constitute the point of reference in the perturbation analysis (the topic of the next

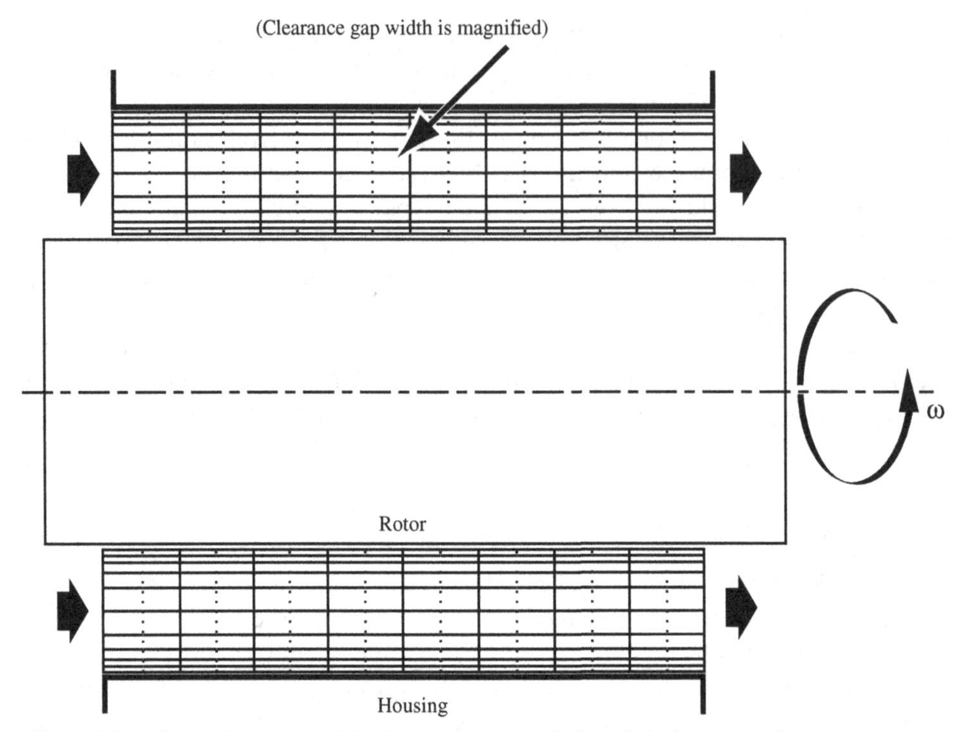

Figure 16.6. Finite element model of the axisymmetric flow field in an annular seal.

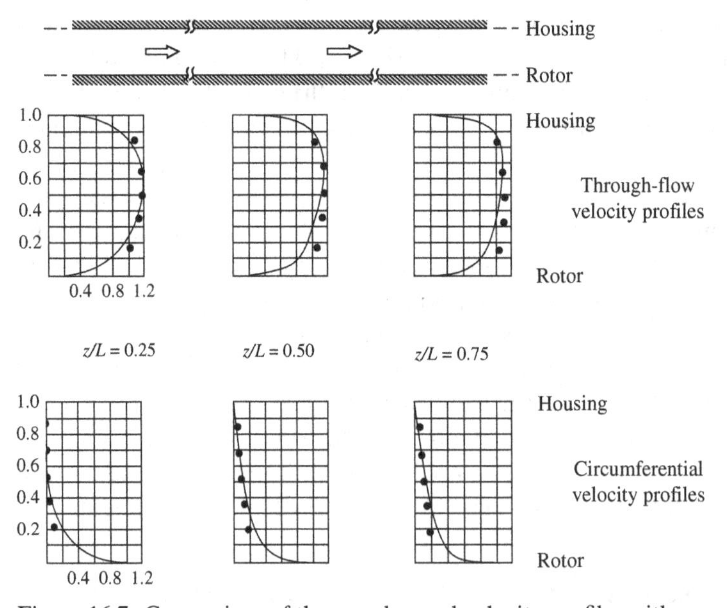

Figure 16.7. Comparison of the annular-seal velocity profiles with experimental work.

section). However, because the rotor disturbance presents itself and affects the flow field in a three-dimensional sense, a transition phase is needed whereby the governing equations and the flow solution are cast in a three-dimensional frame of reference. This phase involves defining an alternate frame of reference for casting

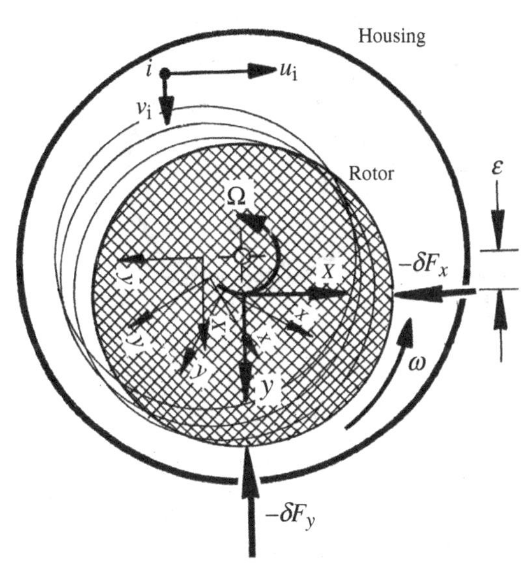

Figure 16.8. A rotating-translating frame of reference eliminates the flow time dependency.

the three-dimensional flow equations and modifying the axisymmetric flow solution accordingly. With the rotor eccentricity being the sole independent variable in the perturbation model, the other whirl component, which is the whirl frequency Ω, has to be part of the zeroth-order flow field. To this end, the alternate rotor-attached coordinate axes are defined to be rotating at an angular speed that is numerically identical to the whirl frequency, leaving the lateral eccentricity to be the focus of the perturbation analysis. These steps ensure compatibility of the zeroth-order flow field with the three-dimensional perturbed flow field once the off-center operation mode is encountered.

Strategy

As discussed earlier, the rotor whirling motion is composed of two components: the rotor displacement ϵ and the whirl frequency Ω. Of these, only the former is the perturbation variable. In building the zeroth-order flow model, therefore, it is essential to include the whirl frequency, as well as its effects on the flow-governing equations and boundary conditions. To this end, an unconventional frame of reference is used, one that is attached to the rotor and is rotating with it at a frequency that is numerically identical to the whirl frequency (Figure 16.8) and not the rotor speed. Note that this frequency is preset, as explained earlier in the rotordynamic analysis, with the rotordynamic coefficients requiring the fluid reaction forces in response to, at least, three magnitudes of it.

In view of the preceding, the perturbation model (in the next section) is centered only around the virtual, infinitesimally small displacement ϵ of the rotor axis (see Figure 16.8). This displacement is perceived to cause varied "virtual" deformations in the entire assembly of finite elements occupying the rotor-to-housing gap (Figure 16.9). The equations associated with each element are then revisited, with

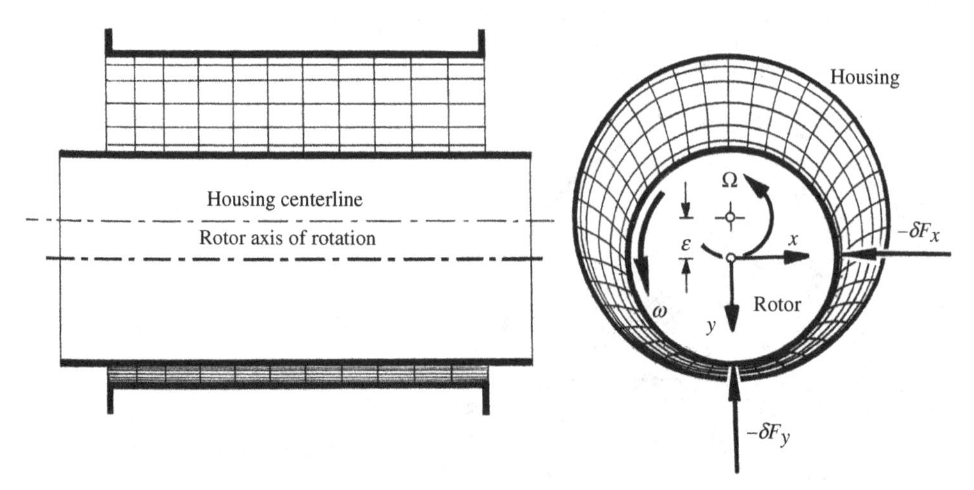

Figure 16.9. Geometric distortions of the finite elements due to the rotor whirl.

the rotor eccentricity now being part of them. These equations are then expanded to separate the eccentricity-related terms and compared with their zeroth-order counterparts (the topic of this section) to find the rates at which the flow variables change with respect to ϵ. The rate at which the rotor-surface pressure varies with ϵ is then separated and integrated over the surface to find the fluid-exerted forces, which dictate the rotordynamic characteristics of the fluid-rotor interaction system.

Transition to an Alternate Frame of Reference

A close look at the rotor-to-housing flow passage (see Figures 16.8 and 16.9) reveals that the displaced-rotor flow field is time-dependent. The time dependency here is clearly a result of the rotor-axis whirling motion. To remove the time dependency of the flow field in the displaced-rotor operation mode, a rotating-translating frame of reference that is attached to the rotor and is whirling with it at a precession (or whirl) frequency Ω (Figure 16.10) is used. At this preparatory step, only the rotation of these coordinate axes is invoked in the manner depicted in Figure 16.10. In this case, two key acceleration contributors, namely, the centripetal and Coriolis acceleration components, become part of the flow-governing equations and are modeled as such.

Regardless of the shift in coordinate axes and their in-plane $(x - y)$ rotation in Figure 16.10, it should be emphasized that the already obtained axisymmetric-flow solution vector $\{\Psi\}$ in Equation (16.20) remains conceptually valid. The fact is that just as we selected the rotating frame of reference, in this section we could have assigned any of a theoretically infinite number of coordinate axes (rotating or stationary) to the same physical problem, and the flow solution would be the same. This explains the next computational step, where the zeroth-order flow solution is deduced from the already existing axisymmetric-flow solution through simple adaptation means in consistency with the newly-introduced rotating frame of reference. However, the need remains to recast the flow-governing equations and their finite element counterparts.

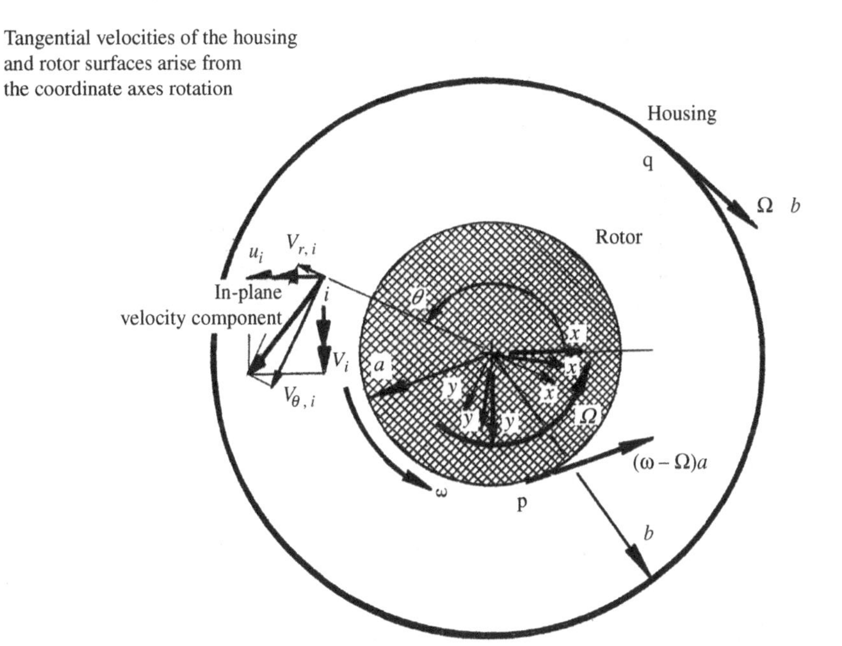

Figure 16.10. Conversion into a rotor-attached frame of reference.

Adaptation of the Axisymmetric Flow Solution

Throughout the next step, the three-dimensional flow model is created and analyzed based on the axisymmetric flow solution plus some other physical extensions. The twenty-noded isoparametric element in Figure 16.11 will again be used.

In creating the three-dimensional version of the flow domain, the meridional plane in Figure 16.11 is repeated around the circumference as many times as deemed sufficient. Correspondingly, a consistent flow solution vector $\{\Psi_o\}$ is constructed by repeating the axisymmetric-flow solution vector $\{\Psi\}$ [Equation 16.20)] as many times as there are meridional planes in Figure 16.12. Next, a transition step is performed whereby the nodal velocity components V_z, V_r, and V_θ are adjusted (the static pressure is axes-independent) in accordance with the change in coordinate system indicated earlier. It is with no lack of generality in Figure 16.10 that the instantaneous position of the axes labeled x and y is chosen for reference purposes. Completing the frame of reference is the z axis (not shown in the figure), which is coincident with both the housing centerline and the rotor axis at this stage of the computational procedure.

Figure 16.10 shows the velocity components $V_{r,i}$ and $V_{\theta,i}$ at a typical node i in a typical cross-flow section of the rotor-to-housing gap. Referring to the x and y components of relative velocity in the rotating frame of reference at the same node by u_i and v_i, respectively, the following velocity conversion relationships apply:

$$u_i = V_{r,i}\cos\theta_i - (V_{\theta,i} - \Omega r_i)\sin\theta_i \tag{16.21}$$

$$v_i = -V_{r,i}\sin\theta_i - (V_{\theta,i} - \Omega r_i)\cos\theta_i \tag{16.22}$$

where ω is the axes rotational frequency, chosen to be numerically identical to the rotor whirl frequency Ω in magnitude and sign. Equations (16.21) and (16.22) apply

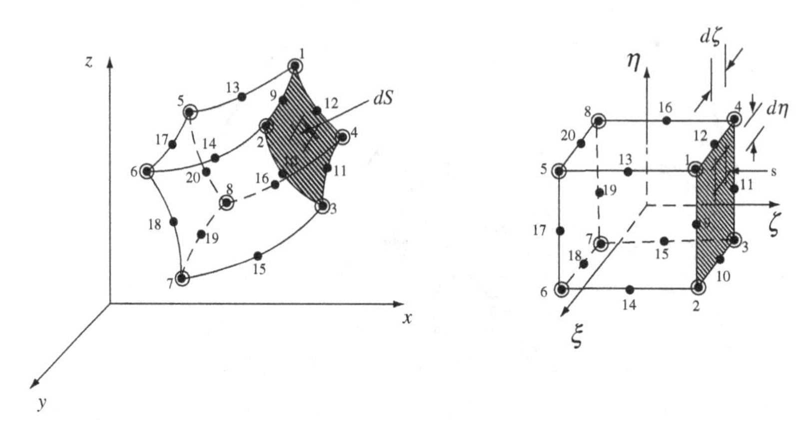

Figure 16.11. Curve-sided isoparametric element for the zeroth order flow field.

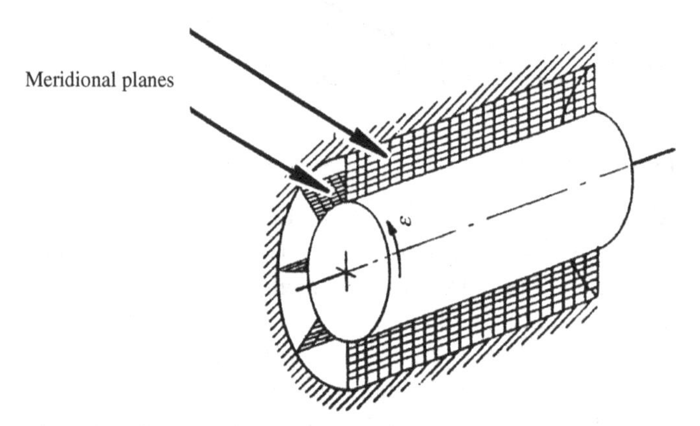

Figure 16.12. Creating a three-dimensional finite element model from the two-dimensional axisymmetric flow-analysis model.

to all computational nodes in the rotor-to-housing flow passage. A special case, however, concerns the nodes existing on the housing surface, such as node q in Figure 16.10. Being physically stationary, the only relative velocity that node q can possess is solely due to the rotation of the coordinate axes and is in the tangential direction, as shown in Figure 16.10. On the other hand, a node such as p in this figure initially possesses the solid-body tangential velocity ωa for existing on the rotor surface, where a is the local rotor radius at this particular axial location. Assuming, as indicated in Figure 16.10, that the two frequencies ω and Ω are indeed in the same direction, an observer who is located at the origin of the rotating axes (and is rotating with them) will register a net relative velocity of $[(\omega - \Omega)a]$, for node p that is also in the tangential direction. Note, by reference to Figure 16.10, that the two angular speeds ω and Ω are in the same direction but don't have to always be. After all, backward (negative) whirl is as common in turbomachinery as forward (positive) whirl. Recall, by reference to the rotordynamic analysis segment earlier in this chapter, that the choice of the whirl frequency (both magnitude and direction) is purely our own. It was then indicated that determination of the rotordynamic coefficients requires knowledge of the fluid-induced forces associated with, at least, three different whirl frequencies. Perhaps the only restriction in selecting these frequencies is to cover a sufficiently large span of the whirl frequency ratio (Ω/ω).

The absolute-to-relative velocity conversion step ends with the axial-velocity component, which remains the same in both frames of reference, that is,

$$w_i = V_{z,i} \tag{16.23}$$

Applied to the flow solution vector $\{\Psi_o\}$, which is simply a repetition of the axisymmetric flow-solution vector, the velocity conversion defined by Equations (16.21) through (16.23) leads to the definition of an equivalent vector $\{\Phi_o\}$ with relative, rather than absolute, velocity components at all computational nodes.

Flow-Governing Equations in the Rotating Frame of Reference

At a given point in the rotor-to-housing gap, let the relative velocity components be u, v, and w in the x, y and z directions, respectively (see Figure 16.10). Using the same flow assumptions cited in the axisymmetric flow analysis section, the flow-conservation equations can be written in the rotating frame of reference (see Figure 16.10) as follows:

Continuity Equation

$$\frac{\partial u}{\partial x} + \frac{\partial v}{\partial y} + \frac{\partial w}{\partial z} = 0 \tag{16.24}$$

x-Momentum Equation

$$\begin{aligned} u\frac{\partial u}{\partial x} + v\frac{\partial u}{\partial y} + w\frac{\partial u}{\partial z} + 2\Omega v - \Omega^2 x = -\frac{1}{\rho}\frac{\partial p}{\partial x} + \nu_e \nabla^2 u + \frac{\partial}{\partial y}\left(\nu_t \frac{\partial u}{\partial y}\right) + \frac{\partial}{\partial z}\left(\nu_t \frac{\partial u}{\partial z}\right) \\ + 2\frac{\partial}{\partial x}\left(\nu_t \frac{\partial u}{\partial x}\right) + \frac{\partial}{\partial y}\left(\nu_t \frac{\partial v}{\partial x}\right) + \frac{\partial}{\partial z}\left(\nu_t \frac{\partial w}{\partial x}\right) \end{aligned} \tag{16.25}$$

y-Momentum Equation

$$\begin{aligned} u\frac{\partial v}{\partial x} + v\frac{\partial v}{\partial y} + w\frac{\partial v}{\partial z} - 2\Omega u - \Omega^2 y = -\frac{1}{\rho}\frac{\partial p}{\partial y} + \nu_e \nabla^2 v + \frac{\partial}{\partial x}\left(\nu_t \frac{\partial v}{\partial x}\right) + \frac{\partial}{\partial z}\left(\nu_t \frac{\partial v}{\partial z}\right) \\ + 2\frac{\partial}{\partial y}\left(\nu_t \frac{\partial v}{\partial y}\right) + \frac{\partial}{\partial x}\left(\nu_t \frac{\partial u}{\partial y}\right) + \frac{\partial}{\partial z}\left(\nu_t \frac{\partial w}{\partial y}\right) \end{aligned} \tag{16.26}$$

z-Momentum Equation

$$\begin{aligned} u\frac{\partial w}{\partial x} + v\frac{\partial w}{\partial y} + w\frac{\partial w}{\partial z} = -\frac{1}{\rho}\frac{\partial p}{\partial z} + \nu_e \nabla^2 w + \frac{\partial}{\partial x}\left(\nu_t \frac{\partial w}{\partial x}\right) + \frac{\partial}{\partial y}\left(\nu_t \frac{\partial w}{\partial y}\right) \\ + 2\frac{\partial}{\partial z}\left(\nu_t \frac{\partial w}{\partial z}\right) + \frac{\partial}{\partial x}\left(\nu_t \frac{\partial u}{\partial z}\right) + \frac{\partial}{\partial y}\left(\nu_t \frac{\partial v}{\partial z}\right) \end{aligned} \tag{16.27}$$

In reviewing the preceding set of equations, note that terms such as $2\Omega v$ and $\Omega^2 x$ represent, respectively, the added Coriolis and centripetal acceleration components associated with the rotation of the coordinate axes.

The boundary conditions here, as well as the turbulence closure, are similar to those in the axisymmetric flow analysis, with the exception that the velocity-related boundary conditions, as well as the vorticity definition (in the turbulence closure), should now be written in the rotating frame of reference (see Figure 16.10), beginning with relationships (16.21) through (16.23). Appendix D is devoted to the topic of the finite element formulation of the *zeroth-order* flow field. This differs from the axisymmetric flow solution in the sense that the former carries in it a finite magnitude of the "whirl" frequency but not yet rotor lateral displacement. This is followed by Appendix E, where the rotor eccentricity is introduced. This Appendix highlights the fundamentals of this chapter's topic, being the finite element–based perturbation analysis.

Calculation of the Force-Related Rotordynamic Coefficients

Of all the nodal values in the subvector $\{P\}$, define a (smaller) subvector $\{\chi\}$ to contain the pressure values only at the subset of nodes existing on the rotor surface, that is,

$$\{\chi\} = \{p_i, i = 1, N_s\} \subset \{P\}$$

where N_s is the total number of corner nodes that exist on the rotor surface (recall that the pressure is a degree of freedom only at the element corner nodes, as shown in Figure 16.12). Next, a subvector $\{\partial\chi/\partial\epsilon\}$ is consistently defined as follows:

$$\left\{\frac{\partial\chi}{\partial\epsilon}\right\} = \left\{\frac{\partial p_i}{\partial\epsilon}, i = 1, N_s\right\}$$

It is precisely this vector of pressure derivatives that is known at this computational step because it is part of the global vector $\partial\{\Phi\}/\partial\epsilon$, which becomes known once the matrix equation is solved.

The remainder of this section involves the integration of these pressure derivatives (Figure 16.13) over the rotor surface in order to obtain the fluid-induced force components $\partial F_x/\partial\epsilon$ and $\partial F_y/\partial\epsilon$. These, as indicated earlier in this chapter, are used to determine the rotordynamic coefficients [Equations (16.6) and (16.7)]. To this end, consider a typical twenty-nodded finite element that shares the cross-hatched face (where $\zeta = 1$) with the rotor surface, as shown in Figure 16.14. Recalling that the pressure is interpolated linearly throughout the element using the linear shape functions M_i, the derivative $\partial p/\partial\epsilon$ can consistently be interpolated over the surface area s as follows:

$$\left(\frac{\partial p}{\partial\epsilon}\right)^{(s)} = \sum_{i=1}^{4} M_i(1, \eta, \xi)\, \frac{\partial p_i}{\partial\epsilon}$$

where $$\frac{\partial p_i}{\partial\epsilon} \in \left\{\frac{\partial\chi}{\partial\epsilon}\right\} \subset \left\{\frac{\partial P}{\partial\epsilon}\right\} \qquad \text{(by definition)}$$

Integration of $(\partial p/\partial\epsilon)^{(s)}$ over the surface area s yields the contribution of that element, namely, $(\partial F/\partial\epsilon)^{(s)}$, to the overall rate of change of the fluid-exerted force,

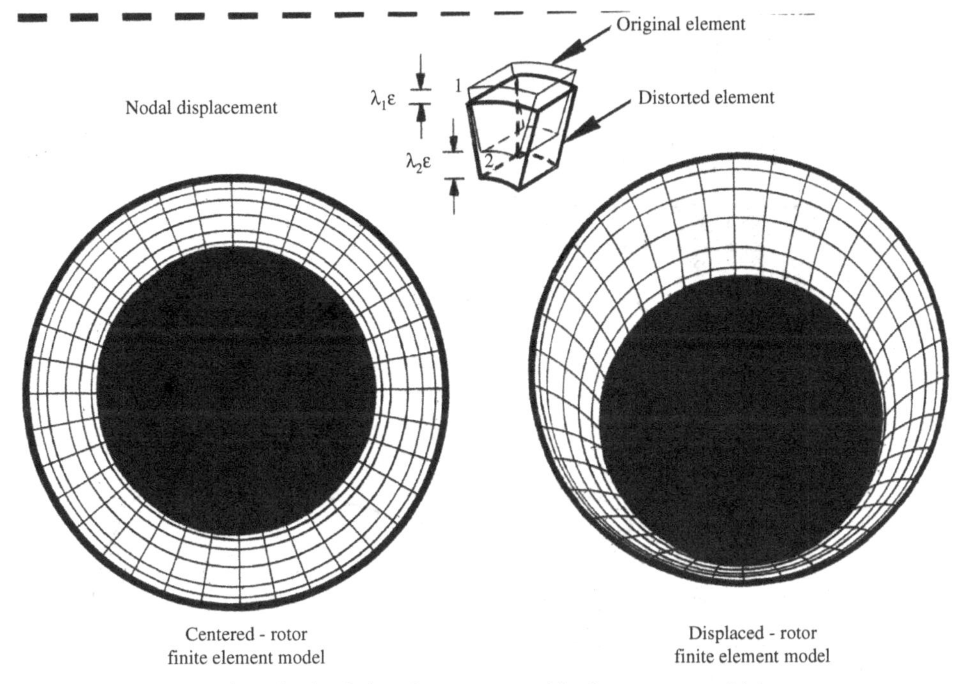

Figure 16.13. Distortions in the finite element assembly due to rotor whirl.

that is, $\partial F/\partial\epsilon$. Now let n_x and n_y denote the components of the local unit vector $\vec{n}$ that is perpendicular to the rotor surface and is pointing inward toward the rotor axis. The components of $(\partial F/\partial\epsilon)^{(s)}$ now can be expressed as follows:

$$\left(\frac{\partial F_x}{\partial\epsilon}\right)^{(s)} = \int_{-1}^{+1}\int_{-1}^{+1} n_x \left(\frac{\partial p}{\partial\epsilon}\right)^{(s)} G(\eta,\xi)\, d\eta d\xi \tag{16.28}$$

$$\left(\frac{\partial F_y}{\partial\epsilon}\right)^{(s)} = \int_{-1}^{+1}\int_{-1}^{+1} n_y \left(\frac{\partial p}{\partial\epsilon}\right)^{(s)} G(\eta,\xi)\, d\eta d\xi \tag{16.29}$$

where $G(\eta,\xi)$ is a Jacobian-like function of Cartesian-to-local-area transformation [13] and can be expressed for the element geometry in Figure 16.14 as follows:

$$G(\eta,\xi) = \left\{\left[\left(\frac{\partial y}{\partial\eta}\frac{\partial z}{\partial\xi} - \frac{\partial y}{\partial\xi}\frac{\partial z}{\partial\eta}\right)^2 + \left(\frac{\partial x}{\partial\xi}\frac{\partial z}{\partial\eta} - \frac{\partial x}{\partial\eta}\frac{\partial z}{\partial\xi}\right)^2 + \left(\frac{\partial x}{\partial\eta}\frac{\partial y}{\partial\xi} - \frac{\partial x}{\partial\xi}\frac{\partial y}{\partial\eta}\right)^2\right]^{1/2}\right\}_{\zeta=1}$$

Referring back to Figure 16.14, note that the element face that exists on the rotor surface (in the physical frame of reference) corresponds to the $\zeta = 1$ face of the conjugate element in its local coordinate system.

Summing up the contributions of all surface areas, the rates at which the fluid-induced force components are developed on the rotor surface can be expressed as follows:

$$\frac{\partial F_x}{\partial\epsilon} = \sum_{s=1}^{M_s}\left(\frac{\partial F_x}{\partial\epsilon}\right)^{(s)}$$

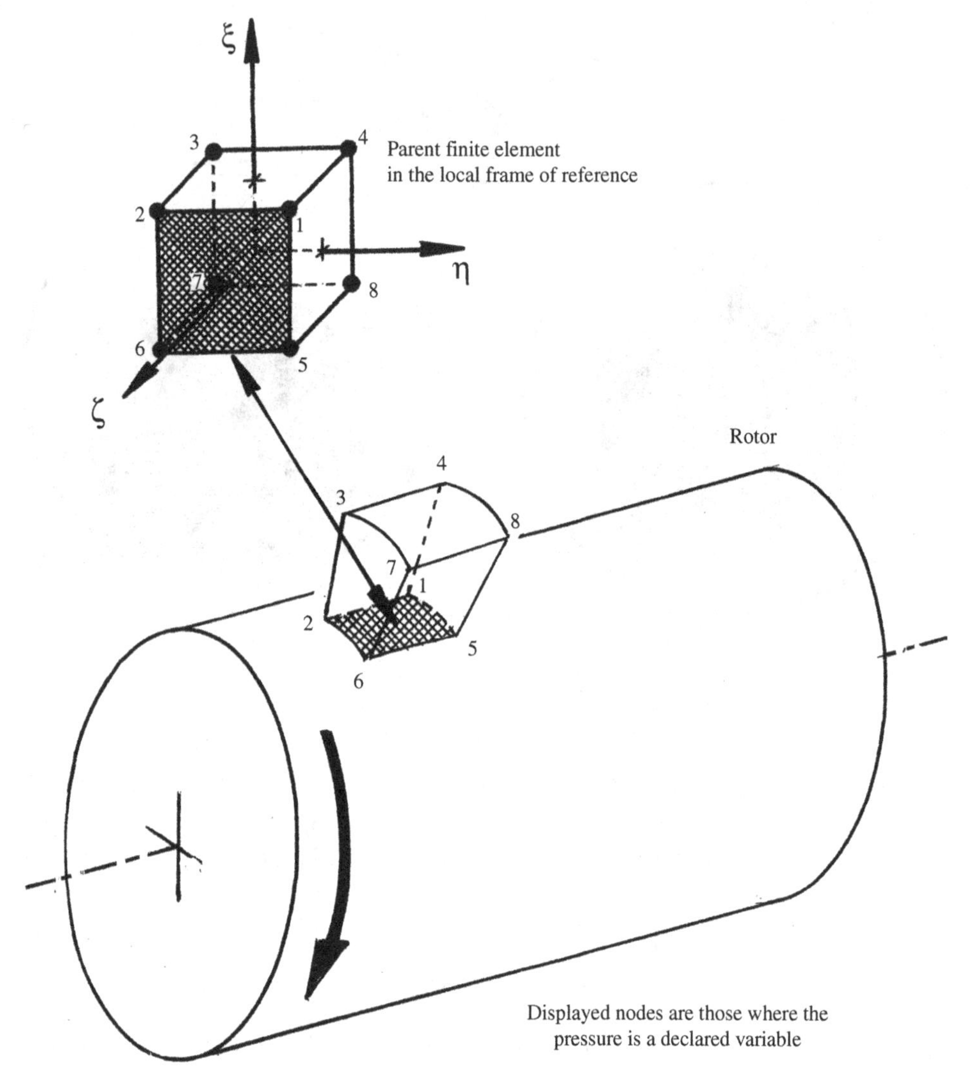

Figure 16.14. Integration of pressure perturbations over the rotor surface.

$$\frac{\partial F_y}{\partial \epsilon} = \sum_{s=1}^{M_s} \left(\frac{\partial F_y}{\partial \epsilon} \right)^{(s)}$$

where M_s is the total number of surface areas that coincide with the rotor surface.

Applications: Benchmark Test Case—Comparison with Cal Tech's Experimental Work

Background

The experimental results reported by Guinzberg et al. [14] are used in this section to validate the finite element–based perturbation model in this chapter. These were obtained in a special rotor force test facility, where the secondary (or leakage) passage in a typical hydraulic pump stage (see Figure 16.1) was simulated. The

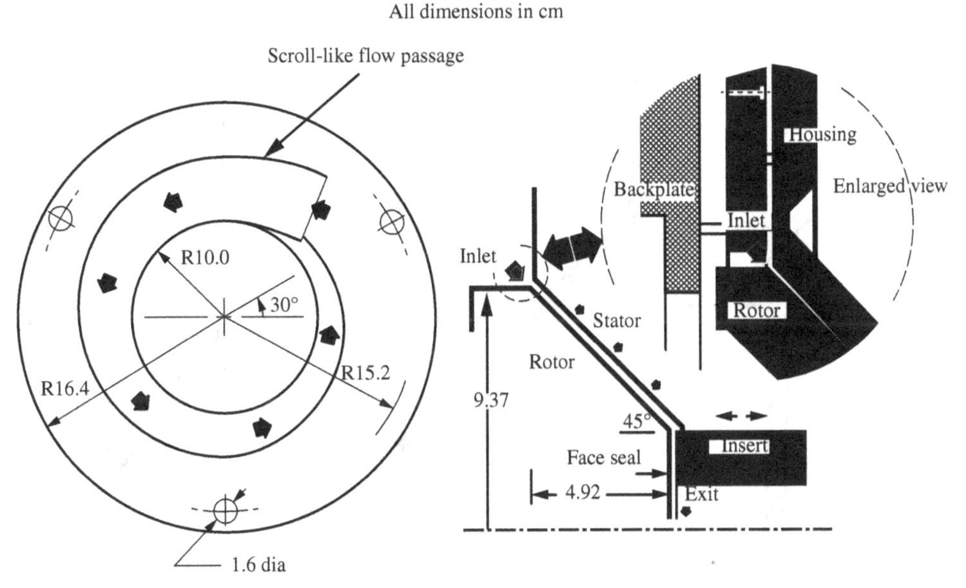

Figure 16.15. Cal Tech's rotordynamic-measurements test rig.

experimental objective in doing so was to investigate the rotordynamic influence of this passage on the shrouded impeller in a typical pump stage.

A schematic of the test facility's relevant features is provided in Figure 16.15. The pump leakage passage is simulated by a long and narrow annular gap ending, as is customarily the case, with a tight-clearance seal that resembles the face seal in the computational domain of Baskharone and Hensel [15]. Typical values of the clearance width, mass-flow rate, rotor speed, and inlet preswirl all were varied in what came out as a comprehensive parametric study. In the test rig, the rotor was externally forced into a whirling-motion type of excitation. Under varied magnitudes of the whirl frequency ratio (whirl-frequency/rotor-speed ratio), the fluid-induced forces were measured.

In an effort to assess the accuracy of the current perturbation model, two different sets of operating conditions were selected for the comparative study. The difference between these was in the inlet-swirl/tip-velocity ratio. This ratio was chosen to be zero (no inlet swirl at the entrance station) and 2.0. The rest of the operating conditions were as follows:

- Volumetric flow rate $\dot{V} = 0.001262 \text{m}^3/\text{s}$
- Rotor speed $\omega = 1{,}000$ rpm
- Seal clearance $t = 1.4$ mm

Features of the Centered-Rotor Flow Field

A finite element discretization model was carefully established throughout the computational domain. Attention was paid especially to the prevention of element overlapping in such a long and narrow flow passage, particularly in the vicinity of the face-seal inlet station. General features of this field-discretization model are displayed in Figure 16.16.

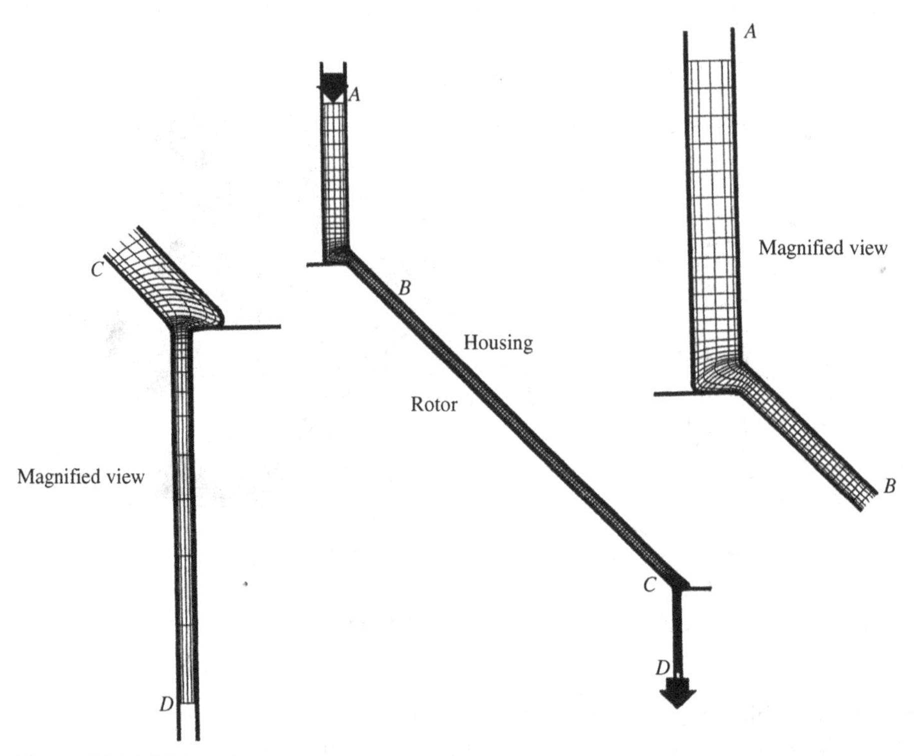

Figure 16.16. Finite element discretization model of the rig test section.

Figure 16.17 shows a plot of the computationally attained velocity vectors in critical subdomains of the flow passage under no preswirl of the inlet flow stream. Examination of this figure reveals a flow separation point (marked "S" in the figure) and a recirculation zone along the bottom of the rotor-side endwall. The through-flow structure seems, otherwise, to exhibit characteristically-fixed fully developed velocity profiles. The flow recirculation region (magnified in Figure 16.17) is caused by the flow tendency to migrate radially outward, particularly over the high-radius rotor-surface segment due to centrifugal-force effects, on one hand, and the tendency to proceed inward due to the streamwise decline in static pressure, on the other. Despite the lack of flow measurements in the study by Guinzberg et al. [14], it was essential to critically examine and try to "reason" the centered-rotor flow structure in this figure because this paves the way to the implementation of the much involved perturbation analysis.

Assessment of the Fluid-Induced Force Components

Figures 16.18 and 16.19 show the fluid-induced force components $F_r{}^*$ and $F_\theta{}^*$ in the radial and tangential directions, respectively, for the two magnitudes of preswirl velocity ratios. These force components are in response to the rotor whirling motion and are nondimensionalized as follows:

$$F_r{}^* = \frac{\dfrac{\partial F_r}{\partial \epsilon}}{\pi R_2{}^2 L \rho \omega^2}$$

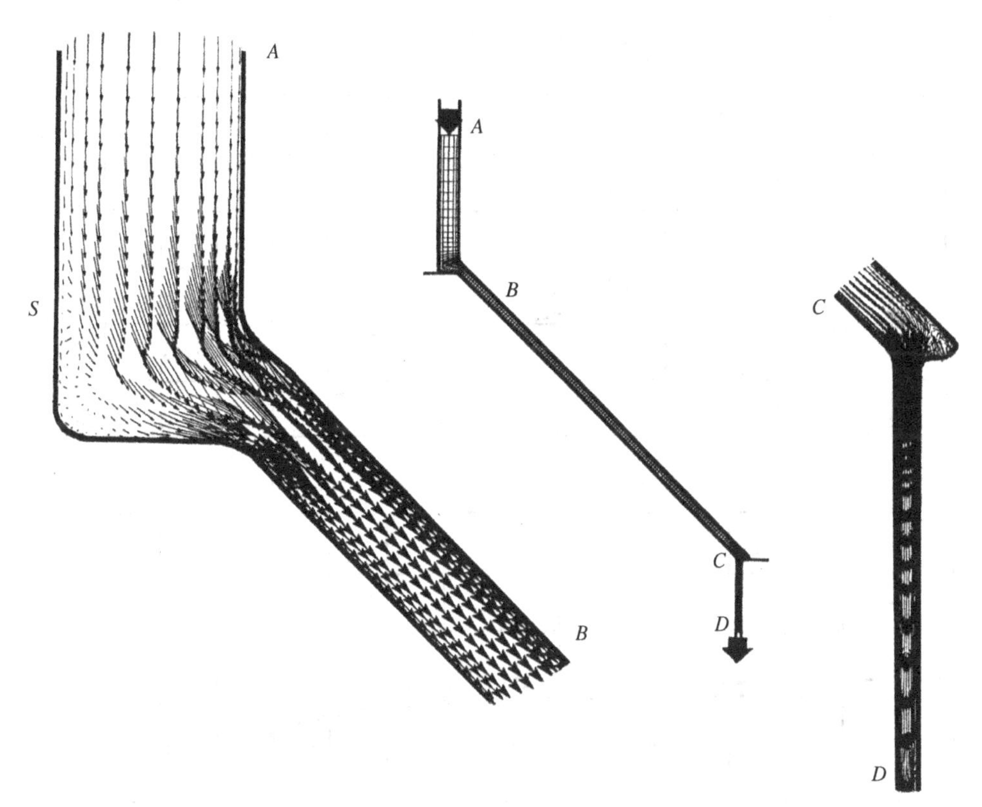

Figure 16.17. Velocity-vector plot for the centered-rotor flow domain.

$$F_\theta{}^* = \frac{\dfrac{\partial F_\theta}{\partial \epsilon}}{\pi R_2{}^2 L \rho \omega^2}$$

where R_2, L, and ω and are the rotor tip radius, axial extension, and running speed, respectively, with ρ being the fluid static density.

Referring to Figure 16.18, the radial force component trend is clearly well predicted by the current perturbation analysis. In fact, a nearly uniform force magnitude seems to practically be the only difference between the two sets of results over the entire range of whirl frequency ratio. The fluid-exerted tangential force component in the same figure is notably different only in the backward-whirl range, where the rotor-axis whirl direction is opposite to the rotor speed. The force magnitudes nevertheless become, virtually identical as the whirl frequency picks up in the positive range. Referring to the $F_\theta{}^*$ distribution in Figure 16.18, the offset between the two sets of results is seen to continually shrink, up to a whirl-frequency ratio of 0.4. The figure also suggests "engineeringly negligible" discrepancies thereafter.

Figure 16.19 shows similar sets of results, but for an entrance tangential velocity that is twice the rotor-tip velocity. Similar to the zero-prerotation case, discussed earlier, the computational force-component distributions seem to follow the experimental trends but continue to mildly underpredict the experimental measurements. Comparing the results in Figure 16.18 with those in Figure 16.19, it is clear that the inlet preswirl has the effect of elevating the fluid-exerted radial force component for the entire range of whirl-frequency ratios under investigation. More importantly,

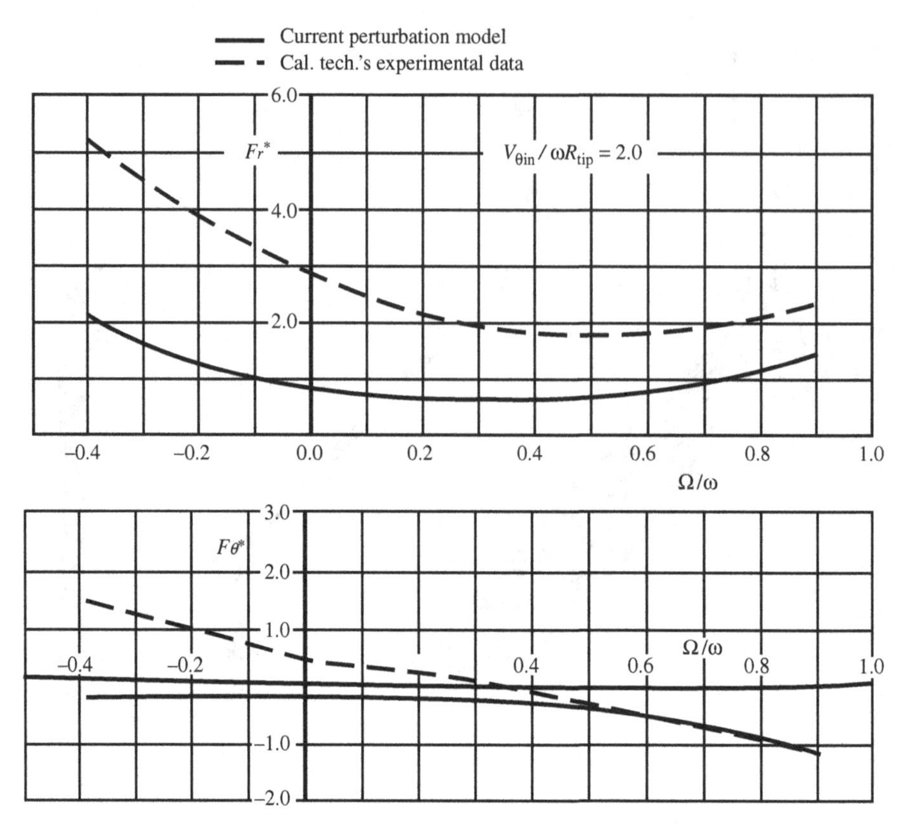

Figure 16.18. Comparison of the computed results with experimental data for no inlet swirl.

the high preswirl ratio creates a wide range of whirl frequencies within which the tangential force component acts as a destabilizing agent.

Applications: Perturbed Flow Structure due to Synchronous Whirl

Overview

Validation of the current model in the preceding test case was done in a "macroscopic" sense, with the net integrated rotor forces being the only variable under examination. In this example, the fine details of the perturbed rotor-to-housing flow field are rigorously verified. The intention here is to provide a better understanding of the source of rotordynamic forces, which, in turn, should serve the design process itself.

Choice of this test case was motivated by the availability of LDA flow measurements in an annular seal (Figure 16.20) with a synchronously whirling rotor (where the whirl frequency is identical to the rotor speed) by Morrison et al. [16]. The perturbed flow variables, composed of the velocity components and static pressure, in the clearance gap are subjects of comparison in this case. Next, the rotor surface is unwrapped, and the pressure perturbation distribution is examined. Results of the latter step had to be outside the comparison with experimental data, for the latter did not include any surface-pressure perturbation measurements.

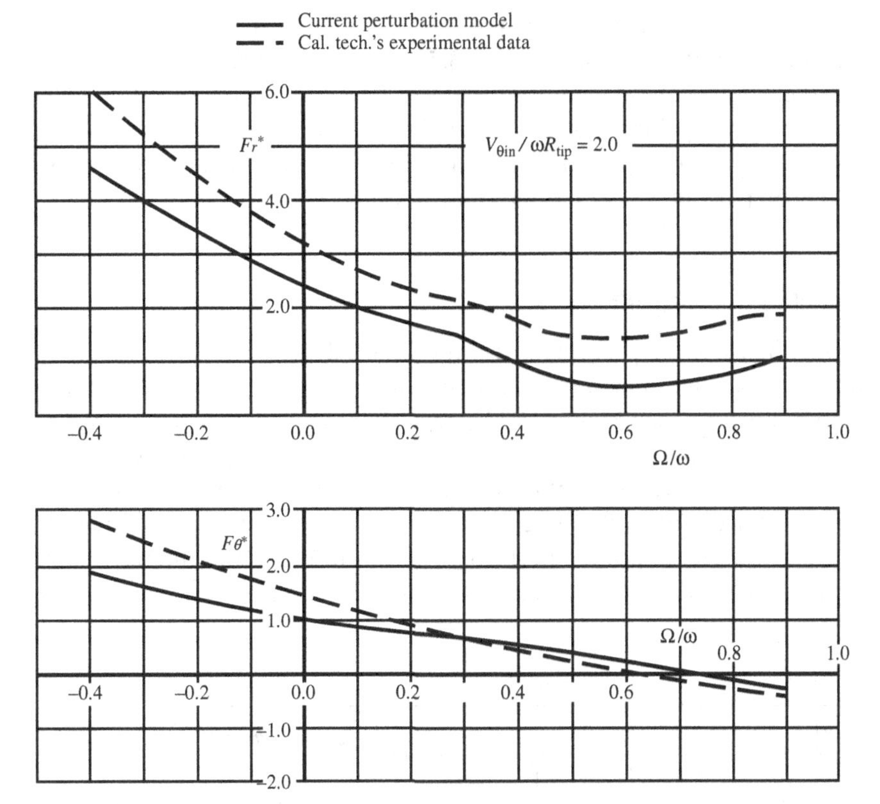

Figure 16.19. Comparison of the computed results with experimental data for nonzero inlet swirl.

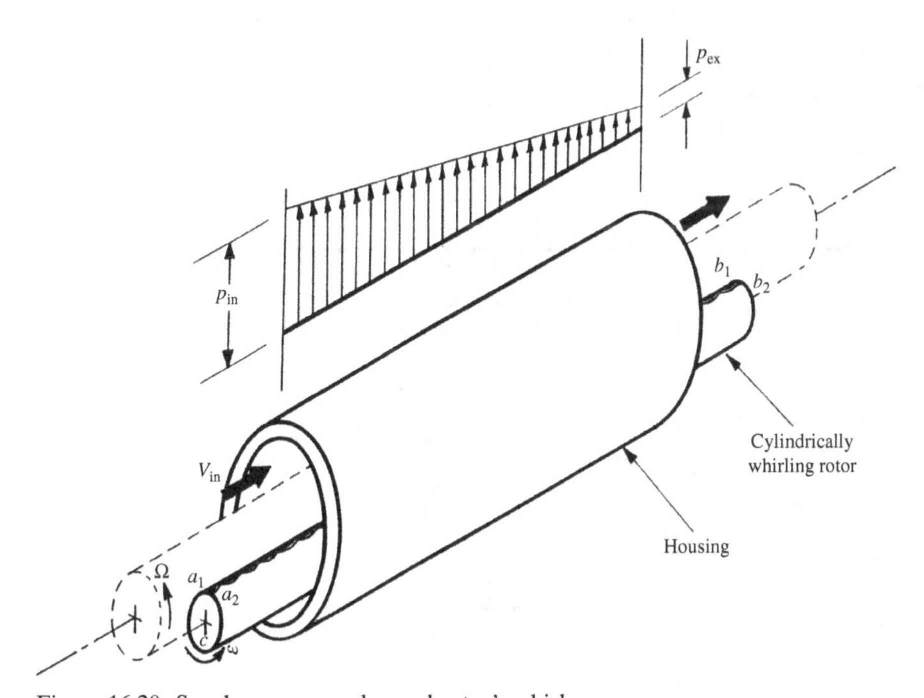

Figure 16.20. Synchronous annular-seal rotor's whirl.

The seal geometry and operating conditions in this study are identical to those of Morrison et al. [16]. With a seal length of 3.73 cm, a rotor diameter of 16.41 cm, and a nominal clearance of 1.27 mm, this seal was tested at a rotor speed of 3,600 rpm using water as the working medium. At all times, the rotor whirl frequency was identical (in magnitude and direction) to the rotor speed, and the eccentricity ratio ϵ/c was fixed at 50 percent, where c is the rotor-housing clearance width.

Grid Dependency Investigation

As is naturally the case, the perturbed flow resolution dependens on the grid resolution, which is a strong function of the number of computational planes in the circumferential direction (see Figure 16.11). Investigation of this dependency was carried out by varying this number and computing the fluid-exerted forces on the whirling rotor each time. Results of this preliminary step are shown in Figure 16.21, where the fluid-induced force components ($F_r{}^*$ and $F_\theta{}^*$) are obtained by integrating the rotor-surface pressure-force perturbations (on resolution in the radial and tangential directions) over all contributing finite element surfaces.

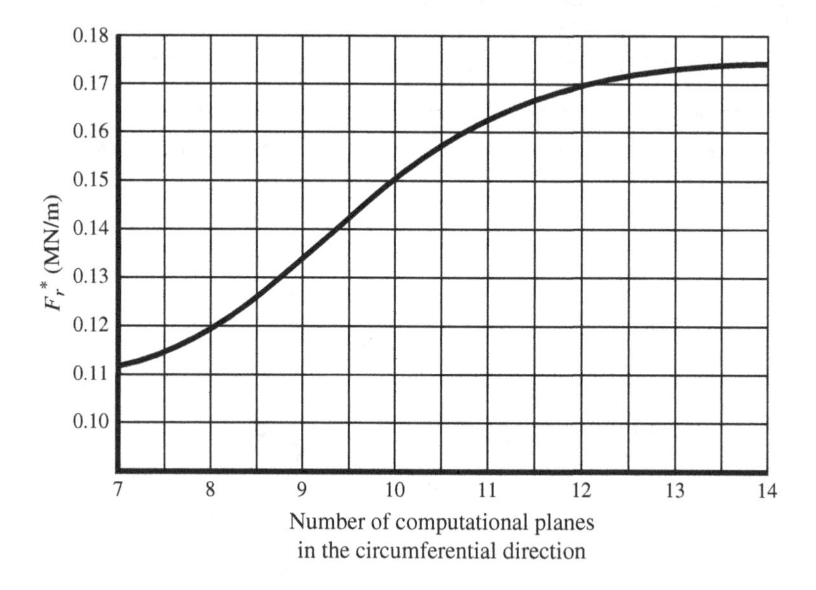

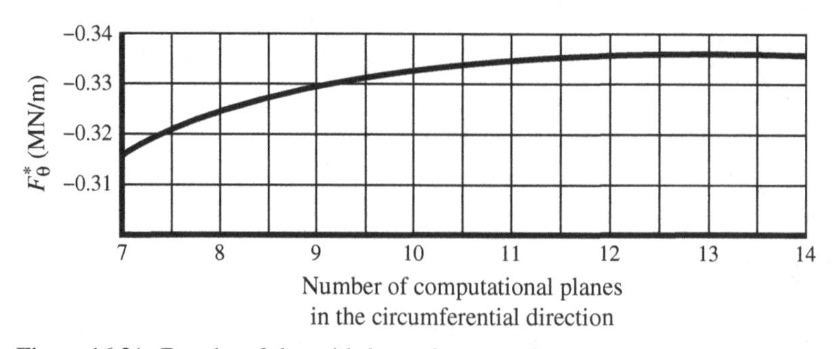

Figure 16.21. Results of the grid-dependency study.

The force trends in Figure 16.21 suggest that the radial-force component is more sensitive to the grid tangential resolution by comparison. The figure also shows that changes in the force magnitudes become acceptably small for a tangential computational-station count that is in excess of 11. In the current study, this parameter was selected to be 13, as a compromise between a sufficiently high resolution, on the one hand, and the CPU time consumption, on the other

Samples of the Computational Results

Examples of the computational results are shown in Figure 16.22 for the seal middle cross section as a representative location. The figure shows contours of the nondimensionalized perturbations in the velocity components (i.e., $V_z{}^*$, $V_r{}^*$, and $V_\theta{}^*$), which are defined as follows:

$$V_z{}^* = \frac{c}{V_{in}} \frac{\partial V_z}{\partial \epsilon}$$

$$V_r{}^* = \frac{c}{V_{in}} \frac{\partial V_r}{\partial \epsilon}$$

$$V_\theta{}^* = \frac{c}{\omega r_i} \frac{\partial V_\theta}{\partial \epsilon}$$

where

- V_{in} is the seal-inlet average through-flow velocity,
- ω is the rotor spinning speed,
- r_i is the rotor radius, and
- c is the seal nominal clearance,

with the whirl frequency being identical to the rotor speed, giving rise to the synchronous whirl, the topic of this study. There is, however, no possible comparison between the results in Figure 16.22 and Morrison's flow measurements because the latter involved the perturbed and centered-rotor flow variables combined.

Examination of Figure 16.22 reveals that the maximum axial-velocity perturbation occurs in the vicinity of the rotor "pressure" side (where the clearance is smaller than the nominal value) and at a tangential location that lags the minimum-clearance position by reference to the whirl direction. The radial velocity perturbations in the same figure are clearly smaller than those of the axial-velocity component (recall that this is a straight annular seal) and attains its maximum value near the rotor pressure side as well. The difference, however, is that the radial-velocity perturbation peaks ahead of the minimum-clearance tangential position. As for the tangential-velocity perturbations in Figure 16.22, the peak value appears closer to the housing and in the region of the rotor suction side.

Figure 16.23 shows contours of the nondimensionalized pressure perturbations $p^*c/\rho V_{in}{}^2$, as well as those of the combined (centered-rotor plus whirl-related) pressure values in the annulus and at the same (50 percent seal length) axial location. Of these, the combined-pressure values are non dimensionalized using the seal-inlet static pressure p_{in}. As expected, the peak values in each plot occur on the rotor pressure side but not at the minimum-clearance position, as may be intuitively anticipated.

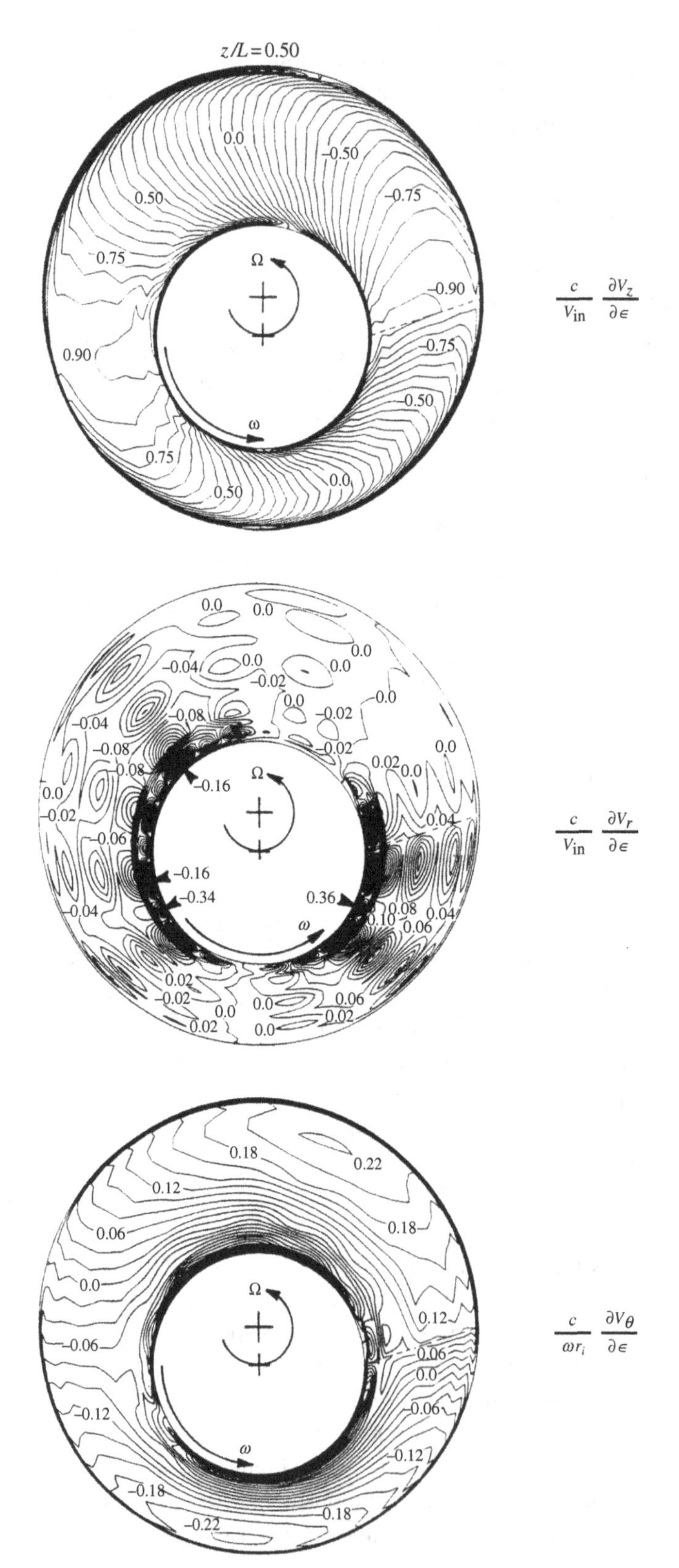

Figure 16.22. Perturbations of the velocity components at 50 percent of the seal's length.

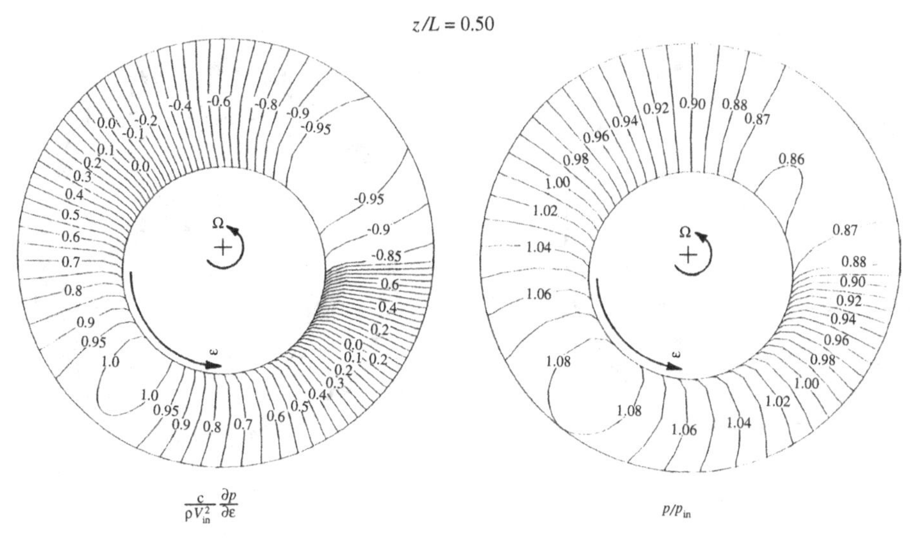

Figure 16.23. Pressure and pressure-perturbation distributions at 50 percent of the seal's length.

Comparison with Experimental Data

One-to-one comparisons with Morrison's flow measurements are presented in Figures 16.24 through 16.28. In these figures, the computational results were obtained by superimposing the perturbations of each flow variable on the variable's centered-rotor magnitude.

In assessing the perturbation-model accuracy, it was important to verify a fact that was solidly stated by Morrison et al. [16], that the maximum through-flow velocity magnitude occurs on the pressure side of the rotor over the early seal section but then moves toward the suction side as the flow progresses to the seal's downstream sections. Comparison of the computed velocity contours in Figures 16.24 through 16.28 reveals that the computational results are indeed in agreement with Morrison's documented observation.

It is also important, in comparing the numerical and experimental sets of data, to point out that early sections of the seal (perhaps up to the midseal axial location) are those where no significant agreement would be anticipated. This is largely due to the flow admission losses in the actual test rig, whereas the computational model treats this early seal segment as systematically, in the sense of the turbulence model and near-wall analysis, as does the entire seal. This is primarily why the velocity contours become more and more in agreement with the experimental data as the seal exit station is approached (e.g., Figures 16.27 and 16.28). Included, in particular, is the tangential shift of the maximum axial-velocity component from a tangential location in the pressure (small-clearance) region gradually toward the suction side. In fact, one would go even further in reviewing the flow measurements, indicating that the measurements themselves do not exhibit a characteristically similar axial-velocity pattern until the 77 percent seal-length section is reached

An interesting phenomenon that is confirmed by both the numerical and experimental sets of results is that of flow separation and recirculation near the seal exit station. These actions, by reference to Figures 16.27 and 16.28, take place on the

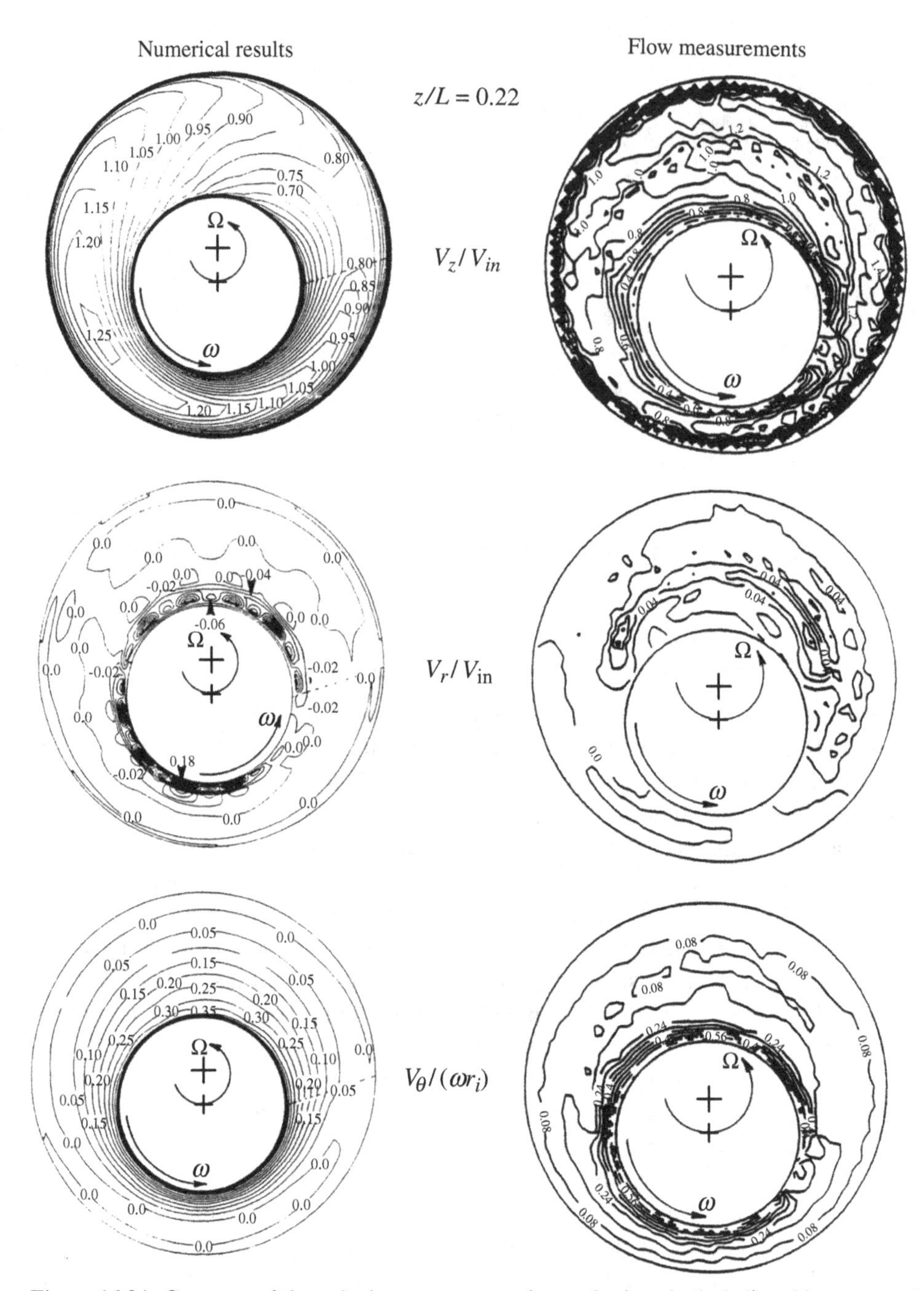

Figure 16.24. Contours of the velocity components (perturbations included) at 22 percent of the seal's length.

rotor suction side and are more pronounced at the tangential location marked *s* in these two figures. It is also in these figures that the numerical results yield shapes and magnitudes of the velocity contours that are very much consistent with the flow measurements.

Another flow structure that is equally interesting involves the tangential velocity distribution and pertains, in particular, to the seal sections at 49 percent and 77

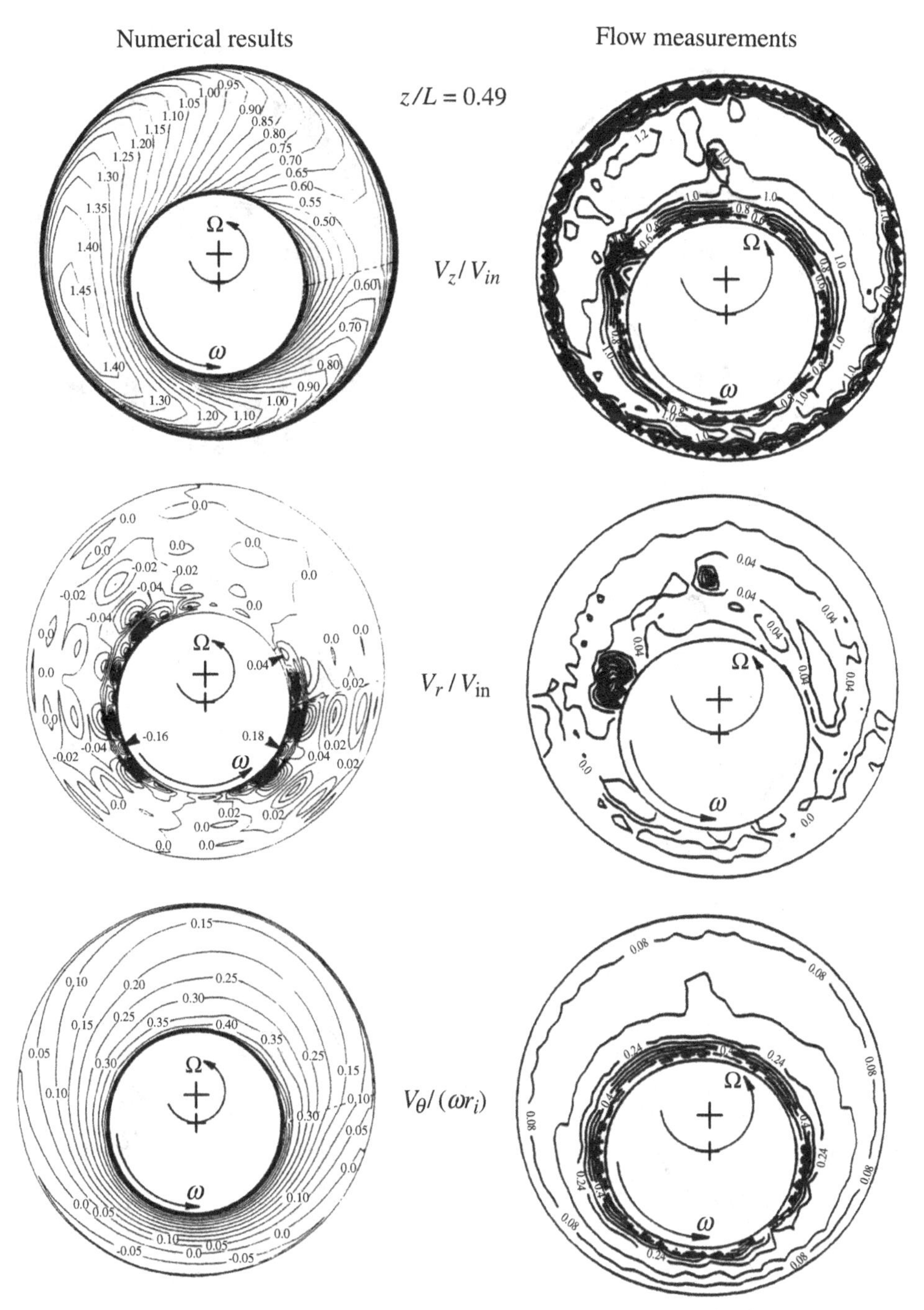

Figure 16.25. Contours of the velocity components (perturbations included) at 49 percent of the seal's length.

percent of seal length (see Figures 16.25 and 16.26). According to these two figures, there exists a region of reversed tangential velocity in each of these two sections near the housing and at the minimum-clearance location. This implies local in-plane flow separation and recirculation within this subregion. Such flow behavior was apparently suspected (but not confirmed) by Morrison, who stated, "If a tangential

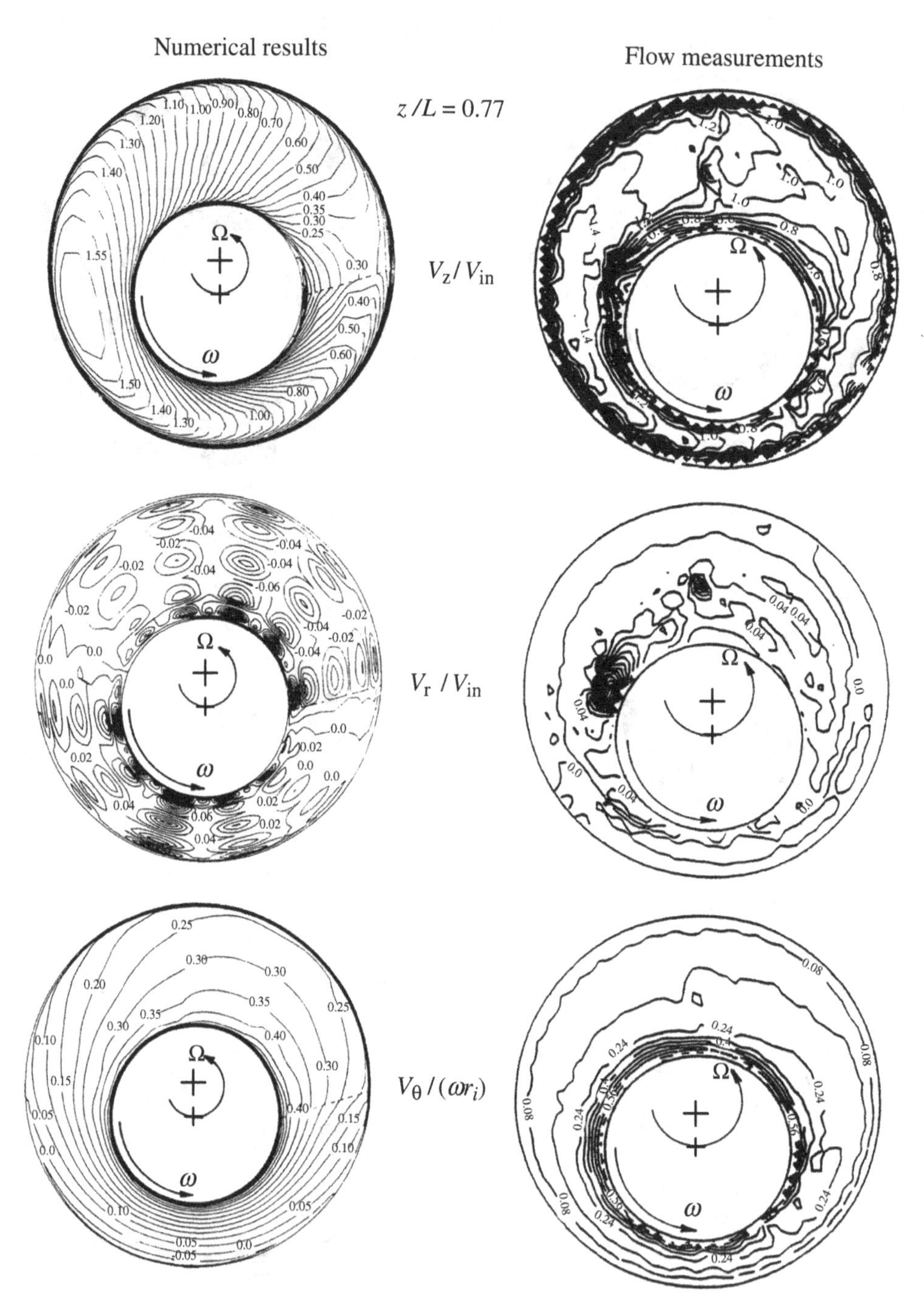

Figure 16.26. Contours of the velocity components (perturbations included) at 77 percent of the seal's length.

recirculation zone exists, it must be between the stator (meaning the housing) and the first radial grid line."

Figure 16.29 shows a contour plot of the nondimensionalized pressure perturbations, namely, $p^*c/\rho V_{in}{}^2$, and the resulting (centered-rotor and whirl-related) magnitudes of the nondimensionalized pressure p/p_{in} over the rotor surface.

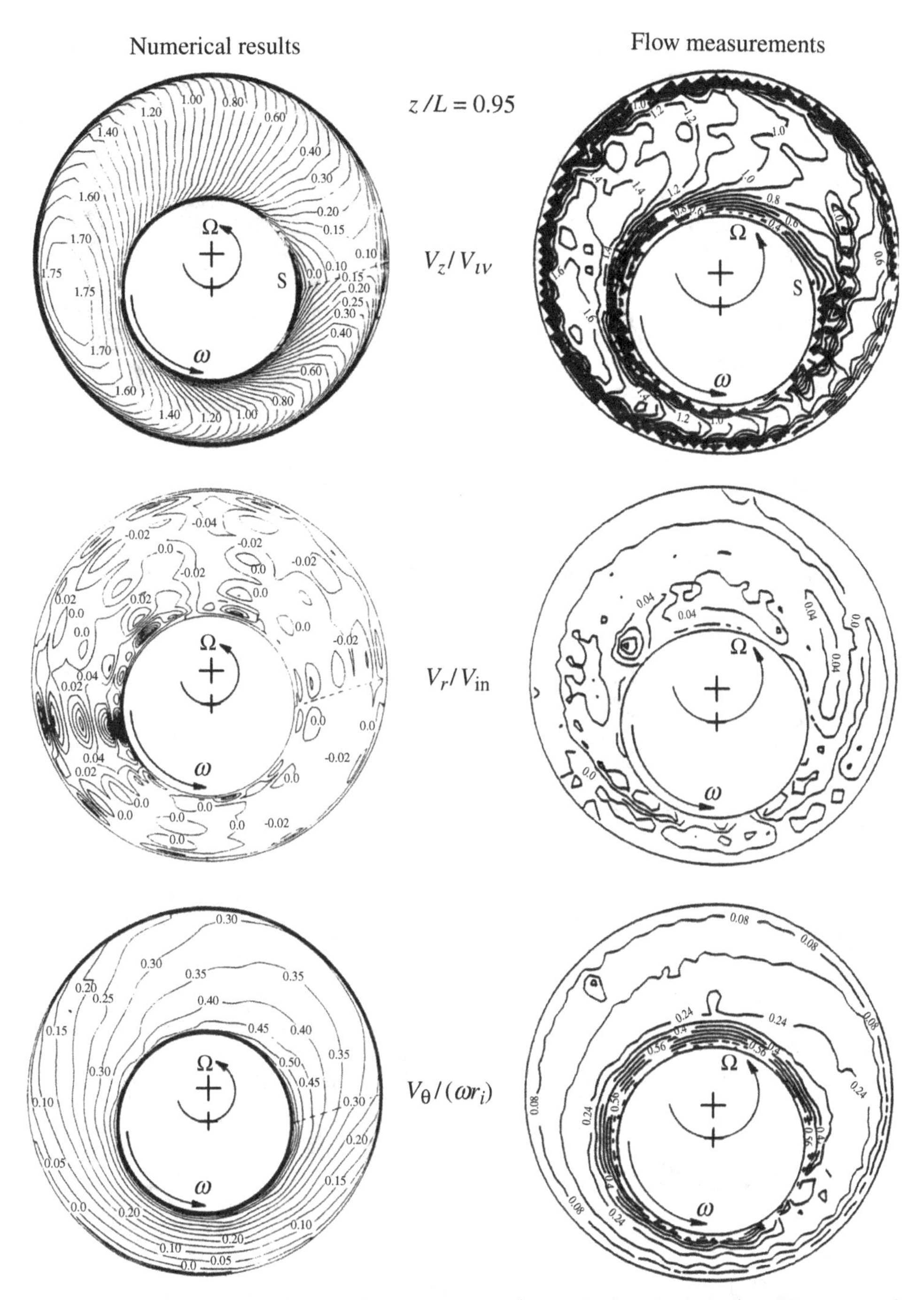

Figure 16.27. Contours of the velocity components (perturbations included) at 95 percent of the seal's length.

These pressure contours are shown on the "unwrapped" rotor surface, which is obtained by "splitting" the surface along the lines labeled $[a_1 - b_1]$ and $[a_2 - b_2]$ in Figure 16.20. Given the lack of pressure measurements in Morrison's study, no experimental verification of these results was possible. Examination of the pressure

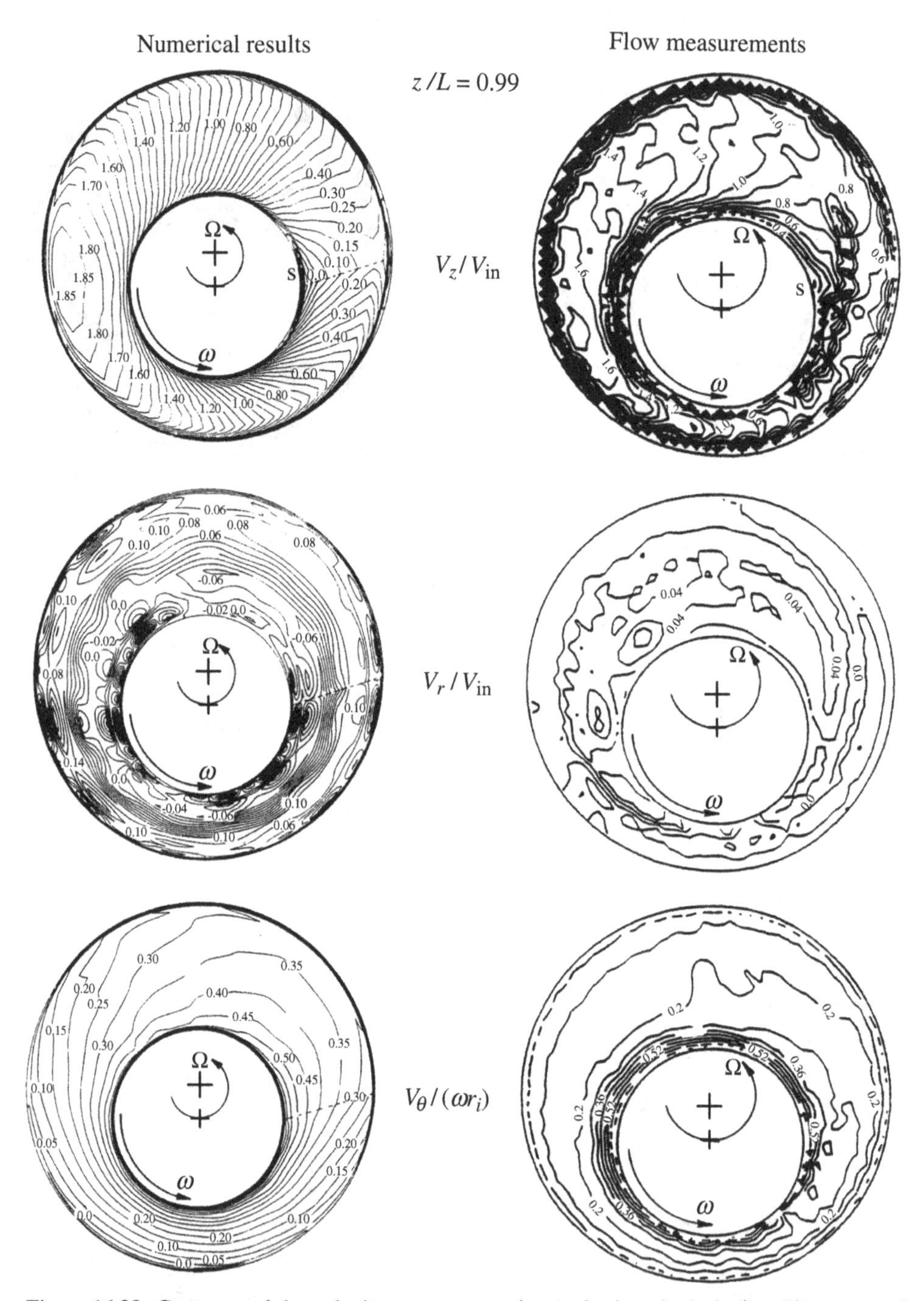

Figure 16.28. Contours of the velocity components (perturbations included) at 99 percent of the seal's length.

contours in this figure reveals that the peak value occurs at a tangential location that lags the minimum-clearance position by reference to the whirl direction. It is conceivable, and indeed likely, that the maximum-pressure location is a function of the whirl direction (forward or backward) and/or the whirl frequency ratio Ω/ω. For the synchronous-whirl case under consideration, Figure 11.29 shows that the rotor

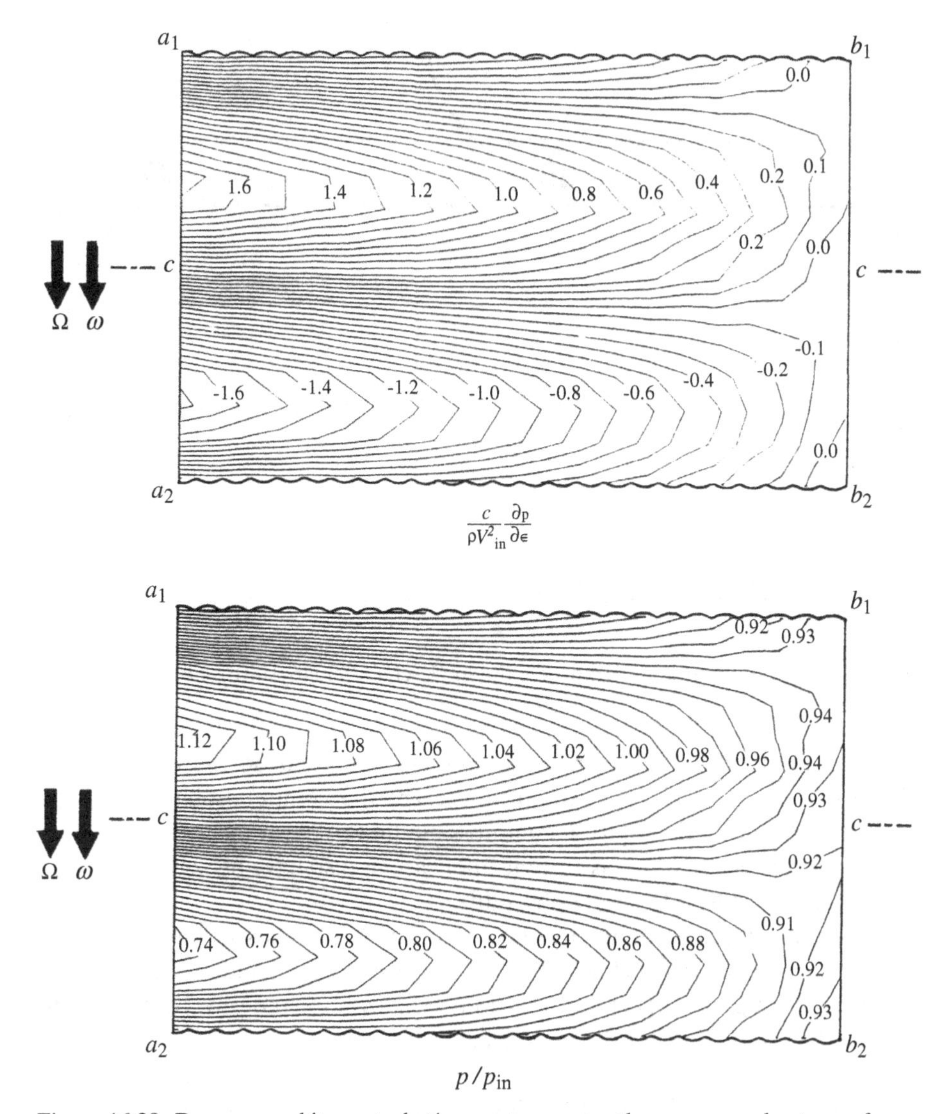

Figure 16.29. Pressure and its perturbation contours over the unwrapped rotor surface.

pressure peaks at a value of approximately 1.12 of the inlet pressure magnitude and that with the streamwise accumulation of total pressure losses (due to friction), this magnitude gradually declines to approximately 0.93 at the seal exit station. Figure 16.29 also shows that the maximum and minimum rotor-surface pressure locations hardly change their tangential locations along the overwhelming majority of the seal length.

Applications: Rotordynamic Analysis of Labyrinth Seals

Literature Survey

Labyrinth seals are generally regarded as efficient leakage-control devices in turbomachinery applications (Figure 16.30). Because leakage flow is considered to be one of the primary loss mechanisms and, subsequently, causes performance degradation,

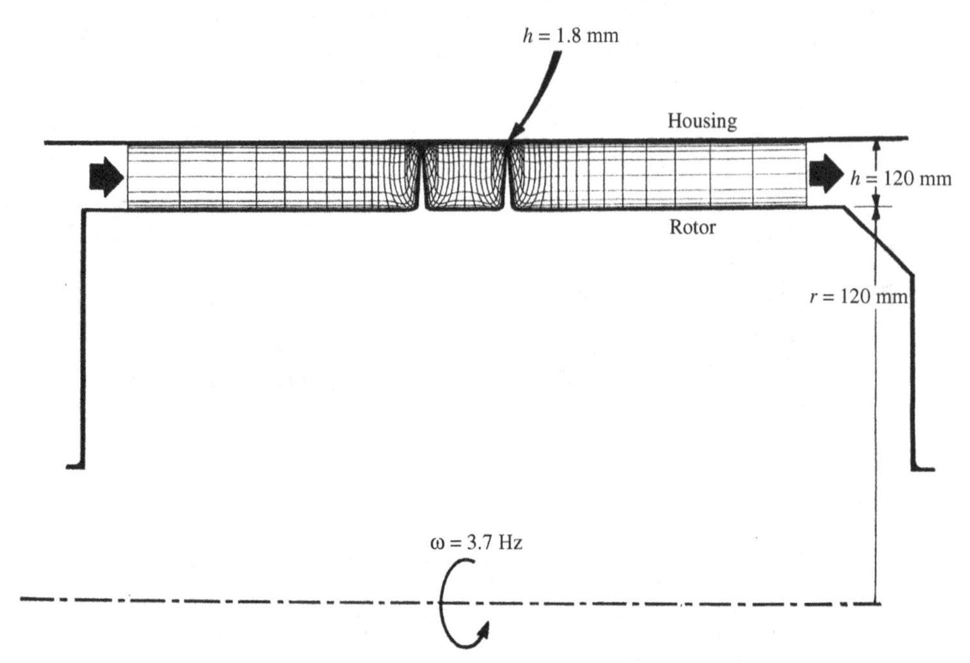

Figure 16.30. Finite element discretization of the single-chamber labyrinth-seal configuration.

much of the numerical studies on seals have been focused on the effectiveness of this particular seal category. Among these are the computational models by Stoff [17] and Rhode et al. [18] for incompressible-flow applications. Compressible-flow labyrinth-seal models for gas-turbine applications also have been devised (e.g., Wittig et al., [19] and Rhode and Hibbs, [20]).

Experimental measurements and flow-visualization studies in labyrinth seals have also been reported in the literature. Examples of these include the LDA measurements by Wittig et al. [19] and the flow-visualization study by Iwatsubo and Kawai [21]. The complexity of the flow structure in labyrinth seals has also encouraged researchers to devise and continually update empirical correlations (e.g., Zimmermann and Wolf, [22]) for inclusion during a preliminary design procedure.

Despite the significant performance improvements attributed to labyrinth seals, many rotordynamicists blame them for serious fluid-induced vibration problems. Among those, Alford [23] was the first to point out the destabilizing effects of labyrinth seals. Iwatsubo and Kawai [21] later reported significantly poor direct damping coefficients for a hydraulic labyrinth seal under a matrix of different operating conditions. In a comparative study involving several seal categories, Childs and Elrod [24] also concluded that the flow swirl at the seal inlet station makes a bad rotordynamic situation even worse and that an effective swirl "brake" (a special cascade of deswirl vanes) would be required in this case.

Theoretical developments in this area have generally been either too simplified to account for the real flow effects within the seal section or so tailored as to handle simple geometries and/or uniform lateral eccentricity of the whirling rotor. Under the first category are the "one-volume" model by Iwatsubo [25] and the model by Childs and Scharrer [26]. To the author's knowledge, the first attempt to apply rigorous computational fluid-dynamics tools was that of Nordmann and

Weiser [27]. This finite-difference-based model relies on the transformation of the entire rotor-to-housing computational domain into a fictitious frame of reference, whereby the rotor eccentricity (assumed uniform along the seal axis) is eliminated. Despite the apparent generality and complex nature of this approach, the method is virtually limited to perfectly-rectangular tooth-to-tooth chambers and is inherently inapplicable to rotor excitations in the form of conical whirl (discussed later in this chapter), which would be of serious mechanical consequences for long seals with an appreciable streamwise pressure differential.

Centered-Rotor Flow Field

Two labyrinth seals composed of the same tooth-to-tooth chamber geometry but different in the number of chambers were selected. These one- and five-chamber hydraulic seals were the subject of a flow-visualization study and rotordynamic testing by Iwatsubo and Kawai [21]. Out of the experimental matrix of operating conditions, a set of operational modes, where the seal-inlet static pressure and rotor speed are 147.0 KPa and 3.7 Hz, respectively, were arbitrarily chosen for the current rotordynamic analysis. The finite element discretization models created for the two seal configurations are shown in Figure 16.31, where the discretization unit is the biquadratic nine-nodded curve-sided Lagrangian element in Figure 16.4.

Figures 16.32 and 16.33 show the centered-rotor flow solution for the one-chamber seal configuration. Of these, Figure 16.32 illustrates the meridional flow behavior within critical subregions, which are magnified to clarify the flow rcirculation within the seal chamber and the vortex breakdown in the "dump" region downstream from the chamber. The flow structure in Figure 16.33 was qualitatively compared with its counterpart in the flow-visualization study by Iwatsubo and Kawai, and the computational results were found to be consistent with the experimental findings. The swirl velocity contours are shown in Figure 16.33

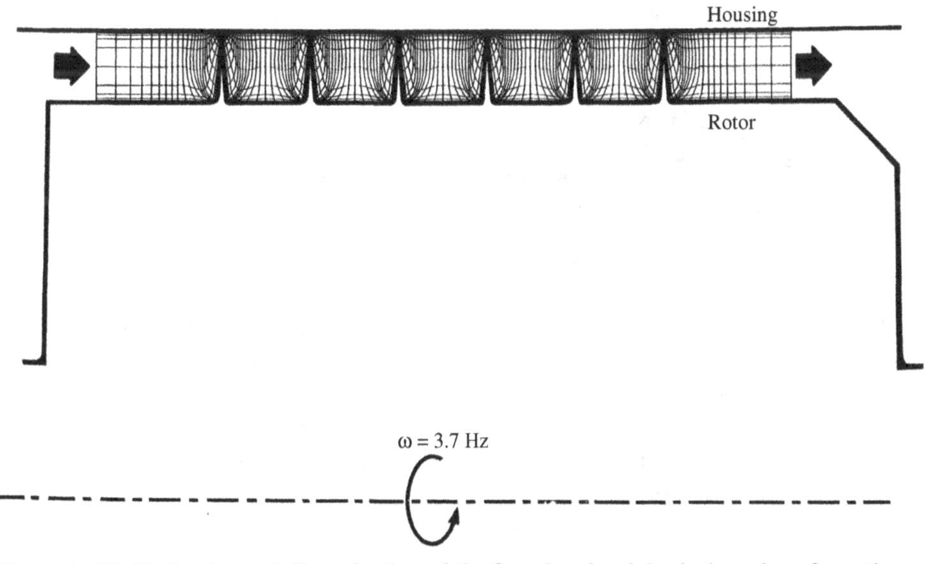

Figure 16.31. Finite element discretization of the five-chamber labyrinth-seal configuration.

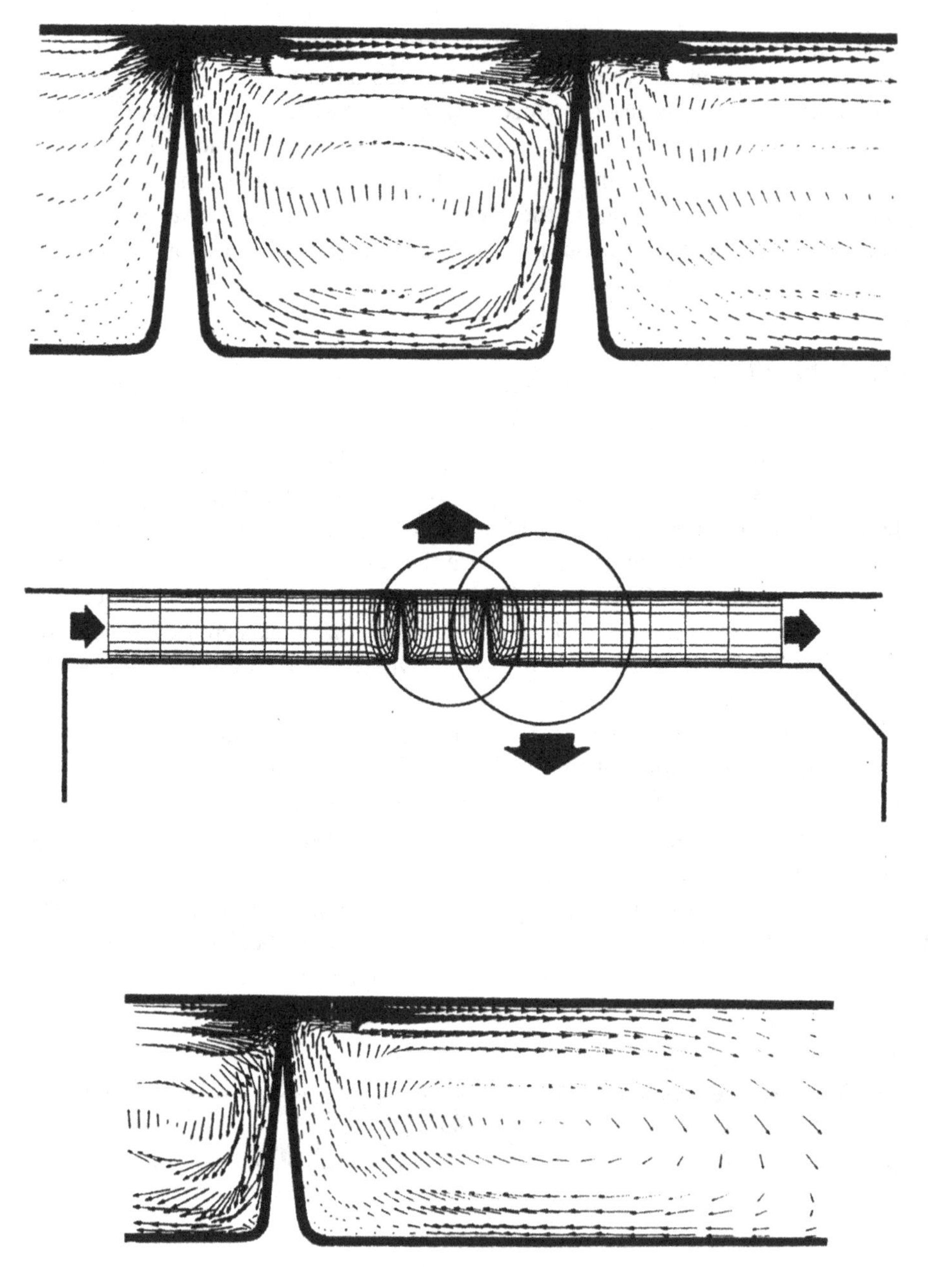

Figure 16.32. Velocity-vector plot associated with the centered-rotor operation mode.

within the same magnified segments. Magnitudes of V_θ, in this figure are non-dimensionalized using the average inlet through-flow velocity $V_{z,in}$. Note that the figure shows no inlet preswirl, which is consistent with Iwatsubo's seal-inlet conditions. Also note the swirl-velocity boundary-layer-profile development over both the rotor and housing surfaces. Figure 16.33 also shows that the swirl velocity near the housing surface appears within the tooth-to-tooth chamber region but steadily declines away from the chamber, up to the location where the flow recirculation zone ends.

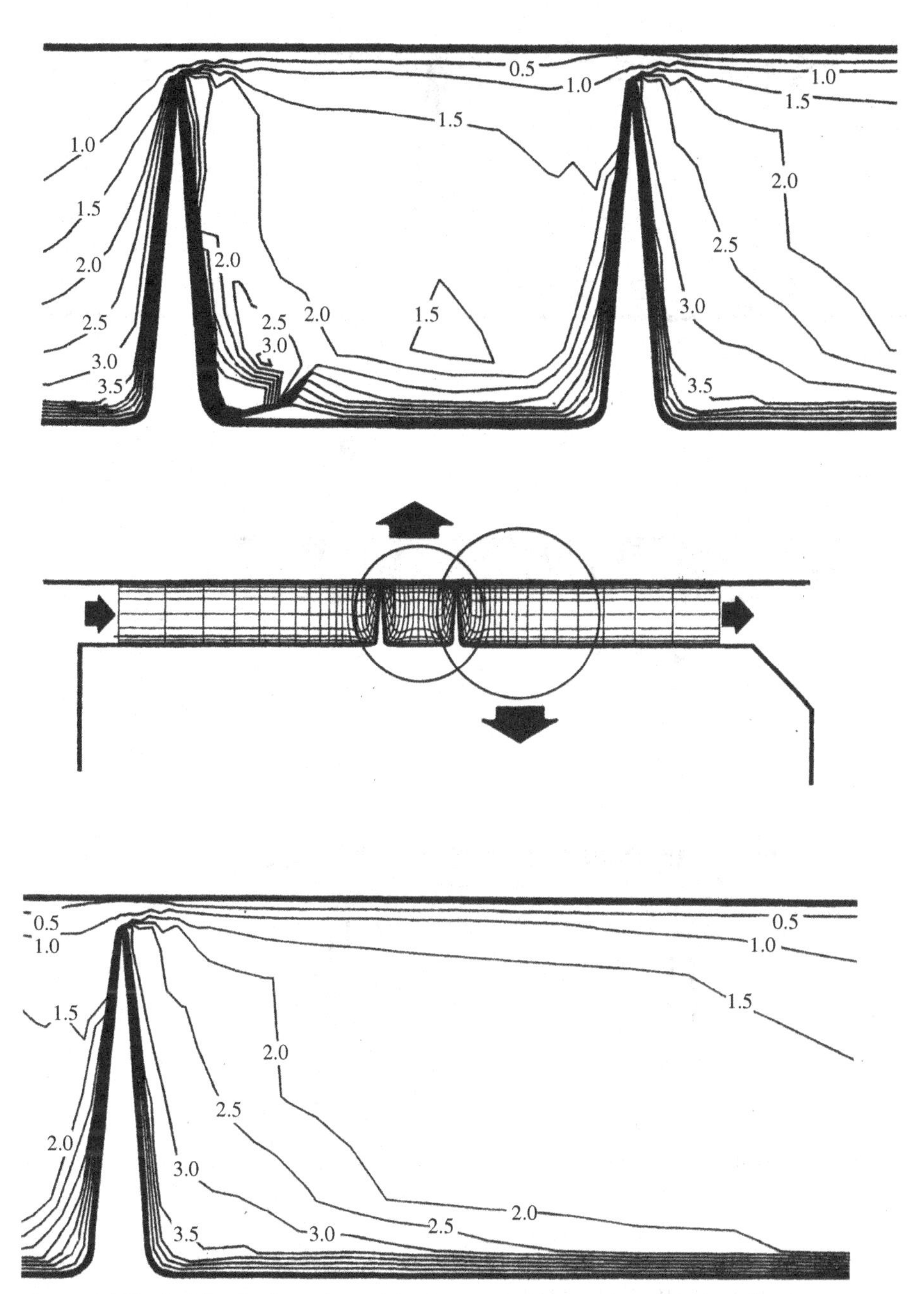

Figure 16.33. Swirl-velocity contours in the centered-rotor operation mode.

The centered-rotor flow pattern in the five-chamber seal configuration is displayed in Figures 16.34 and 16.35, with particular emphasis on the flow structure in the first and last seal chambers. Examination of Figure 16.34 reveals that the center of the cavity recirculatory motion gradually moves toward the tooth-pressure side (the lagging side) as the flow progresses from the first to the last cavity. The most significant difference between the swirl-velocity distribution in Figure 16.35 and that in Figure 16.33 is the continuous rise in V_θ magnitude in the through-flow direction this time. This characteristic simply means that the five-chamber seal configuration gives

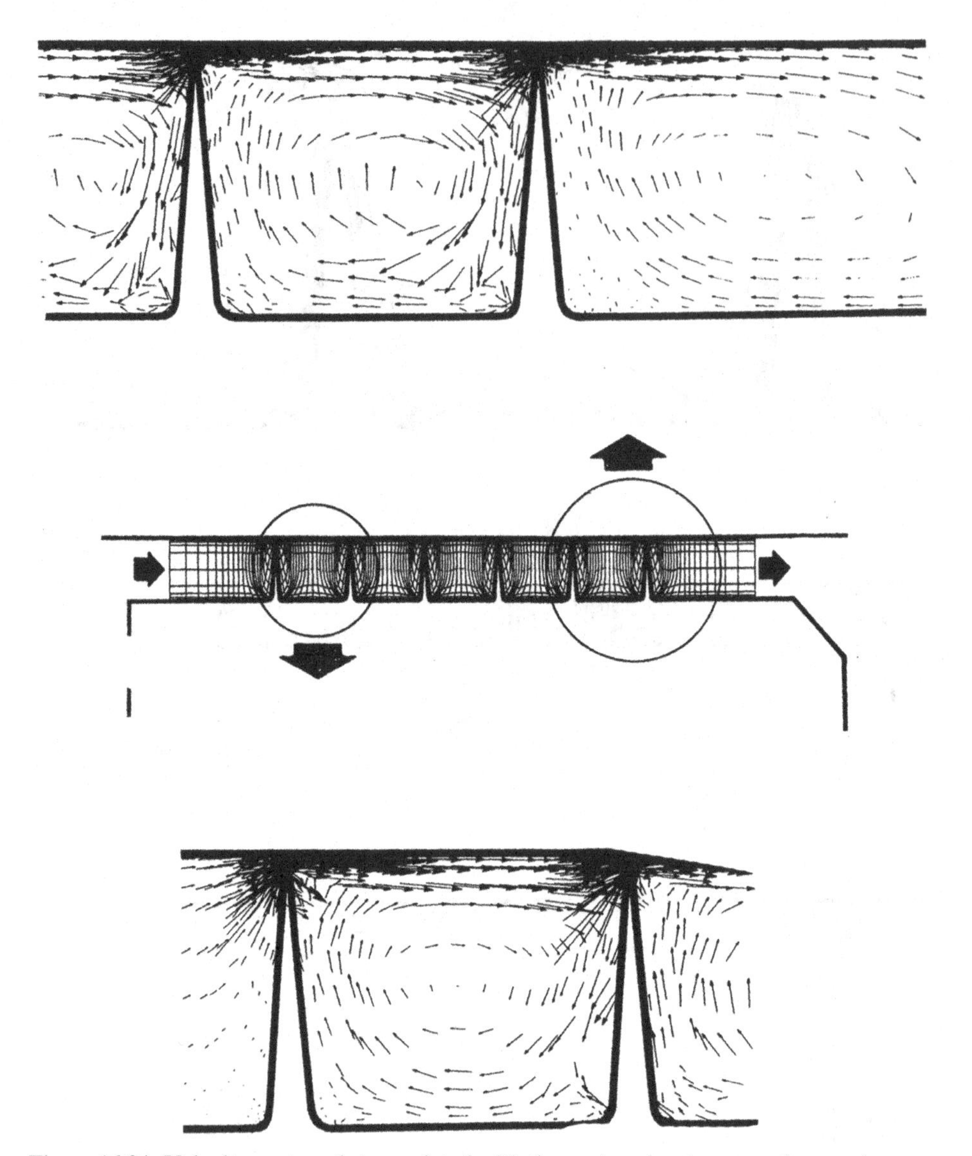

Figure 16.34. Velocity-vector plot associated with the centered-rotor operation mode.

rise to "elongated" flow trajectories over the last few chambers. One would expect, as a result, a substantial total pressure decline across these chambers, which is consistent with the well-known fact that the leakage-suppression capability of labyrinth seals is drastically enhanced by increasing the number of tooth-to-tooth chambers.

Investigation of the Grid Dependency

As would be naturally anticipated, the perturbed flow precision depends on the grid resolution, which, at this procedural step, translates into the number of computational stations in the circumferential direction (see Figures 16.30 and 16.31). Investigation of this dependency was carried out by varying this number and repeatedly computing the fluid-exerted forces on the rotor for an arbitrarily chosen

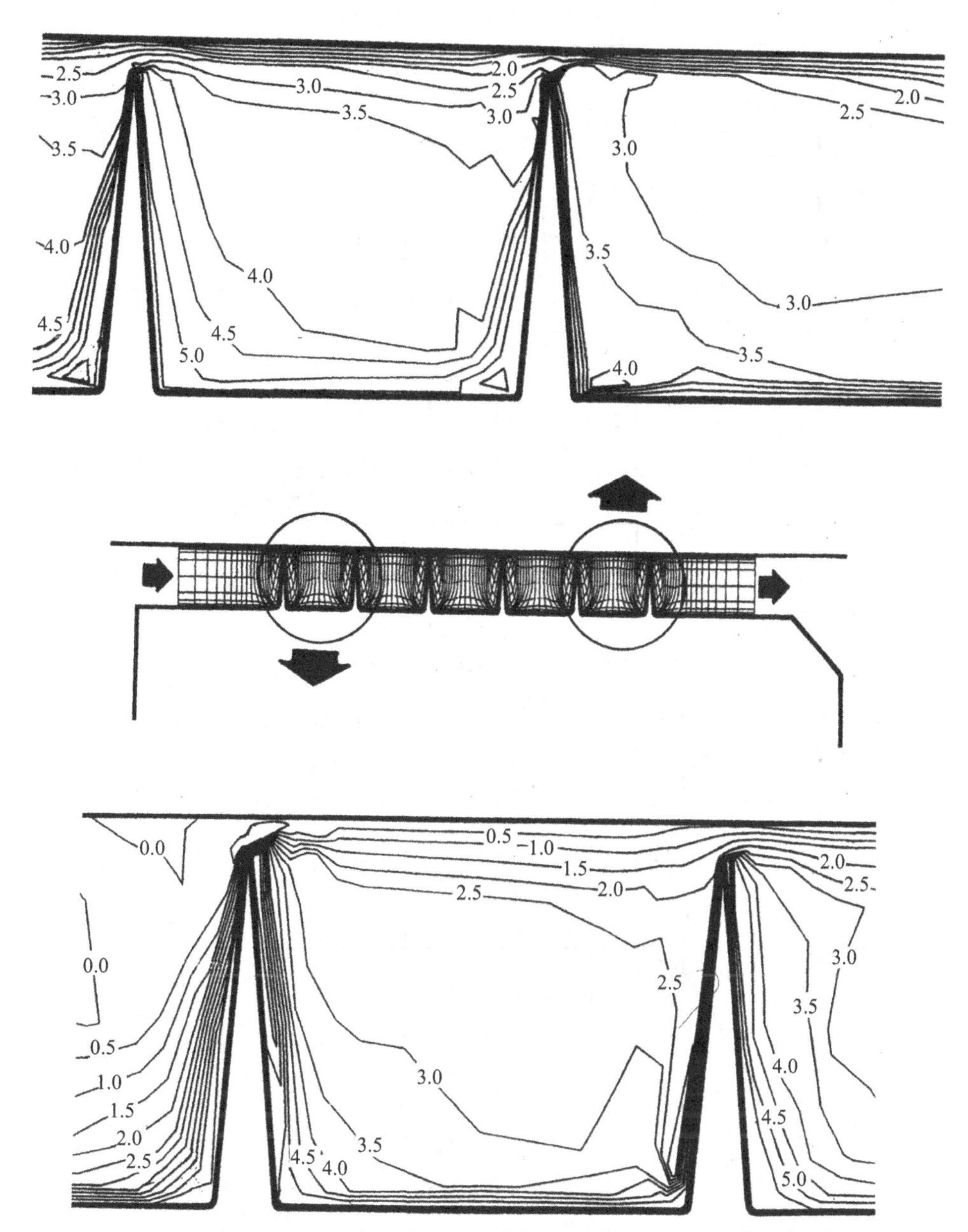

Figure 16.35. Contours of the swirl velocity in the centered-rotor operation mode.

whirl frequency Ω that is identical to the operating speed ω [28]. Results of this preliminary step are shown in Figure 16.36.

Examination of the force trends in Figure 16.36 reveals that the radial-force component $F_r{}^*$ is more sensitive to the tangential grid resolution, by comparison. The figure also shows that changes in the force magnitudes become acceptably small as the tangential computational-station count approaches 11. Given the fact that $F_r{}^*$ itself varies by as little as 2.7 percent by increasing the number of these stations from 10 to 11, and in an attempt to maintain a "manageable" CPU time consumption, the number of these computational stations was fixed at 11 thereafter.

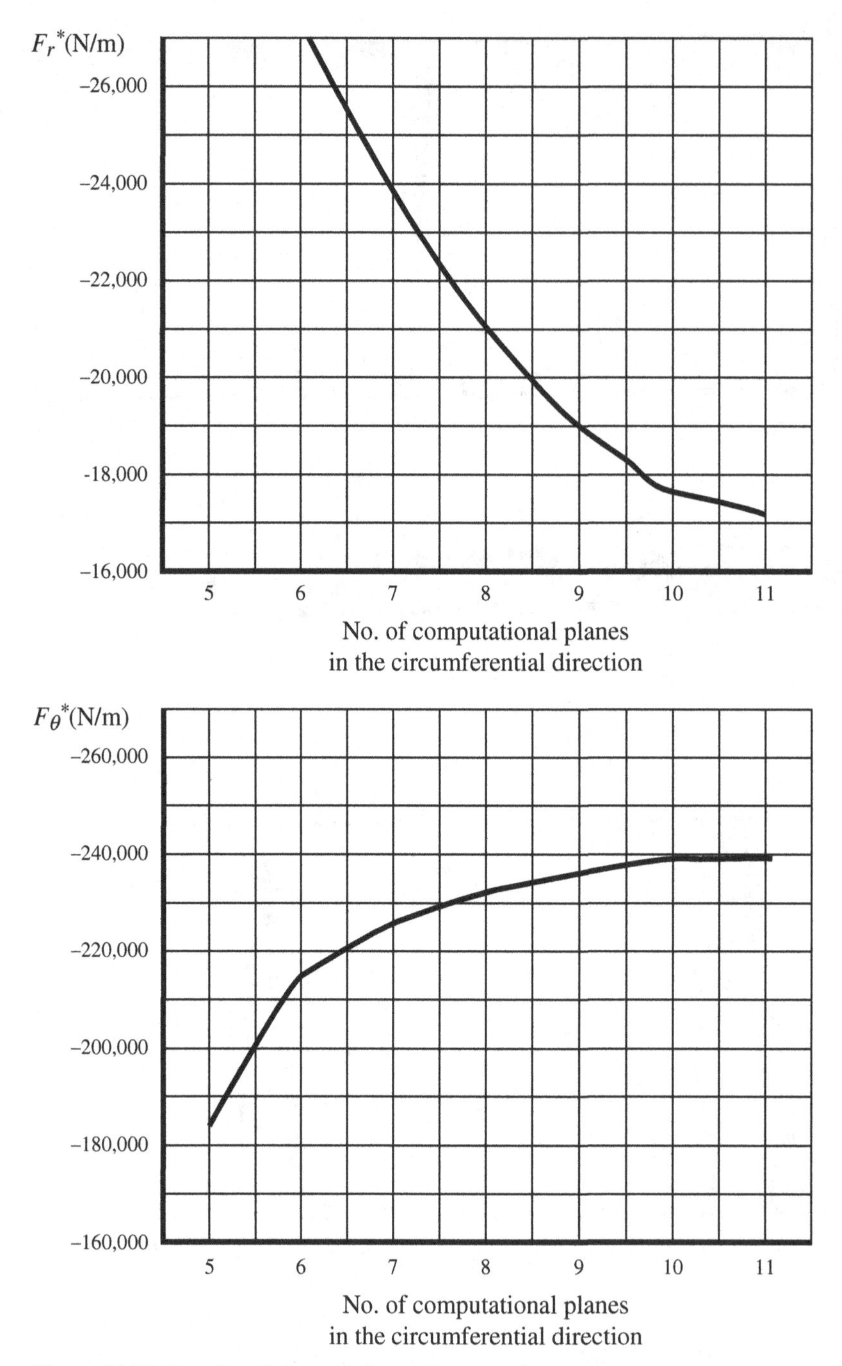

Figure 16.36. Results of the grid-dependency study.

Fluid-Induced Forces and Rotordynamic Coefficients

The fluid-induced force components (F_r^* and F_θ^*) are shown in Figure 16.37 for the five-chamber seal configuration over a range of whirl frequency ratio Ω/ω between -200 percent and $+200$ percent, with the negative magnitudes indicating a backward whirl. One reassuring feature of the force-component trends in this figure is

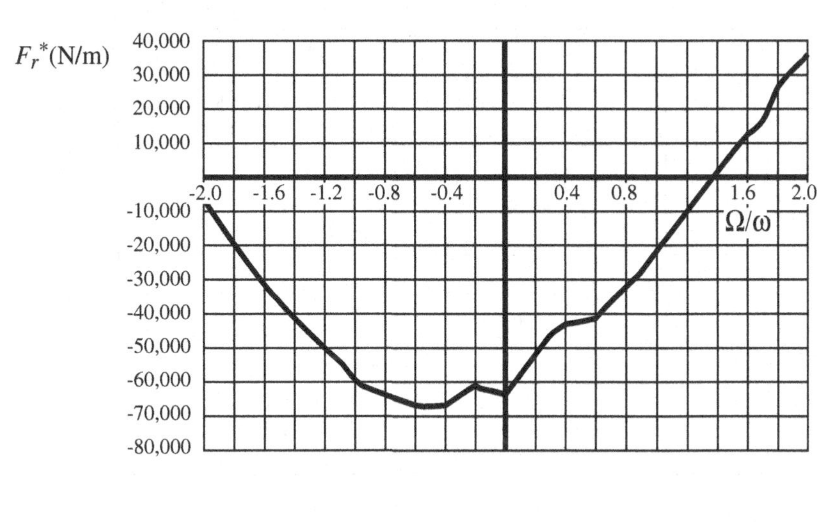

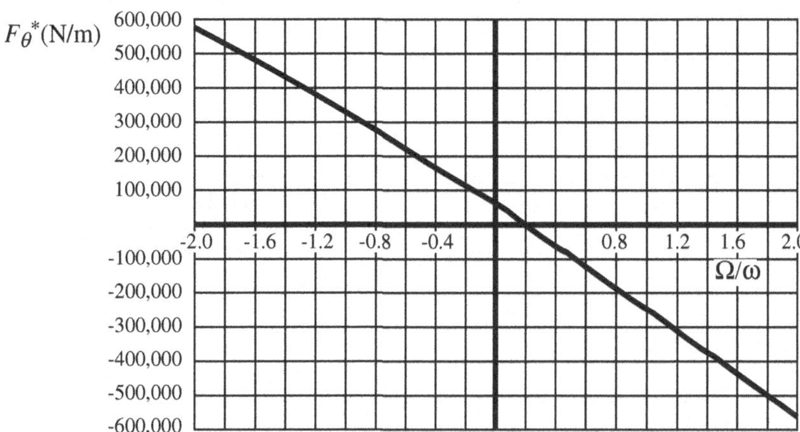

Figure 16.37. Fluid-induced force components for the five-chamber labyrinth seal.

that they are more or less parabolic, a commonly known characteristic of seals in general. Of the two force components in Figure 16.37, the tangential component is exclusively responsible for restoring or aggravating the rotor whirling motion. With this in mind, the impression one would have, by examining this figure, is that of rotordynamic instability in the range of positive (forward) whirl between the whirl-frequency ratios of zero and 200 percent, judging by the positive tangential force within this range.

The computed fluid-induced force components were then used to find the rotordynamic coefficients of the fluid-rotor interaction system in the manner outlined earlier in this chapter. Results of this computational step are contained in Figure 16.38 for the one- and five-chamber seal configurations. Reproduced in the same figure are the experimental magnitudes reported by Iwatsubo and Kawai [21] for comparison. Perhaps the most alarming feature in this figure is the poor values of the direct damping coefficient C for both seal configurations. In fact, the negative value of this coefficient for the single-chamber configuration implies a highly destabilizing effect, particularly in view of the positive (also destabilizing) value of the cross-coupled stiffness coefficient k in this case.

Rotor dynamic coefficients	Current perturbation model	Experimental measurements by Iwatsubo and Kawai (1984)
K (N/m)	6.60*10^4	1.0*10^4
k (N/m)	2.20*10^4	0.94*10^4
C (NS/m)	-3.49*10^2	-1.01*10^2
C (NS/m)	-9.19*10^2	-7.75*10^2

Rotor dynamic coefficients	Current perturbation model	Experimental measurements by Iwatsubo and Kawai (1984)
K (N/m)	5.87*10^4	2.30*10^4
k (N/m)	6.19*10^4	2.25*10^3
C (NS/m)	1.22*10^4	1.25*10^2
C (NS/m)	-6.60*10^2	-1.75*10^3

Figure 16.38. Comparison of the five-chamber-seal rotordynamic coefficients with experimental data.

A useful rotordynamic-stability indicator in the general fluid-encompassed-rotor problems is the so-called relative stability ratio f, which is defined as follows:

$$f = \frac{k}{\omega C}$$

where ω is the rotor speed. This factor, in effect, carries the ratio of the destabilizing-to-stabilizing forces acting on the rotor. A given seal would have a stabilizing influence for positive f magnitudes below 1.0 [26]. Application of this criterion to the five-chamber seal configuration leads to the conclusion that this particular seal imparts rotordynamic stability to the shaft (in a "global" sense), for it gives rise to a relative stability ratio f of approximately 0.22 on an average basis. However, this same "relative stability" criterion is inapplicable to the single-chamber seal configuration. This is due to the negative value of the direct damping coefficient C in this case, a fact that simply renders the seal unstable regardless of the positive cross-coupled stiffness coefficient magnitude.

Applications: Rotordynamic Behavior of a Shrouded Pump Impeller

The focus of this section is a shrouded-impeller pump stage, which was the focus of rotordynamic measurements by Bolleter et al. [29]. This is schematically shown, with the flow domain of concern highlighted, in Figure 16.39. The stage dimensions and operating conditions are as follows:

- Impeller tip radius = 17.5 cm
- Impeller-outlet axial width = 2.8 cm
- Impeller speed = 2000 rpm

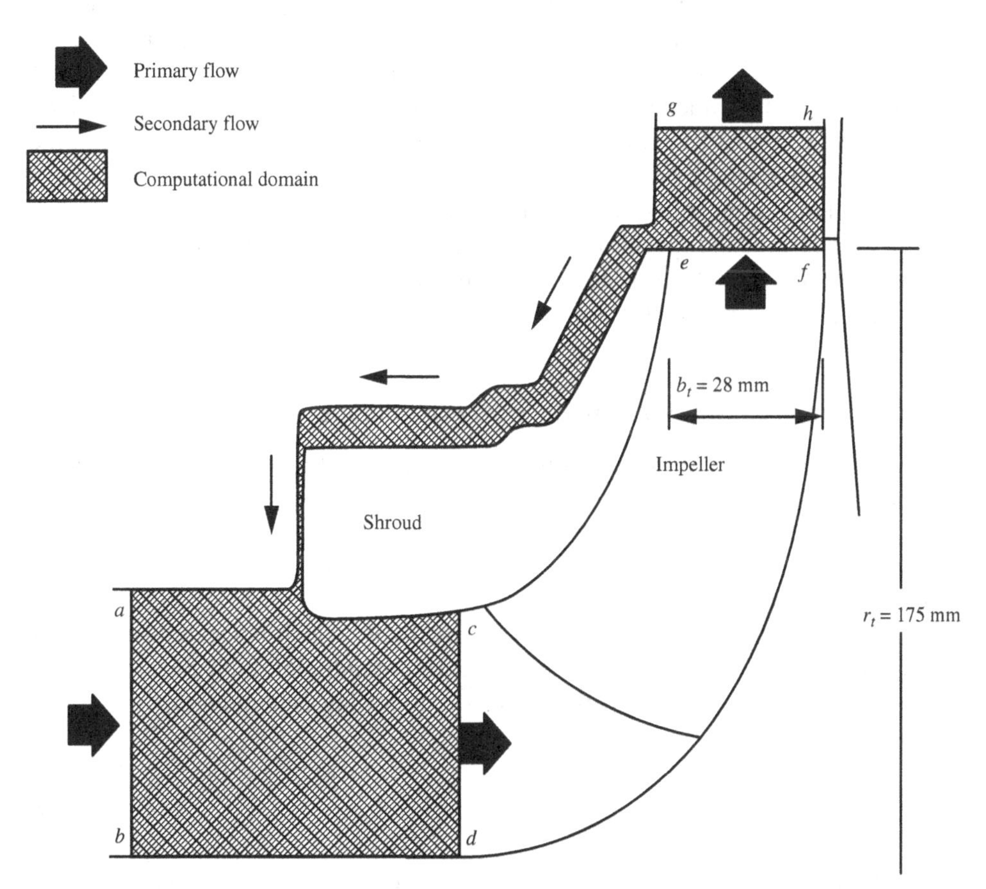

Figure 16.39. Computational domain for the shroud-to-housing secondary passage in a shrouded-impeller pump.

- Working medium is water at 30° C
- Volumetric flow rate = 130 L/s
- Total head = 68 m
- Reynolds no. (based on the impeller radius and tip velocity) = 8.02×10^6

Centered-Impeller Subproblem: Contouring the Flow Domain

The problem with the geometry and the drastically different sizes of the different subdomains pose real difficulties in contouring and discretizing a realistic flow passage. With the primary flow confined between the hub and shroud surfaces, both parts of the impeller component, it was realized early on that this particular flow stream will have a minimum effect on the impeller rotordynamic behavior depending on how the computational domain is defined. The decision therefore was made to extract the primary-flow passage from the computational domain (see Figure 16.39). In doing so, attention was particularly focused on a realistic means to reflect the natural continuation of the primary and secondary passages into one another. In other words, the option of beginning and ending the secondary passage where the two streams separate and then merge is now out of the picture. In practical terms, there is no realistic way to accurately specify inlet and exit boundary conditions at these complex-flow stations. Despite its obvious popularity, such an option was viewed as simply too problematic from an analytical viewpoint.

Figure 16.39 shows a computational domain that is composed of the secondary-flow passage that is attached to two (inlet and exit) primary-passage segments [30]. The objective here is to allow a more or less natural flow interaction between the secondary-flow passage and these two primary-flow subregions, with their interface boundaries now being internal surfaces. From an analytical standpoint, this choice also permits the specification of straightforward boundary conditions that, if wrongly chosen, would overspecify the problem. These flow constraints (now over a set of entry, reentry and discharge boundary segments) are discussed next.

Centered-Impeller Subproblem: Boundary Conditions

Referring to the flow-permeable boundary segments in Figure 16.39, the following boundary conditions are imposed:

Stage Inlet Station

This is the boundary segment $a - b$ in Figure 16.39, which is located sufficiently far upstream of the impeller, from an engineering standpoint. Fully developed flow is assumed at this location, giving rise to the following Neumann-type boundary condition:

$$\frac{\partial V_r}{\partial z} = \frac{\partial V_z}{\partial z} = \frac{\partial V_\theta}{\partial z} = 0$$

In addition, the stage inlet static pressure is specified at the node midway between the endwalls on this station.

Impeller Inlet and Exit Stations

These are labeled $c-d$ and $e-f$ in Figure 16.39. Fixed profiles of the velocity components corresponding to the stage operating conditions are specified over these boundary segments.

Stage Exit Station

The flow behavior at this station (designated $g-h$ in Figure 16.39) is viewed as predominantly confined to satisfying the mass and angular-momentum conservation principles in a "global" sense. In their derivative forms, these can be expressed as follows:

$$\frac{\partial V_r}{\partial r} = -\frac{V_r}{r}$$

$$\frac{\partial V_\theta}{\partial r} = -\frac{V_\theta}{r}$$

These are linear boundary conditions and are, therefore, introduced non-iteratively in the finite element equations. Moreover, considering the geometry of the endwalls in the exit flow segment, a zero normal derivative of V_z is specified over this station. Comparing this with a zero V_z boundary condition, this derivative-type constraint is less restrictive, for it allows V_z to exist (regardless of its much smaller magnitude) as a result of the boundary-layer blockage effect along the endwalls and the subsequent shifting of the streamlines away from these walls at the boundary layer edge. In addition, the static pressure is prescribed at the computational node midway between the endwalls at this station, in accordance with the pump operating conditions.

Solid Boundary Segments

These involve the housing and shroud surfaces in contact with the primary and secondary flow streams, as well as the endwall segments $b-d$ and $f-h$. Over these boundary segments, the no-slip boundary condition applies as follows:

$$V_r = V_z = 0$$

$$V_\theta = C$$

where C is equal to zero and ωr for stationary and rotating boundary segments, respectively, with r being the local radius.

Flow Structure

In this section, the flow fields corresponding to two different seal configurations are compared. These are the face-seal and wear-ring stage configurations shown in Figures 16.40 and 16.43, respectively, with the rest of the stage geometry aspects being otherwise identical. In comparing the two flow structures, two criteria are used in preferring one over the other. First, the seal performance as an effective leakage-control device is assessed, with the leakage/primary volumetric flow ratio being the decisive factor in this case. However, the anticipated rotordynamic characteristics are judged in each case by examining the swirl-velocity component magnitude

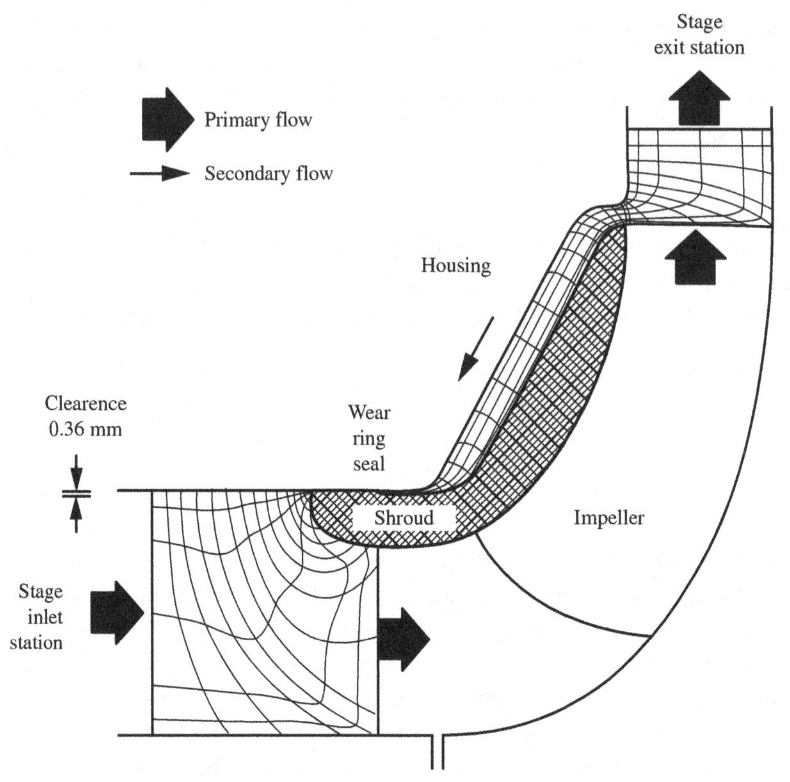

Figure 16.40. Finite element model for the wear-ring-seal configuration.

throughout the entire secondary-flow passage, including that at its inlet station. The logic behind this choice is the established fact that the swirl-velocity component will always act as a destabilizing factor as the impeller enters its whirling-motion excitation. In fact, the effort by Baskharone [30] is focused on virtually eliminating this component altogether at the secondary-passage inlet station through what was generically termed a *swirl brake*. This is a radial cascade of deswirl vanes that is located immediately downstream from the impeller exit station and was shown (in this study) to significantly improve the rotordynamic characteristics of the face-seal pump configuration in Figure 16.43.

Figures 16.41 and 16.42, for the wear-ring seal configuration, and Figures 16.43 through 16.45, for its face-seal counterpart, present the most significant features of the centered-impeller flow structure in the secondary (or leakage) flow passage. The complexity of this flow field is most apparent in Figures 16.41 and 16.43, both of which are vector plots of the meridional velocity component throughout the computational domain. The flow patterns in these two figures are characterized by massive separation and recirculation zones in virtually all segments of the secondary-flow passage. This feature is the result of the flow tendency to migrate radially outward near the (spinning) shroud surface due to the locally high centrifugal force, which is opposed by the tendency to proceed radially inward due to the streamwise static pressure decline across the secondary passage, a decline that creates the secondary-flow stream in the first place. Equally important are the swirl-velocity profiles in

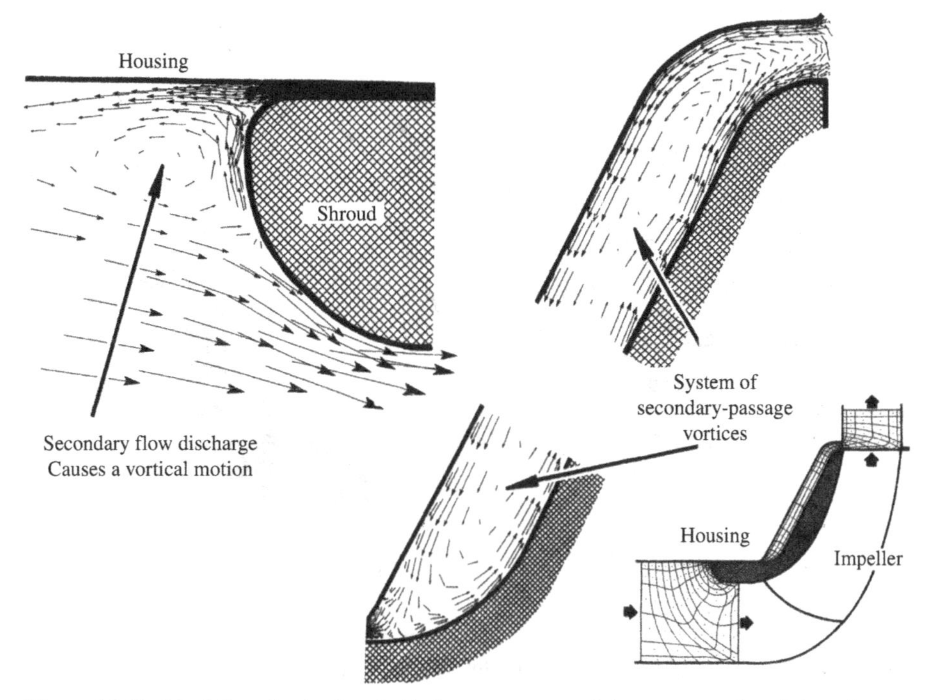

Figure 16.41. Meridional velocity plot in key segments of the computational domain.

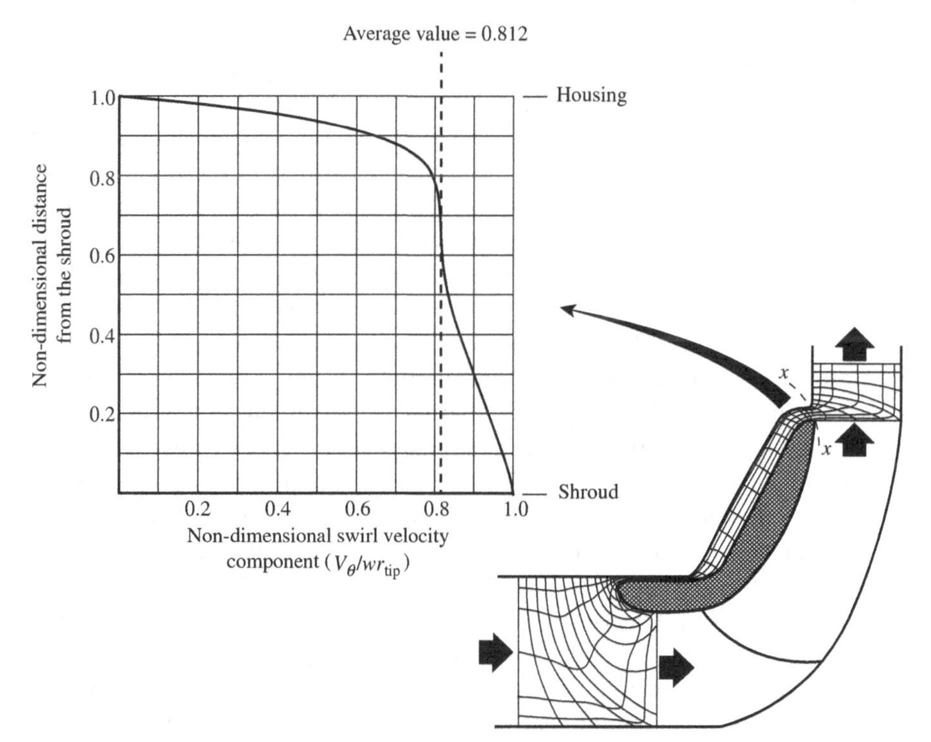

Figure 16.42. Swirl-velocity-component profile at the secondary-passage inlet station.

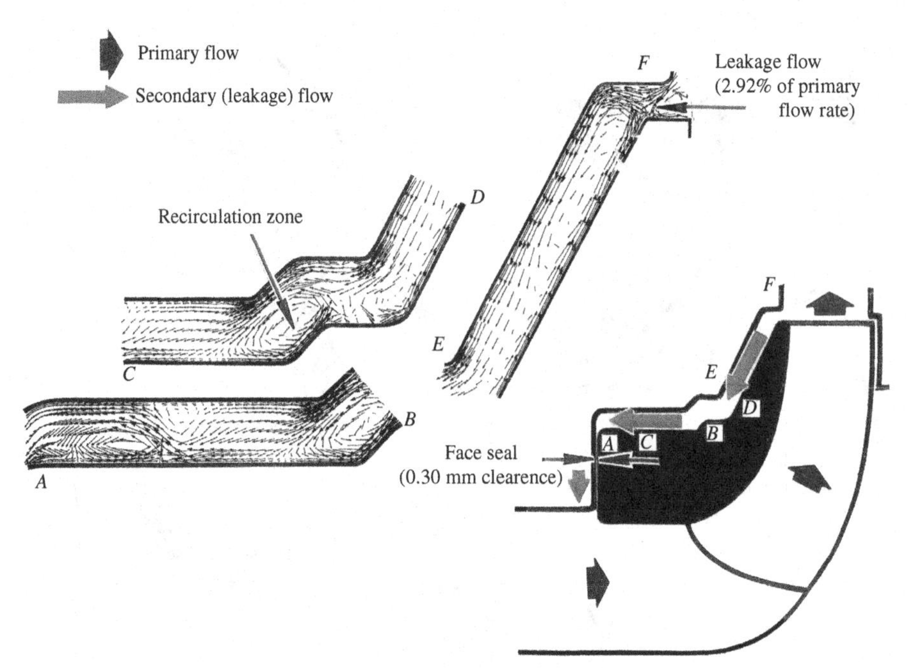

Figure 16.43. Leakage-flow behavior under a face-seal configuration of the leakage discourager.

Figure 16.42 and those in Figures 16.44 and 16.45 owing to the previously discussed rotordynamic implications of the swirl velocity in the secondary-flow passage.

The leakage mass-flow rate through the secondary-flow passage was obtained by integrating the mass flux over the passage inlet station. Results of this step came to establish the superiority of the wear-ring seal as a leakage-control device because it tolerated only 0.85 percent of the primary mass-flow rate in the secondary passage compared with 2.92 percent for its face-seal counterpart. This, in part, is due to the highly favorable streamwise pressure gradient across the poorly performing face seal as a result of radius diminishment (between the seal inlet and exit stations), which suppresses the boundary layer buildup over the solid walls in this region (see Figure 16.43), an important factor with its own flow-blockage effect. However, the tendency of the fluid particles to migrate radially outward near the constant-radius shroud surface bearing the wear-ring seal added to the large-scale recirculatory motion just upstream the seal inlet station created an environment for an early flow separation over the shroud surface at a location that is approximately one-quarter of the seal length downstream of the seal inlet station. Despite the rapid flow reattachment in this case, the recirculation region following the separation point has the effect of enhancing the seal frictional resistance and therefore the sealing effectiveness of the wear-ring seal configuration by comparison. Another contributing factor behind the wear-ring seal effectiveness is the existence of a substantial recirculation zone near the seal exit station (not shown in Figure 16.41), with no comparable seal exit flow behavior in the case of its face-seal counterpart.

As for the seal capability, as a swirl-velocity dissipator, the computed results came to favor the face-seal pump configuration instead. A thorough examination

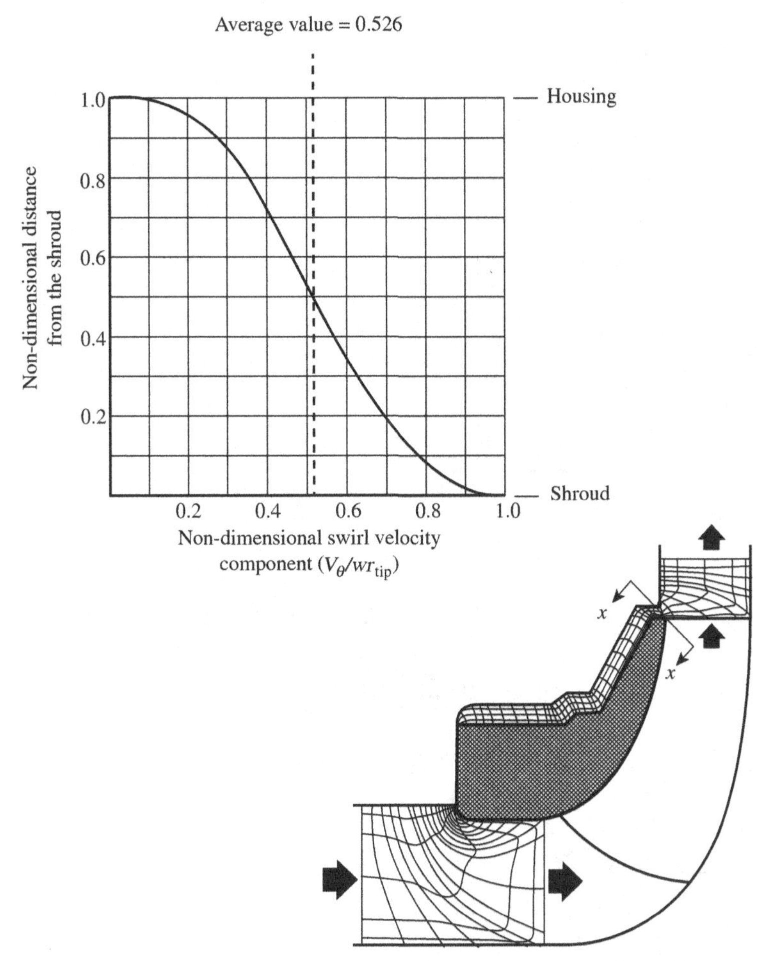

Figure 16.44. Distribution of the swirl velocity at the secondary-passage inlet station.

of the secondary-passage swirl-velocity distribution revealed that the "kink" in the shape of the secondary passage of the face-seal pump configuration (see Figure 16.43) is partially responsible for this result because it causes a sudden and significant reduction in the swirl-velocity component in the horizontal-passage segment with no equivalent swirl-velocity reduction mechanism in the wear-ring-seal pump configuration. More importantly the average nondimensional swirl-velocity component at the secondary-passage inlet station is shown in Figure 16.42 to be as large as 0.812 compared with a significantly lower magnitude of 0.526 for the face-seal pump configuration (see Figure 16.44). In fact, the face-seal exit magnitude of the same variable, by reference to Figure 16.46, is as low as 0.351, which also goes to establish the superiority of the face-seal configuration as a swirl-velocity suppressant.

An interesting issue at this point may be meaningfully raised. The question is whether the computational domain can now be "shrunk" to contain only the secondary-flow passage between its inlet and exit stations in Figures 16.44 and 16.45 prior to executing the more-involved perturbation procedure. Legitimacy of the question rests on the basis that all three velocity components are now known in

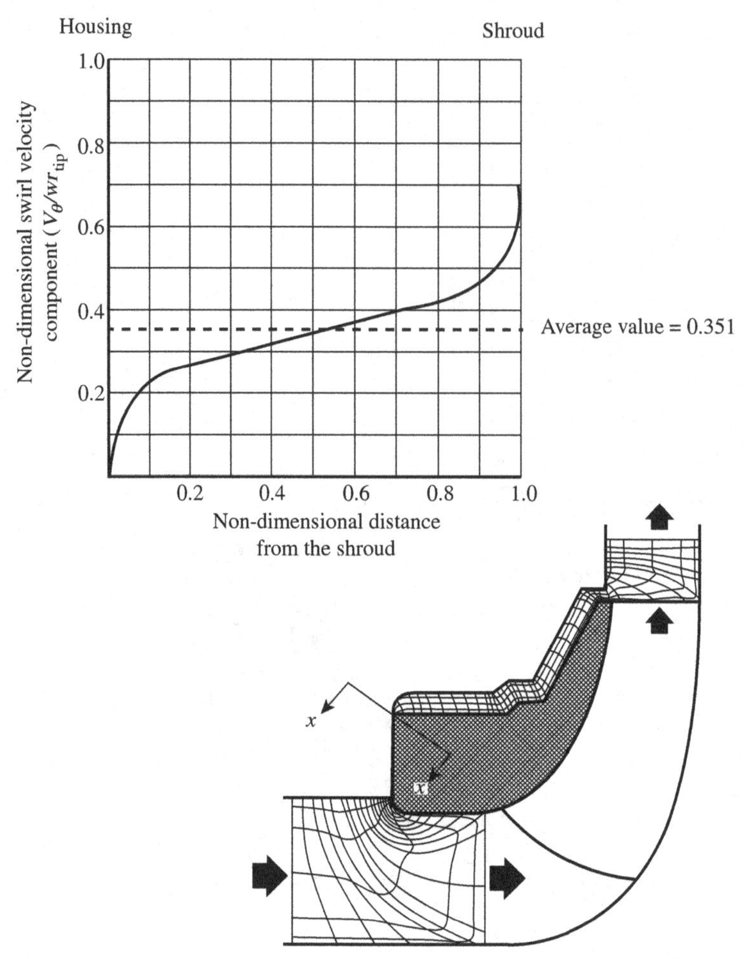

Figure 16.45. Extracting versus manipulating the primary-flow passage in defining the solution domain.

the form of profiles over these two limiting stations. In many existing flow models, the answer to this question is yes. In fact, the strategy of isolating the flow subregion of interest, with either experimentally available or previously attained boundary conditions, is more common than not.

However, it is the author's experience with this issue that indicates that such a strategy would create its own inaccuracies, particularly near the two limiting boundaries. As is well known, the Dirichlet-type (specified-variable) boundary conditions that would be employed in this case are the strongest and most restrictive. As a result, even the slightest inconsistency of the specified magnitudes (relative to the outcome of analyzing the smaller flow domain with *perfectly accurate* boundary conditions) can, and indeed do, produce fictitious flow-behavioral characteristics near these two boundaries. In fact, such a subdomain-isolation approach could impart the serious problem of analytical overspecification unless one of these boundaries is assigned derivative-type boundary conditions. Needless to say, the choice of placing the limiting boundaries partially or totally inside a flow recirculation subregion would be doubly senseless.

Simulation of the Impeller Subdomain Effects

In addition to the computational domain in Figure 16.39, Figure 16.46 also shows an alternate flow-domain choice, which is a full-scale flow domain because it combines both the primary- and secondary-flow passages in the face-seal pump configuration with the impeller-mounted shroud separating the two subdomains. In terms of boundary conditions, this choice makes it much simpler to find and impose them over sufficiently far upstream and downstream of the inlet and exit stations. As large a numerical model as it obviously is, this domain-contouring method was also investigated, and the centered-impeller flow field was reported by Baskharone and Wyman [31].

In the interest of retaining a manageable-size finite element model, only the flow turning effect by the impeller blades was simulated, without any physical inclusion of the blades themselves. Assuming full guidedness of the relative flow field (in the rotating-impeller passage) by these blades, the relative flow angle β was piecewise equated to the blade angle β' at the same meridional location (meaning the same z and r coordinates) in Figure 16.46. To simplify the process even more, the blade angle was assumed to vary along the "master" streamline (meaning the streamline midway between the hub and shroud lines) in the manner depicted in Figure 16.47. Representation of the impeller-blade angle in this figure follows a third-order polynomial with the streamwise derivatives at both ends both set to zero. This is a realistic representation because it is consistent with the majority of the impeller-blade design features. Finally, the blade angles at all nodes existing on any cross-flow station are taken to be identical to those at the middle node on the master streamline. The meridional velocity component V_m, also assumed constant over any cross-flow station in Figure 16.46, is obtained by applying the continuity equation over this particular station. Referring to Figure 16.47 and noting that the

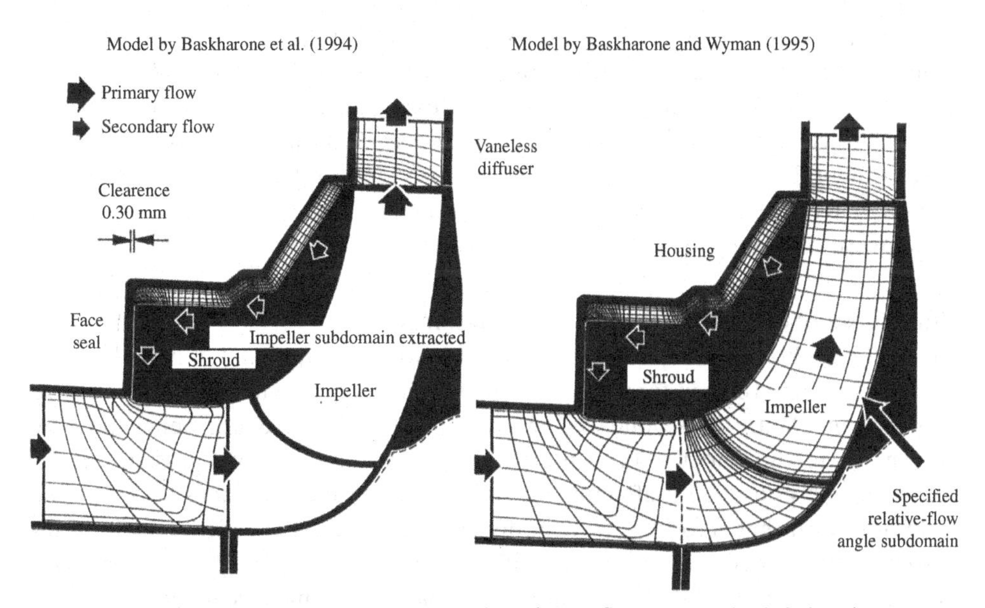

Figure 16.46. Extracting versus maintaining the primary-flow passage in defining the computational domain.

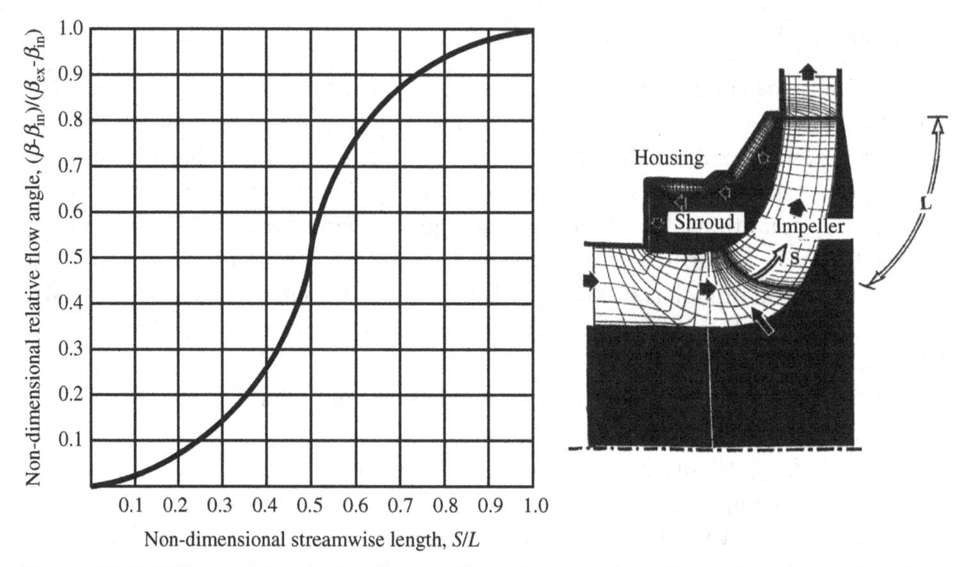

Figure 16.47. Effects of simulating the impeller subdomain in defining the solution domain.

relative magnitude of the meridional velocity (i.e., W_m) is identical to the absolute value V_m, we can compute the absolute magnitude V_θ of the swirl velocity through application of the following well-known velocity-triangle relationship:

$$V_\theta = W_\theta + \omega r = V_m \tan\beta + \omega r$$

where r is the local radius and ω is the rotor speed. With the relative flow angle β and meridional velocity component V_m both assumed to be constant over each cross-flow station in the impeller subdomain, the only variable (in the preceding equation) that will cause V_θ to vary across that station is the local radial coordinate r.

The simplified approach to obtain the nodal magnitudes of V_θ ignores, among other factors, the "deviation" angle between the actual relative flow stream and the blade surface. Of course, the inaccuracy of such a simplification would be more pronounced should the relative flow stream experience any separation over the blade surface. Note that the likelihood of flow separation, owing to the streamwise unfavorable pressure gradient, is worthy of consideration by comparison with the relative flow-stream behavior across the rotor of, say, water turbines, where a streamwise favorable pressure gradient prevails.

Worthiness of Simulating the Impeller Subdomain

Figure 16.48 shows a comparison between the two computational-domain configurations in Figure 16.46 in terms of the swirl-velocity component and static pressure distributions. With the secondary shroud-to-housing flow passage being the only region where the impeller off-center excitation is felt, it is this passage that is exclusively responsible for affecting the rotordynamic behavior and is therefore the only subregion that is focus-worthy in this figure. As far as the swirl-velocity component is concerned, the figure generally reveals an agreement between the two configurations, particularly in the horizontal-passage segment leading to the face seal, in which the fluid-induced forces are of direct impact on the rotordynamic

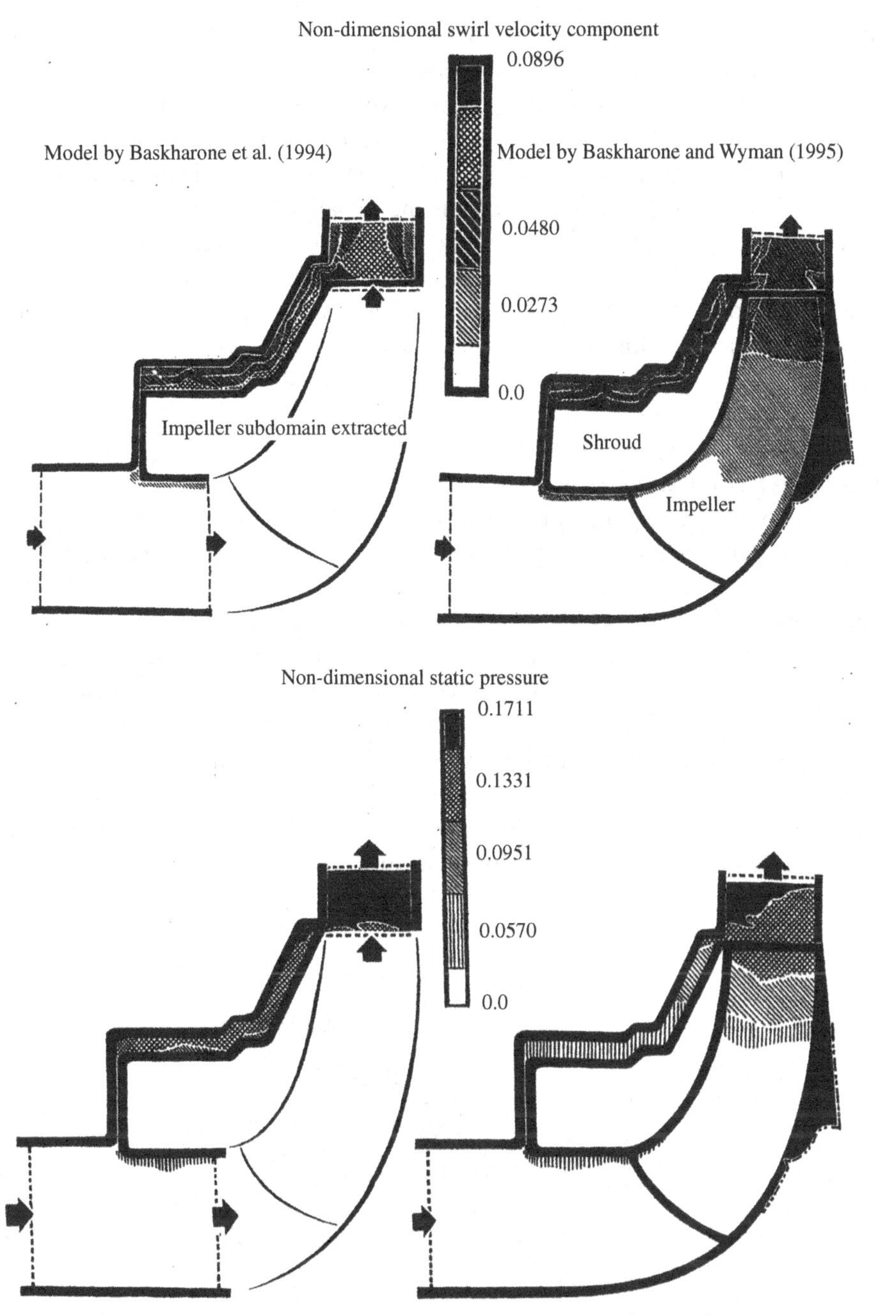

Figure 16.48. Effects of simulating the impeller subdomain in defining the computational domain.

coefficients. The difference, however, appears to be largely confined to the stage exit-flow segment, for it is the closest to the added impeller subdomain and has a decaying influence on the secondary-flow passage up to the knee-shaped segment of it. The difference between the two sets of data seems to be more significant as we compare the static-pressure distribution contours, including differences within the secondary-flow passage. This outcome was as surprising as it was unfortunate, at least initially, for the simple reason that you would primarily think of pressure anytime you think of impeller-applied forces. However, it is the pressure perturbation as the impeller encounters its whirling excitation that determines the fluid reaction forces and not the initial centered-impeller pressure magnitudes. Nevertheless, and in the bigger picture, the issue becomes one in which the increase in the finite element degrees of freedom in the expanded computational domain may very well cause a significant accuracy enhancement as far as the rotordynamic coefficients are concerned. In the following, the rotordynamic analysis will be applied only to the configuration on the left-hand side of Figure 16.46, with the enlarged-domain application left as a worthy extension of the current perturbation analysis.

Results of the Perturbation Analysis

Figure 16.48 shows a comparison of the swirl-velocity component and static pressure. The comparison here is between the two computational-domain choices in Figure 16.46.

The isometric sketch in Figure 16.49 shows the finite element assembly in the pump secondary-flow passage and the varied elemental distortions due to the rotor whirl. Also indicated in the figure, it is the previously discussed rotating-translating frame of reference that is attached to the rotor and whirls with it.

Results of the perturbation model in the form of fluid-induced forces on the shroud are shown in Figure 16.50 for a range of whirl-frequency ratios Ω/ω between -1.25 and +1.25, the same range of investigation as in the experimental study by Bolleter et al. [29]. The fluid-exerted forces in this figure are non-dimensionalized

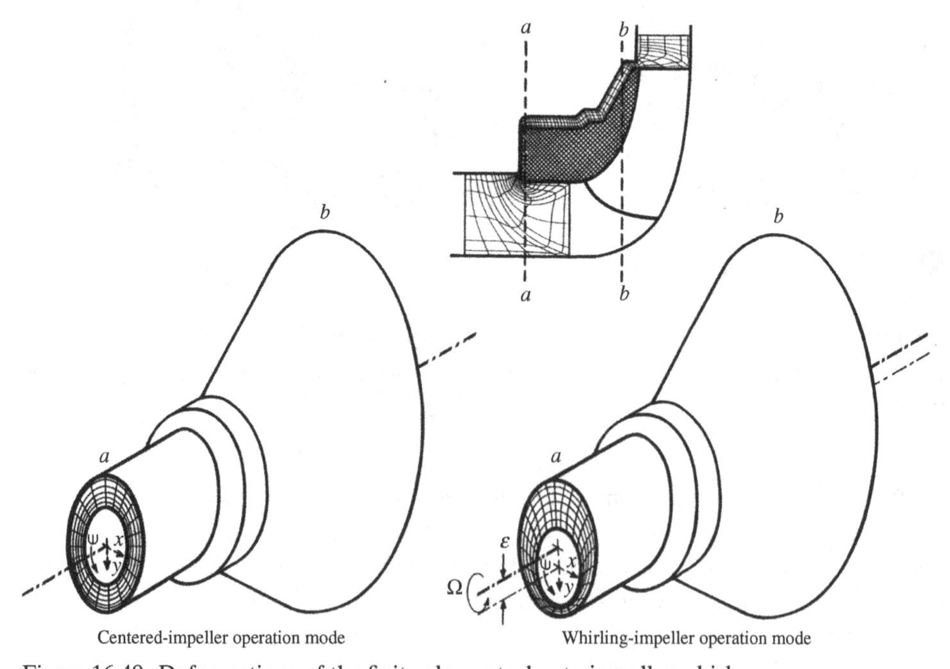

Figure 16.49. Deformations of the finite elements due to impeller whirl.

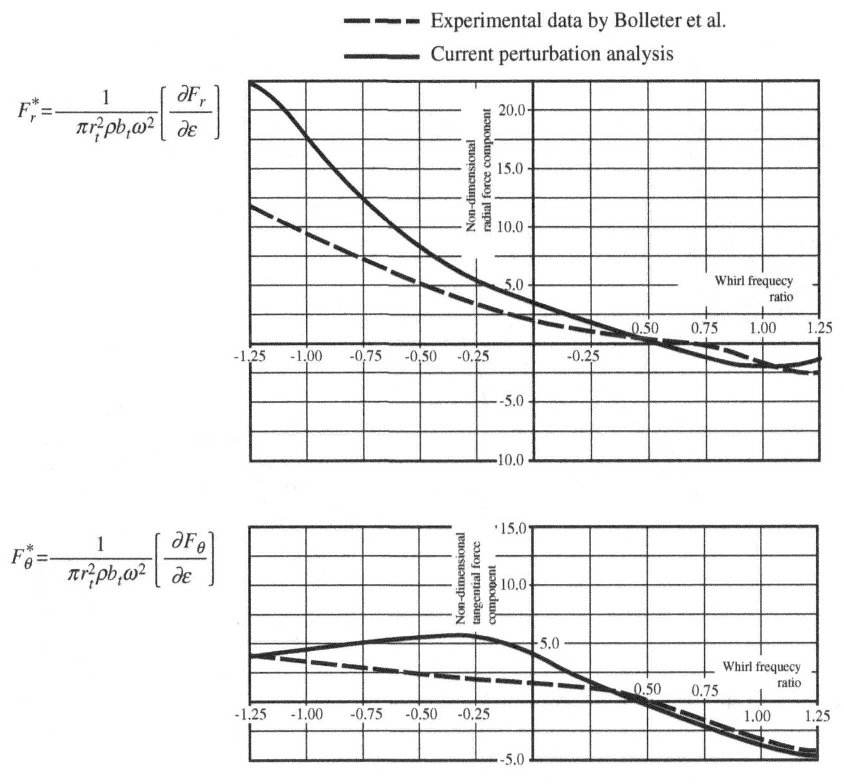

Figure 16.50. Comparison of the fluid-induced force components with experimental data.

as follows:

$$F_r{}^* = \frac{\frac{\partial F_r}{\partial \epsilon}}{\pi r_t{}^2 \rho b_t \omega^2}$$

$$F_\theta{}^* = \frac{\frac{\partial F_\theta}{\partial \epsilon}}{\pi r_t{}^2 \rho b_t \omega^2}$$

where ϵ is the impeller-axis eccentricity, r_t is the impeller tip radius, b_t is the impeller-outlet width, ρ is the fluid static density, and ω is the shaft speed. Reproduced in Figure 16.50 are the previously cited experimental measurements.

Examination of Figure 16.50 reveals a generally satisfactory agreement between the computed force components and their experimental counterparts. Excluded in this conclusion is the trend of tangential force component in the backward-whirl range, where the computed force component is nearly uniform over this entire range. Nevertheless, the trend agreement of the radial force component with the experimental data, within the same range, is perhaps evidence of the perturbation-model validity under a backward impeller whirl. However, the figure displays an excellent agreement between the computational and experimental results in the forward-whirl range in terms of both trend and magnitudes.

Over the unwrapped outer shroud surface (Figure 16.51), the pressure perturbations $\partial p/\partial \epsilon$ are displayed in Figure 16.52 for backward, forward, and synchronous

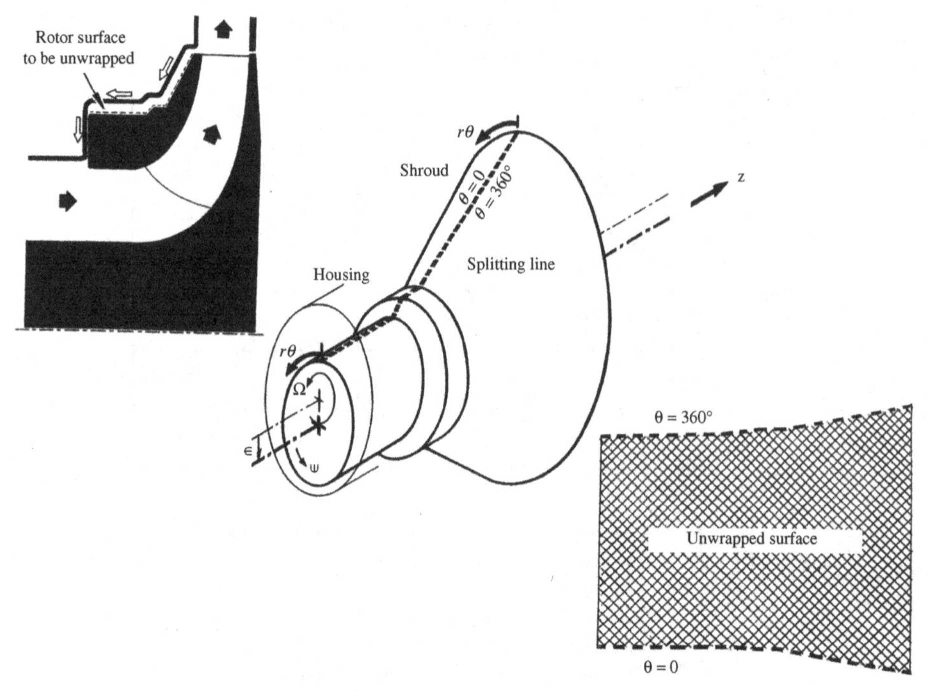

Figure 16.51. Unwrapping the secondary-flow-exposed shroud surface.

whirl frequency ratios of -0.2, 0.2, and 1.0, respectively. Corresponding to the first (backward-whirl) frequency ratio, the pressure-perturbations peak magnitude is seen to occur at a tangential location that is lagging the minimum-clearance position (marked $c - c$ in the figure) by reference to the direction of the whirling motion. Referring back to Figure 16.50, specifically the positive sign of $F_\theta{}^*$, this surface-pressure perturbation pattern gives rise to a tangential force that is whirl-sustaining and is therefore destabilizing.

As the impeller-shroud assembly enters the forward-whirl excitation mode, an abrupt redistribution of the pressure perturbation contours takes place, with the peak value occurring at an angular location that is ahead of the minimum-clearance position, as shown in Figure 16.52. Note that the impeller forward whirl gives rise to a negative tangential-force component, a restoring (or stabilizing) effect for a whirl-frequency ratio that is in excess of 0.3, by reference to Figure 16.50.

Assessment of the Single-Harmonic Perturbation Assumption

With the shroud-surface pressure-perturbation distribution already computed, the opportunity was there to assess the analytical consequences of limiting the circumferential variation of this variable to a single harmonic. This assumption is common among lower-order (so-called bulk-flow) perturbation models such as that of Childs [3], where perturbations of all variables were confined to the same assumption. To this end, a Fourier expansion (around the circumference) was carried out using the computed pressure perturbations at several streamwise locations, all corresponding to an arbitrarily selected whirl-frequency ratio of 0.3. Introduced in each expansion were two harmonics, with the lower-order harmonic amplitudes closely monitored.

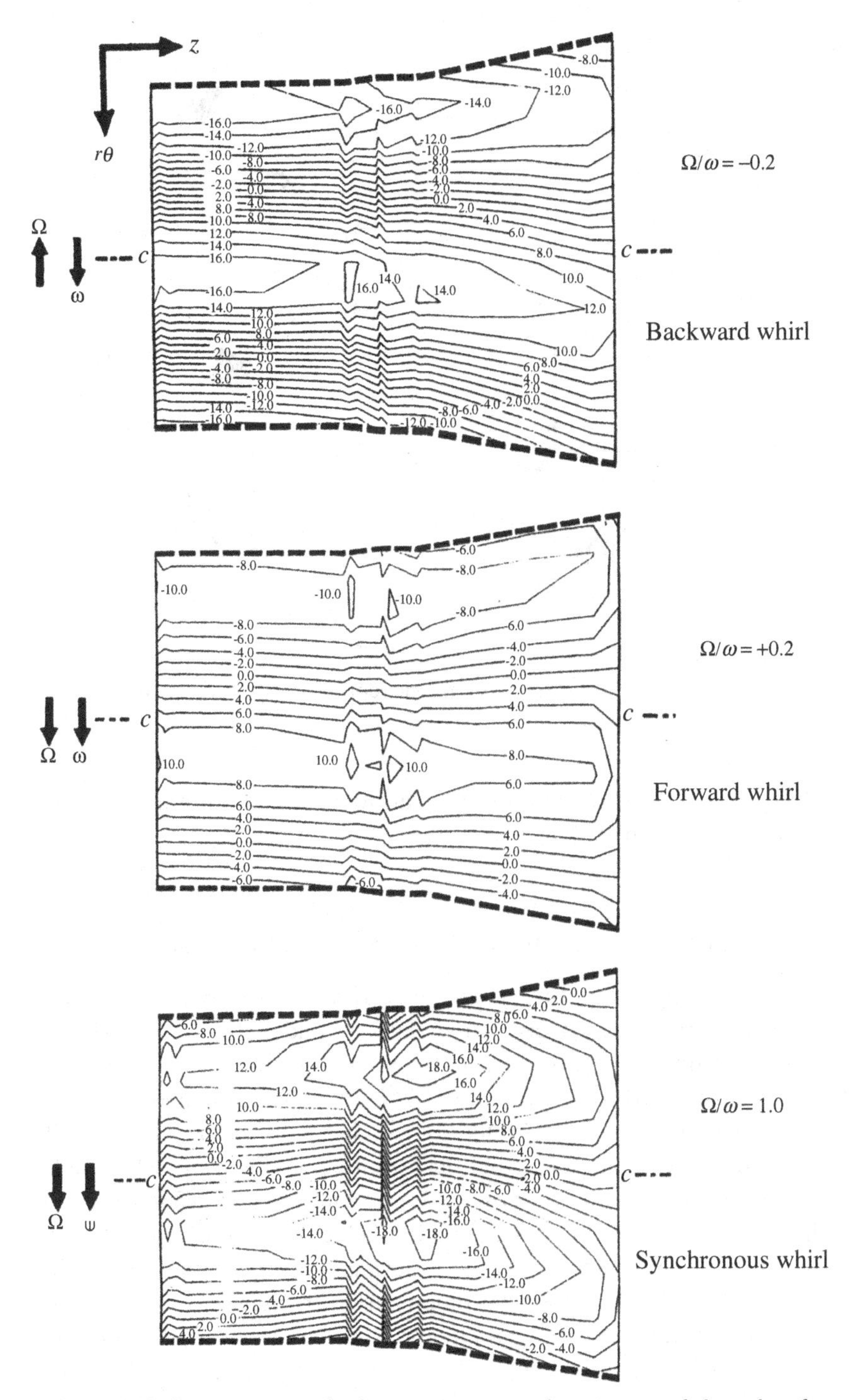

Figure 16.52. Pressure-perturbation contours over the unwrapped shroud surface.

Results of this analysis indicated that these amplitudes were hardly ignorable fractions of the primary-harmonic amplitudes, only over the shroud segments where massive flow separation and recirculation zones exist (see Figure 16.43). These findings are indicated in Figure 16.53, where an error magnitude as high as 36 percent

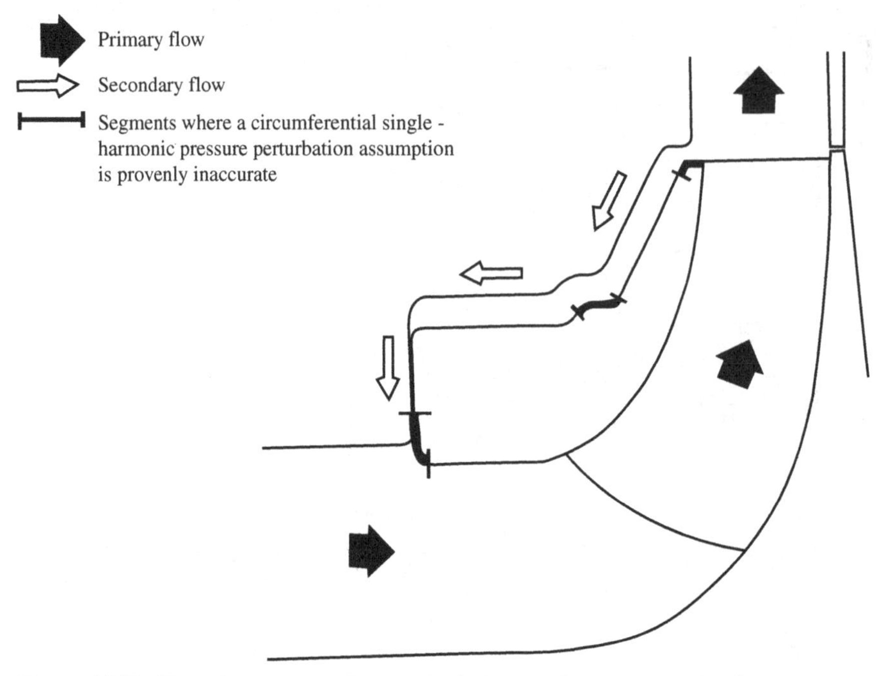

Figure 16.53. Shroud segments where a single-harmonic representation is erroneous.

was detected as a result of assuming a single-harmonic circumferential variation pattern of the pressure perturbation.

Applications: Investigation of Annular Seals under Conical Whirl

Not particularly surprising, the rotordynamic analysis of annular seals, both analytically and experimentally, has been overwhelmingly under the cylindrical type of whirl. The nontraditional study by Childs [35] was the first analytical attempt to quantify the moment-related rotordynamic coefficients in annular seals. This was a perturbation analysis based on Hirs' bulk-flow model, [39], where the rotor tilting motion was coupled with a purely-cylindrical rotor whirl (Figure 16.54). Childs' results were later compared with those by Kanemori and Iwatsubo [36] for a long seal where the fluid-induced moments would naturally be of predominant influence on the seal dynamic behavior. However, the experimental study was itself incomplete, for it was only concerned with the fluid-exerted moment resulting from a uniform lateral eccentricity, constituting a categorically cylindrical whirling motion. This type of rotor excitation, by reference to Figure 16.54, indeed leads to tilting moments, especially in long seals. It is in the next section that the effects of these two rotor excitations are treated as interrelated and the matrices of direct and cross-coupled rotordynamic coefficients are accordingly expanded.

In the current application of the finite element–based perturbation model, the task is to compute the direct effects of an angular rotor-axis excitation associated with a purely conical type of rotor whirl in the manner depicted in Figure 16.54. Recall that there was no uniform-eccentricity assumption in deriving the current

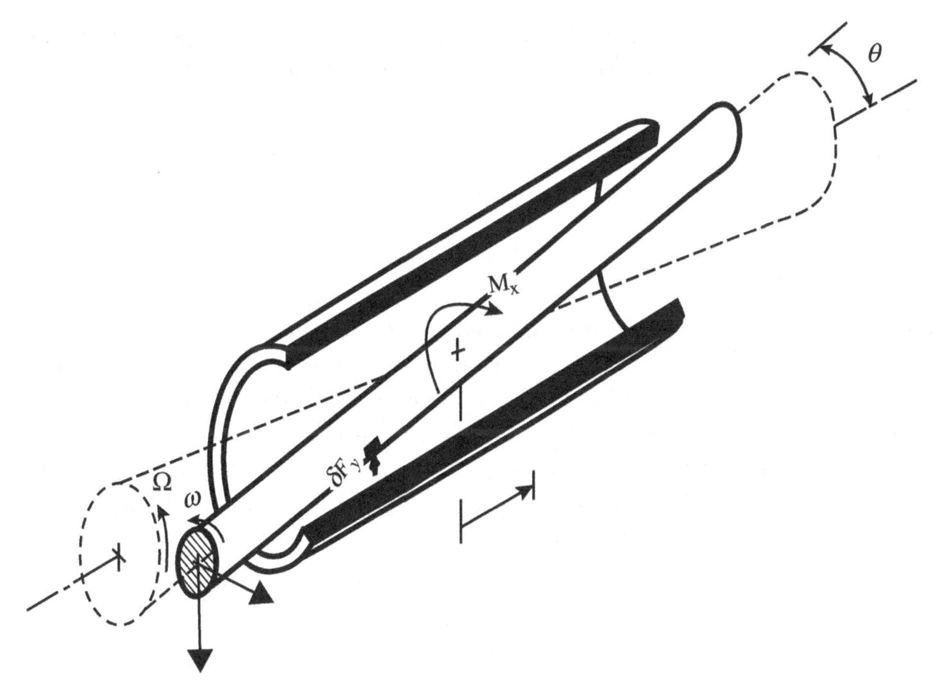

Figure 16.54. Conical whirl of an annular-seal rotor.

perturbation model, a versatility feature that makes it applicable to conical as well as cylindrical types of rotor-axis whirl.

Rotordynamic Analysis of the Fluid/Rotor Interaction System

Figure 16.54 shows a whirling rotor that is experiencing an infinitesimally small angular eccentricity θ around the center of the tilting motion coupled with a finite whirl frequency around the housing centerline. Eccentricity of the rotor axis in this case can generally be resolved at any given time t into two angular-displacement components around the x and y axes as follows:

$$\theta_x = \frac{e}{z}\cos(\Omega t)$$

$$\theta_y = \frac{e}{z}\sin(\Omega t)$$

where e is the local magnitude of lateral eccentricity, and z is the axial distance from the center of tilting motion. The fluid reaction in this case is in the form of differential moments δM_x and δM_y around the x and y axes, for which the following linearized relationship applies:

$$-\begin{Bmatrix} \delta M_x \\ \delta M_y \end{Bmatrix} = \begin{bmatrix} \bar{K} & \bar{k} \\ -\bar{k} & \bar{K} \end{bmatrix} \begin{Bmatrix} \theta_x \\ \theta_y \end{Bmatrix} + \begin{bmatrix} \bar{C} & \bar{c} \\ -\bar{c} & \bar{C} \end{bmatrix} \begin{Bmatrix} \dot{\theta}_x \\ \dot{\theta}_y \end{Bmatrix} + \begin{bmatrix} \bar{M} & \bar{m} \\ -\bar{m} & \bar{M} \end{bmatrix} \begin{Bmatrix} \ddot{\theta}_x \\ \ddot{\theta}_y \end{Bmatrix}$$

The direct ($\bar{K}$, $\bar{C}$, and $\bar{M}$) and cross-coupled ($\bar{k}$, $\bar{c}$, and $\bar{m}$) rotordynamic coefficients in the square matrices are the moment-related stiffness, damping, and inertia rotordynamic coefficients of the fluid-rotor interaction system. The overbar here is meant to simply distinguish these coefficients from their force-related counterparts under a purely cylindrical type of rotor whirl. The fluid reaction moments δM_x and δM_y are defined as follows:

$$\delta M_x = \int_z z\delta F_y$$

$$\delta M_y = \int_z z\delta F_x$$

with δF_x and δF_y being the differential changes in the force components that are exerted on the rotor due to the infinitesimally small shift in the pressure on the rotor surface as a result of the angular eccentricity (see Figure 16.54). With no lack of generality, consider the special location of the displaced rotor axis in a plane that is perpendicular to the housing centerline and at a distance z from the center of the tilting motion. At this location, we can express the angular location, velocity, and acceleration of the rotor axis as follows:

$$\theta_x \approx \frac{e}{z} = \theta \qquad \text{and} \qquad \theta_y = 0$$

$$\dot{\theta}_x = 0 \qquad \text{and} \qquad \dot{\theta}_y \approx \frac{\dot{e}}{z} = \Omega\theta$$

$$\ddot{\theta}_x \approx \frac{\ddot{e}}{z} = -\Omega^2\theta \qquad \text{and} \qquad \ddot{\theta}_y = 0$$

where e is the local lateral eccentricity of the rotor axis. Use of these three expressions in the preceding matrix equation gives rise to the following two relationships:

$$\frac{\partial M_x}{\partial \theta} = -\bar{K} - \Omega\bar{c} + \Omega^2\bar{M}$$

$$\frac{\partial M_y}{\partial \theta} = \bar{k} - \Omega\bar{C} + \Omega^2\bar{m}$$

Determination of the stiffness, damping, and inertia coefficients in these two equations requires computation of $\partial M_x/\partial\theta$ and $\partial M_y/\partial\theta$ at three or more different values of the whirl frequency Ω. Interpolation of these two derivatives as quadratic functions of Ω using curve-fitting techniques leads to the magnitudes of the moment coefficients $\bar{K}$, $\bar{k}$, $\bar{C}$, and so on simply by equating corresponding terms on both sides of each equation.

The procedure to determine $\partial M_x/\partial\theta$ and $\partial M_y/\partial\theta$ is initiated by securing a finite element solution of the undisturbed flow field in the rotor-to-housing annular gap for the centered-rotor operation mode. Next, a virtual angular eccentricity θ is introduced. As a result, the finite element nodal points are permitted to undergo virtual lateral displacements of different magnitudes depending on the spatial location of each node including, this time, the axial coordinate away from the center of tilting motion (see Figure 16.54). Implementation of the current perturbation analysis yields, among other variables, the rate at which the rotor-surface pressure varies with the angular eccentricity (i.e., $\partial p/\partial\theta$). The quantities $\partial M_x/\partial\theta$ and $\partial M_y/\partial\theta$ are, in the end, obtained by integrating the moment produced by each differential force (around the center of the tilting motion) from one end of the seal to the other.

Computational Results

A hydraulic annular seal with a clearance/length ratio of 0.85 percent was selected for model-verification purposes. The force coefficients of this seal were determined by Dietzen and Nordmann [32], and later verified by Baskharone and Hensel [33]. This seal has a rotor radius of 23.5 mm and a Reynolds number (based on the clearance and inlet through-flow velocity) of 4,700. Operating speeds of 2,000, 4,000, and 6,000 rpm were considered, together with preswirl–inlet-velocity ratios of 0.04, 0.10, and 0.17.

Shown in Figure 16.55 is an assessment of the zeroth-order centered-rotor flow solution. Compared in this figure is the computed nondimensional pressure drop across the seal (referred to as the *seal resistance*) with that obtained through Yamada's experimental correlation [34]. This particular correlation is probably the most accurate empirical means of determining the seal resistance in terms of the seal geometry, rotor speed, and preswirl velocity ratio. As seen in the figure, the computed pressure drop is in good agreement with Yamada's predictions because the average deviation between the two data sets within the seal operation range is approximately 5 percent.

The computed moment-related rotordynamic coefficients are compared with those obtained through Childs' bulk-flow analysis [35] for the same seal configuration and operating conditions. Noteworthy is the fact that Childs' simplified model has reportedly been successful when applied to simple clearance gaps, where zones of flow separation and recirculation, such as those in labyrinth seals, are

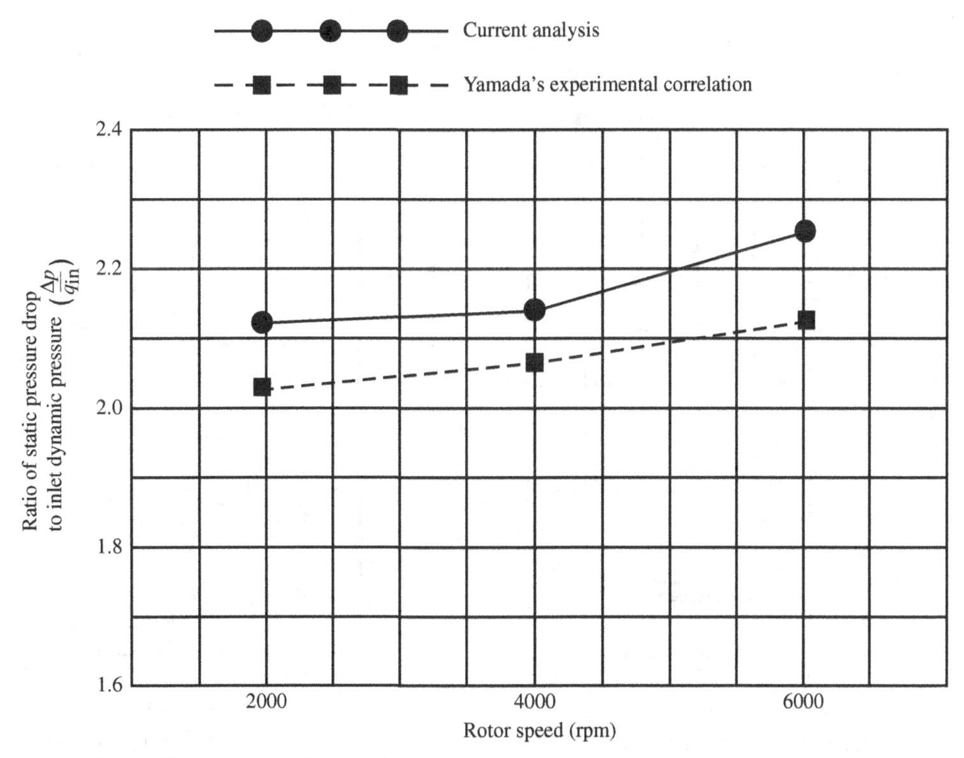

Figure 16.55. Comparison of the seal pressure drop with Yamada's correlation.

between rare and nonexisting. Absence of these complicating factors provides an environment where Childs' idealization of the clearance-gap flow, including parabolized flow equations and unsimulated shear stresses away from the solid walls, have a minimum impact on the numerical results. In fact, even the assumption of a single-harmonic perturbation representation would have an insignificant impact in this case in the absence of any notable recirculation zones in the computational domain.

Figure 16.56 shows a comparison between the computed conical-whirl-related rotordynamic coefficients and those obtained through Childs' bulk-flow model. These are the direct ($\bar{K}$, $\bar{C}$) and cross-coupled ($\bar{k}$, $\bar{c}$) stiffness and damping coefficients. The magnitude of direct inertia coefficient $\bar{M}$, in both sets of results was extremely small, making it comparable with the truncation errors during the numerical procedure. On the other hand, it is known that straight annular seals such as the present are incapable of displaying any significant cross-coupled inertia coefficient $\bar{m}$ whatsoever.

In the numerical procedure leading to the moment coefficients in Figure 16.57, it was realized that the current eccentricity mechanism, namely, conical whirl, is created and sustained by the moment component M_y and that a positive value of $\partial M_y/\partial\theta$ creates an aggravating effect in the case of forward whirl. With this in mind, and referring back to the definition of moment coefficients, such a destabilizing effect would be created by large and positive values of $\bar{k}$ and would diminish at sufficiently large and positive magnitudes of $\bar{C}$. Examination of the results in Figure 16.56 in light of this discussion reveals that the seal under consideration becomes more stable as the rotor speed ω is increased.

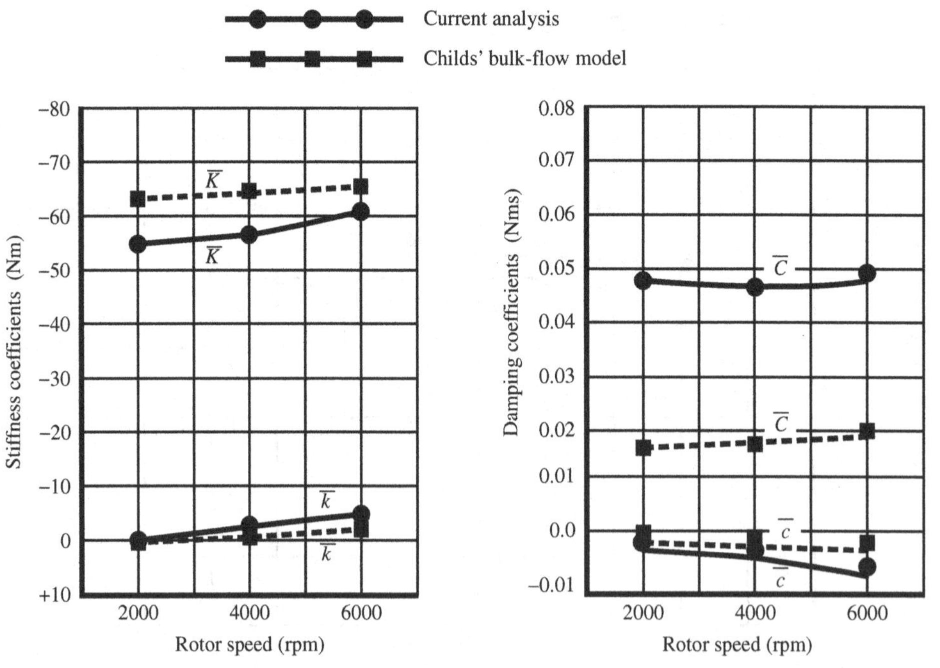

Figure 16.56. Comparison of the rotordynamic coefficients with those of Childs' simplified bulk-flow model.

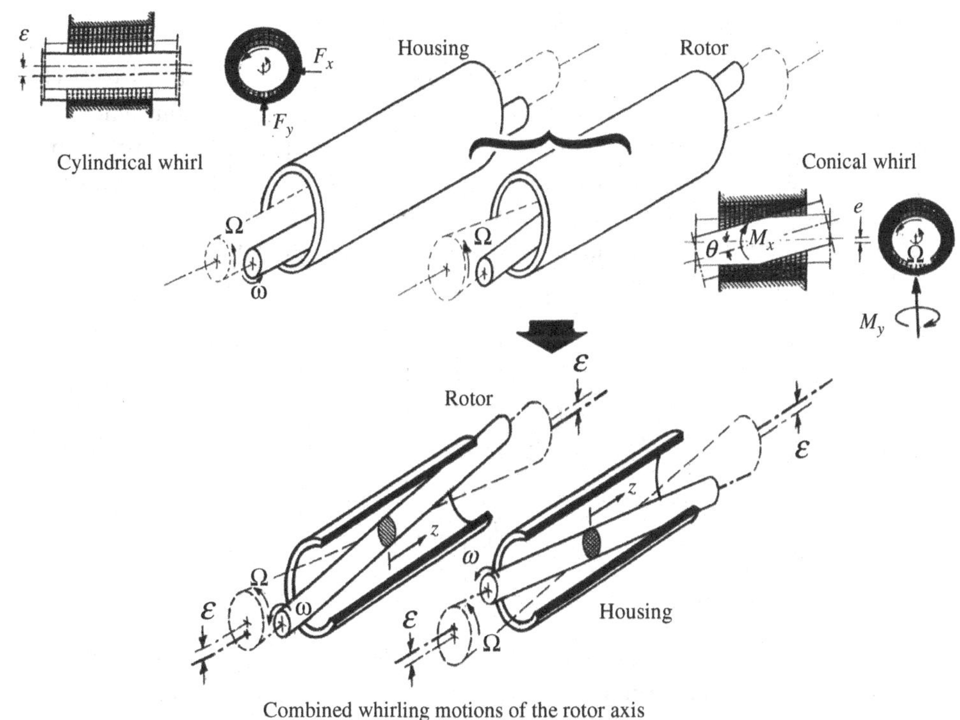

Figure 16.57. Interrelated rotordynamic effects on cylindrical and conical whirl.

Applications: Interrelated Effects of the Cylindrical/Conical Rotor Whirl

Under focus in this section are the mutual interaction effects of the simultaneous cylindrical and conical whirling motions in annular seals. It is common, as well as natural, to always associate a fluid reaction force with a uniform lateral eccentricity and a reaction moment with an angular eccentricity. The fact, however, is that each of these rotor excitations prompts a fluid response that is composed of a rotor-applied force components as well as moments. As a result, it is legitimate to speak of a rotordynamic coefficient such as $K_{\theta\epsilon}$, defined as the direct stiffness coefficient of moment production as a result of a lateral rotor displacement ϵ, and vice versa.

Among other existing models of rotordynamics, only the simplified model by Childs [35] is capable of handling the problem at hand. Other models, by comparison, are limited not only to simple seal geometries but also have been tailored around a single rotor excitation, namely, a uniform lateral eccentricity, with a fluid-induced force being the only rotordynamic response. Nevertheless, shortcomings of Childs' so-called bulk-flow model are significant from an accuracy viewpoint. These include a fictitiously parabolized flow field (prohibiting such real flow features as separation and recirculation) and unsimulated shear stress away from the immediate vicinity of a solid boundary segment. It is therefore important to point out that comparing the results of the current rotordynamic analysis with those of Childs' model is an exercise that has a rather limited value because the latter would understandably yield less accurate results.

Expanded Rotordynamic Analysis

The rotor's virtual eccentricity under consideration is shown in Figure 16.57. This consists of a uniform lateral eccentricity ϵ on which an angular displacement θ is superimposed. These infinitesimally small deviations from the centered-rotor operation mode are accompanied by a rotor whirl at a finite frequency Ω, as shown in this figure. The fluid reaction in this case is in the form of forces and moments in the x and y directions. These can be related to the displacement components of the whirling rotor as follows:

$$\begin{Bmatrix} \delta F_x \\ \delta F_y \\ \delta M_y \\ \delta M_x \end{Bmatrix} = \begin{bmatrix} K_\epsilon & k_\epsilon & K_{\epsilon\theta} & -k_{\epsilon\theta} \\ -k_\epsilon & K_\epsilon & k_{\epsilon\theta} & -K_{\epsilon\theta} \\ K_{\theta\epsilon} & k_{\theta\epsilon} & K_\theta & -k_\theta \\ k_{\theta\epsilon} & -K_{\theta\epsilon} & k_\theta & K_\theta \end{bmatrix} \begin{Bmatrix} x \\ y \\ \theta_y \\ \theta_x \end{Bmatrix} + \begin{bmatrix} C_\epsilon & c_\epsilon & C_{\epsilon\theta} & -c_{\epsilon\theta} \\ -c_\epsilon & C_\epsilon & -c_{\epsilon\theta} & -C_{\epsilon\theta} \\ C_{\theta\epsilon} & c_{\theta\epsilon} & C_\theta & -c_\theta \\ c_{\theta\epsilon} & -C_{\theta\epsilon} & c_\theta & C_\theta \end{bmatrix} \begin{Bmatrix} \dot{x} \\ \dot{y} \\ \dot{\theta}_y \\ \dot{\theta}_x \end{Bmatrix}$$

$$+ \begin{bmatrix} M_\epsilon & m_\epsilon & M_{\epsilon\theta} & -m_{\epsilon\theta} \\ -m_\epsilon & M_\epsilon & -m_{\epsilon\theta} & -M_{\epsilon\theta} \\ M_{\theta\epsilon} & m_{\theta\epsilon} & M_\theta & -m_\theta \\ m_{\theta\epsilon} & -M_{\theta\epsilon} & m_\theta & M_\theta \end{bmatrix} \begin{Bmatrix} \ddot{x} \\ \ddot{y} \\ \ddot{\theta}_y \\ \ddot{\theta}_x \end{Bmatrix}$$

where the symbols (K, k), (C, c) and (M, m) are the direct and cross-coupled stiffness, damping, and added mass (inertia) coefficients of the fluid-rotor interaction system. A single subscript of these coefficients implies force-displacement or moment-tilt types of correspondence. Double subscripts, on the other hand, tie fluid-exerted moments to a lateral eccentricity and forces to an angular eccentricity. Note that a rotor eccentricity, whether lateral or angular, in the x direction (for instance) will generally produce a restoring or aggravating fluid reaction in the y direction. Also note that straight annular seals such as the present are categorically incapable of displaying any significant magnitude of the cross-coupled inertia coefficients, that is,

$$m_\epsilon = m_\theta = m_{\epsilon\theta} = m_{\theta\epsilon} = 0$$

The procedure to compute the rotordynamic coefficients in the preceding matrix equation is initiated by expressing the local radius δ of the rotor-axis orbit (see Figure 16.52) as follows:

$$\delta = \epsilon + \epsilon_\theta$$

where the lateral eccentricity ϵ_θ due to the rotor tilting motion can be expressed as follows:

$$\epsilon_\theta = \theta z$$

a relationship that is always correct because the angular eccentricity θ is infinitesimally small, by definition. Motion of the rotor axis can be decoupled into the cylindrical and conical whirl modes depicted in Figure 16.57. Referred to the coordinate axes in the same figure, the lateral displacement components x and y of the tilting-motion center can be written for the whirling axis at any given time t as follows:

$$x = \epsilon \cos(\Omega t)$$

$$y = \epsilon \sin(\Omega t)$$

where Ω is the rotor-axis whirl frequency. Similarly, the angular displacement can be expressed in terms of the whirl frequency (see Figure 16.57) as follows:

$$\theta_x = \frac{\epsilon}{z} \cos(\Omega t)$$

$$\theta_y = \frac{\epsilon}{z} \sin(\Omega t)$$

With no lack of generality, consider the location of the displaced rotor axis, where $x=0$ and $\theta_y=0$, that is shown in Figure 16.57. Using the expressions of x, y, θ_x, and θ_y, the linear and angular coordinates of the rotor axis at the axial location z, as well as the velocity and acceleration components, can be written as follows:

$$x=0 \qquad y=\epsilon$$

$$\dot{x}=\Omega\epsilon \qquad \dot{y}=0$$

$$\ddot{x}=0 \qquad \ddot{y}=-\Omega^2\epsilon$$

$$\theta_x=\theta \qquad \theta_y=0$$

$$\dot{\theta}_x=0 \qquad \dot{\theta}_y=\Omega\theta$$

$$\ddot{\theta}_x=-\Omega^2\theta \qquad \ddot{\theta}_y=0$$

Substituting these expressions into the preceding matrix equation, and taking the limit as both the rotor-axis eccentricity components tend to zero, the following equations are achieved:

$$\frac{\partial F_x}{\partial \epsilon} = -k_\epsilon - \Omega C_\epsilon$$

$$\frac{\partial F_y}{\partial \epsilon} = -K_\epsilon + \Omega c_\epsilon + \Omega^2 M_\epsilon$$

$$\frac{\partial F_x}{\partial \theta} = k_{\epsilon\theta} - \Omega C_{\epsilon\theta} + \Omega^2 M_{\epsilon\theta}$$

$$\frac{\partial F_y}{\partial \theta} = K_{\epsilon\theta} + \Omega c_{\epsilon\theta}$$

$$\frac{\partial M_y}{\partial \epsilon} = -k_{\theta\epsilon} - \Omega C_{\theta\epsilon}$$

$$\frac{\partial M_x}{\partial \epsilon} = K_{\theta\epsilon} - \Omega c_{\theta\epsilon} - \Omega^2 M_{\theta\epsilon}$$

$$\frac{\partial M_y}{\partial \theta} = k_\theta - \Omega C_\theta$$

$$\frac{\partial M_x}{\partial \theta} = -K_\theta - \Omega c_\theta + \Omega^2 M_\theta$$

where the cross-coupled inertia coefficients were set to zero, as previously explained. Determination of the rotordynamic coefficients now reduces to the problem of computing the derivatives on the left-hand side of the preceding equations at a minimum of three whirl frequencies. Interpolation of these derivatives as quadratic expressions of the whirl frequency Ω using curve-fitting techniques leads to these coefficients by simply equating the different terms in these expressions to the right-hand sides of the preceding equations. With this in mind, the computational procedure becomes one of applying the current perturbation analysis to calculate the rate at which a fluid response (force or moment) varies with respect to an eccentricity (linear or angular) of the rotor axis.

Computational Results

The hydraulic seal previously used in conjunction with a purely conical whirl in the preceding section is again used to compute the interrelated rotordynamic coefficients in the current problem. The results are compared with those obtained through Childs' bulk-flow-model code [35].

Shown in Figure 16.58 is a comparison between the moment coefficients $K_{\theta\epsilon}$, $k_{\theta\epsilon}$, $C_{\theta\epsilon}$, and $c_{\theta\epsilon}$ associated with the rotor cylindrical whirl and those obtained through Childs' code. Of these, the cross-coupled stiffness and direct damping coefficients $k_{\theta\epsilon}$ and $C_{\theta\epsilon}$ represent the moments around the y axis (see Figure 16.57) that arise from the rotor cylindrical whirl. These moments conceptually suppress or aggravate the rotor whirling motion and are, therefore, of particular interest. As seen in Figure 16.58, the trends of these two coefficients with the rotor spinning speed ω are basically similar in both sets of results. However, the bulk-flow model seems to produce magnitudes that are significantly smaller than those produced by the current perturbation analysis.

The rotordynamic coefficients $K_{\epsilon\theta}$, $k_{\epsilon\theta}$, $C_{\epsilon\theta}$, and $c_{\epsilon\theta}$ are shown in Figure 16.59. This figure represents the fluid-exerted forces as a result of conical whirl. Aside from the moderate agreement between the computed values of $K_{\epsilon\theta}$ and $c_{\epsilon\theta}$ and those corresponding to the bulk-flow model, the figure shows that the respective values of $k_{\epsilon\theta}$ and $C_{\epsilon\theta}$ are far apart. In fact, results of the current perturbation analysis indicate that the rotor is much more stable by comparison.

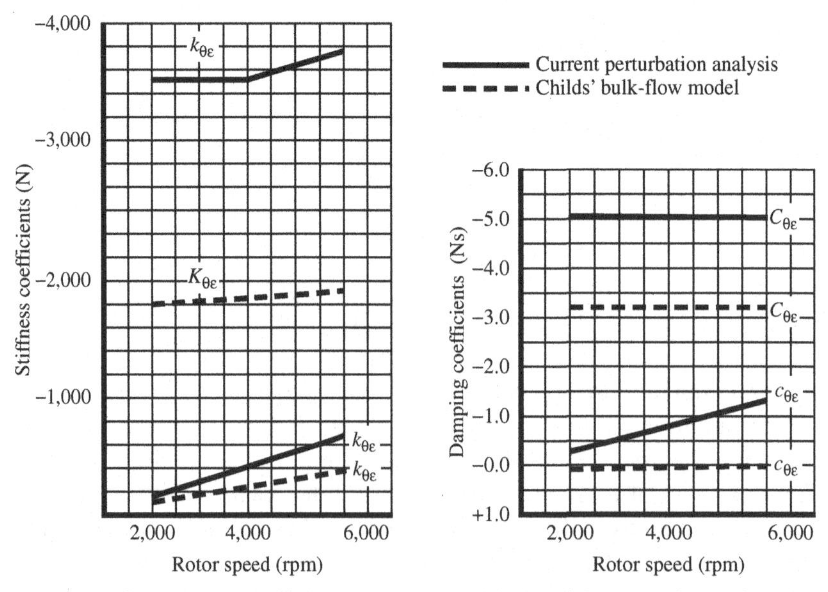

Figure 16.58. Moment coefficients due to cylindrical whirl: comparison with Childs' results.

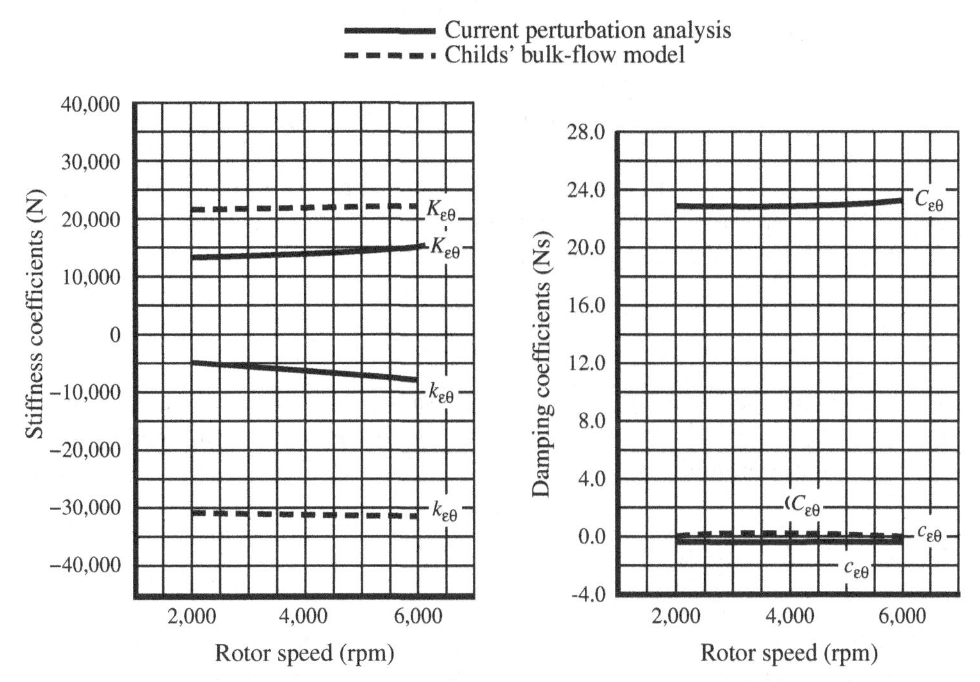

Figure 16.59. Force coefficients due to conical whirl: comparison with Childs' results.

Applications: Compressible-Flow Gas Seals Using a Simplified Adiabatic-Flow Approach

A more involved extension of the new perturbation model involves the gas-seal category of the fluid-induced vibration problems. This seal is frequently used in gas turbine engine applications, with the objective of minimizing the through-flow leakage, which is proportional to the streamwise static-pressure differential created by the primary flow stream. As for the rotordynamics aspect, the gas-seal problem lends itself to the same perturbation approach once the flow compressibility feature is added to the flow model. The gas-seal application would be a good starting point toward consideration of the larger problem of the shrouded-impeller centrifugal compressor problem as an example.

Leaving the rigorous analysis of this problem to Appendix F as a proposed analysis upgrade, the analytical model in this section and the computational results are based on the assumption of an adiabatic flow throughout the computational domain. In this case, and within an inner loop, the static-density magnitude is continually updated every time a new flow field is computed. In preparation to implement the equation of state, recall that the static-pressure nodal magnitudes constitute part of the flow solution. Remaining, then, is the magnitude of static temperature at the same computational nodes, which is where the flow adiabaticity presents itself. In an adiabatic flow, the total (or stagnation) temperature is constant everywhere in the solution domain and is equal to its magnitude T_t at the seal inlet station, which is assumed known. For a typical computational node where the velocity components have already been computed as V_z, V_r, and V_θ, the static temperature T can be calculated as follows:

$$T = T_t - \left[\frac{(\gamma - 1)}{2\gamma R}\right]\left(V^2{}_z + V^2{}_r + V^2{}_\theta\right)$$

where R is the gas constant, and γ is the specific heat ratio. Once the static temperature T is computed at all nodes where the static pressure has been obtained, the equation of state can be used to compute the static density ρ, as follows:

$$\rho = \frac{p}{RT}$$

The process just outlined is to be repeated toward convergence once a newly computed flow field is within a sufficiently small preset margin when compared with that of the previous iteration.

Computational Results

Given the approximate nature of this analysis, the relative stability ratio was identified to be the appropriate variable for comparison with experimental data because it is based on the ratio between two rotordynamic coefficients, deviating from the individual magnitudes of these coefficients. To this end, the gas seal investigated by Childs et al. [37] was chosen, and three values of the inlet preswirl were selected for investigation.

The relative stability ratio f in Figure 16.61 is defined as follows:

$$f = \frac{k}{\omega C}$$

where ω is the rotor speed. A large magnitude of f, as discussed earlier, is a sign of instability in the fluid-rotor interaction system.

As seen in Figure 16.60, the trend of computational results agrees with the experimental findings. As is well known in this engineering discipline, the seal inlet preswirl will always act as a destabilizing factor, and Figure 16.60 simply quantifies this fact.

Comment

The simplicity of this approach lies in the fact that it requires no additional degrees of freedom in the finite element model. This is a much-appreciated feature in a computational model that consumes a great deal of computational resources, including the large core size it demands. At the end of this text (specifically in Appendix F), a rigorous approach to compressible-flow-model adaptation is proposed. Inclusion of the energy equation in this case and the static temperature as a field variable will unavoidably increase the size of the computational model in terms of the total number of degrees of freedom. Worthiness of the upgrade, however, lies in the vast number of gas-turbine applications that the model is able to embrace. Specific issues that are unique to gas seals, such as friction choking, are part of the proposed upgrade in Appendix F.

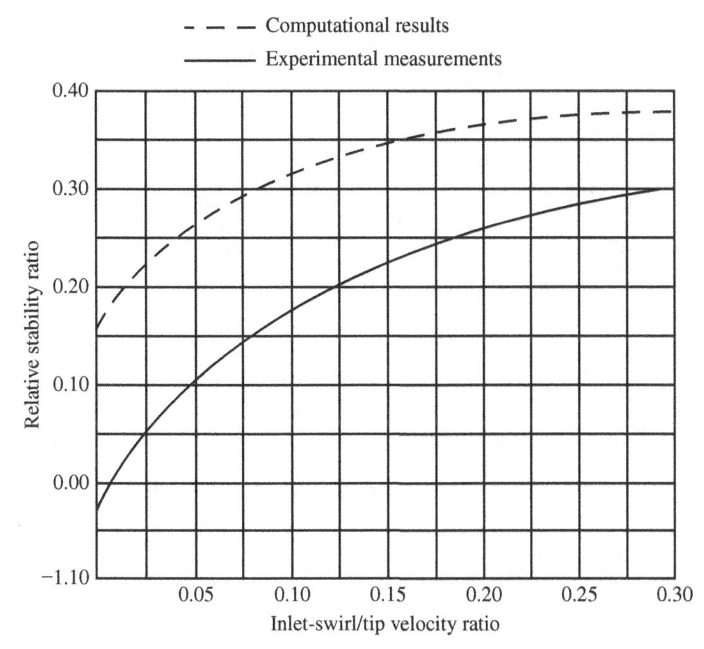

Figure 16.60. Relative stability ratio versus inlet swirl ratio for an annular gas seal.

Proposed Upgrades of the Perturbation Analysis

Inclusion of the Shear-Stress Perturbations in Computing the Fluid-Induced Forces

In computing the fluid-induced force-perturbation components [$\partial F_x/\partial\epsilon$ and $\partial F_y/\partial\epsilon$] in Equations (16.28) and (16.29)], only the rotor surface-pressure perturbations were taken into account. Ignored, in doing so, is the surface-applied shear-stress contributions, which, depending on the Reynolds number, would generally be of a comparable order of magnitude.

With the required variables being readily available at this point of the computational procedure, the recommended addition only involves an added post-processing step. This is where perturbations of all flow variables are known on computing the vector $\partial\{\Phi\}/\partial\epsilon$ once the final set of equations is solved. In the following, a subset of this vector, composed of perturbations of the velocity components in the vicinity of the rotor surface, is isolated and then used in computing the shear-stress contributions to $\partial F_x/\partial\epsilon$ and $\partial F_y/\partial\epsilon$. In the end, the process becomes one of simply summing up all contributions of the differential surface areas of the type shown in Figure 16.61.

Referring to Figure 16.61, the differential force δF_s due to the surface-exposed shear stress can be expressed as follows:

$$\delta F_s = \mu_e \left[\frac{\partial}{\partial n} (\vec{e}_\theta . \vec{V}_R) \right] dS \tag{16.30}$$

where $\vec{e}_\theta$ is the local unit vector in the tangential direction, and $\vec{n}$ is the local outward unit vector perpendicular to the differential surface area dS, as illustrated in Figure 16.61. The relative velocity vector $\vec{V}_R$ (relative to the whirling frame of

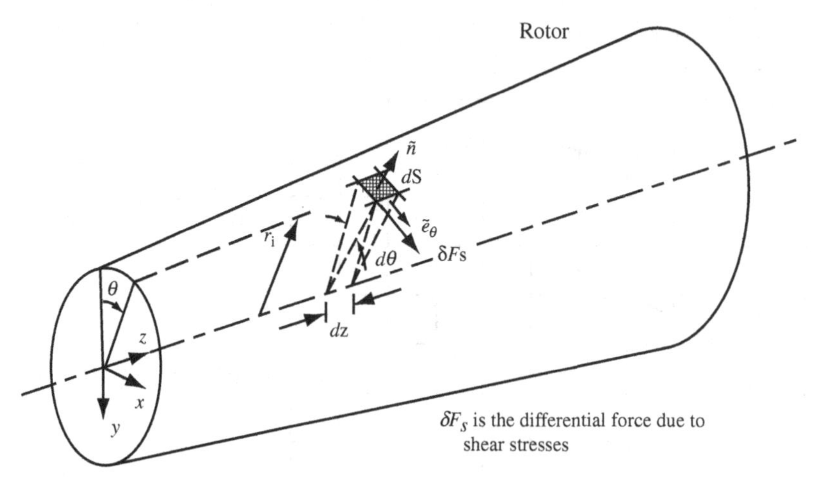

Figure 16.61. Including the rotor-surface shear stress in computing the fluid-induced forces.

reference in Figure 16.8) can be written in the following form:

$$\vec{V}_R = u\vec{e}_x + v\vec{e}_y + w\vec{e}_z \tag{16.31}$$

Dividing Equation (16.30) through by the rotor lateral eccentricity ϵ (see Figure 16.8), and taking the limit as ϵ tends to zero, we obtain the shear-stress-caused force perturbation $\partial F_s/\partial\epsilon$ as follows:

$$\frac{\partial F_s}{\partial \epsilon} \equiv \lim_{\epsilon \to 0} \frac{\delta F_s}{\epsilon} = \frac{\partial}{\partial \epsilon}\left\{\mu_e \left[\frac{\partial}{\partial n}(\vec{e}_\theta . \vec{V}_R)\right] dS\right\} \tag{16.32}$$

Using expression (16.31) for the relative-velocity vector $\vec{V}_{6R}$, Equation (16.32) can be rewritten as follows:

$$\begin{aligned}\frac{\partial F_s}{\partial \epsilon} &= \frac{\partial}{\partial \epsilon}\left\{\mu_e \left[(\vec{e}_\theta . \vec{e}_x)\frac{\partial u}{\partial n} + (\vec{e}_\theta . \vec{e}_y)\frac{\partial v}{\partial n} + (\vec{e}_\theta . \vec{e}_z)\frac{\partial w}{\partial n}\right]\right\} dS \\ &= \left\{\mu_e \left[\sin\theta \frac{\partial}{\partial n}\left(\frac{\partial u}{\partial \epsilon}\right) + \cos\theta \frac{\partial}{\partial n}\left(\frac{\partial v}{\partial \epsilon}\right)\right]\right\} dS\end{aligned} \tag{16.33}$$

where the last term on the right-hand side was dropped, because $\vec{e}_\theta$ and $\vec{e}_z$ are perpendicular to one another. The effective viscosity coefficient μ_e is the result of applying the Baldwin and Lomax turbulence closure, combined with the Benim and Zinser near-wall analysis discussed in Chapter 13. Note that all three derivatives in parentheses are known at this stage of the computational procedure because the vector $\partial\{\Phi\}/\partial\epsilon$ is now known, where

$$\frac{\partial\{\Phi\}}{\partial \epsilon} = \begin{Bmatrix} \dfrac{\partial\{U\}}{\partial \epsilon} \\ \dfrac{\partial\{V\}}{\partial \epsilon} \\ \dfrac{\partial\{W\}}{\partial \epsilon} \\ \dfrac{\partial\{P\}}{\partial \epsilon} \end{Bmatrix} \tag{16.34}$$

with $\{U\}$, $\{V\}$, $\{W\}$, and $\{P\}$ being the vectors of all nodal values of u, v, w, and p, respectively.

Referring to Figure 16.61, the shear-stress-related force-perturbation components contributed by the differential rotor-surface area dS can be resolved in the x and y directions as follows:

$$\frac{\partial F_{s,x}}{\partial \epsilon} = \frac{\partial F_s}{\partial \epsilon}\sin\theta \tag{16.35}$$

$$\frac{\partial F_{s,y}}{\partial \epsilon} = \frac{\partial F_s}{\partial \epsilon}\cos\theta \tag{16.36}$$

Noting that expressions (16.35) and (16.36) concern the contributions of the differential area dS and that:

$$dS = r_i d\theta\, dz \tag{16.37}$$

where r_i is the rotor local radius (see Figure 16.61), the net force perturbation components due to shear stresses over the rotor surface can now be obtained by summing up all such contributions as follows:

$$\left(\frac{\partial F_{s,x}}{\partial \epsilon}\right)_{net} = \sum_{N_s}\left\{\int\int \mu_e\left[\frac{\partial}{\partial n}\left(\frac{\partial u}{\partial \epsilon}\right)\right]\sin\theta + \mu_e\left[\frac{\partial}{\partial n}\left(\frac{\partial v}{\partial \epsilon}\right)\right]\right\}\cos\theta r_i\, d\theta dz \tag{16.38}$$

$$\left(\frac{\partial F_{s,y}}{\partial \epsilon}\right)_{net} = \sum_{N_s}\left\{\int\int \mu_e\left[\frac{\partial}{\partial n}\left(\frac{\partial u}{\partial \epsilon}\right)\right]\sin\theta + \mu_e\left[\frac{\partial}{\partial n}\left(\frac{\partial v}{\partial \epsilon}\right)\right]\right\}\sin\theta r_i\, d\theta dz \tag{16.39}$$

where N_s is the total number of the elemental surface areas existing on the rotor surface. It is certain that adding the shear-stress-related contributions to the pressure-caused force component perturbations in Equations (16.28) and (16.29) will significantly enhance the accuracy of the perturbation analysis.

Rigorous Adaptation to Compressible-Flow Problems

This detailed segment in the book is separated to be the subject of Appendix F which carries the same title.

Relevant Remarks

Not before the development of the current perturbation model were finite-elements treated as variable-geometry (deformable) subdomains. The model is founded on the idea that deformations of the flow domain can be transmitted to the individual elements through varied displacements of the computational nodes once the rotor enters its off-center operation mode. Away from the lengthy process of switching the reference frame, derivation of the distorted-element equations, and the analytical effort to separate the rotor-eccentricity components, the overall objective is as simple as absorbing the element distortions in the volume and surface integration steps that ultimately gives rise to the set of finite element equations. In effect, therefore, the finite element–based perturbation model exploits this very fact that the finite element equations are obtained by integrating different functionals over the element and its boundary. This feature, for instance, is nonexistent in a typical finite-difference procedure, whereas the equations reflect functional relationships relating the flow variable(s) at a given computational point to the same at neighboring grid points. Details of the model itself reveal the fact that the model's significant versatility is but a natural inheritance of the adaptability and versatility of the finite element–based perturbation approach.

In validating the current perturbation model, and out of seven problems in this chapter, the first is considered a real test case. This, as presented earlier, involves the rotordynamic testing of a special rotor configuration under a cylindrical type of whirling motion. The relevant feature in this case is that the flow passage ending with the face seal is rather long, making the admission losses small enough not to cause a grave concern in the analytical simulation of this passage. Moreover, the streamwise declining radius of the flow passage creates a continually accelerating flow where the friction-related losses are conceptually small. In all, this test case takes away most of the factors that would create substantial deviations of the zeroth-order flow solution relative to that in the test rig. It is under such a favorable environment that the accuracy of the final results is a true reflection of the perturbation model accuracy. Recall that investigation of this particular test case produced results that are significantly consistent with the experimental data.

Other examples where comparison with experimental data is, in reality, unfair to the perturbation model results concern situations where the rotor eccentricity is too large to be modeled via perturbation means. Suspected to be sufficiently small, the rotor-axis orbit radius ϵ in many cases was not provided. The exception here concerns the problem of synchronous whirl (second test problem), where the LDA measurements of the disturbed flow field were taken for a whirl-orbit radius that is as large as 50 percent of the clearance-gap width. Comparing these measurements with the outcome of this or any other perturbation model (where the rotor

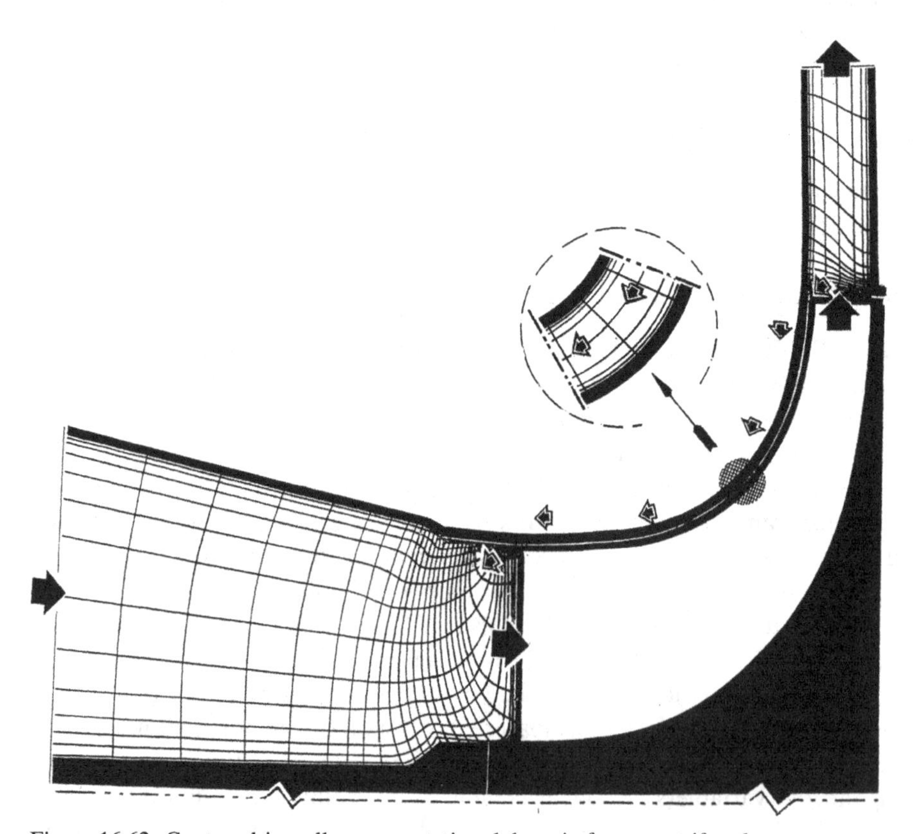

Figure 16.62. Centered-impeller computational domain for a centrifugal compressor.

eccentricity ϵ is virtual and the results are obtained in the limit as it tends to zero) clearly raises the issue of inconsistency. The exception, however, is the situation where dependency of the fluid-rotor interaction system on ϵ is linear. Judging by the trend agreements of the two sets of results in this case, this linearity appears to be more or less satisfied in this particular problem. A conclusive judgment, nevertheless, remains unreachable in this case because no other rotor-eccentricity magnitudes were experimentally investigated.

Figure 16.62 shows the centered-impeller finite element model for determining the leakage through the secondary passage of a centrifugal compressor.

PROBLEMS

P.1 Figure 16.63 shows the geometric arrangements and coordinate axes associated with the zeroth-order flow field in a simple annular seal. The rotor in this figure is about to undergo a backward whirling motion at a whirl frequency, Ω that is numerically equal to the rotor speed. The coordinate axes are rotating at the same magnitude and in the same direction of Ω. Determine, in terms of the rotor speed ω, the velocity magnitudes and directions at the two computational nodes i and j on the rotor and housing surfaces, respectively.

P.2 Figure 16.64 shows a shrouded radial-turbine rotor that is synchronously whirling ($\Omega = \omega$) around the housing centerline. In defining the fluid-rotor interaction domain, the figure shows three different domain configurations beginning with the most rigorous, where the primary-flow passage (in its entirety) is added to the secondary (whirl-impacting) passage. Next is the domain definition where only two (inlet and exit) segments of the primary-flow passage are connected to both ends of the secondary passage. The third configuration in the figure is the least desirable because it confines the fluid-structure interaction effects to the secondary-flow passage, requiring the specification of boundary conditions at both ends of this passage. Recalling that the rotor eccentricity is virtual, and viewing the flow problem from the usual rotating-translating frame of reference:

a. Determine the magnitude and direction of swirl-velocity components at the two computational nodes i and j in the figure.

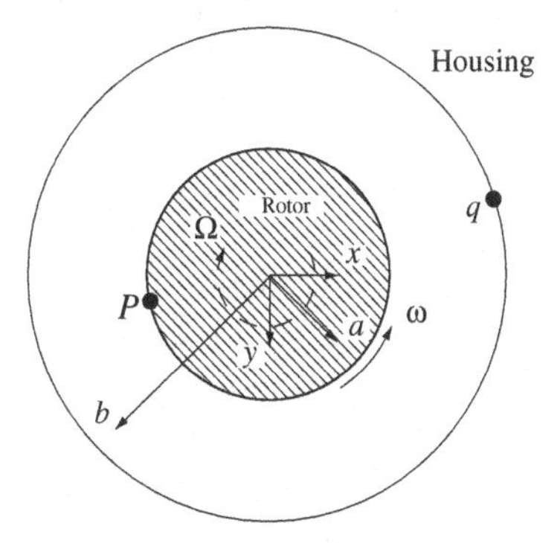

Figure 16.63. Backward whirl where the whirl frequency is equal but opposite to the rotor speed.

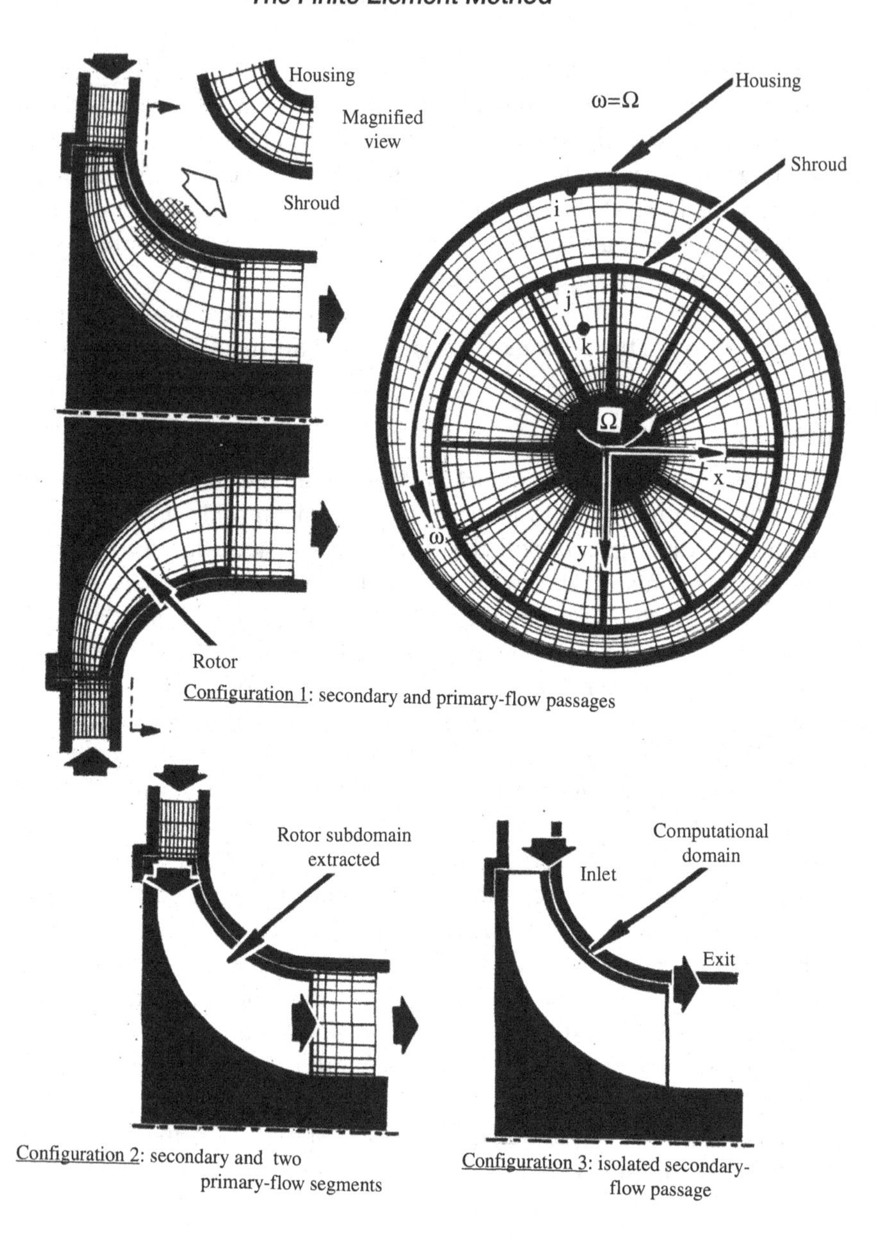

Figure 16.64. A radial-turbine shrouded rotor with three configurations for defining the computational domain around the secondary-flow passage.

b. Indicate the magnitudes of the node-displacement fraction λ at the three nodes i, j, and k.

c. Considering separately each of the three domain configurations in Figure 16.64, identify (in each case) the flow-permeable boundary segments, and list the appropriate boundary condition(s) you would choose to apply at these segments. You may assume a fully developed flow at the exit of any flow passage in any of the three computational-domain configurations.

P.3 Figure 16.65 shows two configurations of the same pump stage, the first with a casing-mounted labyrinth seal and the other with a shroud-mounted

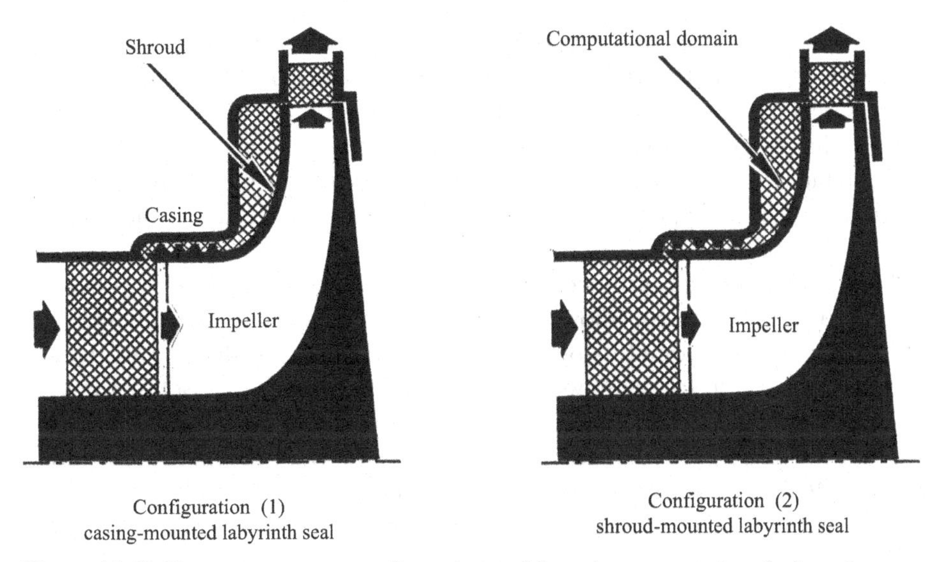

Figure 16.65. Two pump-stage configurations with casing-mounted and shroud-mounted labyrinth seals.

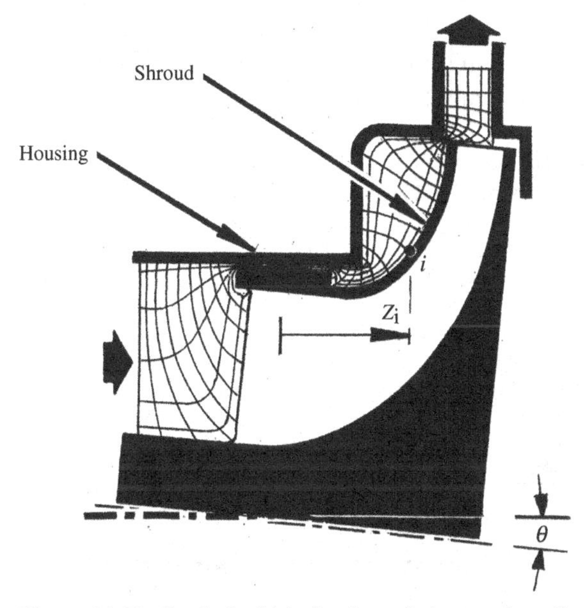

Figure 16.66. Conical whirl of a shrouded pump impeller.

labyrinth seal. Considering the centered-rotor axisymmetric flow field governed by Equations (16.9), (16.10), and (16.11),

a. Write down the boundary conditions over the shroud and labyrinth seal in each of the two pump configurations.
b. By integrating the pressure forces in the axial direction, show how you can calculate the thrust force acting on the impeller.

P.4 Figure 16.66 shows a shrouded pump impeller that is undergoing a conical whirl with a cone half-angle of θ. In this figure, note that the shroud lateral displacement e_i at the typical node i is equal to θz_i. Referring to the conical whirl section in this chapter as well as expressions (16.28), (16.29), (16.38), and (16.39), write down expressions for $\partial F_x/\partial\theta$, $\partial F_y/\partial\theta$, $\partial M_x/\partial\theta$ and $\partial M_y/\partial\theta$, taking both the pressure and shear-stress contributions into account.

P.5 Referring to Figure 16.2, let us assume an entirely laminar flow field through the annular seal. Now consider the submatrices of influence coefficients $a_1)_{i,j}$ through $a_{15})_{i,j}$ in Appendix D. Re-express these coefficients under the laminar-flow assumption.

P.6 As a general rule, a rotational flow field (one with nonzero vorticity) does not have to involve a viscous-flow stream. In fact, the vorticity (the fluid-particle tendency to spin around its own axis) can be imparted to the flow domain as the effect of an upstream flow component and would present itself in the inlet velocity profiles. Under the simplification of an inviscid-flow field, rewrite the $\bar{a}_1)_{i,j}$ through $\bar{a}_{15})_{i,j}$ submatrices of influence coefficients in Appendix F.

P.7 Figure 16.67 shows the centered-rotor operation mode of a simple annular seal, where the flow is swirling (due to the rotor spinning speed ω) and axisymmetric. Under this operation mode, the set of finite element equations associated with the cross-hatched element can be expressed as follows:

$$\begin{bmatrix} [A] & [B] & [0] & [0] \\ [E] & [F] & [0] & [G] \\ [H] & [0] & [0] & [0] \\ [R] & [S] & [0] & [0] \end{bmatrix} \begin{Bmatrix} \{V_z\} \\ \{V_r\} \\ \{V_\theta\} \\ \{P\} \end{Bmatrix} = \begin{Bmatrix} \{0\} \\ \{I_z\} \\ \{I_r\} \\ \{I_\theta\} \end{Bmatrix}$$

where the submatrices [A] and [B] are 4×8 arrays, [R] and [S] are 8×4 arrays, and the submatrices [C], [D], [E], [G], and [H] are 8×8 square submatrices. The

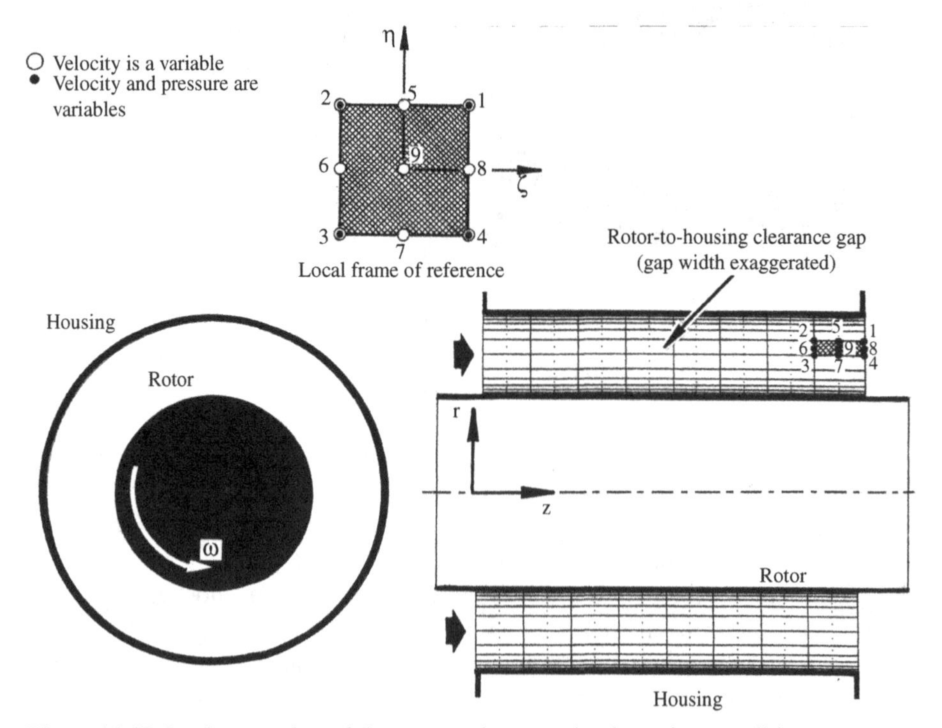

Figure 16.67. Implementation of the zero-surface-traction boundary conditions.

submatrices and subvectors in this expression are the same as those introduced in Equations (16.16) through (16.19). The exit boundary conditions over the element side 1-8-4 are those termed *zero surface tractions*, which are defined as follows:

$$\frac{\partial V_z}{\partial z} - \left(\frac{1}{2\rho \nu_{eff}}\right)p = 0$$

$$\frac{\partial V_z}{\partial r} + \frac{\partial V_r}{\partial z} = 0$$

$$\frac{\partial V_\theta}{\partial z} = 0$$

Limiting the imposition of these three conditions to computational node 8 of the highlighted element, prove that this step will alter the entries of the eighth row in the submatrices [C], [R], [E], [F], [G], and [H], as well as the eighth entry of the subvectors $\{I_z\}$, $\{I_r\}$, and $\{I_\theta\}$ in the following manner:

$$C_{8,i} = V_{1,1}\frac{\partial N_i}{\partial \zeta} + V_{1,2}\frac{\partial N_i}{\partial \eta} \qquad \text{where} i = 1,2,3,\ldots,9$$

$$R_{8,k} = -\frac{1}{2\rho \nu_{eff}}M_k \qquad \text{where} k = 1,2,3,4$$

$$E_{8,i} = V_{2,1}\frac{\partial N_i}{\partial \zeta} + V_{2,2}\frac{\partial N_i}{\partial \eta} \qquad \text{where} i = 1,2,3,\ldots,9$$

$$F_{8,i} = V_{1,1}\frac{\partial N_i}{\partial \zeta} + V_{1,2}\frac{\partial N_i}{\partial \eta} \qquad \text{where} i = 1,2,3,\ldots,9$$

$$H_{8,i} = V_{1,1}\frac{\partial N_i}{\partial \zeta} + V_{1,2}\frac{\partial N_i}{\partial \eta} \qquad \text{where} i = 1,2,3,\ldots,9$$

$$(I_z)_8 = 0$$

$$(I_r)_8 = 0$$

$$(I_\theta)_8 = 0$$

where the 2×2 coordinate-transformation matrix [V] is defined as follows:

$$[V] = \frac{1}{T}\begin{bmatrix} \left(\sum_{i=1}^{9}\frac{\partial N_i}{\partial \eta}r_i\right) & -\left(\sum_{i=1}^{9}\frac{\partial N_i}{\partial \eta}z_i\right) \\ \left(\sum_{i=1}^{9}\frac{\partial N_i}{\partial \zeta}r_i\right) & \left(\sum_{i=1}^{9}\frac{\partial N_i}{\partial \zeta}z_i\right) \end{bmatrix}$$

where

$$T = \left(\sum_{i=1}^{9}\frac{\partial N_i}{\partial \zeta}z_i\right)\left(\sum_{i=1}^{9}\frac{\partial N_i}{\partial \eta}r_i\right) - \left(\sum_{i=1}^{9}\frac{\partial N_i}{\partial \eta}z_i\right)\left(\sum_{i=1}^{9}\frac{\partial N_i}{\partial \zeta}r_i\right)$$

REFERENCES

[1] Baskharone, E. A., and Hensel, S. J., "A Finite Element Perturbation Approach to Fluid/Rotor Interaction in Turbomachinery Elements. 1: Theory," *Journal of Fluids Engineering*, Vol. 113, No. 3, 1991, pp. 353–361.

[2] Baskharone, E. A., and Hensel, S. J., "A Finite Element Perturbation Approach to Fluid/Rotor Interaction in Turbomachinery Elements. 2: Applications," *Journal of Fluids Engineering*, Vol. 113, No. 3, 1991, pp. 362–367.
[3] Childs, D., *Turbomachinery Rotordynamics: Phenomena, Modeling and Analysis*, Wiley, New York, 1993.
[4] Baskharone, E. A., and Hensel, S. J., "A New Model for Leakage Prediction in Shrouded-Impeller Turbopumps," *Journal of Fluids Engineering*, Vol. 111, No. 2, 1989, pp. 118–123.
[5] Baskharone, E. A., and Hensel, S. J, "Flow Field in the Secondary, Seal-Containing Passages of Centrifugal Pumps," *Journal of Fluids Engineering*, Vol. 115, No. 4, 1993, pp. 702–709.
[6] Baldwin, B. S., and Lomax, H., "Thin Layer Approximation and Algebraic Model for Separated Turbulent Flows," *AIAA Paper No. 78–257*, 1978.
[7] Benim, A. C., and Zinser, W., "Investigation of the Finite Element Analysis of Confined Turbulent Flows Using a $\kappa - \epsilon$ Model of Turbulence," *Computer Methods in Applied Mechanics and Engineering*, Vol. 51, 1985, pp. 507–523.
[8] Heinrich, J. C., and Zieniewicz, O. C., "Quadratic Finite Element Schemes for Two-Dimensional Convective Transport Problems," *International Journal of Numerical Methods in Engineering*, Vol. 11, 1977, pp. 1831–1834.
[9] Carey, G. F., and Oden, J. T., *Finite Elements: Fluid Mechanics*, Vol. 6, Prentice-Hall, Englewood Cliffs, NJ, 1986.
[10] Hood, P., "Frontal Solution Program for Unsymmetric Matrices," *International Journal of Numerical Methods in Engineering*, Vol. 10, 1976, pp. 379–399.
[11] Morrison, G. L., DeOtte, R. E., and Thomas, H. D., "Turbulence Measurements of High Shear Flow Fields in Turbomachine Seal Configuration," presented at the Advanced Earth-to-Orbit Propulsion Technology Conference, Huntsville, AL, 1992.
[12] Zienkiewicz, O. C., *The Finite Element Method in Engineering Science*, McGraw-Hill, New York, 1971.
[13] Baskharone, E., and Hamed, A., "A New Approach in Cascade Flow Analysis Using the Finite Element Method," *AIAA Journal*, Vol. 19, No. 1, 1981, pp. 65–71.
[14] Guinzberg, A., Brennen, C. A., Acosta, A. J., and Caughey, T. K., "The Effect of Inlet Swirl on the Rotordynamic Shrouded Forces in Centrifugal Pumps," *Journal of Engineering for Gas Turbine and Power*, Vol. 115, 1993, pp. 287–293.
[15] Baskharone, E. A., Daniel, A. S., and Hensel, S. J., "Rotordynamic Effects of the Shroud-to-Housing Leakage Flow in Centrifugal Pumps," *Journal of Fluids Engineering*, Vol. 116, No. 3, 1994, pp. 558–563.
[16] Morrison, G. L, Johnson, M. C., and Tatterson, G. B., "Three-Dimensional Laser Anemometer Measurements in an Annular Seal," *Journal of Tribology*, Vol. 113, 1991, pp. 421–427.
[17] Stoff, H., Incompressible Flow in a Labyrinth Seal," *Journal of Fluid Mechanics*, Vol. 100, 1980. pp. 817–829.
[18] Rhode, D. L., Demko, J. A., Morrison, G. L., Traegner, U. K., and Sobolic, S. R., "On the Prediction of Incompresible Flow in Labyrinth Seals," *Journal of Fluids Engineering*, Vol. 108, 1984, pp. 19–25.
[19] Wittig, S., Scheling, U., Kim, S., and Jacobsen, K., "Numerical Predictions and Measurements of Discharge Coefficients in Labyrinth Seals," *ASME Paper 87-GT-188*, 1987.
[20] Rhode, D. L., and Hibbs, R. I., "New Model for Flow Over Open Cavities, II: Assessment for Seal Leakage," *AIAA Journal of Propulsion and Power*, Vol. 8, No. 2, 1992, pp. 398–402.

[21] Iwatsubo, T., and Kawai, R., "Analysis of Dynamic Characteristics of Fluid Force Induced by Labyrinth Seal," *NASA Conference Publication 2338*, Rotordynamic Instability in High Performance Turbomachinery, 1984, pp. 211–234.

[22] Zimmermann, H., and Wolf, K. H., "Comparison between Emperical and Numerical Labyrinth Seal Correlations," *ASME Paper 87-GT-86*, 1987.

[23] Alford, J. S., "Protecting Turbomachinery from Self-Excited Rotor Whirl," *Journal of Engineering for Power*, 1985, pp. 333–344.

[24] Childs, D., and Elrod, D., "Annular Honeycomb Seals: Test Results for Leakage and Rotordynamic Coefficients; Comparison to Labyrinth and Smooth Configurations," *NASA Conference Publication 3026*, Rotordynamic Instability Oroblems in High-Performance Turbomachinery, 1988, pp. 143–159.

[25] Iwatsubo, T., "Evaluation of Instability Forces of Labyrinth Seals in Turbines or Compressors," *NASA Conference Publication 2133*, Rotordynamic Insability Problems in High-Performance Turbomachinery, 1980, pp. 139–167.

[26] Childs, D., and Scharrer, J., "An Iwatsubo-Based Solution for Labyrinth Seals, Comparison with Experimental Results," *NASA Conference Publication 2338*, Rotordynamic Instability Problems in High-Performance Turbomachinery, 1984, pp. 427–434.

[27] Nordmann, R., and Weiser, P., "Rotordynamic Coefficients for Labyrinth Seals Calculated by Means of a Finite Difference Technique," *NASA Conference Publication 3026*, Rotordynamic Instability in High-Performance Turbomachinery, 1988, pp. 161–175.

[28] Baskharone, E. A., and Ghali, A., "Theoretical versus Experimental Rotordynamic Coefficients of Incompressible Flow Labyrinth Seals," *AIAA Journal of Propulsion and Power*, Vol. 10, No. 5, 1994, pp. 721–728.

[29] Bolleter, U., Leibundgut, E., and Sturchler, R., "Hydraulic Interaction and Excitation Forces of High Head Pump Impellers," presented at the Third Joint ASCE/ASME Mechanics Conference, La Jolla, CA, 1989.

[30] Baskharone, E. A., "Swirl Brake Effect on the Rotordynamic Stability of a Shrouded Impeller," *Journal of Turbomachinery*, Vol. 121, No. 1, 1999, pp. 127–133.

[31] Baskharone, E. A., and Wyman, N. J., "Primary/Leakage Flow Interaction in a Pump Stage," *Journal of Fluids Engineering*, Vol. 121, No. 1, 1999, pp. 133–138.

[32] Dietzen, F. J., and Nordmann, R. "Calculation of Rotordynamic Coefficients of Seals by Finite Difference Techniques," *Journal of Tribology*, Vol. 109, 1987.

[33] Hensel, S. J., "A New Method for Determining Rotordynamic Forces on Rotating Mechanical Components," Ph.D. dissertation, Texas A&M University, 1989.

[34] Yamada, Y., "Resistance of a Flow Through an Annulus with an Inner Rotating Cylinder," *Bulletin of the Japanese Society of Mechanical Engineering*, Vol. 5, 1962.

[35] Childs, D. W., "Rotordynamic Moment Coefficients for Finite-Length Turbulent Seals," *Proceedings of the International Conference on Rotordynamic Problems in Power Plants*, Rome, Italy, 1982, pp. 371–378.

[36] Kanemori, Y., and Iwatsubo, T., "Experimental Study of Dynamic Fluid Forces and Moments for a Long Annular Seal: Machinery Dynamics Applications and Vibration Control Problems," *ASME Publication No. DE-Vol. 18-2*, 1989, pp. 141–147.

[37] Childs, D. W., Nelson, C. C., Nicks, C., Scharrer, J., Elrod, D., and Hale, K., "Theory versus Experimental for the Rotordynamic Coefficients of Annular Gas Seals. 1: Test Facility and Apparatus," *Journal of Tribology*, Vol. 108, 1986, pp. 426–432.

[38] Baskharone, E. A., "Finite Element Analysis of Turbulent Flow in Annular Exhaust Diffusers of Gas Turbine Engines," *Journal of Fluids Engineering*, Vol. 113, No. 1, 1991, pp. 104–110.
[39] Hirs, G.G., "A bulk-flow theory for turbulence in lubricant film," *J. Lubric., Tech.* 137, 1973.

APPENDIX A

Natural Coordinates for Three-Dimensional Surface Elements

Evaluation of the surface integrals in the problem of three-dimensional (3D) heat conduction corresponding to the different boundary conditions requires the discretization of the body surfaces into surface finite elements that take the form of surface triangles. A set of natural coordinates would be advantageous in defining these elements, especially if these surfaces are curved.

The natural coordinates are local coordinates that vary in a range between zero and unity. At any of the element's vertices, one of these coordinates has a value of unity, whereas the others are all zeros. Use of these coordinates simplifies the evaluation of integrals in the element's equations. This additional advantage is a consequence of the existing closed-form integration formulas that evaluate these integrals.

The derivation given in this appendix generalizes the natural coordinates' definition for two-dimensional (2D) plane elements, all lying in one plane, to the case in which these plane elements exist in a 3D space. Such generalization was essential because the elements dealt with in the analysis lie on the 3D body-surface segments, which are, in turn, 3D.

Let A represent the area of the triangular element in Figure A.1, with $i(x_i, y_i, z_i)$, $j(x_j, y_j, z_j)$, and $k(x_k, y_k, z_k)$ denoting its vertices. The area A can be expressed in terms of the position vectors r_i, r_j, and r_k of the vertices i, j, and k as follows:

$$A = \frac{1}{2} \mid (\vec{r}_j - \vec{r}_i) \times (\vec{r}_k - \vec{r}_i) \mid$$

In terms of the vertices' Cartesian coordinates, the preceding expression becomes

$$A = \{[(y_j - y_i)(z_k - z_i) - (y_k - y_i)(z_j - z_i)]^2 + [(x_j - x_i)(z_k - z_i) - (x_k - x_i)(z_j - x_i)]^2 + [(x_j - x_i)(y_k - y_i) - (x_k - x_i)(y_j - y_i)]^2\}^{1/2} \tag{A.1}$$

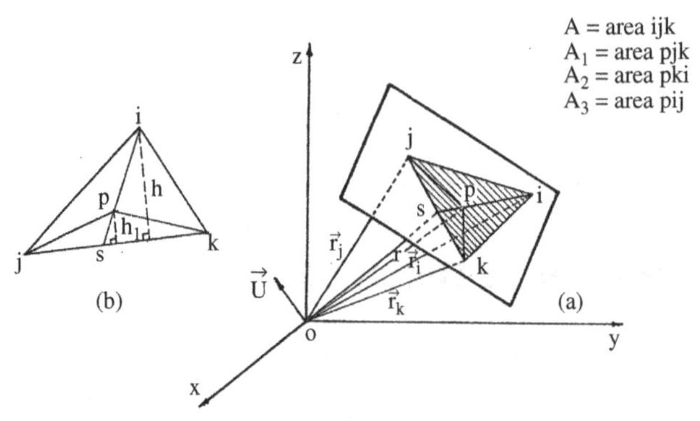

Figure A.1. Three-dimensional natural coordinates.

The Cartesian coordinates of an arbitrary interior point $P(x,y,z)$ within the triangle are linearly related to the natural coordinates L_1, L_2, and L_3 as follows:

$$x = L_1 x_i + L_2 x_j + L_3 x_k \tag{A.2a}$$

$$y = L_1 y_i + L_2 y_j + L_3 y_k \tag{A.2b}$$

$$z = L_1 z_i + L_2 z_j + L_3 z_k \tag{A.2c}$$

In order to reveal the physical interpretation of the natural coordinates, the set of Equations (A.2) can be solved for L_1, L_2, and L_3, giving

$$L_1 = \begin{vmatrix} x & x_j & x_k \\ y & y_j & y_k \\ z & z_j & z_k \end{vmatrix} \Big/ \begin{vmatrix} x_i & x_j & x_k \\ y_i & y_j & y_k \\ z_i & z_j & z_k \end{vmatrix}$$

or

$$L_1 = \begin{vmatrix} x & y & z \\ x_j & y_j & z_j \\ x_k & y_k & z_k \end{vmatrix} \Big/ \begin{vmatrix} x_i & y_i & z_i \\ x_j & y_j & z_j \\ x_k & y_k & z_k \end{vmatrix}$$

or

$$L_1 = \frac{\vec{r}.(\vec{r}_j \times \vec{r}_k)}{\vec{r}_i.(\vec{r}_j \times \vec{r}_k)} \tag{A.3a}$$

where $\vec{r}$ represents the position vector of the arbirary point P.

Similar expressions can be obtained for L_2 and L_3. Denoting the vector product $(\vec{r}_j \times \vec{r}_k)$ by $\vec{u}$, Equation (A.3a) can be rewritten in the following form:

$$L_1 = \frac{\vec{r}.\vec{u}}{\vec{r}_i.\vec{u}} \tag{A.3b}$$

Referring to Figure A.1, the vectors $\vec{r}$ and $\vec{r}_i$ can be expressed as follows:

$$\vec{r} = \vec{os} + \vec{sp} \tag{A.4a}$$

$$\vec{r}_i = \vec{os} + \vec{si} \tag{A.4b}$$

where s refers to the point at which ip intersects jk. Substituting Equations (A.4a) and (A.4b) into Equatin (A.3b), we get

$$L_1 = \frac{(\vec{os} + \vec{sp})}{(\vec{os} + \vec{u}) + (\vec{si}.\vec{u})} \bar{u}(\vec{os} + \vec{si}).\vec{u} = \frac{(\vec{os}.\vec{u}) + (\vec{sp}.\vec{u})}{(\vec{os}.\vec{u}) = (\vec{si}.\vec{u})} \tag{A.5}$$

Recalling that $\vec{u}$ is the vector product of $\vec{r}_j$ and $\vec{r}_k$, which means that $\vec{u}$ is perpendicular to the plane *ojk*, and noting that the vector $\vec{os}$ lies in this plane, the following can be concluded:

$$\vec{os}.\vec{u} = 0 \tag{A.6}$$

Substituting Equation (A.6) into Equation (A.5), we get

$$L_1 = \frac{\vec{sp}.\vec{u}}{\vec{si}.\vec{u}} \tag{A.7}$$

Denoting the angle between $\vec{si}$ and $\vec{u}$ by ϕ, Equation (A.7) can be rewritten in the following form:

$$L_1 = \frac{|\vec{sp}||\vec{u}|\cos\phi}{|\vec{si}||\vec{u}|\cos\phi} = \frac{sp}{si} \tag{A.8}$$

Equation (A.8) can be rewritten as follows:

$$L_1 = \frac{h_1}{h} = \frac{A_1}{A} \tag{A.9}$$

where

- A_1 is the area of the triangle *pjk*, and
- A is the area of the triangle *ijk*.

Following the same procedure, the following relationships are attained:

$$L_2 = \frac{A_2}{A} \tag{A.10}$$

$$L_3 = \frac{A_3}{A} \tag{A.11}$$

From Equations (A.9), (A.10), and (A.11), it is clear that

$$L_1 + L_2 + L_3 = 1 \tag{A.12}$$

From Equation (A.12), it can be seen that only two of the natural coordinates are independent. This was expected because the points p, i, j, and k are coplanar.

Because the definitions of natural coordinates (A.9), (A.10), and (A.11) are independent of the number of Cartesian coordinates describing the problem, it can be concluded that defining natural coordinates on and for a triangle existing in the 3D space x, y, z can be reduced to the usual problem of defining these coordinates for a triangle lying in the x-y plane.

APPENDIX B

Classification and Finite Element Formulation of Viscous Flow Problems

Introduction

Viscous effects in a flow field are present in all real fluids. As a consequence of the assumption defining an ideal flow, two adjacent portions of a fluid can move at different relative velocities, provided that the interface is a streamline. The assumption that there are no viscous effects will provide an adequate mathematical model for certain flow classes. However, we would not expect this to be reasonable for flows such as those encountered in polymer processing. Even for more standard fluids and gases, viscous effects may be important in many applications. Finally, important viscosity-related effects may be confined to regions close to the solid boundary, whereas flows far from the boundary may essentially be inviscid. Such cases provide the basic foundation of boundary-layer theory, introduced by Prandtl and since applied extensively in fluid mechanics.

Steady Navier-Stokes Problems: The "Primitive" Formulation

The procedure here is to apply Galerkin's method of weighted residuals. Consider a two-dimensional (2D) flow domain Ω that is bounded by the curve C. For a general element in this domain, we select u, v, and p as nodal variables as follows:

$$u^{(e)} = \sum N_i^u(x,y)u_i = [N]\{u\}$$

$$v^{(e)} = \sum N_i^u(x,y)v_i = [N]\{v\}$$

$$p^{(e)} = \sum N_i^p(x,y)p_i = [N]\{p\}$$

where N_i^u, N_i^v, and N_i^p are the interpolation functions, which need not necessarily be of the same order.

The Galerkin's procedure, applied at node i of an element, becomes

$$\int_{\Omega^{(e)}} \left[\frac{\partial}{\partial x}(\sigma_x - p) + \frac{\partial \tau_{xy}}{\partial y}\right] W_i \, d\Omega = 0$$

$$\int_{\Omega^{(e)}} \left[\frac{\partial \tau_{xy}}{\partial x} + \frac{\partial}{\partial y}(\sigma_y - p)\right] W_i \, d\Omega = 0$$

$$\int_{\Omega^{(e)}} \left(\frac{\partial u}{\partial x} + \frac{\partial v}{\partial y} \right) H_i \, d\Omega = 0$$

where W_i and H_i are the weighing (or simply weight) functions, which we take as

$$W_i = N_i$$

$$H_i = N_i^p$$

Because we have chosen the weight functions for the momentum equations as the interpolation functions for the velocity components, we are indeed using a Bobnov-Galerkin approach. In an alternate approach, some authors use a Petrov-Galerkin approach to introduce the concept of "upwind" finite-elements.

If we integrate the two preceding equations using Gauss theorem and then introduce the velocity components, we get

$$\int_{\Omega^{(e)}} [\left(2\mu \frac{\partial u}{\partial x} - p \right) \frac{\partial N_i}{\partial x} + \mu \left(\frac{\partial u}{\partial y} + \frac{\partial v}{\partial x} \frac{\partial N_i}{\partial y} \right] d\Omega = \int_{S_2} \bar{\sigma}_x N_i \, dS_2$$

$$\int_{\Omega^{(e)}} \left[\mu \left(\frac{\partial u}{\partial y} + \frac{\partial v}{\partial x} + (2\mu \frac{\partial v}{\partial y} - p) \frac{\partial N_i}{\partial y} \right) \right] d\Omega = \int_{S_2} \bar{\sigma}_y N_i \, dS_2$$

We note that by starting with the stress-component form of the momentum equations, the natural boundary conditions, that is, the surface tractions, appear directly in the "load" vectors on the right-hand side.

When the approximations of u, v, and p are substituted into the preceding equations, the matrix equations for node i result. These are

$$2 \int_{\Omega^{(e)}} \mu \frac{\partial N_i}{\partial x} \left[\frac{\partial N}{\partial x} \right] d\Omega + \int_{\Omega^{(e)}} \mu \frac{\partial N_i}{\partial y} \left[\frac{\partial N}{\partial y} \right] d\Omega \{u\}$$

$$+ \int_{\Omega^{(e)}} \frac{\partial N_i}{\partial y} \frac{\partial N}{\partial x}] \, d\Omega \{v\} - \int_{\Omega^{(e)}} \frac{\partial N_i}{\partial x} [N^p] d\Omega \{p\} = \int_{S_2} \bar{\sigma}_x N_i \, dS_2$$

$$\int_{\Omega^{(e)}} \mu \frac{\partial N_i}{\partial x} \left[\frac{\partial N}{\partial y} \right] d\Omega \{u\} + 2 \int_{\Omega^{(e)}} \mu \frac{\partial N_i}{\partial y} \left[\frac{\partial N}{\partial y} \right] d\Omega + \int_{\Omega^{(e)}} \mu \frac{\partial N_i}{\partial x} \left[\frac{\partial N}{\partial x} \right] d\Omega \{v\}$$

$$- \int \Omega^{(e)} \frac{\partial N_i}{\partial y} [N^p] \, d\omega \{p\} = \int_{S_2} \bar{\sigma}_y N_i \, dS_2$$

$$\int_{\Omega^{(e)}} N_i^p \left[\frac{\partial N}{\partial x} \, d\Omega \{u\} + \int_{\Omega^{(e)}} N_i^p \left[\frac{\partial N}{\partial y} \right] d\Omega \{v\} = 0 \right.$$

From these equations we can write the element's matrix equation by inspection. Suppose that the velocity components are interpolated at r nodes of the element, whereas the pressure is interpolated at s nodes where, in general, $r > s$. Then the matrix equation will take the following form:

$$\begin{bmatrix} [2K_{11}+K_{22}] & [K_{12}] & [L_1] \\ [K_{12}] & [K_{11}+2K_{22}] & [L_2] \\ [L_1]^* & [L_2]^* & [0] \end{bmatrix} \begin{Bmatrix} u_1 \\ u_2 \\ \cdot \\ \cdot \\ \cdot \\ u_r \\ v_1 \\ v_2 \\ \cdot \\ \cdot \\ \cdot \\ v_r \\ p_1 \\ p_2 \\ \cdot \\ \cdot \\ \cdot \\ p_s \end{Bmatrix} = \begin{Bmatrix} R_{u1} \\ R_{u2} \\ \cdot \\ \cdot \\ \cdot \\ R_{ur} \\ R_{v1} \\ R_{v2} \\ \cdot \\ \cdot \\ \cdot \\ R_{vr} \\ 0 \\ 0 \\ \cdot \\ \cdot \\ \cdot \\ 0 \\ 0 \\ \cdot \\ \cdot \\ \cdot \\ 0 \end{Bmatrix}$$

where

$$[K_{11}] = \int_{\Omega^{(e)}} \mu \left\{\frac{\partial N}{\partial x}\right\}\left\{\frac{\partial N}{\partial x}\right\} d\Omega$$

$$[K_{22}] = \int_{\Omega^{(e)}} \mu \left\{\frac{\partial N}{\partial y}\right\}\left\{\frac{\partial N}{\partial y}\right\} d\Omega$$

$$[K_{12}] = \int_{\Omega^{(e)}} \mu \left\{\frac{\partial N}{\partial y}\right\}\left\{\frac{\partial N}{\partial x}\right\} d\Omega$$

$$[L_1] = -\int_{\Omega^{(e)}} \left\{\frac{\partial N}{\partial x}\right\} [N^p]\, d\Omega$$

$$[L_2] = -\int_{\Omega^{(e)}} \left\{\frac{\partial N}{\partial y}\right\} [N^p]\, d\Omega$$

$$\{R_u\} = \int_{S_2} \bar{\sigma}_x \{N\}\, dS_2$$

$$\{R_v\} = \int_{S_2} \bar{\sigma}_y \{N\}\, dS_2$$

We note that the complete coefficients matrix on the left-hand side of the matrix equation is unsymmetric, although some of the submatrices are symmetric.

Serious attention must be paid to the choice of interpolation functions for the velocity components and pressure. Several different approaches have established that the interpolation functions for the velocity components should be one order higher than those of the pressure components. This was reached through consideration of a variational formula. Typical finite-elements for viscous flow maintain C^0 continuity, meaning continuity of the field variable(s) across the elements' interfaces.

Once the element equations have been evaluated, they can be assembled in the usual manner to form the system (or global) set of equations. On one portion of the boundary, the velocity components are specified, and those are handled in the manner previously discussed in the main text. On the remaining part of the boundary, the surface tractions are prescribed, and these boundary conditions comprise the load vectors $\{R_u\}$ and $\{R_v\}$ appearing in the preceding matrix equation.

Incompressible Viscous Flow without Inertia (Creeping Flow)

At the beginning of the second part (fluid flow applications) of the main text we covered a special case of inviscid irrotational (potential) flow with and without multiconnectivity. In this section we consider the simplest of viscous flow problems, namely, creeping flow (also called *Stokes flow*). If the full Navier-Stokes equations are made dimensionless, there results a dimensionless group known as the *Reynolds number Re*, which represents the ratio of inertia to viscous forces of a fluid in motion. When the Reynolds number is very small, the inertia forces are insignificant compared with the viscous forces and can be omitted from the governing momentum equations. Small Reynolds numbers characterize slow-moving flows and flows of very viscous fluids. These types of flows occur, for example, in viscometry and polymer processing.

A steady 2D isoviscous flow is governed by the following equations.

Continuity Equation

$$\frac{\partial u}{\partial x}+\frac{\partial v}{\partial y}=0$$

Momentum Equations

$$\frac{\partial}{\partial x}(\sigma_x-p)+\frac{\partial \tau_{xy}}{\partial y}=0$$

$$\frac{\partial \tau_{xy}}{\partial x}+\frac{\partial}{\partial y}(\sigma_y-p)=0$$

where

$$\sigma_x=2\mu\frac{\partial u}{\partial x}$$

$$\sigma_y=2\mu\frac{\partial v}{\partial y}$$

$$\tau_{xy}=\mu\left(\frac{\partial u}{\partial y}+\frac{\partial v}{\partial x}\right)$$

An alternative to the stress-component form of the momentum equations (given earlier) is to substitute the velocity components into the momentum equations to express the latter in terms of u and v. We choose to start from the stress-component form because in a velocity-pressure formulation of the finite element equations, the boundary conditions are handled in a consistent manner.

The boundary conditions consist of specifying the velocity components on a portion of the boundary S_1 and the surface tractions on the remainder of the surface S_2. Thus

$$u = g(x,y), \qquad v = h(x,y) \qquad \text{on } S_1$$
$$\bar{\sigma}_x = (\sigma - p)n_x + \tau_{xy}n_y \qquad \bar{\sigma}_y = \tau_{xy}n_x + (\sigma_y - p)n_y \qquad \text{on } S_2$$

where $\bar{\sigma}_x$ and $\bar{\sigma}_y$ denote x and y components of the total surface tractions, with n_x and n_y standing for the direction cosines of the unit outward vector normal to the surface S_2.

To solve these equations by the finite element method, we may use one of two different formulations,

- We can introduce a stream function and work with one governing equation of the fourth order, or
- We can work with the velocity and pressure as field variables.

The second of these two choices will now be explored.

Incompressible Viscous Flow with Inertia: Velocity-Pressure (Primitive) Formulation

Several authors favor the velocity-pressure formulation as the most straightforward finite element procedure for solution of nonlinear Navier-Stokes equations. In addition to avoiding some of the difficulties associated with the stream function/vorticity formulations, the following reasons are cited in favor of the velocity-pressure formulations:

- The formulation is readily extendible to three-dimensional (3D) applications.
- Only C^0 continuity is required of the element's interpolation functions.
- Pressure, velocity, velocity gradients, and stress boundary conditions can be incorporated directly into the matrix equations.
- Free-surface problems are tractable.
- The formulation appears to require less computational time than such other formulations as the stream function/vorticity formulation.

There are two different velocity-pressure formulations currently in use. The first treats the velocity components and pressure as unknown flow variables and develops the finite element equations from simultaneous solution of the continuity and momentum equations. In a second, and more recent, approach, the penalty-function formulation eliminates the pressure as an unknown field variable through the use of a "penalty" parameter and solves modified momentum equations for the velocity components. We will now proceed with the first of these two methods.

Primitive Formulation

The approach to derive the element's equation relies on the Bobnov-Galerkin method. Let the velocity and static pressure fields be interpolated over an element, and choose these interpolation functions as the weight functions. Development of the element's equations follows that used in conjunction with Stokes (creeping) flow, except for the acceleration terms on the left-hand side. The unsteady acceleration terms $\rho\partial u/\partial t$ and $\rho\partial v/\partial t$ cause no difficulty, but the nonlinear convection acceleration terms require special consideration. The Galerkin's method can be used if we linearize the governing equations by approximating the nonlinear convection terms. Suppose that (u_n, v_n) is some approximate solution to the flow problem. For example, (u_n, v_n) could be the result of Stokes solution. Then the momentum equations can be written as follows:

$$\rho\left(\frac{\partial u}{\partial t}+u_n\frac{\partial u}{\partial x}+v_n\frac{\partial u}{\partial y}\right)=\frac{\partial}{\partial x}(\sigma_x-p)+\frac{\partial\tau_{xy}}{\partial y}$$

$$\rho\left(\frac{\partial v}{\partial t}+u_n\frac{\partial v}{\partial x}+v_n\frac{\partial v}{\partial y}\right)=\frac{\partial\tau_{xy}}{\partial x}+\frac{\partial}{\partial y}(\sigma_y-p)$$

The Galerkin's criterion, applied to these equations, leads to

$$\int_{\Omega^{(e)}}\left[-\rho\left(\frac{\partial u}{\partial t}+u_n\frac{\partial u}{\partial x}\right]N_i\,d\Omega+v_n\frac{\partial u}{\partial y}\right)+\frac{\partial\tau_{xy}}{\partial x}(\sigma_x-p)+\frac{\partial\tau_{xy}}{\partial y}]N_i\,d\Omega=0$$

$$\int_{\Omega^{(r)}}[-\rho\left(\frac{\partial v}{\partial t}+u_n\frac{\partial v}{\partial x}+v_n\frac{\partial\tau_{xy}}{\partial x}+\frac{\partial}{\partial y}(\sigma_y-p)\right]N_i\,d\Omega=0$$

$$\int_{\Gamma^{(e)}}\left(\frac{\partial u}{\partial x}+\frac{\partial v}{\partial y}\right)N_i^P\,d\Omega=0$$

Now we integrate the stress terms using Gauss theorem and then introduce the natural boundary conditions in the resulting line integral. The element's mâtrices arising from the stress terms are identical to the terms we obtained for Stokes flow. Now terms arise from the acceleration terms, but the derivation is straightforward, and for brevity, we will omit the details. The resulting element equations are as follows:

$$\begin{bmatrix}[M] & [0] & [0]\\ [0] & [M] & [0]\\ [0] & [0] & [M]\end{bmatrix}\begin{Bmatrix}\{\dot{u}\}\\ \{\dot{v}\}\\ \{\dot{p}\}\end{Bmatrix}+\begin{bmatrix}[C_{11}+C_{22}] & [0] & [0]\\ [0] & [C_{11}+C_{22}] & [0]\\ [0] & [0] & [0]\end{bmatrix}\begin{Bmatrix}\{u\}\\ \{v\}\\ \{p\}\end{Bmatrix}$$

$$+\begin{bmatrix}[2K_{11}+K_{22}] & [K_{12}] & [L_1]\\ [K_12]^T & [K_{11}+2K_{22}] & [L_2]\\ [L_1]^T & [L_2]^T & [0]\end{bmatrix}\begin{Bmatrix}\{u\}\\ \{v\}\\ \{p\}\end{Bmatrix}=\begin{Bmatrix}\{R_u\}\\ \{R_v\}\\ \{0\}\end{Bmatrix}$$

where $$[M]=\int_{\Omega^{(e)}}\rho[N][N]d\omega$$

$$[C_{11}]=\int_{\Omega^{(e)}}\rho u_n[N]\left[\frac{\partial N}{\partial x}\right]d\Omega$$

$$[C_{22}]=\int_{\Omega^{(e)}}\rho v_n[N]\left[\frac{\partial N}{\partial y}\right]d\Omega$$

and the remainder of the submatrices have earlier been defined. To discuss the finite element equations, we write the preceding matrix equation in a more compact form as follows:

$$[M]\begin{Bmatrix} \dot{u} \\ \dot{v} \\ \dot{p} \end{Bmatrix} + [C(u_n, v_n)]\begin{Bmatrix} u \\ v \\ p \end{Bmatrix} + [K]\begin{Bmatrix} u \\ v \\ p \end{Bmatrix} = \begin{Bmatrix} R_u \\ R_v \\ 0 \end{Bmatrix}$$

where the coefficient matrices $[M]$, $[C(u_n, v_n)]$, and $[K]$ were defined earlier.

Once the element matrices are assembled into the "global" system and the boundary conditions have been implemented, we get the formidable task of solving the final set of equations. For steady flow, we can solve the (now steady) nonlinear algebraic equations through an iterative procedure. For unsteady flow, we have a set of nonlinear ordinary differential equations that we can solve by combining a transient time-integration algorithm with an iterative scheme at each step. Solution of the nonlinear equations either for steady or time-dependent flow is a significant computational challenge. Important considerations for high-quality solutions include the proper element choice, effective solution algorithms, correct use of the boundary conditions, and proper discretization of the solution domain. These and other important aspects of the numerical solutions have been studied and evaluated in the literature. Briefly stated, the computational experience of these researchers is that an element with a nine-node quadratic velocity interpolation (a Lagrangian element) combined with a four-node pressure interpolation generally gives the best performance. Consistent use of the velocity and surface-traction boundary conditions is necessary, and pressure boundary conditions are to be avoided. Finally, in regions of high velocity and/or pressure gradients, refined meshes are required to avoid spatial oscillations in the flow variables.

APPENDIX C

Numerical Integration

Consider the problem of numerically evaluating a one-dimensional integral of the following form:

$$I = \int_{-1}^{+1} f\xi \, d\xi$$

The Gauss quadrature approach for evaluating I is given below. This method has proven most useful in the finite element framework. Extension to integrals in two dimensions will follow.

Consider the n-point approximation

$$I = \int_{-1}^{+1} f(\zeta) \, d\zeta = w_1 f(\zeta_1) + w_2 f(\zeta_2) + \cdots, + w_n f(\zeta_n)$$

where w_1, w_2, ..., w_n are the *weights*, and ζ_1, ζ_2, ..., ζ_n are the sampling points or *Gauss* points. The idea behind the Gaussian quadrature is to select the n Gauss points and n weights (tabulated in this Appendix) in such a way that the preceding expression provides an exact answer for polynomials $f(\zeta)$ of as large a degree as possible. In other words, the idea is that if the n-point integration formula is exact for all polynomials up to as high a degree as possible, then the formula will work well even if f is not a polynomial. To get some intuition for the method, the one-point and two-point approximations are discussed next.

One-Point Formula

Consider the formula with $n = 1$ as

$$\int_{-1}^{+1} f(\zeta) \, d\zeta = w_1 f(\zeta_1)$$

Because there are two parameters, w_1 and ζ_1, we consider requiring the formula in the preceding expression to be exact when $f(\zeta)$ is a polynomial of order 1. Thus, if

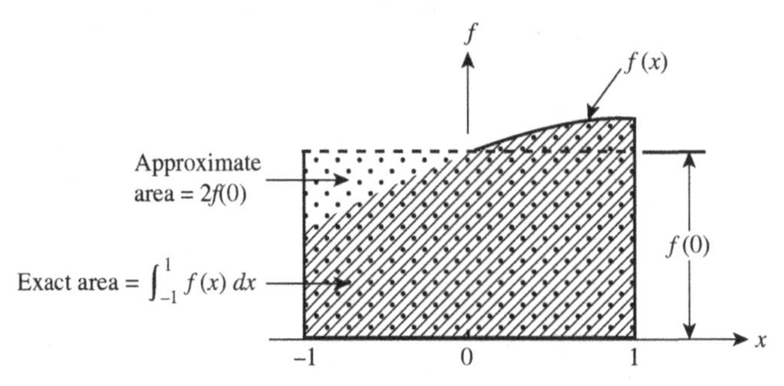

Figure C.1. One-point Gauss quadrature.

$f(\zeta) = a_0 + a_1\zeta$, then we require that

$$\text{Error} = \int_{-1}^{+1} (a_0 + a_1\zeta)\,d\zeta - w_1 f(\zeta_1) = 0$$

or

$$\text{Error} = a_0(2 - w_1) - w_1(a_0 + a_1\zeta_1) = 0$$

or

$$\text{Error} = a_0(2 - w_1) - w_1 a_1 \zeta_1 = 0$$

From this equation we see that the error is zero if

$$w_1 = 2$$
$$\zeta_1 = 0$$

For any general f, then, we have

$$I = \int_{-1}^{+1} f(\zeta)\,d\zeta = 2f(0)$$

which is seen to be the familiar *midpoint rule* (Figure C.1)

Two-Point Formula

Consider the formula with $n = 2$ as

$$\int_{-1}^{+1} f(\zeta)\,d\zeta = w_1 f(\zeta_1) + w_2 f(\zeta_2)$$

We now have four parameters to choose. We can, therefore, expect the formula to be exact for a cubic polynomial. Thus, choosing $f(\zeta) = a_0 + a_1\zeta + a_2\zeta^2 + a_3\zeta^3$ yields

$$\text{Error} = \left[\int_{-1}^{+1} (a_0 + a_1\zeta + a_2\zeta^2 + a_3\zeta^3)\,d\zeta\right] - [w_1 f(\zeta_1) + w_2 f(\zeta_2)]$$

$$\int_{-1}^{1} f(\xi)\, d\xi \approx \sum_{i=1}^{n} W_i f(\xi_i)$$

Number of points, n	Location, ξ_i	Weights, w_i
1	0.0	2.0
2	$\pm 1/\sqrt{3} = \pm 0.5773502692$	1.0
3	± 0.7745966692	0.5555555556
	0.0	0.8888888889
4	± 0.8611363116	0.3478548451
	± 0.3399810436	0.6521451549
5	± 0.9061798459	0.2369268851
	± 0.5384693101	0.4786286705
	0.0	0.5688888889
6	± 0.9324695142	0.1713244924
	± 0.6612093865	0.3607615730
	± 0.2386191861	0.4679139346

Figure C.2. Gauss points and weights for Gauss quadrature.

Requiring zero error yields

$$w_1 + w_2 = 2$$

$$w_1\zeta_1 + w_2\zeta_2 = 0$$

$$w_1\zeta_1^2 + w_2\zeta_2^2 = \frac{2}{3}$$

$$w_1\zeta_1^3 + w_2\zeta_2^3 = 0$$

These nonlinear equations have the following unique solution:

$$w_1 = w_2 = 1$$

$$-\zeta_1 = \zeta_2 = \frac{1}{\sqrt{3}} = 0.5773502$$

From the preceding derivations we can conclude that the n-point Gaussian quadrature will provide an exact answer if f is a polynomial of order $(2n - 1)$ or less. Figure C.2 gives the values of w_i and ζ_i for Gauss quadrature formulas of order $n = 1$ through $n = 6$. Note that the Gauss points are located symmetrically with respect to the origin and that symmetrically placed points have the same weights. Moreover, the large number of digits given in Figure C.2 should be used in the calculations for accuracy; that is, it would be wise to select double precision when employing the method in a computer program.

Example

Evaluate the following integral:

$$I = \int_{-1}^{+1} \left[3e^x + x^2 + \frac{1}{(x+2)} \right] dx$$

using one and two-point Gauss quadrature.

$$\int_0^1 \int_{-1}^{1-\xi} f(\xi,\eta)\, d\eta\, d\xi \approx \sum_{i=1}^{n} W_i f(\xi_i\, \eta_i)$$

No of points, n	Weights, w_i	Multi-plicity	ξ_i	η_i	ζ_i
One	$\frac{1}{2}$	1	$\frac{1}{3}$	$\frac{1}{3}$	$\frac{1}{3}$
Three	$\frac{1}{6}$	3	$\frac{1}{3}$	$\frac{1}{6}$	$\frac{1}{6}$
Three	$\frac{1}{6}$	3	$\frac{1}{2}$	$\frac{1}{2}$	0
Four	$-\frac{9}{32}$	1	$\frac{1}{3}$	$\frac{1}{3}$	$\frac{1}{3}$
	$\frac{25}{96}$	3	$\frac{3}{5}$	$\frac{1}{5}$	$\frac{1}{5}$
Six	$\frac{1}{12}$	6	0.6590276223	0.2319333685	0.1090390090

Figure C.3. Gauss quadrature for a triangle.

Solution

For $n = 1$, we have $w_1 = 2$, $x_1 = 0$, and

$$I = 2f(0) = 7.0$$

For $n = 2$, $w_1 = w_2 = 1$, $x_1 = -0.57735$, $x_2 = +0.57735$, and $I = 8.7857$

This may be compared with the exact solution, which is:

$$I_{exact} = 8.8165$$

Two-Dimensional Integrals

Extension of the Gaussian quadrature to two-dimensional (2D) integrals of the form:

$$I = \int_{-1}^{+1} \int_{-1}^{+1} f(\zeta,\eta)\, d\zeta\, d\eta$$

follows readily because

$$I = \int_{-1}^{+1} \left[\sum_{i=1}^{n} w_i f(\zeta,\eta) \right] d\eta = \sum_{j=1}^{n} w_j \left[\sum_{i=1}^{n} w_i f(\zeta,\eta) \right]$$

or

$$I = \sum_{i=1}^{n} \sum_{j=1}^{n} w_i w_j f(\zeta,\eta)$$

The different Gauss points and weight functions for a triangle integration are shown in Figure C.3.

For integrated functions with practically any order over domains with arbitrary shapes, the reader is directed to the books by Stroud [1], as well as that by Baker and Pepper [2].

REFERENCES

[1] Stroud, A. H., *Approximate Calculation of Multiple Integrals*, Prentice-Hall, Englewood Cliffs, NJ, 1971.

[2] Baker, A. J., and Pepper, D. W., *Finite Elements One-Two-Three*, McGraw-Hill, New York, 1991.

APPENDIX D

Finite Element–Based Perturbation Analysis: Formulation of the Zeroth-Order Flow Field

The undistorted flow region, which is viewed as a three-dimensional (3D) domain, is replaced by an assembly of nonoverlapping twenty-noded isoparametric finite-elements of the type shown in Figure 16.12 while retaining the same meridional-projection computational nodes (see Figure 16.11) used in the earlier axisymmetric flow analysis. Within a typical element (e), let the velocity components, static pressure, and spatial coordinates be interpolated as follows:

$$u^{(e)} = \sum_{i=1}^{20} N_i(\zeta, \eta, \xi) u_i$$

$$v^{(e)} = \sum_{i=1}^{20} N_i(\zeta, \eta, \xi) v_i$$

$$w^{(e)} = \sum_{i=1}^{20} N_i(\zeta, \eta, \xi) w_i$$

$$p^{(e)} = \sum_{k=1}^{8} M_k(\zeta, \eta, \xi) p_k$$

$$x^{(e)} = \sum_{i=1}^{20} N_i(\zeta, \eta, \xi) x_i$$

$$y^{(e)} = \sum_{i=1}^{20} N_i(\zeta, \eta, \xi) y_i$$

$$z^{(e)} = \sum_{i=1}^{20} N_i(\zeta, \eta, \xi) z_i$$

where N_i refers to a set of quadratic shape functions associated with all corner and midside nodes [12], M_k refers to a set of linear shape functions associated with corner nodes [12].

Using these interpolating functions and following (as we previously did) the Petrov-Galerkin weighted-residual approach, the following set of finite element equations is obtained for a typical element (e):

$$\begin{bmatrix} [a_1] & [a_2] & [a_3] & [a_4] \\ [a_5] & [a_6] & [a_7] & [a_8] \\ [a_9] & [a_{10}] & [a_{11}] & [a_{12}] \\ [a_{13}] & [a_{14}] & [a_{15}] & 0 \end{bmatrix} \begin{Bmatrix} \{u\} \\ \{v\} \\ \{w\} \\ \{p\} \end{Bmatrix} = \begin{Bmatrix} \{q_1\} \\ \{q_2\} \\ \{q_3\} \\ 0 \end{Bmatrix}$$

which can be expanded in the following form:

$$[a_1]\{u\} + [a_2]\{v\} + [a_3]\{w\} + [a_4]\{p\} = \{q_1\}$$

$$[a_5]\{u\} + [a_6]\{v\} + [a_7]\{w\} + [a_8]\{p\} = \{q_2\}$$

$$[a_9]\{u\} + [a_{10}]\{v\} + [a_{11}]\{w\} + [a_{12}]\{p\} = \{q_3\}$$

$$[a_{13}]\{u\} + [a_{14}]\{v\} + [a_{15}]\{w\} = 0$$

in which the first three correspond to the three momentum equations, and the fourth is the finite element equivalent of the continuity equation. The vectors $\{u\}$, $\{v\}$, and $\{w\}$ contain the twenty nodal values of each velocity component u, v, and w, respectively, whereas the vector $\{p\}$ is composed of the eight pressure values at the element corner nodes (see Figure 16.11). The submatrices $[a_1]$ through $[a_{15}]$ and the subvectors $\{q_1\}$, $\{q_2\}$, and $\{q_3\}$ are defined as follows:

$$\begin{aligned} a_1)_{i,j} = &\int_{V^{(e)}} \bar{\nu}_e \left(\frac{\partial N_i}{\partial x}\frac{\partial N_j}{\partial x} + \frac{\partial N_i}{\partial y}\frac{\partial N_j}{\partial y} + \frac{\partial N_i}{\partial z}\frac{\partial N_j}{\partial z} + \frac{\partial N_i}{\partial z}\frac{\partial N_j}{\partial z} \right) dV \\ &+ \int_{V^{(e)}} N_i \left(\frac{\partial \bar{\nu}_t}{\partial y}\frac{\partial N_j}{\partial y} + \frac{\partial \bar{\nu}_t}{\partial z}\frac{\partial N_j}{\partial z} + 2\frac{\partial \bar{\nu}_t}{\partial x}\frac{\partial N_j}{\partial x} \right) dV \\ &+ \int_{V^{(e)}} \left(\bar{u} W_i \frac{\partial N_j}{\partial x} + \bar{v} W_i \frac{\partial N_j}{\partial y} + \bar{w} W_i \frac{\partial N_j}{\partial z} \right) dV \end{aligned}$$

$$a_2)_{i,j} = \int_{V^{(e)}} -\left(N_i \frac{\partial \nu_t}{\partial z}\frac{\partial N_j}{\partial z} - 2\Omega N_i N_j \right) dV$$

$$a_3)_{i,j} = \int_{V^{(e)}} -\left(N_i \frac{\partial \nu_t}{\partial z}\frac{\partial N_j}{\partial s} \right) dV$$

$$a_4)_{i,k} = \int_{V^{(e)}} \left(\frac{1}{\rho} N_i \frac{\partial M_k}{\partial x} \right) dV$$

$$a_5)_{i,j} = \int_{V^{(e)}} -\left(N_i \frac{\partial \bar{\nu}_t}{\partial x} + 2\Omega N_i N_j \right) dV$$

$$\begin{aligned} a_6)_{i,j} = &\int_{V^{(e)}} \bar{\nu}_e \left(\frac{\partial N_i}{\partial x}\frac{\partial N_j}{\partial x} + \frac{\partial N_i}{\partial y}\frac{\partial N_j}{\partial y} + \frac{\partial N_i}{\partial z}\frac{\partial N_j}{\partial z} \right) dV \\ &+ \int_{V^{(e)}} N_i \left(\frac{\partial \bar{\nu}_t}{\partial x}\frac{\partial N_j}{\partial x} + \frac{\partial \bar{\nu}_t}{\partial z}\frac{\partial N_j}{\partial z} + 2\frac{\partial \bar{\nu}_t}{\partial y}\frac{\partial N_j}{\partial y} \right) dV \\ &+ \int_{V^{(e)}} \left(\bar{u} W_i \frac{\partial N_j}{\partial x} + \bar{v} W_i \frac{\partial N_j}{\partial y} + \bar{w} W_i \frac{\partial N_j}{\partial z} \right) dV \end{aligned}$$

$$a_7)_{i,j} = \int_{V^{(e)}} -N_i \frac{\partial \bar{\nu}_t}{\partial z}\frac{\partial N_j}{\partial y}\, dV$$

$$a_{8})_{i,k} = \int_{V^{(e)}} \frac{1}{\rho} N_i \frac{\partial M_k}{\partial y}\, dV$$

$$a_{9})_{i,j} = \int_{V^{(e)}} -N_i \frac{\partial \bar{v}_t}{\partial x} \frac{\partial N_j}{\partial z}\, dV$$

$$a_{10})_{i,j} = \int_{V^{(e)}} -N_i \frac{\partial \bar{v}_t}{\partial y} \frac{\partial N_j}{\partial z}\, dV$$

$$a_{11})_{i,j} = \int_{V^{(e)}} \bar{v}_e \left(\frac{\partial N_i}{\partial x}\frac{\partial N_j}{\partial x} + \frac{\partial N_i}{\partial y}\frac{\partial N_j}{\partial y} + \frac{\partial N_i}{\partial z}\frac{\partial N_j}{\partial z} \right) dV$$
$$+ \int_{V^{(e)}} N_i \left(\frac{\partial \bar{v}_t}{\partial x}\frac{\partial N_j}{\partial x} + \frac{\partial \bar{v}_t}{\partial y}\frac{\partial N_j}{\partial y} + 2\frac{\partial \bar{v}_t}{\partial z}\frac{\partial N_j}{\partial z} \right) dV$$
$$+ \int_{V^{(e)}} \left(\bar{u} W_i \frac{\partial N_j}{\partial x} + \bar{v} W_i \frac{\partial N_j}{\partial y} + \bar{w} W_i \frac{\partial N_j}{\partial z} \right) dV$$

$$a_{12})_{i,k} = \int_{V^{(e)}} \frac{1}{\rho} N_i \frac{\partial M_k}{\partial z}\, dV$$

$$a_{13})_{k,j} = \int_{V^{(e)}} M_k \frac{\partial N_j}{\partial x}\, dV$$

$$a_{14})_{k,j} = \int_{V^{(e)}} M_k \frac{\partial N_j}{\partial y}\, dV$$

$$a_{15})_{k,j} = \int_{V^{(e)}} M_k \frac{\partial N_j}{\partial z}\, dV$$

$$q_{1})_i = \int_{S^{(e)}} \bar{v}_e N_i \frac{\partial u}{\partial n} dS + \Omega^2 \int_{V^{(e)}} N_i \left(\sum_{l=1}^{20} N_l x_l \right) dV$$

$$q_{2})_i = \int_{S^{(e)}} \bar{v}_e N_i \frac{\partial v}{\partial n} dS + \Omega^2 \int_{V^{(e)}} N_i \left(\sum_{l=1}^{20} N_l y_l \right) dV$$

$$q_{3})_i = \int_{S^{(e)}} \bar{v}_e N_i \frac{\partial w}{\partial n}\, dS$$

where

- $i = 1, 2, 3, \ldots, 20$,
- $j = 1, 2, 3, \ldots, 20$,
- $k = 1, 2, 3, \ldots, 8$,
- $\vec{n}$ is the local outward unit vector perpendicular to the element surface,
- dS is the differential area on the element surface,
- $\{u\}$ is the finite element vector of u nodal magnitudes (20 entries),
- $\{v\}$ is the element vector of v nodal magnitudes (20 entries),
- $\{w\}$n is the element vector of w nodal magnitudes (20 entries),
- $\{p\}$ is the element vector of p nodal magnitudes (8 entries),
- $dV = dxdydz = \mid J_0 \mid d\zeta d\eta d\xi$, and
- $\mid J_0 \mid$ is the Jacobian of Cartesian-to-local coordinate conversion.

The subscript 0 in the Jacobian symbol identifies it as associated with the zeroth-order (undisplaced-rotor) domain geometry. Again, the “upwind” weight functions, explained earlier in conjunction with the axisymmetric flow model, are used in weighing the inertia (or convection) terms in the three momentum equations.

As a transitional step toward the derivation of the perturbation model, and in consistency with its details, the influence coefficients $a_1)_{i,j}$ through $a_{15})_{i,j}$ are re-expressed in terms of a transformation matrix $[T]$ where

$$\begin{Bmatrix} \frac{\partial}{\partial x} \\ \frac{\partial}{\partial y} \\ \frac{\partial}{\partial z} \end{Bmatrix} = \begin{bmatrix} T_{1,1} & T_{1,2} & T_{1,3} \\ T_{2,1} & T_{2,2} & T_{2,3} \\ T_{3,1} & T_{3,2} & T_{3,3} \end{bmatrix} \begin{Bmatrix} \frac{\partial}{\partial \zeta} \\ \frac{\partial}{\partial \eta} \\ \frac{\partial}{\partial \xi} \end{Bmatrix}$$

in which

$$T_{1,1} = \frac{1}{|J_0|}\left(\sum_{i=1}^{20}\frac{\partial N_i}{\partial \eta}y_i\sum_{i=1}^{20}\frac{\partial N_i}{\partial \xi}z_i - \sum_{i=1}^{20}\frac{\partial N_i}{\partial \eta}z_i\sum_{i=1}^{20}\frac{\partial N_i}{\partial \xi}y_i\right)$$

$$T_{1,2} = \frac{1}{|J_0|}\left(\sum_{i=1}^{20}\frac{\partial N_i}{\partial \xi}y_i\sum_{i=1}^{20}\frac{\partial N_i}{\partial \zeta}z_i - \sum_{i=1}^{20}\frac{\partial N_i}{\partial \zeta}y_i\sum_{i=1}^{20}\frac{\partial N_i}{\partial \xi}z_i\right)$$

$$T_{1,3} = \frac{1}{|J_0|}\left(\sum_{i=1}^{20}\frac{\partial N_i}{\partial \zeta}y_i\sum_{i=1}^{20}\frac{\partial N_i}{\partial \eta}z_i - \sum_{i=1}^{20}\frac{\partial N_i}{\partial \eta}y_i\sum_{i=1}^{20}\frac{\partial N_i}{\partial \xi}z_i\right)$$

$$T_{2,1} = \frac{1}{|J_0|}\left(\sum_{i=1}^{20}\frac{\partial N_i}{\partial \xi}x_i\sum_{i=1}^{20}\frac{\partial N_i}{\partial \eta}z_i - \sum_{i=1}^{20}\frac{\partial N_i}{\partial \eta}x_i\sum_{i=1}^{20}\frac{\partial N_i}{\partial \xi}z_i\right)$$

$$T_{2,2} = \frac{1}{|J_0|}\left(\sum_{i=1}^{20}\frac{\partial N_i}{\partial \zeta}x_i\sum_{i=1}^{20}\frac{\partial N_i}{\partial \xi}z_i - \sum_{i=1}^{20}\frac{\partial N_i}{\partial \xi}x_i\sum_{i=1}^{20}\frac{\partial N_i}{\partial \zeta}z_i\right)$$

$$T_{2,3} = \frac{1}{|J_0|}\left(\sum_{i=1}^{20}\frac{\partial N_i}{\partial \eta}x_i\sum_{i=1}^{20}\frac{\partial N_i}{\partial \zeta}z_i - \sum_{i=1}^{20}\frac{\partial N_i}{\partial \zeta}x_i\sum_{i=1}^{20}\frac{\partial N_i}{\partial \eta}z_i\right)$$

$$T_{3,1} = \frac{1}{|J_0|}\left(\sum_{i=1}^{20}\frac{\partial N_i}{\partial \eta}x_i\sum_{i=1}^{20}\frac{\partial N_i}{\partial \xi}y_i - \sum_{i=1}^{20}\frac{\partial N_i}{\partial \xi}x_i\sum_{i=1}^{20}\frac{\partial N_i}{\partial \eta}y_i\right)$$

$$T_{3,2} = \frac{1}{|J_0|}\left(\sum_{i=1}^{20}\frac{\partial N_i}{\partial \xi}x_i\sum_{i=1}^{20}\frac{\partial N_i}{\partial \zeta}y_i - \sum_{i=1}^{20}\frac{\partial N_i}{\partial \zeta}x_i\sum_{i=1}^{20}\frac{\partial N_i}{\partial \xi}y_i\right)$$

$$T_{3,3} = \frac{1}{|J_0|}\left(\sum_{i=1}^{20}\frac{\partial N_i}{\partial \xi}x_i\sum_{i=1}^{20}\frac{\partial N_i}{\partial \zeta}y_i - \sum_{i=1}^{20}\frac{\partial N_i}{\partial \zeta}x_i\sum_{i=1}^{20}\frac{\partial N_i}{\partial \xi}y_i\right)$$

The zeroth-order transformation Jacobian $|J_0|$ can itself be written as follows:

$$|J_0\} = \begin{vmatrix} \sum_{i=1}^{20}\frac{\partial N_i}{\partial \zeta}x_i & \sum_{i=1}^{20}\frac{\partial N_i}{\partial \zeta}y_i & \sum_{i=1}^{20}\frac{\partial N_i}{\partial \zeta}z_i \\ \sum_{i=1}^{20}\frac{\partial N_i}{\partial \eta}x_i & \sum_{i=1}^{20}\frac{\partial N_i}{\partial \eta}y_i & \sum_{i=1}^{20}\frac{\partial N_i}{\partial \eta}z_i \\ \sum_{i=1}^{20}\frac{\partial N_i}{\partial \xi}x_i & \sum_{i=1}^{20}\frac{\partial N_i}{\partial \xi}y_i & \sum_{i=1}^{20}\frac{\partial N_i}{\partial \xi}z_i \end{vmatrix}$$

With the definition of the transformation matrix $[T]$, the influence coefficients $a_1)_{i,j}$ through $a_{15})_{i,j}$ can now be expressed as follows:

$$
\begin{aligned}
a_1)_{i,j} = & \int_{\xi=-1}^{+1}\int_{\eta=-1}^{+1}\int_{\zeta=-1}^{+1} \bar{\nu}_e\left(T_{1,1}\frac{\partial N_i}{\partial \zeta}+T_{1,2}\frac{\partial N_i}{\partial \eta}+T_{1,3}\frac{\partial N_i}{\partial \xi}\right) \\
& \times\left(T_{1,1}\frac{\partial N_j}{\partial \zeta}+T_{1,2}\frac{\partial N_j}{\partial \eta}+T_{1,3}\frac{\partial N_j}{\partial \xi}\right)| J_0 |d\zeta\, d\eta d\xi \\
& +\int_{\xi=-1}^{+1}\int_{\eta=-1}^{+1}\int_{\zeta=-1}^{+1} \bar{\nu}_e\left(T_{2,1}\frac{\partial N_i}{\partial \zeta}+T_{2,2}\frac{\partial N_i}{\partial \eta}+T_{2,3}\frac{\partial N_i}{\partial \xi}\right) \\
& \times\left(T_{2,1}\frac{\partial N_j}{\partial \zeta}+T_{2,2}\frac{\partial N_j}{\partial \eta}+T_{2,3}\frac{\partial N_j}{\partial \xi}\right)| J_0 |d\zeta\, d\eta d\xi \\
& +\int_{\xi=-1}^{+1}\int_{\eta=-1}^{+1}\int_{\zeta=-1}^{+1} \bar{\nu}_e\left(T_{3,1}\frac{\partial N_i}{\partial \zeta}+T_{3,2}\frac{\partial N_i}{\partial \eta}+T_{3,3}\frac{\partial N_i}{\partial \xi}\right) \\
& \times\left(T_{3,1}\frac{\partial N_j}{\partial \zeta}+T_{3,2}\frac{\partial N_j}{\partial \eta}+T_{3,3}\frac{\partial N_j}{\partial \xi}\right)| J_0 |d\zeta\, d\eta d\xi \\
& -\int_{\xi=-1}^{+1}\int_{\eta=-1}^{+1}\int_{\zeta=-1}^{+1} N_i\left(T_{2,1}\frac{\partial \bar{\nu}_t}{\partial \zeta}+T_{2,2}\frac{\partial \bar{\nu}_t}{\partial \eta}+T_{2,3}\frac{\partial \bar{\nu}_t}{\partial \xi}\right) \\
& \times\left(T_{2,1}\frac{\partial N_j}{\partial \zeta}+T_{2,2}\frac{\partial N_j}{\partial \eta}+T_{2,3}\frac{\partial N_j}{\partial \xi}\right)| J_0 |d\zeta\, d\eta d\xi \\
& -\int_{\xi=-1}^{+1}\int_{\eta=-1}^{+1}\int_{\zeta=-1}^{+1} N_i\left(T_{3,1}\frac{\partial \bar{\nu}_t}{\partial \zeta}+T_{3,2}\frac{\partial \bar{\nu}_t}{\partial \eta}+T_{3,3}\frac{\partial \bar{\nu}_t}{\partial \xi}\right) \\
& \times\left(T_{3,1}\frac{\partial N_j}{\partial \zeta}+T_{3,2}\frac{\partial N_j}{\partial \eta}+T_{3,3}\frac{\partial N_j}{\partial \xi}\right)| J_0 |d\, \zeta d\eta d\xi \\
& -2\int_{\xi=-1}^{+1}\int_{\eta=-1}^{+1}\int_{\zeta=-1}^{+1} N_i\left(T_{1,1}\frac{\partial \bar{\nu}_t}{\partial \zeta}+T_{1,2}\frac{\partial \bar{\nu}_t}{\partial \eta}+T_{1,3}\frac{\partial \bar{\nu}_t}{\partial \xi}\right) \\
& \times\left(T_{1,1}\frac{\partial N_j}{\partial \zeta}+T_{1,2}\frac{\partial N_j}{\partial \eta}+T_{1,3}\frac{\partial N_j}{\partial \xi}\right)| J_0 |d\zeta\, d\eta d\xi \\
& +\int_{\xi=-1}^{+1}\int_{\eta=-1}^{+1}\int_{\zeta=-1}^{+1} W_i\bar{u}\left(T_{1,1}\frac{\partial N_j}{\partial \zeta}+T_{1,2}\frac{\partial N_j}{\partial \eta}+T_{1,3}\frac{\partial N_j}{\partial \xi}\right)| J_0 |d\zeta\, d\eta d\xi \\
& +\int_{\xi=-1}^{+1}\int_{\eta=-1}^{+1}\int_{\zeta=-1}^{+1} W_i\bar{v}\left(T_{2,1}\frac{\partial N_j}{\partial \zeta}+T_{2,2}\frac{\partial N_j}{\partial \eta}+T_{2,3}\frac{\partial N_j}{\partial \xi}\right)| J_0 |d\zeta\, d\eta d\xi \\
& +\int_{\xi=-1}^{+1}\int_{\eta=-1}^{+1}\int_{\zeta=-1}^{+1} W_i\bar{w}\left(T_{3,1}\frac{\partial N_j}{\partial \zeta}+T_{3,2}\frac{\partial N_j}{\partial \eta}+T_{3,3}\frac{\partial N_j}{\partial \xi}\right)| J_0 \mid d\zeta d\eta d\xi
\end{aligned}
$$

$$
\begin{aligned}
a_2)_{i,j} = & \int_{\xi=-1}^{+1}\int_{\eta=-1}^{+1}\int_{\zeta=-1}^{+1} -N_i\left(T_{2,1}\frac{\partial \bar{\nu}_t}{\partial \zeta}+T_{2,2}\frac{\partial \bar{\nu}_t}{\partial \eta}+T_{2,3}\frac{\partial \bar{\nu}_t}{\partial \xi}\right) \\
& \times\left(T_{1,1}\frac{\partial N_j}{\partial \zeta}+T_{1,2}\frac{\partial N_j}{\partial \eta}+T_{1,3}\frac{\partial N_j}{\partial \xi}\right)| J_0 \mid d\zeta d\eta d\xi \\
& +\int_{\xi=-1}^{+1}\int_{\eta=-1}^{+1}\int_{\zeta=-1}^{+1} 2\Omega N_i N_j| J_0 \mid d\zeta d\eta d\xi
\end{aligned}
$$

$$a_{3})_{i,j} = \int_{\xi=-1}^{+1}\int_{\eta=-1}^{+1}\int_{\zeta=-1}^{+1} -N_i\left(T_{3,1}\frac{\partial \bar{v}_t}{\partial \zeta} + T_{3,2}\frac{\partial \bar{v}_t}{\partial \eta} + T_{3,3}\frac{\partial \bar{v}_t}{\partial \xi}\right)$$
$$\times\left(T_{1,1}\frac{\partial N_j}{\partial \zeta} + T_{1,2}\frac{\partial N_j}{\partial \eta} + T_{1,3}\frac{\partial N_j}{\partial \xi}\right)| J_0 |\, d\zeta\, d\eta\, d\xi$$

$$a_{4})_{i,k} = \int_{\xi=-1}^{+1}\int_{\eta=-1}^{+1}\int_{\zeta=-1}^{+1} \frac{1}{\rho}N_i\left(T_{1,1}\frac{\partial M_k}{\partial \zeta} + T_{1,2}\frac{\partial M_k}{\partial \eta} + T_{1,3}\frac{\partial M_k}{\partial \xi}\right)| J_0 |\, d\zeta\, d\eta\, d\xi$$

$$a_{5})_{i,j} = -\int_{\xi=-1}^{+1}\int_{\eta=-1}^{+1}\int_{\zeta=-1} +1N_i\left(T_{1,1}\frac{\partial \bar{v}_t}{\partial \zeta} + T_{1,2}\frac{\partial \bar{v}_t}{\partial \eta} + T_{1,3}\frac{\partial \bar{v}_t}{\partial \xi}\right)$$
$$\times\left(T_{2,1}\frac{\partial N_j}{\partial \zeta} + T_{2,2}\frac{\partial N_j}{\partial \eta} + T_{2,3}\frac{\partial N_j}{\partial \xi}\right)| J_0 |d\zeta\, d\eta\, d\xi$$
$$-2\int_{\xi=-1}^{+1}\int_{\eta=-1}^{+1}\int_{\zeta=-1}^{+1} (\Omega N_i N_j)| J_0 |\, d\zeta\, d\eta\, d\xi$$

$$a_{6})_{i,j} = \int_{\xi=-1}^{+1}\int_{\eta=-1}^{+1}\int_{\zeta=-1}^{+1} \bar{v}_e\left(T_{1,1}\frac{\partial N_i}{\partial \zeta} + T_{1,2}\frac{\partial N_i}{\partial \eta} + T_{1,3}\frac{\partial N_i}{\partial \xi}\right)$$
$$\times\left(T_{1,1}\frac{\partial N_j}{\partial \zeta} + T_{1,2}\frac{\partial N_j}{\partial \eta} + T_{1,3}\frac{\partial N_j}{\partial \xi}\right)| J_0 |\, d\zeta\, d\eta\, d\xi$$
$$+\int_{\xi=-1}^{+1}\int_{\eta=-1}^{+1}\int_{\zeta=-1}^{+1} \bar{v}_e\left(T_{2,1}\frac{\partial N_i}{\partial \zeta} + T_{2,2}\frac{\partial N_i}{\partial \eta} + T_{2,3}\frac{\partial N_i}{\partial \xi}\right)$$
$$\times\left(T_{2,1}\frac{\partial N_j}{\partial \zeta} + T_{2,2}\frac{\partial N_j}{\partial \eta} + T_{2,3}\frac{\partial N_j}{\partial \xi}\right)| J_0 |\, d\zeta\, d\eta\, d\xi$$
$$+\int_{\xi=-1}^{+1}\int_{\eta=-1}^{+1}\int_{\zeta=-1}^{+1} \bar{v}_e\left(T_{3,1}\frac{\partial N_i}{\partial \zeta} + T_{3,2}\frac{\partial N_i}{\partial \eta} + T_{3,3}\frac{\partial N_i}{\partial \xi}\right)$$
$$\times\left(T_{3,1}\frac{\partial N_j}{\partial \zeta} + T_{3,2}\frac{\partial N_j}{\partial \eta} + T_{3,3}\frac{\partial N_j}{\partial \xi}\right)| J_0 |\, d\zeta\, d\eta\, d\xi$$
$$-\int_{\xi=-1}^{+1}\int_{\eta=-1}^{+1}\int_{\zeta=-1}^{+1} N_i\left(T_{1,1}\frac{\partial \bar{v}_t}{\partial \zeta} + T_{1,2}\frac{\partial \bar{v}_t}{\partial \eta} + T_{1,3}\frac{\partial \bar{v}_t}{\partial \xi}\right)$$
$$\times\left(T_{1,1}\frac{\partial N_j}{\partial \zeta} + T_{1,2}\frac{\partial N_j}{\partial \eta} + T_{1,3}\frac{\partial N_j}{\partial \xi}\right)| J_0 |\, d\zeta\, d\eta\, d\xi$$
$$-\int_{\xi=-1}^{+1}\int_{\eta=-1}^{+1}\int_{\zeta=-1}^{+1} N_i\left(T_{3,1}\frac{\partial \bar{v}_t}{\partial \zeta} + T_{3,2}\frac{\partial \bar{v}_t}{\partial \eta} + T_{3,3}\frac{\partial \bar{v}_t}{\partial \xi}\right)$$
$$\times\left(T_{3,1}\frac{\partial N_j}{\partial \zeta} + T_{3,2}\frac{\partial N_j}{\partial \eta} + T_{3,3}\frac{\partial N_j}{\partial \xi}\right)| J_0 |\, d\zeta\, d\eta\, d\xi$$
$$-2\int_{\xi=-1}^{+1}\int_{\eta=-1}^{+1}\int_{\zeta=-1}^{+1} N_i\left(T_{2,1}\frac{\partial \bar{v}_t}{\partial \zeta} + T_{2,2}\frac{\partial \bar{v}_t}{\partial \eta} + T_{2,3}\frac{\partial \bar{v}_t}{\partial \xi}\right)$$
$$\times\left(T_{2,1}\frac{\partial N_j}{\partial \zeta} + T_{2,2}\frac{\partial N_j}{\partial \eta} + T_{2,3}\frac{\partial N_j}{\partial \xi}\right)| J_0 |\, d\zeta\, d\eta\, d\xi$$
$$+\int_{\xi=-1}^{+1}\int_{\eta=-1}^{+1}\int_{\zeta=-1}^{+1} W_i\bar{u}\left(T_{1,1}\frac{\partial N_j}{\partial \zeta} + T_{1,2}\frac{\partial N_j}{\partial \eta} + T_{1,3}\frac{\partial N_j}{\partial \xi}\right)| J_0 |\, d\zeta\, d\eta\, d\xi$$

$$+\int_{\xi=-1}^{+1}\int_{\eta=-1}^{+1}\int_{\zeta=-1}^{+1} W_i\bar{v}\left(T_{2,1}\frac{\partial N_j}{\partial \zeta}+T_{2,2}\frac{\partial N_j}{\partial \eta}+T_{2,3}\frac{\partial N_j}{\partial \xi}\right)\mid J_0\mid d\zeta d\eta d\xi$$

$$+\int_{\xi=-1}^{+1}\int_{\eta=-1}^{+1}\int_{\zeta=-1}^{+1} W_i\bar{w}\left(T_{3,1}\frac{\partial N_j}{\partial \zeta}+T_{3,2}\frac{\partial N_j}{\partial \eta}+T_{3,3}\frac{\partial N_j}{\partial \xi}\right)\mid J_0\mid d\zeta d\eta d\xi$$

$$a_7)_{i,j}=-\int_{\xi=-1}^{+1}\int_{\eta=-1}^{+1}\int_{\zeta=-1}^{+1} N_i\left(T_{3,1}\frac{\partial \bar{v}_t}{\partial \zeta}+T_{3,2}\frac{\partial \bar{v}_t}{\partial \eta}+T_{3,3}\frac{\partial \bar{v}_t}{\partial \xi}\right)$$
$$\times\left(T_{2,1}\frac{\partial N_j}{\partial \zeta}+T_{2,2}\frac{\partial N_j}{\partial \eta}+T_{2,3}\frac{\partial N_j}{\partial \xi}\right)\mid J_0\mid d\zeta d\eta d\xi$$

$$a_8)_{i,k}=\int_{\xi=-1}^{+1}\int_{\eta=-1}^{+1}\int_{\zeta=-1}^{+1} \frac{1}{\rho}N_i\left(T_{2,1}\frac{\partial M_k}{\partial \zeta}+T_{2,2}\frac{\partial M_k}{\partial \eta}+T_{2,3}\frac{\partial M_k}{\partial \xi}\right)\mid J_0\mid d\zeta d\eta d\xi$$

$$a_9)_{i,j}=-\int_{\xi=-1}+1\int_{\eta=-1}^{+1}\int_{\zeta=-1}+1N_i\left(T_{1,1}\frac{\partial \bar{v}_t}{\partial \zeta}+T_{1,2}\frac{\partial \bar{v}_t}{\partial \eta}+T_{1,3}\frac{\partial \bar{v}_t}{\partial \xi}\right)$$
$$\times\left(T_{3,1}\frac{\partial N_j}{\partial \zeta}+T_{3,2}\frac{\partial N_j}{\partial \eta}+T_{3,3}\frac{\partial N_j}{\partial \xi}\right)\mid J_0\mid d\zeta d\eta d\xi$$

$$a_{10})_{i,j}=-\int_{\xi=-1}^{+1}\int_{\eta=-1}^{+1}\int_{\zeta=-1}^{+1} N_i\left(T_{2,1}\frac{\partial \bar{v}_t}{\partial \zeta}+T_{2,2}\frac{\partial \bar{v}_t}{\partial \eta}+T_{2,3}\frac{\partial \bar{v}_t}{\partial \xi}\right)$$
$$\times\left(T_{3,1}\frac{\partial N_j}{\partial \zeta}+T_{3,2}\frac{\partial N_j}{\partial \eta}+T_{3,3}\frac{\partial N_j}{\partial \xi}\right)\mid J_0\mid d\zeta d\eta d\xi$$

$$a_11)_{i,j}=\int_{\xi=-1}^{+1}\int_{\eta=-1}^{+1}\int_{\zeta=-1}^{+1} \bar{v}_e\left(T_{1,1}\frac{\partial N_i}{\partial \zeta}+T_{1,2}\frac{\partial N_i}{\partial \eta}+T_{1,3}\frac{\partial N_i}{\partial \xi}\right)$$
$$\times\left(T_{1,1}\frac{\partial N_j}{\partial \zeta}+T_{1,2}\frac{\partial N_j}{\partial \eta}+T_{1,3}\frac{\partial N_j}{\partial \xi}\right)\mid J_0\mid d\zeta d\eta d\xi$$
$$+\int_{\xi=-1}^{+1}\int_{\eta=-1}^{+1}\int_{\zeta=-1}^{+1} \bar{v}_e\left(T_{2,1}\frac{\partial N_i}{\partial \zeta}+T_{2,2}\frac{\partial N_i}{\partial \eta}+T_{2,3}\frac{\partial N_i}{\partial \xi}\right)$$
$$\times\left(T_{2,1}\frac{\partial N_j}{\partial \zeta}+T_{2,2}\frac{\partial N_j}{\partial \eta}+T_{2,3}\frac{\partial N_j}{\partial \xi}\right)\mid J_0\mid d\zeta d\eta d\xi$$
$$+\int_{\xi=-1}^{+1}\int_{\eta=-1}^{+1}\int_{\zeta=-1}^{+1} \bar{v}_e\left(T_{3,1}\frac{\partial N_i}{\partial \zeta}+T_{3,2}\frac{\partial N_i}{\partial \eta}+T_{3,3}\frac{\partial N_i}{\partial \xi}\right)$$
$$\times\left(T_{3,1}\frac{\partial N_j}{\partial \zeta}+T_{3,2}\frac{\partial N_j}{\partial \eta}+T_{3,3}\frac{\partial N_j}{\partial \xi}\right)\mid J_0\mid d\zeta d\eta d\xi$$
$$-\int_{\xi=-1}^{+1}\int_{\eta=-1}^{+1}\int_{\zeta=-1}^{+1} N_i\left(T_{1,1}\frac{\partial \bar{v}_t}{\partial \zeta}+T_{1,2}\frac{\partial \bar{v}_t}{\partial \eta}+T_{1,3}\frac{\partial \bar{v}_t}{\partial \xi}\right)$$
$$\times\left(T_{1,1}\frac{\partial N_j}{\partial \zeta}+T_{1,2}\frac{\partial N_j}{\partial \eta}+T_{1,3}\frac{\partial N_j}{\partial \xi}\right)\mid J_0\mid d\zeta d\eta d\xi$$
$$-\int_{\xi=-1}^{+1}\int_{\eta=-1}^{+1}\int_{\zeta=-1}^{+1} N_i\left(T_{2,1}\frac{\partial \bar{v}_t}{\partial \zeta}+T_{2,2}\frac{\partial \bar{v}_t}{\partial \eta}+T_{2,3}\frac{\partial \bar{v}_t}{\partial \xi}\right)$$
$$\times\left(T_{2,1}\frac{\partial N_j}{\partial \zeta}+T_{2,2}\frac{\partial N_j}{\partial \eta}+T_{2,3}\frac{\partial N_j}{\partial \xi}\right)\mid J_0\mid d\zeta d\eta d\xi$$

$$-2\int_{\xi=-1}^{+1}\int_{\eta=-1}^{+1}\int_{\zeta=-1}^{+1} N_i\left(T_{3,1}\frac{\partial \bar{v}_t}{\partial \zeta}+T_{3,2}\frac{\partial \bar{v}_t}{\partial \eta}+T_{3,3}\frac{\partial \bar{v}_t}{\partial \xi}\right)$$
$$\times\left(T_{3,1}\frac{\partial N_j}{\partial \zeta}+T_{3,2}\frac{\partial N_j}{\partial \eta}+T_{3,3}\frac{\partial N_j}{\partial \xi}\right)\mid J_0\mid d\zeta\, d\eta\, d\xi$$
$$+\int_{\xi=-1}^{+1}\int_{\eta=-1}^{+1}\int_{\zeta=-1}^{+1} W_i\bar{u}\left(T_{1,1}\frac{\partial N_j}{\partial \zeta}+T_{1,2}\frac{\partial N_j}{\partial \eta}+T_{1,3}\frac{\partial N_j}{\partial \xi}\right)\mid J_0\mid d\zeta\, d\eta\, d\xi$$
$$+\int_{\xi=-1}^{+1}\int_{\eta=-1}^{+1}\int_{\zeta=-1}^{+1} W_i\bar{v}\left(T_{2,1}\frac{\partial N_j}{\partial \zeta}+T_{2,2}\frac{\partial N_j}{\partial \eta}+T_{2,3}\frac{\partial N_j}{\partial \xi}\right)\mid J_0\mid d\zeta\, d\eta\, d\xi$$
$$+\int_{\xi=-1}^{+1}\int_{\eta=-1}^{+1}\int_{\zeta=-1}^{+1} W_i\bar{w}\left(T_{3,1}\frac{\partial N_j}{\partial \zeta}+T_{3,2}\frac{\partial N_j}{\partial \eta}+T_{3,3}\frac{\partial N_j}{\partial \xi}\right)\mid J_0\mid d\zeta\, d\eta\, d\xi$$

$$a_{12})_{i,k}=\int_{\xi=-1}^{+1}\int_{\eta=-1}^{+1}\int_{\zeta=-1}^{+1} N_i\left(T_{3,1}\frac{\partial M_k}{\partial \zeta}+T_{3,2}\frac{\partial M_k}{\partial \eta}+T_{3,3}\frac{\partial M_k}{\partial \xi}\right)\mid J_0\mid d\zeta\, d\eta\, d\xi$$

$$a_{13})_{k,j}=\int_{\xi=-1}^{+1}\int_{\eta=-1}^{+1}\int_{\zeta=-1}^{+1} M_k\left(T_{1,1}\frac{\partial N_j}{\partial \zeta}+T_{1,2}\frac{\partial N_j}{\partial \eta}+T_{1,3}\frac{\partial N_j}{\partial \xi}\right)\mid J_0\mid d\zeta\, d\eta\, d\xi$$

$$a_{14})_{k,j}=\int_{\xi=-1}^{+1}\int_{\eta=-1}^{+1}\int_{\zeta=-1}^{+1} M_k\left(T_{2,1}\frac{\partial N_j}{\partial \zeta}+T_{2,2}\frac{\partial N_j}{\partial \eta}+T_{2,3}\frac{\partial N_j}{\partial \xi}\right)\mid J_0\mid d\zeta\, d\eta\, d\xi$$

$$a_{15})_{k,j}=\int_{\xi=-1}^{+1}\int_{\eta=-1}^{+1}\int_{\zeta=-1}^{+1} M_k\left(T_{3,1}\frac{\partial N_j}{\partial \zeta}+T_{3,2}\frac{\partial N_j}{\partial \eta}+T_{3,3}\frac{\partial N_j}{\partial \xi}\right)\mid J_0\mid d\zeta\, d\eta\, d\xi$$

The load vectors $\{q_1\}$, $\{q_2\}$, and $\{q_3\}$ resulting from the three momentum equations can be similarly expressed as follows:

$$q_1)_i=\int_{S^{(e)}}\bar{v}_e N_i\frac{\partial u}{\partial n}\,dS+\Omega^2\int_{\xi=-1}^{+1}\int_{\eta=-1}^{+1}\int_{\zeta=-1}^{+1} N_i\left(\sum_{l=1}^{20}N_l x_l\right)\mid J_0\mid d\zeta\, d\eta\, d\xi$$

$$q_2)_i=\int_{S^{(e)}}\bar{v}_e N_i\frac{\partial v}{\partial n}\,dS+\Omega^2\int_{\xi=-1}^{+1}\int_{\eta=-1}^{+1}\int_{\zeta=-1}^{+1} N_i\left(\sum_{l=1}^{20}N_l y_l\right)\mid J_0\mid d\zeta\, d\eta\, d\xi$$

$$q_3)_i=\int_{S^{(e)}}\bar{v}_e N_i\frac{\partial w}{\partial n}\,dS$$

In a compact form, the set of equations for the typical finite element (e) can be written for the undisplaced-rotor operation mode (zeroth-order solution) as follows:

$$[a_0]\{\phi_0\}=\{q_0\} \tag{D.1}$$

with $[a_0]$ consisting of the fifteen submatrices $[a_1]$ through $[a_{15}]$, and $\{\phi_0\}$ referring to the vector of the velocity components and pressure at the element nodes. On assembling all contributions of the finite-elements, the "global" system of equations can be written as follows:

$$[A_0]\{\Phi_0\}=\{Q_0\} \tag{D.2}$$

As noted earlier, and this is a rather important point, the zeroth-order solution vector $\{\Phi_0\}$ is already known because it is an expanded version of the axisymmetric flow solution vector $\{\Psi\}$ on adapting the velocity components to fit the rotating 3D frame of reference. This task is attained by simply applying the two relationships (16.21) and (16.22) at all computational nodes, including those on the rotor and housing surfaces.

APPENDIX E

Displaced-Rotor Operation: Perturbation Analysis

Because the whirl frequency is already a built-in parameter in the zeroth-order flow problem, it takes only a lateral eccentricity of the rotor axis for complete whirl excitation to materialize. It is important to understand the physical aspects of the flow field under such disturbance.

Consider the case where an observer is "attached" to the origin of the whirling frame of reference (see Figure 16.8) and is *whirling* with it. To this observer, the distortion of the flow domain is the result of an upward displacement of the housing (and not the rotor) surface. To the same observer, all nodal points in the original finite element discretization model will undergo varied amounts of upward displacement, except for nodes on the rotor surface which will remain undisplaced. As a result, varied amounts of geometric deformations will be experienced by the entire assembly of finite-elements in the rotor-to-housing flow passage (see Figure 16.8). Finally, the observer, whose rotation is at the rate of the whirl frequency Ω, will register a "relative" linear velocity of the housing surface whose magnitude is Ωr_h, where r_h is the local radius of the housing inner surface at this particular axial location, as shown in Figure 16.8.

With no lack of generality, consider the rotor position in Figure 16.13, where the nodes of the typical element (e) have been displaced by different amounts. The nodal displacements, as shown in the figure, depend on the original y coordinate of each individual node. A relationship in this case can be generally established between the nodal coordinates before and after the distortion as follows:

$$X_i = x_i$$

$$Y_i = y_i + \lambda_i \epsilon$$

$$Z_i = z_i$$

where the index i varies from 1 to 20, the total number of nodes per element, and λ_i is a fraction that depends on the node's original location. Distribution of the factor λ among all nodes is basically arbitrary, provided that it attains a value of zero over the rotor surface and -1.0 over the housing surface. Recall that the flow field is now viewed from a reference frame that is attached to the rotor, with which the housing surface will appear as displaced in the negative y direction by an amount that is numerically equal to the rotor virtual displacement ϵ. As for the inner nodes, in the rotor-to-housing gap, the fraction λ was taken to vary linearly with the radial distance from the housing inner surface, where

$$0 \geq \lambda \geq -1$$

Referred to the rotor-attached Cartesian system, the new coordinates for the typical element (e) can now be interpolated as follows:

$$x^{(e)} = \sum_{i=1}^{20} N_i(\zeta,\eta,\xi)X_i = \sum_{i=1}^{20} N_i(\zeta,\eta,\xi)x_i \tag{E.1}$$

$$y^{(e)} = \sum_{i=1}^{20} N_i(\zeta,\eta,\xi)Y_i = \sum_{i=1}^{20} N_i(\zeta,\eta,\xi)y_i + \epsilon \sum_{i=1}^{20} N_i(\zeta,\eta,\xi)\lambda_i \tag{E.2}$$

$$z^{(e)} = \sum_{i=1}^{20} N_i(\zeta,\eta,\xi)Z_i = \sum_{i=1}^{20} N_i(\zeta,\eta,\xi)z_i \tag{E.3}$$

where N_i refers to the element quadratic shape functions.

Referring to the general expressions of the distorted-element nodal coordinates [expressions (16.30) through (16.32)], the following derivative-transformation equation is valid:

$$\begin{Bmatrix} \dfrac{\partial}{\partial\zeta} \\ \dfrac{\partial}{\partial\eta} \\ \dfrac{\partial}{\partial\xi} \end{Bmatrix} = \begin{bmatrix} \left(\sum\limits_{i=1}^{20} \dfrac{\partial N_i}{\partial\zeta} x_i\right) & \left(\sum\limits_{i=1}^{20} \dfrac{\partial N_i}{\partial\zeta} y_i + \epsilon \sum\limits_{i=1}^{20} \dfrac{\partial N_i}{\partial\zeta} \lambda_i\right) & \left(\sum\limits_{i=1}^{20} \dfrac{\partial N_i}{\partial\zeta} z_i\right) \\ \left(\sum\limits_{i=1}^{20} \dfrac{\partial N_i}{\partial\eta} x_i\right) & \left(\sum\limits_{i=1}^{20} \dfrac{\partial N_i}{\partial\eta} y_i + \epsilon \sum\limits_{i=1}^{20} \dfrac{\partial N_i}{\partial\eta} \lambda_i\right) & \left(\sum\limits_{i=1}^{20} \dfrac{\partial N_i}{\partial\eta} z_i\right) \\ \left(\sum\limits_{i=1}^{20} \dfrac{\partial N_i}{\partial\xi} x_i\right) & \left(\sum\limits_{i=1}^{20} \dfrac{\partial N_i}{\partial\xi} y_i + \epsilon \sum\limits_{i=1}^{20} \dfrac{\partial N_i}{\partial\xi} \lambda_i\right) & \left(\sum\limits_{i=1}^{20} \dfrac{\partial N_i}{\partial\xi} z_i\right) \end{bmatrix} \begin{Bmatrix} \dfrac{\partial}{\partial x} \\ \dfrac{\partial}{\partial y} \\ \dfrac{\partial}{\partial z} \end{Bmatrix}$$

which, on inversion, yields

$$\begin{Bmatrix} \dfrac{\partial}{\partial x} \\ \dfrac{\partial}{\partial y} \\ \dfrac{\partial}{\partial z} \end{Bmatrix} = ([T(\zeta,\eta,\xi)] + \epsilon[P(\zeta,\eta,\xi)]) \begin{Bmatrix} \dfrac{\partial}{\partial\zeta} \\ \dfrac{\partial}{\partial\eta} \\ \dfrac{\partial}{\partial\xi} \end{Bmatrix} \tag{E.4}$$

with the second- and higher-order terms being ignored. The matrices $[T]$ and $[P]$ in expression (E.4) can be written as follows:

$$[T] = \frac{[B]}{|J_0|} \tag{E.5}$$

$$[P] = \frac{[C]}{|J_0|} - \frac{|J_1|[B]}{|J_0|^2} \tag{E.6}$$

$$|J| = |J_0| + \epsilon|J_1| \tag{E.7}$$

The determinants $| J_0 |$ and $| J_1 |$ are defined as follows:

$$| J_0 | = \begin{vmatrix} \sum_{i=1}^{20} \frac{\partial N_i}{\partial \zeta} x_i & \sum_{i=1}^{20} \frac{\partial N_i}{\partial \zeta} y_i & \sum_{i=1}^{20} \frac{\partial N_i}{\partial \zeta} z_i \\ \sum_{i=1}^{20} \frac{\partial N_i}{\partial \eta} x_i & \sum_{i=1}^{20} \frac{\partial N_i}{\partial \eta} y_i & \sum_{i=1}^{20} \frac{\partial N_i}{\partial \eta} z_i \\ \sum_{i=1}^{20} \frac{\partial N_i}{\partial \xi} x_i & \sum_{i=1}^{20} \frac{\partial N_i}{\partial \xi} y_i & \sum_{i=1}^{20} \frac{\partial N_i}{\partial \xi} z_i \end{vmatrix}$$

$$| J_1 | = \begin{vmatrix} \sum_{i=1}^{20} \frac{\partial N_i}{\partial \zeta} x_i & \sum_{i=1}^{20} \frac{\partial N_i}{\partial \zeta} \lambda_i & \sum_{i=1}^{20} \frac{\partial N_i}{\partial \zeta} z_i \\ \sum_{i=1}^{20} \frac{\partial N_i}{\partial \eta} x_i & \sum_{i=1}^{20} \frac{\partial N_i}{\partial \eta} \lambda_i & \sum_{i=1}^{20} \frac{\partial N_i}{\partial \eta} z_i \\ \sum_{i=1}^{20} \frac{\partial N_i}{\partial \xi} x_i & \sum_{i=1}^{20} \frac{\partial N_i}{\partial \xi} \lambda_i & \sum_{i=1}^{20} \frac{\partial N_i}{\partial \xi} z_i \end{vmatrix}$$

The matrices $[B]$ and $[C]$ in expressions (E.5) and (E.6) are both 3×3 arrays and are defined on an entry-by-entry basis as follows:

$$B_{11} = \sum_{i=1}^{20} \frac{\partial N_i}{\partial \eta} y_i \sum_{i=1}^{20} \frac{\partial N_i}{\partial \xi} z_i - \sum_{i=1}^{20} \frac{\partial N_i}{\partial \eta} z_i \sum_{i=1}^{20} \frac{\partial N_i}{\partial \xi} y_i$$

$$B_{12} = \sum_{i=1}^{20} \frac{\partial N_i}{\partial \xi} y_i \sum_{i=1}^{20} \frac{\partial N_i}{\partial \zeta} z_i - \sum_{i=1}^{20} \frac{\partial N_i}{\partial \zeta} y_i \sum_{i=1}^{20} \frac{\partial N_i}{\partial \xi} z_i$$

$$B_{13} = \sum_{i=1}^{20} \frac{\partial N_i}{\partial \zeta} y_i \sum_{i=1}^{20} \frac{\partial N_i}{\partial \eta} z_i - \sum_{i=1}^{20} \frac{\partial N_i}{\partial \eta} y_i \sum_{i=1}^{20} \frac{\partial N_i}{\partial \zeta} z_i$$

$$B_{21} = \sum_{i=1}^{20} \frac{\partial N_i}{\partial \xi} x_i \sum_{i=1}^{20} \frac{\partial N_i}{\partial \eta} z_i - \sum_{i=1}^{20} \frac{\partial N_i}{\partial \eta} x_i \sum_{i=1}^{20} \frac{\partial N_i}{\partial \xi} z_i$$

$$B_{22} = \sum_{i=1}^{20} \frac{\partial N_i}{\partial \zeta} x_i \sum_{i=1}^{20} \frac{\partial N_i}{\partial \xi} z_i - \sum_{i=1}^{20} \frac{\partial N_i}{\partial \xi} x_i \sum_{i=1}^{20} \frac{\partial N_i}{\partial \zeta} z_i$$

$$B_{23} = \sum_{i=1}^{20} \frac{\partial N_i}{\partial \eta} x_i \sum_{i=1}^{20} \frac{\partial N_i}{\partial \zeta} z_i - \sum_{i=1}^{20} \frac{\partial N_i}{\partial \zeta} x_i \sum_{i=1}^{20} \frac{\partial N_i}{\partial \eta} z_i$$

$$B_{31} = \sum_{i=1}^{20} \frac{\partial N_i}{\partial \eta} x_i \sum_{i=1}^{20} \frac{\partial N_i}{\partial \xi} y_i - \sum_{i=1}^{20} \frac{\partial N_i}{\partial \xi} x_i \sum_{i=1}^{20} \frac{\partial N_i}{\partial \eta} y_i$$

$$B_{32} = \sum_{i=1}^{20} \frac{\partial N_i}{\partial \xi} x_i \sum_{i=1}^{20} \frac{\partial N_i}{\partial \eta} y_i - \sum_{i=1}^{20} \frac{\partial N_i}{\partial \zeta} x_i \sum_{i=1}^{20} \frac{\partial N_i}{\partial \xi} y_i$$

$$B_{33} = \sum_{i=1}^{20} \frac{\partial N_i}{\partial \zeta} x_i \sum_{i=1}^{20} \frac{\partial N_i}{\partial \eta} y_i - \sum_{i=1}^{20} \frac{\partial N_i}{\partial \eta} x_i \sum_{i=1}^{20} \frac{\partial N_i}{\partial \zeta} y_i$$

In a similar fashion, the matrix $[C]$ in expression (E.6) is defined as follows:

$$C_{11} = \sum_{i=1}^{20} \frac{\partial N_i}{\partial \eta}\lambda_i \sum_{i=1}^{20} \frac{\partial N_i}{\partial \xi} z_i - \sum_{i=1}^{20} \frac{\partial N_i}{\partial \zeta}\lambda_i \sum_{i=1}^{20} \frac{\partial N_i}{\partial \eta} z_i$$

$$C_{12} = \sum_{i=1}^{20} \frac{\partial N_i}{\partial \xi}\lambda_i \sum_{i=1}^{20} \frac{\partial N_i}{\partial \zeta} z_i - \sum_{i=1}^{20} \frac{\partial N_i}{\partial \zeta}\lambda_i \sum_{i=1}^{20} \frac{\partial N_i}{\partial \xi} z_i$$

$$C_{13} = \sum_{i=1}^{20} \frac{\partial N_i}{\partial \zeta}\lambda_i \sum_{i=1}^{20} \frac{\partial N_i}{\partial \eta} z_i - \sum_{i=1}^{20} \frac{\partial N_i}{\partial \eta}\lambda_i \sum_{i=1}^{20} \frac{\partial N_i}{\partial \zeta} z_i$$

$$C_{21} = 0$$

$$C_{22} = 0$$

$$C_{23} = 0$$

$$C_{31} = \sum_{i=1}^{20} \frac{\partial N_i}{\partial \eta} x_i \sum_{i=1}^{20} \frac{\partial N_i}{\partial \xi}\lambda_i - \sum_{i=1}^{20} \frac{\partial N_i}{\partial \eta}\lambda_i \sum_{i=1}^{20} \frac{\partial N_i}{\partial \eta}\lambda_i$$

$$C_{32} = \sum_{i=1}^{20} \frac{\partial N_i}{\partial \xi} x_i \sum_{i=1}^{20} \frac{\partial N_i}{\partial \zeta}\lambda_i - \sum_{i=1}^{20} \frac{\partial N_i}{\partial \zeta} x_i \sum_{i=1}^{20} \frac{\partial N_i}{\partial \xi}\lambda_i$$

$$C_{33} = \sum_{i=1}^{20} \frac{\partial N_i}{\partial \zeta} x_i \sum_{i=1}^{20} \frac{\partial N_i}{\partial \eta}\lambda_i - \sum_{i=1}^{20} \frac{\partial N_i}{\partial \eta} x_i \sum_{i=1}^{20} \frac{\partial N_i}{\partial \zeta}\lambda_i$$

The finite element equations, compared in Expression D.1 for the zeroth-order flow field, will now assume the following form:

$$([a_0] + \epsilon[\bar{a}])(\{\phi_0\} + \{\delta\phi\}) = (\{q_0\} + \epsilon\{\bar{q}\}) \tag{E.8}$$

In this equation, the zeroth-order stiffness matrix $[a_0]$ and load vector $\{q_0\}$ are precisely those in Equation (D.1). Similar to the manner in which the matrix $[a_0]$ was partitioned earlier, the perturbation matrix $[\bar{a}]$ is broken up into fifteen submatrices and the vector $\{\bar{q}\}$ into three subvectors as follows:

$$\begin{aligned}
\bar{a}_1)_{i,j} = a_1{}^*)_{i,j} &+ \int_{\xi=-1}^{+1}\int_{\eta=-1}^{+1}\int_{\zeta=-1}^{+1} \bar{\nu}_e \left(T_{11}\frac{\partial N_i}{\partial \zeta} + T_{12}\frac{\partial N_i}{\partial \eta} + T_{13}\frac{\partial N_i}{\partial \xi}\right) \\
&\times \left(P_{11}\frac{\partial N_j}{\partial \zeta} + P_{12}\frac{\partial N_j}{\partial \eta} + P_{13}\frac{\partial N_j}{\partial \xi}\right) | J_0 | \, d\zeta\, d\eta\, d\xi \\
&+ \int_{\xi=-1}^{+1}\int_{\eta=-1}^{+1}\int_{\zeta=-1}^{+1} \bar{\nu}_e \left(P_{11}\frac{\partial N_i}{\partial \zeta} + P_{12}\frac{\partial N_i}{\partial \eta} + P_{13}\frac{\partial N_i}{\partial \xi}\right) \\
&\times \left(T_{11}\frac{\partial N_j}{\partial \zeta} + T_{12}\frac{\partial N_j}{\partial \eta} + T_{13}\frac{\partial N_j}{\partial \xi}\right) | J_0 | \, d\zeta\, d\eta\, d\xi \\
&+ \int_{\xi=-1}^{+1}\int_{\eta=-1}^{+1}\int_{\zeta=-1}^{+1} \bar{\nu}_e \left(T_{21}\frac{\partial N_i}{\partial \zeta} + T_{22}\frac{\partial N_i}{\partial \eta} + T_{23}\frac{\partial N_i}{\partial \xi}\right) \\
&\times \left(P_{21}\frac{\partial N_j}{\partial \zeta} + P_{22}\frac{\partial N_j}{\partial \eta} + P_{23}\frac{\partial N_j}{\partial \xi}\right) | J_0 | \, d\zeta\, d\eta\, d\xi
\end{aligned}$$

$$+\int_{\xi=-1}^{+1}\int_{\eta=-1}^{+1}\int_{\zeta=-1}^{+1}\bar{v}_e\left(P_{21}\frac{\partial N_i}{\partial\zeta}+P_{22}\frac{\partial N_i}{\partial\eta}+P_{23}\frac{\partial N_i}{\partial\xi}\right)$$

$$\times\left(T_{21}\frac{\partial N_j}{\partial\zeta}+T_{22}\frac{\partial N_j}{\partial\eta}+T_{23}\frac{\partial N_j}{\partial\xi}\right)|\,J_0\,|\,d\zeta d\eta d\xi$$

$$+\int_{\xi=-1}^{+1}\int_{\eta=-1}^{+1}\int_{\zeta=-1}^{+1}\bar{v}_e\left(T_{31}\frac{\partial N_i}{\partial\zeta}+T_{32}\frac{\partial N_i}{\partial\eta}+T_{33}\frac{\partial N_i}{\partial\xi}\right)$$

$$\times\left(P_{31}\frac{\partial N_j}{\partial\zeta}+P_{32}\frac{\partial N_j}{\partial\eta}+P_{33}\frac{\partial N_j}{\partial\xi}\right)|\,J_0\,|\,d\zeta d\eta d\xi$$

$$+\int_{\xi=-1}^{+1}\int_{\eta=-1}^{+1}\int_{\zeta=-1}^{+1}\bar{v}_e\left(P_{31}\frac{\partial N_i}{\partial\zeta}+P_{32}\frac{\partial N_i}{\partial\eta}+P_{33}\frac{\partial N_i}{\partial\xi}\right)$$

$$\times\left(T_{31}\frac{\partial N_j}{\partial\zeta}+T_{32}\frac{\partial N_j}{\partial\eta}+T_{33}\frac{\partial N_j}{\partial\xi}\right)|\,J_0\,|\,d\zeta d\eta d\xi$$

$$-\int_{xi=-1}^{+1}\int_{\eta=-1}^{+1}\int_{\zeta=-1}^{+1}N_i\left(T_{21}\frac{\partial\bar{v}_t}{\partial\zeta}+T_{22}\frac{\partial\bar{v}_t}{\partial\eta}+T_{23}\frac{\partial\bar{v}_t}{\partial\xi}\right)$$

$$\times\left(P_{21}\frac{\partial N_j}{\partial\zeta}+P_{22}\frac{\partial N_j}{\partial\eta}+P_{23}\frac{\partial N_j}{\partial\xi}\right)|\,J_0\,|\,d\zeta d\eta d\xi$$

$$-\int_{\xi=-1}^{+1}\int_{\eta=-1}^{+1}\int_{\zeta=-1}^{+1}N_i\left(P_{21}\frac{\partial\bar{v}_t}{\partial\zeta}+P_{22}\frac{\partial\bar{v}_t}{\partial\eta}+P_{23}\frac{\partial\bar{v}_t}{\partial\xi}\right)$$

$$\times\left(T_{21}\frac{\partial N_j}{\partial\zeta}+T_{22}\frac{\partial N_j}{\partial\eta}+T_{23}\frac{\partial N_j}{\partial\xi}\right)|\,J_0\,|d\zeta d\eta d\xi$$

$$-\int_{\xi=-1}^{+1}\int_{\eta=-1}^{+1}\int_{\zeta=-1}^{+1}N_i\left(T_{31}\frac{\partial\bar{v}_t}{\partial\zeta}+T_{32}\frac{\partial\bar{v}_t}{\partial\eta}+T_{33}\frac{\partial\bar{v}_t}{\partial\xi}\right)$$

$$\times\left(P_{31}\frac{\partial N_j}{\partial\zeta}+P_{32}\frac{\partial N_j}{\partial\eta}+P_{33}\frac{\partial N_j}{\partial\xi}\right)|\,J_0\,|d\zeta d\eta d\xi$$

$$-\int_{\xi=-1}^{+1}\int_{\eta=-1}^{+1}\int_{\zeta=-1}^{+1}N_i\left(P_{31}\frac{\partial\bar{v}_t}{\partial\zeta}+P_{32}\frac{\partial\bar{v}_t}{\partial\eta}+P_{33}\frac{\partial\bar{v}_t}{\partial\xi}\right)$$

$$\times\left(T_{31}\frac{\partial N_j}{\partial\zeta}+T_{32}\frac{\partial N_j}{\partial\eta}+T_{33}\frac{\partial N_j}{\partial\xi}\right)|\,J_0\,|d\zeta d\eta d\xi$$

$$-2\int_{\xi=-1}^{+1}\int_{\eta=-1}^{+1}\int_{\zeta=-1}^{+1}N_i\left(T_{11}\frac{\partial\bar{v}_t}{\partial\zeta}+T_{12}\frac{\partial\bar{v}_t}{\partial\eta}+T_{13}\frac{\partial\bar{v}_t}{\partial\xi}\right)$$

$$\times\left(P_{11}\frac{\partial N_j}{\partial\zeta}+P_{12}\frac{\partial N_j}{\partial\eta}+P_{13}\frac{\partial N_j}{\partial\xi}\right)|\,J_0\,|d\zeta d\eta d\xi$$

$$-2\int_{\xi=-1}^{+1}\int_{\eta=-1}^{+1}\int_{\zeta=-1}^{+1}N_i\left(P_{11}\frac{\partial\bar{v}_t}{\partial\zeta}+P_{12}\frac{\partial\bar{v}_t}{\partial\eta}+P_{13}\frac{\partial\bar{v}_t}{\partial\xi}\right)$$

$$\times\left(T_{11}\frac{\partial N_j}{\partial\zeta}+T_{12}\frac{\partial N_j}{\partial\eta}+T_{13}\frac{\partial N_j}{\partial\xi}\right)|\,J_0\,|d\zeta d\eta d\xi+\bar{u}W_i$$

$$\times\left(P_{11}\frac{\partial N_j}{\partial\zeta}+P_{12}\frac{\partial N_j}{\partial\eta}+P_{13}\frac{\partial N_j}{\partial\xi}\right)|\,J_0\,|d\zeta d\eta d\xi+\bar{v}W_i$$

$$\times\left(P_{21}\frac{\partial N_j}{\partial\zeta}+P_{22}\frac{\partial N_j}{\partial\eta}+P_{23}\frac{\partial N_j}{\partial\xi}\right)|\,J_0\,|d\zeta d\eta d\xi+\bar{w}W_i$$

$$\times\left(P_{31}\frac{\partial N_j}{\partial \zeta}+P_{32}\frac{\partial N_j}{\partial \eta}+P_{33}\frac{\partial N_j}{\partial \xi}\right)|\,J_0\,|d\zeta\, d\eta\, d\xi$$

$$\bar{a_2})_{i,j}=a_2{}^*)_{i,j}-\int_{\xi=-1}^{+1}\int_{\eta=-1}^{+1}\int_{\zeta=-1}^{+1}N_i\left(T_{21}\frac{\partial \bar{v}_t}{\partial \zeta}+T_{22}\frac{\partial \bar{v}_t}{\partial \eta}+T_{23}\frac{\partial \bar{v}_t}{\partial \xi}\right)$$

$$\times\left(P_{11}\frac{\partial N_j}{\partial \zeta}+P_{12}\frac{\partial N_j}{\partial \eta}+P_{13}\frac{\partial N_j}{\partial \xi}\right)|\,J_0\,|\,d\zeta\, d\eta\, d\xi$$

$$-\int_{\xi=-1}^{+1}\int_{\eta=-1}^{+1}\int_{\zeta=-1}^{+1}N_i\left(P_{21}\frac{\partial \bar{v}_t}{\partial \zeta}+P_{22}\frac{\partial \bar{v}_t}{\partial \eta}+P_{23}\frac{\partial \bar{v}_t}{\partial \xi}\right)$$

$$\times\left(T_{11}\frac{\partial N_j}{\partial \zeta}+T_{12}\frac{\partial N_j}{\partial \eta}+T_{13}\frac{\partial N_j}{\partial \xi}\right)|\,J_0\,|\,d\zeta\, d\eta\, d\xi$$

$$\bar{a_3})_{i,j}=a_3{}^*)_{i,j}-\int_{\xi=-1}^{+1}\int_{\eta=-1}^{+1}\int_{\zeta=-1}^{+1}N_i\left(T_{31}\frac{\partial \bar{v}_t}{\partial \zeta}+T_{32}\frac{\partial \bar{v}_t}{\partial \eta}+T_{33}\frac{\partial \bar{v}_t}{\partial \xi}\right)$$

$$\times\left(P_{11}\frac{\partial N_j}{\partial \zeta}+P_{12}\frac{\partial N_j}{\partial \eta}+P_{13}\frac{\partial N_j}{\partial \xi}\right)|\,J_0\,|\,d\zeta\, d\eta\, d\xi$$

$$-\int_{\xi=-1}^{+1}\int_{\eta=-1}^{+1}\int_{\zeta=-1}^{+1}N_i\left(P_{31}\frac{\partial \bar{v}_t}{\partial \zeta}+P_{32}\frac{\partial \bar{v}_t}{\partial \eta}+P_{33}\frac{\partial \bar{v}_t}{\partial \xi}\right)$$

$$\times\left(T_{11}\frac{\partial N_j}{\partial \zeta}+T_{12}\frac{\partial N_j}{\partial \eta}+T_{13}\frac{\partial N_j}{\partial \xi}\right)|\,J_0\,|\,d\zeta\, d\eta\, d\xi$$

$$\bar{a_4})_{i,k}=a_4{}^*)_{i,k}+\int_{\xi=-1}^{+1}\int_{\eta=-1}^{+1}\int_{\zeta=-1}^{+1}\frac{1}{\rho}N_i$$

$$\times\left(P_{11}\frac{\partial M_k}{\partial \zeta}+P_{12}\frac{\partial M_k}{\partial \eta}+P_{13}\frac{\partial M_k}{\partial \xi}\right)|\,J_0\,|\,d\zeta\, d\eta\, d\xi$$

$$\bar{a_5})_{i,j}=a_5{}^*)_{i,j}-\int_{\xi=-1}^{+1}\int_{\eta=-1}^{+1}\int_{\zeta=-1}^{+1}N_i\left(T_{11}\frac{\partial \bar{v}_t}{\partial \zeta}+T_{12}\frac{\partial \bar{v}_t}{\partial \eta}+T_{13}\frac{\partial \bar{v}_t}{\partial \xi}\right)$$

$$\times\left(P_{21}\frac{\partial N_j}{\partial \zeta}+P_{22}\frac{\partial N_j}{\partial \eta}+P_{23}\frac{\partial N_j}{\partial \xi}\right)|\,J_0\,|d\zeta\, d\eta\, d\xi$$

$$-\int_{\xi=-1}^{+1}\int_{\eta=-1}^{+1}\int_{\zeta=-1}^{+1}N_i\left(P_{11}\frac{\partial \bar{v}_t}{\partial \zeta}+P_{12}\frac{\partial \bar{v}_t}{\partial \eta}+P_{13}\frac{\partial \bar{v}_t}{\partial \xi}\right)$$

$$\times\left(T_{21}\frac{\partial N_j}{\partial \zeta}+T_{22}\frac{\partial N_j}{\partial \eta}+T_{23}\frac{\partial N_j}{\partial \xi}\right)|\,J_0\,|\,d\zeta\, d\eta\, d\xi$$

$$\bar{a_6})_{i,j}=a_6{}^*)_{i,j}+\int_{\xi=-1}^{+1}\int_{\eta=-1}^{+1}\int_{\zeta=-1}^{+1}\bar{v}_e\left(T_{11}\frac{\partial N_i}{\partial \zeta}+T_{12}\frac{\partial N_i}{\partial \eta}+T_{13}\frac{\partial N_i}{\partial \xi}\right)$$

$$\times\left(P_{11}\frac{\partial N_j}{\partial \zeta}+P_{12}\frac{\partial N_j}{\partial \eta}+P_{13}\frac{\partial N_j}{\partial \xi}\right)|\,J_0\,|\,d\zeta\, d\eta\, d\xi$$

$$+\int_{\xi=-1}^{+1}\int_{\eta=-1}^{+1}\int_{\zeta=-1}^{+1}\bar{v}_e\left(P_{11}\frac{\partial N_i}{\partial \zeta}+P_{12}\frac{\partial N_i}{\partial \eta}+P_{13}\frac{\partial N_i}{\partial \xi}\right)$$

$$\times\left(T_{11}\frac{\partial N_j}{\partial \zeta}+T_{12}\frac{\partial N_j}{\partial \eta}+T_{13}\frac{\partial N_j}{\partial \xi}\right)|\,J_0\,|\,d\zeta\, d\eta\, d\xi$$

$$+\int_{\xi=-1}^{+1}\int_{\eta=-1}^{+1}\int_{\zeta=-1}^{+1}\bar{v}_e\left(T_{21}\frac{\partial N_i}{\partial \zeta}+T_{22}\frac{\partial N_i}{\partial \eta}+T_{23}\frac{\partial N_i}{\partial \xi}\right)$$

$$\times\left(P_{21}\frac{\partial N_j}{\partial \zeta}+P_{22}\frac{\partial N_j}{\partial \eta}+P_{23}\frac{\partial N_j}{\partial \xi}\right)|\,J_0\,|\,d\zeta\,d\eta\,d\xi$$

$$+\int_{\xi=-1}^{+1}\int_{\eta=-1}^{+1}\int_{\zeta=-1}^{+1}\bar{v}_e\left(P_{21}\frac{\partial N_i}{\partial \zeta}+P_{22}\frac{\partial N_i}{\partial \eta}+P_{23}\frac{\partial N_i}{\partial \xi}\right)$$

$$\times\left(T_{21}\frac{\partial N_j}{\partial \zeta}+T_{22}\frac{\partial N_j}{\partial \eta}+T_{23}\frac{\partial N_j}{\partial \xi}\right)|\,J_0\,|\,d\zeta\,d\eta\,d\xi$$

$$+\int_{\xi=-1}^{+1}\int_{\eta=-1}^{+1}\int_{\zeta=-1}^{+1}\bar{v}_e\left(T_{31}\frac{\partial N_i}{\partial \zeta}+T_{32}\frac{\partial N_i}{\partial \eta}+T_{33}\frac{\partial N_i}{\partial \xi}\right)$$

$$\times\left(P_{31}\frac{\partial N_j}{\partial \zeta}+P_{32}\frac{\partial N_j}{\partial \eta}+P_{33}\frac{\partial N_j}{\partial \xi}\right)|\,J_0\,|\,d\zeta\,d\eta\,d\xi$$

$$+\int_{\xi=-1}^{+1}\int_{\eta=-1}^{+1}\int_{\zeta=-1}^{+1}\bar{v}_e\left(P_{31}\frac{\partial N_i}{\partial \zeta}+P_{32}\frac{\partial N_i}{\partial \eta}+P_{33}\frac{\partial N_i}{\partial \xi}\right)$$

$$\times\left(T_{31}\frac{\partial N_j}{\partial \zeta}+T_{32}\frac{\partial N_j}{\partial \eta}+T_{33}\frac{\partial N_j}{\partial \xi}\right)|\,J_0\,|\,d\zeta\,d\eta\,d\xi$$

$$-\int_{xi=-1}^{+1}\int_{\eta=-1}^{+1}\int_{\zeta=-1}^{+1}N_i\left(T_{11}\frac{\partial \bar{v}_t}{\partial \zeta}+T_{12}\frac{\partial \bar{v}_t}{\partial \eta}+T_{13}\frac{\partial \bar{v}_t}{\partial \xi}\right)$$

$$\times\left(P_{11}\frac{\partial N_j}{\partial \zeta}+P_{12}\frac{\partial N_j}{\partial \eta}+P_{13}\frac{\partial N_j}{\partial \xi}\right)|\,J_0\,|\,d\zeta\,d\eta\,d\xi$$

$$-\int_{\xi=-1}^{+1}\int_{\eta=-1}^{+1}\int_{\zeta=-1}^{+1}N_i\left(P_{11}\frac{\partial \bar{v}_t}{\partial \zeta}+P_{12}\frac{\partial \bar{v}_t}{\partial \eta}+P_{13}\frac{\partial \bar{v}_t}{\partial \xi}\right)$$

$$\times\left(T_{11}\frac{\partial N_j}{\partial \zeta}+T_{12}\frac{\partial N_j}{\partial \eta}+T_{13}\frac{\partial N_j}{\partial \xi}\right)|\,J_0\,|d\zeta\,d\eta\,d\xi$$

$$-\int_{\xi=-1}^{+1}\int_{\eta=-1}^{+1}\int_{\zeta=-1}^{+1}N_i\left(T_{31}\frac{\partial \bar{v}_t}{\partial \zeta}+T_{32}\frac{\partial \bar{v}_t}{\partial \eta}+T_{33}\frac{\partial \bar{v}_t}{\partial \xi}\right)$$

$$\times\left(P_{31}\frac{\partial N_j}{\partial \zeta}+P_{32}\frac{\partial N_j}{\partial \eta}+P_{33}\frac{\partial N_j}{\partial \xi}\right)|\,J_0\,|d\zeta\,d\eta\,d\xi$$

$$-\int_{\xi=-1}^{+1}\int_{\eta=-1}^{+1}\int_{\zeta=-1}^{+1}N_i\left(P_{31}\frac{\partial \bar{v}_t}{\partial \zeta}+P_{32}\frac{\partial \bar{v}_t}{\partial \eta}+P_{33}\frac{\partial \bar{v}_t}{\partial \xi}\right)$$

$$\times\left(T_{31}\frac{\partial N_j}{\partial \zeta}+T_{32}\frac{\partial N_j}{\partial \eta}+T_{33}\frac{\partial N_j}{\partial \xi}\right)|\,J_0\,|d\zeta\,d\eta\,d\xi$$

$$-2\int_{\xi=-1}^{+1}\int_{\eta=-1}^{+1}\int_{\zeta=-1}^{+1}N_i\left(T_{21}\frac{\partial \bar{v}_t}{\partial \zeta}+T_{22}\frac{\partial \bar{v}_t}{\partial \eta}+T_{23}\frac{\partial \bar{v}_t}{\partial \xi}\right)$$

$$\times\left(P_{21}\frac{\partial N_j}{\partial \zeta}+P_{22}\frac{\partial N_j}{\partial \eta}+P_{23}\frac{\partial N_j}{\partial \xi}\right)|\,J_0\,|d\zeta\,d\eta\,d\xi$$

$$-2\int_{\xi=-1}^{+1}\int_{\eta=-1}^{+1}\int_{\zeta=-1}^{+1}N_i\left(P_{21}\frac{\partial \bar{v}_t}{\partial \zeta}+P_{22}\frac{\partial \bar{v}_t}{\partial \eta}+P_{23}\frac{\partial \bar{v}_t}{\partial \xi}\right)$$

$$\times\left(T_{21}\frac{\partial N_j}{\partial \zeta}+T_{22}\frac{\partial N_j}{\partial \eta}+T_{23}\frac{\partial N_j}{\partial \xi}\right)|\,J_0\,|d\zeta\,d\eta\,d\xi+\bar{u}W_i$$

$$\times\left(P_{11}\frac{\partial N_j}{\partial \zeta}+P_{12}\frac{\partial N_j}{\partial \eta}+P_{13}\frac{\partial N_j}{\partial \xi}\right)|\,J_0\,|d\zeta\,d\eta\,d\xi+\bar{v}W_i$$

$$\times\left(P_{21}\frac{\partial N_j}{\partial\zeta}+P_{22}\frac{\partial N_j}{\partial\eta}+P_{23}\frac{\partial N_j}{\partial\xi}\right)|J_0|d\zeta d\eta d\xi+\bar{w}W_i$$

$$\times\left(P_{31}\frac{\partial N_j}{\partial\zeta}+P_{32}\frac{\partial N_j}{\partial\eta}+P_{33}\frac{\partial N_j}{\partial\xi}\right)|J_0|d\zeta d\eta d\xi$$

$$\bar{a_7})_{i,j}=a_7{}^*)_{i,j}-\int_{\xi=-1}^{+1}\int_{\eta=-1}^{+1}\int_{\zeta=-1}^{+1}N_i$$

$$\times\left(T_{31}\frac{\partial\bar{v}_t}{\partial\zeta}+T_{32}\frac{\partial\bar{v}_t}{\partial\eta}+T_{33}\frac{\partial\bar{v}_t}{\partial\xi}\right)$$

$$\times\left(P_{21}\frac{\partial N_j}{\partial\zeta}+P_{22}\frac{\partial N_j}{\partial\eta}+P_{23}\frac{\partial N_j}{\partial\xi}\right)|J_0|\,d\zeta d\eta d\xi$$

$$-\int_{\xi=-1}^{+1}\int_{\eta=-1}^{+1}\int_{\zeta=-1}^{+1}N_i\left(P_{31}\frac{\partial\bar{v}_t}{\partial\zeta}+P_{32}\frac{\partial\bar{v}_t}{\partial\eta}+P_{33}\frac{\partial\bar{v}_t}{\partial\xi}\right)$$

$$\times\left(T_{21}\frac{\partial N_j}{\partial\zeta}+T_{22}\frac{\partial N_j}{\partial\eta}+T_{23}\frac{\partial N_j}{\partial\xi}\right)|J_0|\,d\zeta d\eta d\xi$$

$$\bar{a_8})_{i,k}=a_8{}^*)_{i,k}+\int_{\xi=-1}^{+1}\int_{\eta=-1}^{+1}\int_{\zeta=-1}^{+1}\frac{1}{\rho}N_i$$

$$\times\left(P_{21}\frac{\partial M_k}{\partial\zeta}+P_{22}\frac{\partial M_k}{\partial\eta}+P_{23}\frac{\partial M_k}{\partial\xi}\right)|J_0|\,d\zeta d\eta d\xi$$

$$\bar{a_9})_{i,j}=a_9{}^*)_{i,j}-\int_{\xi=-1}^{+1}\int_{\eta=-1}^{+1}\int_{\zeta=-1}^{+1}N_i\left(T_{11}\frac{\partial\bar{v}_t}{\partial\zeta}+T_{12}\frac{\partial\bar{v}_t}{\partial\eta}+T_{13}\frac{\partial\bar{v}_t}{\partial\xi}\right)$$

$$\times\left(P_{31}\frac{\partial N_j}{\partial\zeta}+P_{32}\frac{\partial N_j}{\partial\eta}+P_{33}\frac{\partial N_j}{\partial\xi}\right)|J_0|\,d\zeta d\eta d\xi$$

$$-\int_{\xi=-1}^{+1}\int_{\eta=-1}^{+1}\int_{\zeta=-1}^{+1}N_i\left(P_{11}\frac{\partial\bar{v}_t}{\partial\zeta}+P_{12}\frac{\partial\bar{v}_t}{\partial\eta}+P_{13}\frac{\partial\bar{v}_t}{\partial\xi}\right)$$

$$\times\left(T_{31}\frac{\partial N_j}{\partial\zeta}+T_{32}\frac{\partial N_j}{\partial\eta}+T_{33}\frac{\partial N_j}{\partial\xi}\right)|J_0|\,d\zeta d\eta d\xi$$

$$\bar{a_{10}})_{i,j}=a_{10}{}^*)_{i,j}-\int_{\xi=-1}^{+1}\int_{\eta=-1}^{+1}\int_{\zeta=-1}^{+1}N_i\left(T_{21}\frac{\partial\bar{v}_t}{\partial\zeta}+T_{22}\frac{\partial\bar{v}_t}{\partial\eta}+T_{23}\frac{\partial\bar{v}_t}{\partial\xi}\right)$$

$$\times\left(P_{31}\frac{\partial N_j}{\partial\zeta}+P_{32}\frac{\partial N_j}{\partial\eta}+P_{33}\frac{\partial N_j}{\partial\xi}\right)|J_0|\,d\zeta d\eta d\xi$$

$$-\int_{\xi=-1}^{+1}\int_{\eta=-1}^{+1}\int_{\zeta=-1}^{+1}N_i\left(P_{21}\frac{\partial\bar{v}_t}{\partial\zeta}+P_{22}\frac{\partial\bar{v}_t}{\partial\eta}+P_{23}\frac{\partial\bar{v}_t}{\partial\xi}\right)$$

$$\times\left(T_{31}\frac{\partial N_j}{\partial\zeta}+T_{32}\frac{\partial N_j}{\partial\eta}+T_{33}\frac{\partial N_j}{\partial\xi}\right)|J_0|\,d\zeta d\eta d\xi$$

$$\bar{a}_{11})_{i,j}=a_{11}{}^*)_{i,j}+\int_{\xi=-1}^{+1}\int_{\eta=-1}^{+1}\int_{\zeta=-1}^{+1}\bar{v}_e\left(T_{11}\frac{\partial N_i}{\partial\zeta}+T_{12}\frac{\partial N_i}{\partial\eta}+T_{13}\frac{\partial N_i}{\partial\xi}\right)$$

$$\times\left(P_{11}\frac{\partial N_j}{\partial\zeta}+P_{12}\frac{\partial N_j}{\partial\eta}+P_{13}\frac{\partial N_j}{\partial\xi}\right)|J_0|\,d\zeta d\eta d\xi$$

$$+\int_{\xi=-1}^{+1}\int_{\eta=-1}^{+1}\int_{\zeta=-1}^{+1}\bar{v}_e\left(P_{11}\frac{\partial N_i}{\partial\zeta}+P_{12}\frac{\partial N_i}{\partial\eta}+P_{13}\frac{\partial N_i}{\partial\xi}\right)$$

$$\times\left(T_{11}\frac{\partial N_j}{\partial\zeta}+T_{12}\frac{\partial N_j}{\partial\eta}+T_{13}\frac{\partial N_j}{\partial\xi}\right)|\,J_0\,|\,d\zeta d\eta d\xi$$

$$+\int_{\xi=-1}^{+1}\int_{\eta=-1}^{+1}\int_{\zeta=-1}^{+1}\bar{\nu}_e\left(T_{21}\frac{\partial N_i}{\partial\zeta}+T_{22}\frac{\partial N_i}{\partial\eta}+T_{23}\frac{\partial N_i}{\partial\xi}\right)$$

$$\times\left(P_{21}\frac{\partial N_j}{\partial\zeta}+P_{22}\frac{\partial N_j}{\partial\eta}+P_{23}\frac{\partial N_j}{\partial\xi}\right)|\,J_0\,|\,d\zeta d\eta d\xi$$

$$+\int_{\xi=-1}^{+1}\int_{\eta=-1}^{+1}\int_{\zeta=-1}^{+1}\bar{\nu}_e\left(P_{21}\frac{\partial N_i}{\partial\zeta}+P_{22}\frac{\partial N_i}{\partial\eta}+P_{23}\frac{\partial N_i}{\partial\xi}\right)$$

$$\times\left(T_{21}\frac{\partial N_j}{\partial\zeta}+T_{22}\frac{\partial N_j}{\partial\eta}+T_{23}\frac{\partial N_j}{\partial\xi}\right)|\,J_0\,|\,d\zeta d\eta d\xi$$

$$+\int_{\xi=-1}^{+1}\int_{\eta=-1}^{+1}\int_{\zeta=-1}^{+1}\bar{\nu}_e\left(T_{31}\frac{\partial N_i}{\partial\zeta}+T_{32}\frac{\partial N_i}{\partial\eta}+T_{33}\frac{\partial N_i}{\partial\xi}\right)$$

$$\times\left(P_{31}\frac{\partial N_j}{\partial\zeta}+P_{32}\frac{\partial N_j}{\partial\eta}+P_{33}\frac{\partial N_j}{\partial\xi}\right)|\,J_0\,|\,d\zeta d\eta d\xi$$

$$+\int_{\xi=-1}^{+1}\int_{\eta=-1}^{+1}\int_{\zeta=-1}^{+1}\bar{\nu}_e\left(P_{31}\frac{\partial N_i}{\partial\zeta}+P_{32}\frac{\partial N_i}{\partial\eta}+P_{33}\frac{\partial N_i}{\partial\xi}\right)$$

$$\times\left(T_{31}\frac{\partial N_j}{\partial\zeta}+T_{32}\frac{\partial N_j}{\partial\eta}+T_{33}\frac{\partial N_j}{\partial\xi}\right)|\,J_0\,|\,d\zeta d\eta d\xi$$

$$-\int_{xi=-1}^{+1}\int_{\eta=-1}^{+1}\int_{\zeta=-1}^{+1}N_i\left(T_{11}\frac{\partial\bar{\nu}_t}{\partial\zeta}+T_{12}\frac{\partial\bar{\nu}_t}{\partial\eta}+T_{13}\frac{\partial\bar{\nu}_t}{\partial\xi}\right)$$

$$\times\left(P_{11}\frac{\partial N_j}{\partial\zeta}+P_{12}\frac{\partial N_j}{\partial\eta}+P_{13}\frac{\partial N_j}{\partial\xi}\right)|\,J_0\,|\,d\zeta d\eta d\xi$$

$$-\int_{\xi=-1}^{+1}\int_{\eta=-1}^{+1}\int_{\zeta=-1}^{+1}N_i\left(P_{11}\frac{\partial\bar{\nu}_t}{\partial\zeta}+P_{12}\frac{\partial\bar{\nu}_t}{\partial\eta}+P_{13}\frac{\partial\bar{\nu}_t}{\partial\xi}\right)$$

$$\times\left(T_{11}\frac{\partial N_j}{\partial\zeta}+T_{12}\frac{\partial N_j}{\partial\eta}+T_{13}\frac{\partial N_j}{\partial\xi}\right)|\,J_0\,|\,d\zeta d\eta d\xi$$

$$-\int_{\xi=-1}^{+1}\int_{\eta=-1}^{+1}\int_{\zeta=-1}^{+1}N_i\left(T_{21}\frac{\partial\bar{\nu}_t}{\partial\zeta}+T_{22}\frac{\partial\bar{\nu}_t}{\partial\eta}+T_{23}\frac{\partial\bar{\nu}_t}{\partial\xi}\right)$$

$$\times\left(P_{21}\frac{\partial N_j}{\partial\zeta}+P_{22}\frac{\partial N_j}{\partial\eta}+P_{23}\frac{\partial N_j}{\partial\xi}\right)|\,J_0\,|\,d\zeta d\eta d\xi$$

$$-\int_{\xi=-1}^{+1}\int_{\eta=-1}^{+1}\int_{\zeta=-1}^{+1}N_i\left(P_{21}\frac{\partial\bar{\nu}_t}{\partial\zeta}+P_{22}\frac{\partial\bar{\nu}_t}{\partial\eta}+P_{23}\frac{\partial\bar{\nu}_t}{\partial\xi}\right)$$

$$\times\left(T_{21}\frac{\partial N_j}{\partial\zeta}+T_{22}\frac{\partial N_j}{\partial\eta}+T_{23}\frac{\partial N_j}{\partial\xi}\right)|\,J_0\,|\,d\zeta d\eta d\xi$$

$$-2\int_{\xi=-1}^{+1}\int_{\eta=-1}^{+1}\int_{\zeta=-1}^{+1}N_i\left(T_{31}\frac{\partial\bar{\nu}_t}{\partial\zeta}+T_{32}\frac{\partial\bar{\nu}_t}{\partial\eta}+T_{33}\frac{\partial\bar{\nu}_t}{\partial\xi}\right)$$

$$\times\left(P_{11}\frac{\partial N_j}{\partial\zeta}+P_{12}\frac{\partial N_j}{\partial\eta}+P_{13}\frac{\partial N_j}{\partial\xi}\right)|\,J_0\,|\,d\zeta d\eta d\xi$$

$$-2\int_{\xi=-1}^{+1}\int_{\eta=-1}^{+1}\int_{\zeta=-1}^{+1}N_i\left(P_{31}\frac{\partial\bar{\nu}_t}{\partial\zeta}+P_{32}\frac{\partial\bar{\nu}_t}{\partial\eta}+P_{33}\frac{\partial\bar{\nu}_t}{\partial\xi}\right)$$

$$\times\left(T_{31}\frac{\partial N_j}{\partial\zeta}+T_{32}\frac{\partial N_j}{\partial\eta}+T_{33}\frac{\partial N_j}{\partial\xi}\right)|\,J_0\,|\,d\zeta\,d\eta\,d\xi+\bar{u}W_i$$

$$\times\left(P_{11}\frac{\partial N_j}{\partial\zeta}+P_{12}\frac{\partial N_j}{\partial\eta}+P_{13}\frac{\partial N_j}{\partial\xi}\right)|\,J_0\,|d\zeta\,d\eta\,d\xi+\bar{v}W_i$$

$$\times\left(P_{21}\frac{\partial N_j}{\partial\zeta}+P_{22}\frac{\partial N_j}{\partial\eta}+P_{23}\frac{\partial N_j}{\partial\xi}\right)|\,J_0\,|d\zeta\,d\eta\,d\xi+\bar{w}W_i$$

$$\times\left(P_{31}\frac{\partial N_j}{\partial\zeta}+P_{32}\frac{\partial N_j}{\partial\eta}+P_{33}\frac{\partial N_j}{\partial\xi}\right)|\,J_0\,|\,d\zeta\,d\eta\,d\xi$$

$$\bar{a}_{12})_{i,k}=a_{12}{}^{*})_{i,k}+\int_{\xi=-1}^{+1}\int_{\eta=-1}^{+1}\int_{\zeta=-1}^{+1}\frac{1}{\rho}N_i$$

$$\times\left(P_{31}\frac{\partial M_k}{\partial\zeta}+P_{32}\frac{\partial M_k}{\partial\eta}+P_{33}\frac{\partial M_k}{\partial\xi}\right)|\,J_0\,|\,d\zeta\,d\eta\,d\xi$$

$$\bar{a}_{13})_{k,j}=a_{13}{}^{*})_{k,j}+\int_{\xi=-1}^{+1}\int_{\eta=-1}^{+1}\int_{\zeta=-1}^{+1}M_k$$

$$\times\left(P_{11}\frac{\partial N_j}{\partial\zeta}+P_{12}\frac{\partial N_j}{\partial\eta}+P_{13}\frac{\partial N_j}{\partial\xi}\right)|\,J_0\,|\,d\zeta\,d\eta\,d\xi$$

$$\bar{a}_{14})_{k,j}=a_{14}{}^{*})_{k,j}+\int_{\xi=-1}^{+1}\int_{\eta=-1}^{+1}\int_{\zeta=-1}^{+1}M_k$$

$$\times\left(P_{21}\frac{\partial N_j}{\partial\zeta}+P_{22}\frac{\partial N_j}{\partial\eta}+P_{23}\frac{\partial N_j}{\partial\xi}\right)|\,J_0\,|\,d\zeta\,d\eta\,d\xi$$

$$\bar{a}_{15})_{k,j}=a_{15}{}^{*})_{k,j}+\int_{\xi=-1}^{+1}\int_{\eta=-1}^{+1}\int_{\zeta=-1}^{+1}M_k$$

$$\times\left(P_{31}\frac{\partial N_j}{\partial\zeta}+P_{32}\frac{\partial N_j}{\partial\eta}+P_{33}\frac{\partial N_j}{\partial\xi}\right)|\,J_0\,|\,d\zeta\,d\eta\,d\xi$$

$$\bar{q}_1)_i=q_1{}^{*})_i$$

$$\bar{q}_2)_i=q_2{}^{*})_i+\Omega^2\int_{\xi=-1}^{+1}\int_{\eta=-1}^{+1}\int_{\zeta=-1}^{+1}N_i\left(1+\sum_{l=1}^{20}N_l\lambda_l\right)|\,J_0\,|d\zeta\,d\eta\,d\xi$$

$$\bar{q}_3)_i=q_3{}^{*})_i$$

where $i=1,2,3,\ldots,20, j=1,2,3,\ldots,20$, and $k=1,2,3,\ldots,8$.

In the foregoing expressions, a variable with an asterisk is defined to have the same form as its counterpart in the expressions defining the contents of Equation (D.1), with the exception that the Jacobian $|\,J_0\,|$ is now replaced by $|\,J_1\,|$, which was defined earlier in this appendix. Referring to the expressions of the load subvectors (last three preceding expressions), note that only the middle subvector is influenced by the rotor excitation and that this influence is proportional to the rotor whirl frequency Ω.

On assembling all elemental contributions with reference to [Equation (D.2)], the "global" set of finite element equations can consistently be written as follows:

$$\left([A_0]+\epsilon[\bar{A}]\right)\{\Phi_0+\delta\Phi\}=\{Q_0\}+\epsilon\{\bar{Q}\} \tag{E.9}$$

Comparing this system of equations for the displaced-rotor flow field with that associated with the zeroth-order flow equations (D.2), we can isolate the differential changes in the flow variables as follows:

$$[A_0]\{\delta\Phi\} = \epsilon\left(\{\bar{Q}\} - [\bar{A}]\{\Phi_0\}\right) \tag{E.10}$$

in which the second-order term containing the product $\epsilon\{\delta\Phi\}$ was eliminated. Taking the limit of Equation (E.10) as the rotor virtual eccentricity ϵ tends to zero, the rate of change of nodal variables (velocity components and pressure) with respect to ϵ can be expressed as follows:

$$\frac{\partial\{\Phi\}}{\partial\epsilon} = [A_0]^{-1}\left(\{\bar{Q}\} - [\bar{A}]\{\Phi_0\}\right) \tag{E.11}$$

Note that the "global" vector $\{\Phi\}$ and its derivative $\partial\{\phi\}/\partial\epsilon$ can, by definition, be expressed as follows:

$$\{\Phi\} = \begin{Bmatrix} \{U\} \\ \{V\} \\ \{W\} \\ \{P\} \end{Bmatrix} \quad \text{and} \quad \left\{\frac{\partial\Phi}{\partial\epsilon}\right\} = \begin{Bmatrix} \dfrac{\partial\{U\}}{\partial\epsilon} \\ \dfrac{\partial\{V\}}{\partial\epsilon} \\ \dfrac{\partial\{W\}}{\partial\epsilon} \\ \dfrac{\partial\{P\}}{\partial\epsilon} \end{Bmatrix} \tag{E.12}$$

where

- $\{U\}$ is the subvector containing all nodal values of u,
- $\{V\}$ is the subvector containing all nodal values of v,
- $\{W\}$ is the subvector containing all nodal values of w, and
- $\{P\}$ is the subvector containing all nodal values of p.

In the preceding, note that the three velocity components (u, v and w) are all relative to the whirling frame of reference in Figure 16.8.

Examination of matrix Equation (E.11) reveals that the "global" matrix $[\bar{A}]$ needs only to be assembled but not inverted. However its zeroth-order version $[A_0]$ will have to be inverted at this final stage of the perturbation analysis.

APPENDIX F

Rigorous Adaptation to Compressible-Flow Problems

This model upgrade is intended to embrace a variety of high-speed noncontact seals and bearings traditionally used in gas turbine applications. In this case, the heat-energy exchange between the hardware and the working medium may be far from being negligible. This very fact calls for including a separate energy equation in the set of flow-governing equations cited in Appendix E. The boundary conditions on insertion of this equation involve such variables as the local convection heat transfer coefficient and the local "wall" and flow-stream temperatures.

Given the fact that the rotor-to-housing clearance width is extremely small, the problem of friction choking (in a Fanno-process type of mechanism) is indeed part of this compressible flow problem. Unfortunately, the occurence of this choking type requires external intervention during the flow solution process. During the iterative procedure, where the momentum-equations convection terms are continually linearized, and once the nodal magnitudes of velocity vector are attained, the intervention process begins by computing the corresponding nodal magnitudes of Mach number. These are then examined to see if the Mach number is in excess of unity anywhere in the computational domain (the seal exit station in particular), which is impossible in a subsonic nozzle-like passage. Referring to the simple annular seal in Figure 16.20, the term *nozzle* here is applicable in the sense that the blockage effect of the boundary layer growth over the solid walls causes, in effect, a streamwise reduction in the cross-flow area, turning what is physically a constant-area passage into a subsonic nozzle in the sense of rising displacement thickness (a fraction of the boundary layer thickness that depends on the boundary-layer velocity profile). The corrective action in this case is to set the seal exit Mach number to unity and repeat the preceding computational step.

Once the centered-rotor flow field is covered and the finite element equations are derived, the perturbation analysis is term by term applied to the set of new equations in their discrete finite element form. Given the "bulkiness" of the resulting terms (all being volume or surface integrals), a set of special differential operators is defined and used to shorten the expressions of these terms. From a programming viewpoint, numerical determination of each term becomes much simpler once these operators are individually defined in separate subroutines.

Compressible-Flow Upgrade: Centered-Rotor Axisymmetric Flow Field

Governing Equations

This starting point is where the flow field is axisymmetric, and the equations therefore are free of any tangential derivatives. Recall that the term *centered rotor*, as it comes to defining a flow field, is different from the *zeroth-order* phrase because the latter is where the whirl frequency is built in, as discussed earlier in the main text, when describing the series of computational events comprising the perturbation model.

Through the rotor-to-housing gap, the flow process is governed by the conservation equations of mass, linear momentum, and energy. These can be expressed as follows:

Continuity Equation

$$\frac{\partial(\bar{\rho}V_r)}{\partial r}+\frac{\partial(\bar{\rho}V_z)}{\partial z}+\frac{\bar{\rho}V_r}{r} \tag{F.1}$$

Axial Momentum Equation

$$\bar{V}_r\frac{\partial V_z}{\partial r}+\bar{V}_z\frac{\partial V_z}{\partial z}=-\frac{1}{\bar{\rho}}\frac{\partial p}{\partial z}+\nabla.\left(\nu_{eff}\nabla V_z\right)+\frac{\partial \nu_t}{\partial z}\frac{\partial V_z}{\partial z}+\frac{\partial \nu_t}{\partial r}\frac{\partial V_r}{\partial z} \tag{F.2}$$

Radial Momentum Equation

$$\begin{aligned}\bar{V}_r\frac{\partial V_r}{\partial r}+\bar{V}_z\frac{\partial V_r}{\partial z}-\bar{V}_\theta\frac{V_\theta}{r}&=-\frac{1}{\bar{\rho}}\frac{\partial p}{\partial r}+\nabla.\left(\nu_{eff}\nabla V_r\right)\\&+\frac{\partial \nu_t}{\partial r}\frac{\partial V_r}{\partial r}-\frac{\nu_{eff}}{r^2}V_r+\frac{\partial \nu_t}{\partial z}\frac{\partial V_z}{\partial r}\end{aligned} \tag{F.3}$$

Tangential Momentum Equation

$$\bar{V}_r\frac{\partial V_\theta}{\partial r}+\bar{V}_z\frac{\partial V_\theta}{\partial z}+\frac{\bar{V}_rV_\theta}{r}=\nabla.\left(\nu_{eff}\nabla V_\theta\right)-\left[\frac{1}{r}\frac{\partial \nu_t}{\partial r}+\frac{\nu_{eff}}{r^2}\right]V_\theta \tag{F.4}$$

Energy Equation

$$k\nabla^2T=\bar{\rho}c_p\left(\bar{V}_r\frac{\partial T}{\partial r}+\bar{V}_z\frac{\partial T}{\partial z}\right)-\bar{V}_r\frac{\partial p}{\partial r}-\bar{V}_z\frac{\partial p}{\partial z}+\Phi \tag{F.5}$$

where

- V_z, V_r, and V_θ are the axial, radial, and tangential velocity components, respectively,
- p is the static pressure,
- ρ is the static density,
- k is the gas thermal conductivity,
- ν_t is the eddy viscosity coefficient obtained from the turbulence closure,
- ν_{eff} is the effective (molecular plus eddy) viscosity coefficient, and
- Φ is the viscosity-related dissipation function.

In these equations, the overbar signifies a flow variable that is obtained from the previous iteration or an initial guess. In the energy equation, the dissipation function Φ is defined as follows:

$$\Phi = \bar{\rho}\nu_e \left[2\left\{ \left(\frac{\partial \bar{V}_r}{\partial r}\right)^2 + \left(\frac{\bar{V}_r}{r}\right)^2 + \left(\frac{\partial \bar{V}_z}{\partial z}\right)^2 \right\} \right] + \left(\frac{\partial \bar{V}_\theta}{\partial z}\right)^2 + \left(\frac{\partial \bar{V}_r}{\partial z} + \frac{\partial \bar{V}_z}{\partial r}\right)^2$$
$$+ \left(\frac{\partial \bar{V}_\theta}{\partial r} - \frac{\bar{V}_\theta}{r}\right)^2 + \lambda \left(\frac{\partial \bar{V}_r}{\partial r} + \frac{\bar{V}_r}{r} + \frac{\partial \bar{V}_z}{\partial z}\right)^2 \quad \text{(F.6)}$$

where λ is the gas second viscosity coefficient. Note that the nodal magnitudes of Φ will always be lagging by one iterative step because the velocity components (identified by overbars) arise from the previous iteration or an initial guess, as indicated earlier. Introduction of Φ into the finite element equivalent of the governing equations is in the form of a sourcelike term contributing only to the free term of the energy equation.

The turbulence closure to be implemented in this problem is the Baldwin and Lomax vorticity-based model, subject to the near-wall analysis of Benim and Zinser, both discussed in Chapter 13 of the main text (references [6] and [7]).

Boundary Conditions

These are categorized by the type of boundary segments to which they belong and by reference to Figure 16.20 as follows:

Flow Inlet Station

The distribution of velocity components and temperature at this station is assumed to be known a priori. Normally, the average magnitudes of these variables are available. The analyst in this case would superimpose adequate velocity and thermal boundary-layer profiles or otherwise assume a fully developed flow depending on the manner in which the seal is naturally connected to the gas-turbine component containing it.

Exit Station

Over this station, the velocity components are confined to satisfying the boundary condition of zero surface tractions (reference [38]) as follows:

$$\frac{\partial V_z}{\partial z} - \frac{p}{2\rho\nu_{eff}} = 0 \quad \text{(F.7)}$$

$$\frac{\partial V_r}{\partial z} + \frac{\partial V_z}{\partial r} = 0 \quad \text{(F.8)}$$

$$\frac{\partial V_\theta}{\partial z} = 0 \quad \text{(F.9)}$$

In addition, a zero streamwise temperature gradient is imposed, provided that the seal width/length aspect ratio is sufficiently small, a condition that is, by the mere definition of a seal, overwhelmingly applicable. Finally, an arbitrary "datum" magnitude of the pressure is specified at an arbitrary, usually the middle, computational node on this station. This, of course, is a remedial action to a pressure multivaluedness that would, otherwise, prevail.

Solid Boundary Segments

These include the rotor and housing surfaces in contact with the working medium. The velocity components over these surfaces are subject to the *no-slip* boundary conditions as follows:

$$V_z = V_r = 0$$

$$V_\theta = K$$

where K is equal to zero on the housing surface and ωr on the rotor surface, with ω being the rotor speed.

As for the temperature field, the working medium is viewed as steadily transferring heat energy to the solid boundary segments through the convection mechanism of heat transfer, a condition that the following boundary condition reflects:

$$\frac{\partial T}{\partial n} = \frac{h}{k}(T - T_w) \quad \text{(F.10)}$$

where

- n is the direction of the local outward normal unit vector,
- T_w is the local wall temperature, and
- h is the local coefficient of convection heat transfer.

In the preceding, the coefficient h is continually updated during the iterative procedure in terms of the velocity components using the well-known Chilton-Colburn analogy, together with existing empirical correlations involving the relative surface roughness.

Finite Element Formulation

The introductory steps of this section have largely been covered earlier, including the mixed velocity/pressure interpolation representations in the nine-noded isoparametric element in Figure 16.4, the need for and use of the upwind weight functions W_i, and the physical-to-local spatial coordinates' transformation. The difference, however, has to do with the newly added energy equation and, in particular, the functional representation of the static temperature T within a typical finite element. This variable is piecewise interpolated using the element's quadratic shape functions as follows:

$$T^{(e)} = \sum_{i=1}^{9} N_i(\zeta,\eta)T_i$$

Following the guidelines of the Petrov-Galerkin finite element approach, the finite element equivalent of the flow-governing equations is produced as follows:

Continuity Equation

$$\left[\int_{A^{(e)}} \left(M_k \frac{\partial(\rho N_j)}{\partial r} + \frac{\bar{\rho}}{r} M_k N_j\right) r dA\right] V_{r,j} + \left[\int_{A^{(e)}} M_k \frac{\partial(\rho N_j)}{\partial z} r dA\right] V_{z,j} = 0 \tag{F.11}$$

Axial-Momentum Equation

$$\left[\int_{A^{(e)}} \left\{\nu_{eff}\left(\frac{\partial W_i}{\partial r}\frac{\partial N_j}{\partial r} + \frac{\partial W_i}{\partial z}\frac{\partial N_j}{\partial z}\right) + W_i\left(\bar{V}_r \frac{\partial N_j}{\partial r} + \bar{V}_z \frac{\partial N_j}{\partial z}\right) + W_i \frac{\partial \nu_t}{\partial z}\frac{\partial N_j}{\partial z}\right\} r\, dA\right] V_{z,j}$$
$$- \left[\int_{A^{(e)}} W_i \frac{\partial \nu_t}{\partial r}\frac{\partial N_j}{\partial z}\right] V_{r,j} + \left[\int_{A^{(e)}} \frac{1}{\bar{\rho}} W_i \frac{\partial M_k}{\partial r} r\, dA\right] p_k = \oint_{L^{(e)}} \nu_{eff} r W_i \frac{\partial V_z}{\partial n}\, dL \tag{F.12}$$

Radial-Momentum Equation

$$\left[\int_{A^{(e)}} \left\{\nu_{eff}\left(\frac{\partial W_i}{\partial r}\frac{\partial N_j}{\partial r} + \frac{\partial W_i}{\partial z}\frac{\partial N_j}{\partial z}\right) + W_i\left(\bar{V}_r \frac{\partial N_j}{\partial r} + \bar{V}_z \frac{\partial N_j}{\partial z}\right)\right.\right.$$
$$\left.\left. - W_i \frac{\partial \nu_t}{\partial r}\frac{\partial N_j}{\partial r} + \frac{\nu_{eff}}{r^2} W_i N_j\right\} r\, dA\right] V_{r,j} - \left[\int_{A^{(e)}} \frac{\bar{V}_\theta}{r} W_i N_j r\, dA\right] V_{\theta,j}$$
$$- \left[\int_{A^{(e)}} W_i \frac{\partial \nu_t}{\partial z}\frac{\partial N_j}{\partial r} r\, dA\right] V_{z,j} + \left[\int_{A^{(e)}} \frac{1}{\bar{\rho}} W_i \frac{\partial M_k}{\partial r} r dA\right] p_k = \oint_{L^{(e)}} \nu_{eff} r W_i \frac{\partial V_r}{\partial n}\, dL \tag{F.13}$$

Tangential-Momentum Equation

$$\left[\left\{\nu_{eff}\left(\frac{\partial W_i}{\partial r}\frac{\partial N_j}{\partial r} + \frac{\partial W_i}{\partial z}\frac{\partial N_j}{\partial z}\right) + W_i\left(\bar{V}_r \frac{\partial N_j}{\partial r} + \bar{V}_z \frac{\partial N_j}{\partial z}\right)\right.\right.$$
$$\left.\left. + \left(\frac{1}{r}\frac{\partial \nu_t}{\partial r} + \frac{\nu_{eff}}{r^2} + \frac{\bar{V}_r}{r}\right) W_i N_j\right\} r\, dA\right] V_{\theta,j} = \oint_{L^{(e)}} \nu_{eff} r W_i \frac{\partial V_\theta}{\partial n}\, dL \tag{F.14}$$

Energy Equation

$$\left[\int_{A^{(e)}} \left\{k\left(\frac{\partial W_i}{\partial r}\frac{\partial N_j}{\partial r} + \frac{\partial W_i}{\partial z}\frac{\partial N_j}{\partial z}\right) + W_i \bar{\rho} c_p \left(\bar{V}_r \frac{\partial N_j}{\partial r} + \bar{V}_z \frac{\partial N_j}{\partial z}\right)\right\} r\, dA\right] T_j$$
$$- \left[\int_{A^{(e)}} W_i \left(\bar{V}_r \frac{\partial M_k}{\partial r} + \bar{V}_z \frac{\partial M_k}{\partial z}\right) r\, dA\right] p_k = \oint_{L^{(e)}} k r W_i \frac{\partial T}{\partial n}\, dL$$
$$- \oint_{L^{(e)}} N_i \bar{h} T_w\, dL - \int_{A^{(e)}} W_i \Phi r\, dA \tag{F.15}$$

where

- n is the direction of the local unit vector perpendicular to the element boundary,
- $dA = dz dr = \mid J_0 \mid d\zeta\, d\eta$,

- $| J_0 |$ is the centered-rotor Jacobian of coordinate transformation,
- $\bar{h}$ is the local convection heat transfer coefficient based on the just-computed flow solution or obtained from an initial guess,
- T_w is the local wall temperature,
- $i = 1,2,3,\ldots,9$,
- $j = 1,2,3,\ldots,9$, and
- $k = 1,2,3,4$.

The last term on the right-hand side of the energy equation is proportional to the dissipation function, which is always known from the previous iteration or an initial guess. As discussed earlier in this appendix, this term is treated as a source-like term contributing to the free term that is associated with applying the energy equation at a typical computational node. The term that precedes it will appear only for elements sharing one side with a solid wall, where heat energy is transferred by convection between the working medium and the wall. Also note that

1. The convection heat transfer coefficient h and the viscosity-related dissipation function Φ are both based on an already obtained solution (during the iterative procedure) or an initial guess. As for the density ρ, the equation of state ($\rho = p/(RT)$) is to be applied at the end of each iterative step, which is when the nodal pressure and temperature magnitudes are known.
2. Before proceeding from one iteration to the next, the Mach number should be computed at the seal exit station to detect friction choking, if any, and invoke the previously indicated corrective actions, if warranted. Note that the existence of small supersonic pockets near and at the seal exit station is not indicative of choking as long as a subsonic-flow seal exit segment exists.

Compressible-Flow Upgrade: Zeroth-Order Flow Field

Earlier in the main text and under this same subtitle, details of this computational procedure, including departure from the cylindrical frame of reference to the coordinate axes in Figure 16.10, which are spinning at the whirl-frequency magnitude, insertion of the additional (Coriolis and centripetal) acceleration components, and adaptation of the boundary conditions, were all presented. Applying the same analytical steps to the current set of equations, the zeroth-order set of finite element equations can be cast as follows:

$$\begin{bmatrix} [a_1] & [a_2] & [a_3] & [a_4] & 0 \\ [a_5] & [a_6] & [a_7] & [a_8] & 0 \\ [a_9] & [a_{10}] & [a_{11}] & [a_{12}] & 0 \\ [a_{13}] & [a_{14}] & [a_{15}] & 0 & 0 \\ 0 & 0 & 0 & [a_{16}] & [a_{17}] \end{bmatrix} \begin{Bmatrix} \{u\} \\ \{v\} \\ \{w\} \\ \{P\} \\ \{T\} \end{Bmatrix} = \begin{Bmatrix} \{q_1\} \\ \{q_2\} \\ \{q_3\} \\ 0 \\ \{q_4\} \end{Bmatrix} \tag{F.16}$$

Note that the transition to formulate the zeroth-order flow field, as discussed earlier, is where the alternate (three-dimensional Cartesian) frame of reference is used, the axisymmetric velocity magnitudes consistently adapted, the three-dimensional isoparametric element in Figure 16.12 is instead used, and the whirl frequency Ω introduced (Figure 16.10) as an in-plane rotation of the x and y axes. It follows that

1. The number of unknown variables in the preceding matrix equation, by reference to Figure 16.12, is 88, distributed as 20 nodal values of each of the three velocity components and temperature, as well as 8 nodal values of the pressure at the element corner nodes.
2. Components of the centripetal acceleration ($\omega^2 r$) due to the coordinate axes' rotation will have to appear as free terms in the set of zeroth-order finite element equations (to follow).

Referring to the same earlier section, the following derivative-conversion relationship can be deduced from Equation (16.33) by setting to zero the rotor-axis eccentricity ϵ as follows:

$$\begin{Bmatrix} \frac{\partial}{\partial x} \\ \frac{\partial}{\partial y} \\ \frac{\partial}{\partial z} \end{Bmatrix} = \begin{bmatrix} T_{1,1} & T_{1,2} & T_{1,3} \\ T_{2,1} & T_{2,2} & T_{2,3} \\ T_{3,1} & T_{3,2} & T_{3,3} \end{bmatrix} \begin{Bmatrix} \frac{\partial}{\partial \zeta} \\ \frac{\partial}{\partial \eta} \\ \frac{\partial}{\partial \xi} \end{Bmatrix}$$

where all contents of the matrix $[T]$, meaning $T_{1,1}$, $T_{1,2}$, and so on were also defined.

Now let us define the differential operators $\mathbf{F_1}$, $\mathbf{F_2}$, and $\mathbf{F_3}$ as follows:

$$\mathbf{F_1} = T_{1,1}\frac{\partial}{\partial \zeta} + T_{1,2}\frac{\partial}{\partial \eta} + T_{1,3}\frac{\partial}{\partial \xi}$$

$$\mathbf{F_2} = T_{2,1}\frac{\partial}{\partial \zeta} + T_{2,2}\frac{\partial}{\partial \eta} + T_{1,3}\frac{\partial}{\partial \xi}$$

$$\mathbf{F_3} = T_{3,1}\frac{\partial}{\partial \zeta} + T_{3,2}\frac{\partial}{\partial \eta} + T_{3,3}\frac{\partial}{\partial \xi}$$

With these definitions, the different submatrices and subvectors in Equation (F.16) can be expressed as follows:

$$\begin{aligned} a_1)_{i,j} = & \int_{\xi=-1}^{+1}\int_{\eta=-1}^{+1}\int_{\zeta=-1}^{+1} \nu_e \left\{ \sum_{k=1}^{3} \left[\mathbf{F_k}(N_i)\mathbf{F_k}(N_j)\right] \right\} \\ & - N_i \left\{ \sum_{k=1}^{3} \left[\mathbf{F_k}(\nu_t)\mathbf{F_k}(N_j)\right] \right\} \mid J_0 \mid \, d\zeta \, d\eta \, d\xi \\ & - \int_{\xi=-1}^{+1}\int_{\eta=-1}^{+1}\int_{\zeta=-1}^{+1} N_i \left\{\mathbf{F_1}(\nu_t)\mathbf{F_1}(N_j)\right\} \mid J_0 \mid \, d\zeta \, d\eta \, d\xi \\ & + \int_{\xi=-1}^{+1}\int_{\eta=-1}^{+1}\int_{\zeta=-1}^{+1} N_i \left\{\bar{u}\mathbf{F_1}(N_j) + \bar{v}\mathbf{F_2}(N_j) + \bar{w}\mathbf{F_3}(N_j)\right\} \\ & + \lambda \left\{ \sum_{k=1}^{3} \left[\mathbf{F_k}(N_i)\mathbf{F_k}(N_j)\right] \right\} \mid J_0 \mid \, d\zeta \, d\eta \, d\xi \end{aligned}$$

$$a_2)_{i,j} = \int_{\xi=-1}^{+1}\int_{\eta=-1}^{+1}\int_{\zeta=-1}^{+1} \left(-N_i\left\{\mathbf{F_2}(\nu_t)\mathbf{F_1}(N_j)\right\} - 2\Omega N_i N_j\right) \mid J_0 \mid \; d\zeta\, d\eta d\xi$$

$$a_3)_{i,j} = \int_{\xi=-1}^{+1}\int_{\eta=-1}^{+1}\int_{\zeta=-1}^{+1} \left(-N_i\mathbf{F_3}(\nu_t)\mathbf{F_1}(N_j)\right) \mid J_0 \mid \; d\zeta\, d\eta d\xi$$

$$a_4)_{i,k} = \int_{\xi=-1}^{+1}\int_{\eta=-1}^{+1}\int_{\zeta=-1}^{+1} \left(\frac{1}{\bar{\rho}}N_i\mathbf{F_1}(M_k)\right) \mid J_0 \mid \; d\zeta\, d\eta d\xi$$

$$a_5)_{i,j} = \int_{\xi=-1}^{+1}\int_{\eta=-1}^{+1}\int_{\zeta=-1}^{+1} \left(-N_i\mathbf{F_1}(\nu_t)\mathbf{F_2}(N_j) + 2\Omega N_i N_j\right) \mid J_0 \mid \; d\zeta\, d\eta d\xi$$

$$\begin{aligned}
a_6)_{i,j} = &\int_{\xi=-1}^{+1}\int_{\eta=-1}^{+1}\int_{\zeta=-1}^{+1} \nu_e\left\{\sum_{k=1}^{3}\left[\mathbf{F_k}(N_i)\mathbf{F_k}(N_j)\right]\right\} \\
&- N_i\left\{\sum_{k=1}^{3}\left[\mathbf{F_k}(\nu_t)\mathbf{F_k}(N_j)\right]\right\} \mid J_0 \mid \; d\zeta\, d\eta d\xi \\
&- \int_{\xi=-1}^{+1}\int_{\eta=-1}^{+1}\int_{\zeta=-1}^{+1} N_i\left\{\mathbf{F_2}(\nu_t)\mathbf{F_2}(N_j)\right\} \mid J_0 \mid d\zeta\, d\eta d\xi \\
&+ \int_{\xi=-1}^{+1}\int_{\eta=-1}^{+1}\int_{\zeta=-1}^{+1} N_i\left\{\bar{u}\mathbf{F_1}(N_j) + \bar{v}\mathbf{F_2}(N_j) + \bar{w}\mathbf{F_3}(N_j)\right\} \\
&+ \lambda\left\{\sum_{k=1}^{3}\left[\mathbf{F_k}(N_i)\mathbf{F_k}(N_j)\right]\right\} \mid J_0 \mid \; d\zeta\, d\eta d\xi
\end{aligned}$$

$$a_7)_{i,j} = \int_{\xi=-1}^{+1}\int_{\eta=-1}^{+1}\int_{\zeta=-1}^{+1} \left(-N_i\mathbf{F_3}(\nu_t)\mathbf{F_2}(N_j)\right) \mid J_0 \mid \; d\zeta\, d\eta d\xi$$

$$a_8)_{i,k} = \int_{\xi=-1}^{+1}\int_{\eta=-1}^{+1}\int_{\zeta=-1}^{+1} \left(\frac{1}{\bar{\rho}}N_i\mathbf{F_2}(M_k)\right) \mid J_0 \mid \; d\zeta\, d\eta d\xi$$

$$a_9)_{i,j} = \int_{\xi=-1}^{+1}\int_{\eta=-1}^{+1}\int_{\zeta=-1}^{+1} \left(-N_i\mathbf{F_1}(\nu_t)\mathbf{F_3}(N_j)\right) \mid J_0 \mid \; d\zeta\, d\eta d\xi$$

$$a_{10})_{i,j} = \int_{\xi=-1}^{+1}\int_{\eta=-1}^{+1}\int_{\zeta=-1}^{+1} \left(-N_i\mathbf{F_2}(\nu_t)\mathbf{F_3}(N_j)\right) \mid J_0 \mid \; d\zeta\, d\eta d\xi$$

$$\begin{aligned}
a_{11})_{i,j} = &\int_{\xi=-1}^{+1}\int_{\eta=-1}^{+1}\int_{\zeta=-1}^{+1} \nu_e\left\{\sum_{k=1}^{3}\left[\mathbf{F_k}(N_i)\mathbf{F_k}(N_j)\right]\right\} \\
&- N_i\left\{\sum_{k=1}^{3}\left[\mathbf{F_k}(\nu_t)\mathbf{F_k}(N_j)\right]\right\} \mid J_0 \mid \; d\zeta\, d\eta d\xi \\
&- \int_{\xi=-1}^{+1}\int_{\eta=-1}^{+1}\int_{\zeta=-1}^{+1} N_i\left\{\mathbf{F_3}(\nu_t)\mathbf{F_3}(N_j)\right\} \mid J_0 \mid \; d\zeta\, d\eta d\xi \\
&+ \int_{\xi=-1}^{+1}\int_{\eta=-1}^{+1}\int_{\zeta=-1}^{+1} N_i\left\{\bar{u}\mathbf{F_1}(N_j) + \bar{v}\mathbf{F_2}(N_j) + \bar{w}\mathbf{F_3}(N_j)\right\} \\
&+ \lambda\left\{\sum_{k=1}^{3}\left[\mathbf{F_k}(N_i)\mathbf{F_k}(N_j)\right]\right\} \mid J_0 \mid \; d\zeta\, d\eta d\xi
\end{aligned}$$

$$a_{12})_{i,k} = \int_{\xi=-1}^{+1} \int_{\eta=-1}^{+1} \int_{\zeta=-1}^{+1} (N_i \mathbf{F_3}(M_k)) \mid J_0 \mid \, d\zeta \, d\eta \, d\xi$$

$$a_{13})_{k,j} = \int_{\xi=-1}^{+1} \int_{\eta=-1}^{+1} \int_{\zeta=-1}^{+1} \left(M_k \mathbf{F_1}(N_j)\right) \mid J_0 \mid \, d\zeta \, d\eta \, d\xi$$

$$a_{14})_{k,j} = \int_{\xi=-1}^{+1} \int_{\eta=-1}^{+1} \int_{\zeta=-1}^{+1} \left(M_k \mathbf{F_2}(N_j)\right) \mid J_0 \mid \, d\zeta \, d\eta \, d\xi$$

$$a_{15})_{k,j} = \int_{\xi=-1}^{+1} \int_{\eta=-1}^{+1} \int_{\zeta=-1}^{+1} \left(M_k \mathbf{F_3}(N_j)\right) \mid J_0 \mid \, d\zeta \, d\eta \, d\xi$$

$$a_{16})_{i,j} = \int_{\xi=-1}^{+1} \int_{\eta=-1}^{+1} \int_{\zeta=-1}^{+1} \left(k \left\{ \sum_{k=1}^{3} \left[\mathbf{F_k}(N_i) \mathbf{F_k}(N_j) \right] \right\} \right.$$
$$\left. + \bar{\rho} c_p N_i \left\{ \bar{u} \mathbf{F_1}(N_j) + \bar{v} \mathbf{F_2}(N_j) + \bar{w} \mathbf{F_3}(N_j) \right\} \right) \mid J_0 \mid \, d\zeta \, d\eta \, d\xi - \int_{S^{(e)}} \left(h N_i N_j \right) dS$$

$$a_{17})_{i,k} = \int_{\xi=-1}^{+1} \int_{\eta=-1}^{+1} \int_{\zeta=-1}^{+1} (-N_i \{ \bar{u} \mathbf{F_1}(M_k) + \bar{v} \mathbf{F_2}(M_k) + \bar{w} \mathbf{F_3}(M_k) \}) \mid J_0 \mid \, d\zeta \, d\eta \, d\xi$$

$$q_1)_i = \int_{S^{(e)}} \left(\nu_e N_i \frac{\partial u}{\partial n} \right) dS + \int_{\xi=-1}^{+1} \int_{\eta=-1}^{+1} \int_{\zeta=-1}^{+1}$$
$$\times \left(\Omega^2 N_i \left[\sum_{k=1}^{20} N_k x_k \right] \right) \mid J_0 \mid \, d\zeta \, d\eta \, d\xi$$

$$q_2)_i = \int_{S^{(e)}} \left(\nu_e N_i \frac{\partial v}{\partial n} \right) dS + \int_{\xi=-1}^{+1} \int_{\eta=-1}^{+1} \int_{\zeta=-1}^{+1}$$
$$\times \left(\Omega^2 N_i \left[\sum_{k=1}^{20} N_k y_k \right] \right) \mid J_0 \mid \, d\zeta \, d\eta \, d\xi$$

$$q_3)_i = \int_{S^{(e)}} \nu_e N_i \frac{\partial w}{\partial n} \, dS$$

$$q_4)_i = \int_{S^{(e)}} (-h T_w N_i) \, dS + \int_{\xi=-1}^{+1} \int_{\eta=-1}^{+1} \int_{\zeta=-1}^{+1} (\Phi N_i) \mid J_0 \mid \, d\zeta \, d\eta \, d\xi$$

In reviewing these submatrix and subvector expressions, the following clarifications may be in order:

1. The subscripts i and j vary from 1 to 20, whereas the subscript k varies from 1 to 8.
2. Appearance of the centripetal acceleration components in the "load" vector of free terms was expected to belong to the x- and y-momentum equations (i.e., to the subvectors q_1 and q_2), and they do.
3. Only elements sharing boundaries with uninsulated solid walls will contribute to the last term in the a_{16} expression and the first term in the q_4 expression.
4. Construction of the zeroth-order set of finite element equations (F.11) (by reference to the perturbation analysis) is required, but their solution is not. This, as discussed in the main text, is true in the sense that the flow solution is

independent of the choice of coordinate axes and whether they are stationary or rotating.

5. $| J_0 |$ is the unperturbed Jacobian of physical-to-local coordinate conversion

$$| J_0 |= \begin{vmatrix} \sum_{i=1}^{20} \frac{\partial N_i}{\partial \zeta} x_i & \sum_{i=1}^{20} \frac{\partial N_i}{\partial \zeta} y_i & \sum_{i=1}^{20} \frac{\partial N_i}{\partial \zeta} z_i \\ \sum_{i=1}^{20} \frac{\partial N_i}{\partial \eta} x_i & \sum_{i=1}^{20} \frac{\partial N_i}{\partial \eta} y_i & \sum_{i=1}^{20} \frac{\partial N_i}{\partial \eta} z_i \\ \sum_{i=1}^{20} \frac{\partial N_i}{\partial \xi} x_i & \sum_{i=1}^{20} \frac{\partial N_i}{\partial \xi} y_i & \sum_{i=1}^{20} \frac{\partial N_i}{\partial \xi} z_i \end{vmatrix}$$

The set of Equations (F.16) for the typical element (e) can be rewritten in the following compact form:

$$[a_0]\{\psi\} = \{q_0\} \tag{F.17}$$

Equation (F.17) is similar to its incompressible-flow counterpart. The vector of unknowns $\{\phi\}$ in the latter is replaced by $\{\psi\}$ as a reminder of the fact that it now contains the nodal temperature magnitudes. The subscript 0 in Equation (F.17) signifies the fact that this equation corresponds to the zeroth-order undisplaced-rotor flow field (see Figure 16.10). Assembled among all elements, the "globa" version of Equation (F.17) can be written, in consistency with Equation (16.29), as follows:

$$[A_0]\{\Psi\} = \{Q_0\} \tag{F.18}$$

Compressible-Flow Upgrade: Displaced-Rotor Flow Field: Perturbation Analysis

Once the rotor axis encounters its infinitesimally small lateral eccentricity ϵ, the zeroth-order Equation (F.17) for a typical finite element experiences a small perturbation that is a result of the virtual element-shape distortion in Figure 16.13. The perturbed equation can be expressed as follows:

$$([a_0] + \epsilon[\bar{a}])(\{\psi_0\} + \{\delta\psi\}) = (\{q_0\} + \epsilon\,\{\bar{q}\}) \tag{F.19}$$

Our task, therefore, is to define the perturbation matrix $[\bar{a}]$ and vector $\{\bar{q}\}$ following the same approach explained earlier in the main text.

Because the finite element undergoes the type of deformation depicted in Figure 16.13, the physical-to-local conversion of spatial coordinates is dictated by Equation (E.4), which, for convenience purposes, is repeated here:

$$\begin{Bmatrix} \frac{\partial}{\partial x} \\ \frac{\partial}{\partial y} \\ \frac{\partial}{\partial z} \end{Bmatrix} = ([T(\zeta,\eta,\xi)] + \epsilon[P(\zeta,\eta,\xi)]) \begin{Bmatrix} \frac{\partial}{\partial \zeta} \\ \frac{\partial}{\partial \eta} \\ \frac{\partial}{\partial \xi} \end{Bmatrix}$$

where the matrices $[T]$ and $[P]$ are defined by Equations (E.5) and (E.6). Similar to the derivative operators $\mathbf{F_1}$, $\mathbf{F_2}$, and $\mathbf{F_3}$ in the preceding section, let us define the set of operators $\mathbf{G_1}$, $\mathbf{G_2}$, and $\mathbf{G_3}$ as follows:

$$\mathbf{G_1} = P_{1,1}\frac{\partial}{\partial\zeta} + P_{1,2}\frac{\partial}{\partial\eta} + P_{1,3}\frac{\partial}{\partial\xi}$$
$$\mathbf{G_2} = P_{2,1}\frac{\partial}{\partial\zeta} + P_{2,2}\frac{\partial}{\partial\eta} + P_{2,3}\frac{\partial}{\partial\xi}$$
$$\mathbf{G_3} = P_{3,1}\frac{\partial}{\partial\zeta} + P_{3,2}\frac{\partial}{\partial\eta} + P_{3,3}\frac{\partial}{\partial\xi}$$

With these operators, and considering a typical finite element (e), the perturbation matrix $[\bar{a}]$ and vector $\{\bar{q}\}$ can be written as follows:

$$\begin{bmatrix} [\bar{a_1}] & [\bar{a_2}] & [\bar{a_3}] & [\bar{a_4}] & 0 \\ [\bar{a_5}] & [\bar{a_6}] & [\bar{a_7}] & [\bar{a_8}] & 0 \\ [\bar{a_9}] & [\bar{a_{10}}] & [\bar{a_{11}}] & [\bar{a_{12}}] & 0 \\ [\bar{a_{13}}] & [\bar{a_{14}}] & [\bar{a_{15}}] & 0 & 0 \\ 0 & 0 & 0 & [\bar{a_{16}}] & [\bar{a_{17}}] \end{bmatrix} \begin{Bmatrix} \{U\} \\ \{V\} \\ \{W\} \\ \{P\} \\ \{T\} \end{Bmatrix} = \begin{Bmatrix} \{\bar{q_1}\} \\ \{\bar{q_2}\} \\ \{\bar{q_3}\} \\ 0 \\ \{\bar{q_4}\} \end{Bmatrix} \quad \text{(F.20)}$$

The different submatrices and subvectors in Equation (F.20) are defined as follows:

$$\bar{a_1})_{i,j} = a_1{}^*)_{i,j} + \int_{\xi=-1}^{+1}\int_{\eta=-1}^{+1}\int_{\zeta=-1}^{+1} \nu_e \times \left\{ \sum_{k=1}^{3} \left[\mathbf{F_k}(N_i)\mathbf{G_k}(N_j) + \mathbf{G_k}(N_i)\mathbf{F_k}(N_j)\right] \right\} \mid J_0 \mid \, d\zeta d\eta d\xi$$
$$- \int_{\xi=-1}^{+1}\int_{\eta=-1}^{+1}\int_{\zeta=-1}^{+1} N_i \left\{ \sum_{k=1}^{3} \left[\mathbf{F_k}(\nu_t)\mathbf{G_k}(N_j) + \mathbf{G_k}(\nu_t)\mathbf{F_k}(N_j)\right] \right\} \mid J_0 \mid \, d\zeta d\eta d\xi$$
$$- \int_{\xi=-1}^{+1}\int_{\eta=-1}^{+1}\int_{\zeta=-1}^{+1} N_i \left\{ \mathbf{F_1}(\nu_t)\mathbf{G_1}(N_j) + \mathbf{G_1}(\nu_t)\mathbf{F_1}(N_j) \right\} \mid J_0 \mid \, d\zeta d\eta d\xi$$
$$+ \int_{\xi=-1}^{+1}\int_{\eta=-1}^{+1}\int_{\zeta=-1}^{+1} N_i \left\{ \bar{u}\mathbf{G_1}(N_j) + \bar{v}\mathbf{G_2}(N_j) + \bar{w}\mathbf{G_3}(N_j) \right\} \mid J_0 \mid \, d\zeta d\eta d\xi$$
$$+ \int_{\xi=-1}^{+1}\int_{\eta=-1}^{+1}\int_{\zeta=-1}^{+1} \lambda \left\{ \sum_{k=1}^{3} \left[\mathbf{F_k}(N_i)\mathbf{G_k}(N_j) + \mathbf{G_k}(N_i)\mathbf{F_k}(N_j)\right] \right\} \mid J_0 \mid \, d\zeta d\eta d\xi$$

$$\bar{a_2})_{i,j} = a_2{}^*)_{i,j} + \int_{\xi=-1}^{+1}\int_{\eta=-1}^{+1}\int_{\zeta=-1}^{+1} \left(-N_i\left\{\mathbf{F_2}(\nu_t)\mathbf{G_1}(N_j) + \mathbf{G_2}(\nu_t)\mathbf{F_1}(N_j)\right\}\right) \mid J_0 \mid \, d\zeta d\eta d\xi$$

$$\bar{a_3})_{i,j} = a_3{}^*)_{i,j} + \int_{\xi=-1}^{+1}\int_{\eta=-1}^{+1}\int_{\zeta=-1}^{+1} \left(-N_i\left\{\mathbf{F_3}(\nu_t)\mathbf{G_1}(N_j) + \mathbf{G_3}(\nu_t)\mathbf{F_1}(N_j)\right\}\right) \mid J_0 \mid \, d\zeta d\eta d\xi$$

$$\bar{a_4})_{i,k} = a_4{}^*)_{i,k} + \int_{\xi=-1}^{+1}\int_{\eta=-1}^{+1}\int_{\zeta=-1}^{+1} \left(\frac{1}{\bar{\rho}} N_i\{\mathbf{G_1}(M_k)\}\right) \mid J_0 \mid \, d\zeta d\eta d\xi$$

$$\bar{a_5})_{i,j} = a_5{}^*)_{i,j} + \int_{\xi=-1}^{+1}\int_{\eta=-1}^{+1}\int_{\zeta=-1}^{+1} \left(-N_i\left\{\mathbf{F_1}(\nu_t)\mathbf{G_2}(N_j) + \mathbf{G_1}(\nu_t)\mathbf{F_2}(N_j)\right\}\right) \mid J_0 \mid \, d\zeta d\eta d\xi$$

$$\bar{a_6})_{i,j} = a_6{}^*)_{i,j} + \int_{\xi=-1}^{+1}\int_{\eta=-1}^{+1}\int_{\zeta=-1}^{+1} \nu_e$$

$$\times \left\{ \sum_{k=1}^{3} \left[\mathbf{F_k}(N_i)\mathbf{G_k}(N_j) + \mathbf{G_k}(N_i)\mathbf{F_k}(N_j) \right] \right\} \mid J_0 \mid \; d\zeta\, d\eta d\xi$$

$$- \int_{\xi=-1}^{+1}\int_{\eta=-1}^{+1}\int_{\zeta=-1}^{+1} N_i \left\{ \sum_{k=1}^{3} \left[\mathbf{F_k}(\nu_t)\mathbf{G_k}(N_j) + \mathbf{G_k}(\nu_t)\mathbf{F_k}(N_j) \right] \right\} \mid J_0 \mid \; d\zeta\, d\eta d\xi$$

$$- \int_{\xi=-1}^{+1}\int_{\eta=-1}^{+1}\int_{\zeta=-1}^{+1} N_i \left\{ \mathbf{F_2}(\nu_t)\mathbf{G_2}(N_j) + \mathbf{G_2}(\nu_t)\mathbf{F_2}(N_j) \right\} \mid J_0 \mid \; d\zeta\, d\eta d\xi$$

$$+ \int_{\xi=-1}^{+1}\int_{\eta=-1}^{+1}\int_{\zeta=-1}^{+1} N_i \left\{ \bar{u}\mathbf{G_1}(N_j) + \bar{v}\mathbf{G_2}(N_j) + \bar{w}\mathbf{G_3}(N_j) \right\} \mid J_0 \mid \; d\zeta\, d\eta d\xi$$

$$+ \int_{\xi=-1}^{+1}\int_{\eta=-1}^{+1}\int_{\zeta=-1}^{+1} \lambda \left\{ \sum_{k=1}^{3} \left[\mathbf{F_k}(N_i)\mathbf{G_k}(N_j) + \mathbf{G_k}(N_i)\mathbf{F_k}(N_j) \right] \right\} \mid J_0 \mid \; d\zeta\, d\eta d\xi$$

$$\bar{a_7})_{i,j} = a_7{}^*)_{i,j} + \int_{\xi=-1}^{+1}\int_{\eta=-1}^{+1}\int_{\zeta=-1}^{+1} \left(-N_i \left\{ \mathbf{F_3}(\nu_t)\mathbf{G_2}(N_j) + \mathbf{G_3}(\nu_t)\mathbf{F_2}(N_j) \right\} \right) \mid J_0 \mid \; d\zeta\, d\eta d\xi$$

$$\bar{a_8})_{i,k} = a_8{}^*)_{i,k} + \int_{\xi=-1}^{+1}\int_{\eta=-1}^{+1}\int_{\zeta=-1}^{+1} \left(\frac{1}{\bar{\rho}} N_i \{\mathbf{G_2}(M_k)\} \right) \mid J_0 \mid \; d\zeta\, d\eta d\xi$$

$$\bar{a_9})_{i,j} = a_9{}^*)_{i,j} + \int_{\xi=-1}^{+1}\int_{\eta=-1}^{+1}\int_{\zeta=-1}^{+1} \left(-N_i \left\{ \mathbf{F_1}(\nu_t)\mathbf{G_3}(N_j) + \mathbf{G_1}(\nu_t)\mathbf{F_3}(N_j) \right\} \right) \mid J_0 \mid \; d\zeta\, d\eta d\xi$$

$$\bar{a_{10}})_{i,j} = a_{10}{}^*)_{i,j} + \int_{\xi=-1}^{+1}\int_{\eta=-1}^{+1}\int_{\zeta=-1}^{+1}$$

$$\times \left(-N_i \left\{ \mathbf{F_2}(\nu_t)\mathbf{G_3}(N_j) + \mathbf{G_2}(\nu_t)\mathbf{F_3}(N_j) \right\} \right) \mid J_0 \mid \; d\zeta\, d\eta d\xi$$

$$\bar{a_{11}})_{i,j} = a_{11}{}^*)_{i,j} + \int_{\xi=-1}^{+1}\int_{\eta=-1}^{+1}\int_{\zeta=-1}^{+1} \nu_e$$

$$\times \left\{ \sum_{k=1}^{3} \left[\mathbf{F_k}(N_i)\mathbf{G_k}(N_j) + \mathbf{G_k}(N_i)\mathbf{F_k}(N_j) \right] \right\} \mid J_0 \mid \; d\zeta\, d\eta d\xi$$

$$- \int_{\xi=-1}^{+1}\int_{\eta=-1}^{+1}\int_{\zeta=-1}^{+1} N_i \left\{ \sum_{k=1}^{3} \left[\mathbf{F_k}(\nu_t)\mathbf{G_k}(N_j) + \mathbf{G_k}(\nu_t)\mathbf{F_k}(N_j) \right] \right\} \mid J_0 \mid \; d\zeta\, d\eta d\xi$$

$$- \int_{\xi=-1}^{+1}\int_{\eta=-1}^{+1}\int_{\zeta=-1}^{+1} N_i \left\{ \mathbf{F_3}(\nu_t)\mathbf{G_3}(N_j) + \mathbf{G_3}(\nu_t)\mathbf{F_3}(N_j) \right\} \mid J_0 \mid \; d\zeta\, d\eta d\xi$$

$$+ \int_{\xi=-1}^{+1}\int_{\eta=-1}^{+1}\int_{\zeta=-1}^{+1} N_i \left\{ \bar{u}\mathbf{G_1}(N_j) + \bar{v}\mathbf{G_2}(N_j) + \bar{w}\mathbf{G_3}(N_j) \right\} \mid J_0 \mid \; d\zeta\, d\eta d\xi$$

$$+ \int_{\xi=-1}^{+1}\int_{\eta=-1}^{+1}\int_{\zeta=-1}^{+1} \lambda \left\{ \sum_{k=1}^{3} \left[\mathbf{F_k}(N_i)\mathbf{G_k}(N_j) + \mathbf{G_k}(N_i)\mathbf{F_k}(N_j) \right] \right\} \mid J_0 \mid \; d\zeta\, d\eta d\xi$$

$$\bar{a_{12}})_{i,k} = a_{12}{}^*)_{i,k} + \int_{\xi=-1}^{+1}\int_{\eta=-1}^{+1}\int_{\zeta=-1}^{+1} \left(\frac{1}{\bar{\rho}} N_i \{\mathbf{G_3}(M_k)\} \right) \mid J_0 \mid d\zeta\; d\eta d\xi$$

$$\bar{a_{13}})_{k,j} = a_{13}{}^{*})_{k,j} + \int_{\xi=-1}^{+1}\int_{\eta=-1}^{+1}\int_{\zeta=-1}^{+1} \left(\bar{\rho} M_k\{\mathbf{G_1}(N_j)\}\right) | J_0 | \; d\zeta d\eta d\xi$$

$$\bar{a_{14}})_{k,j} = a_{14}{}^{*})_{k,j} + \int_{\xi=-1}^{+1}\int_{\eta=-1}^{+1}\int_{\zeta=-1}^{+1} \left(\bar{\rho} M_k\{\mathbf{G_2}(N_j)\}\right) | J_0 | \; d\zeta d\eta d\xi$$

$$\bar{a_{15}})_{k,j} = a_{15}{}^{*})_{k,j} + \int_{\xi=-1}^{+1}\int_{\eta=-1}^{+1}\int_{\zeta=-1}^{+1} \left(\bar{\rho} M_k\{\mathbf{G_3}(N_j)\}\right) | J_0 | \; d\zeta d\eta d\xi$$

$$\bar{a_{16}})_{i,j} = a_{16}{}^{*})_{i,j} + \int_{\xi=-1}^{+1}\int_{\eta=-1}^{+1}\int_{\zeta=-1}^{+1} \left(k \left\{ \sum_{k=1}^{3} \left[\mathbf{F_k}(N_i)\mathbf{G_k}(N_j) + \mathbf{G_k}(N_i)\mathbf{F_k}(N_j)\right] \right\} \right.$$
$$\left. + \bar{\rho} c_p N_i \{\bar{u}\mathbf{G_1}(N_j) + \bar{v}\mathbf{G_2}(N_j)\bar{w}\mathbf{G_3}(N_j)\}\right) | J_0 | \; d\zeta d\eta d\xi$$

$$\bar{a_{17}})_{i,k} = a_{17}{}^{*})_{i,k} + \int_{\xi=-1}^{+1}\int_{\eta=-1}^{+1}\int_{\zeta=-1}^{+1}$$
$$\times \left(-N_i\{\bar{u}\mathbf{G_1}(N_j) + \bar{v}\mathbf{G_2}(N_j) + \bar{w}\mathbf{G_3}(N_j)\}\right) | J_0 | \; d\zeta d\eta d\xi$$

$$\bar{q_1})_i = q_1{}^{*})_i$$

$$\bar{q_2})_i = q_2{}^{*})_i + \int_{\xi=-1}^{+1}\int_{\eta=-1}^{+1}\int_{\zeta=-1}^{+1} \omega^2 \left(N_i \left\{ 1 + \sum_{k=1}^{20} [N_k \lambda_k] \right\} \right) | J_0 | \; d\zeta d\eta d\xi$$

$$\bar{q_3})_i = q_3{}^{*})_i$$

$$\bar{q_4})_i = q_4{}^{*})_i$$

In these expressions for $[a_{13}]$, $[a_{14}]$, and $[a_{15}]$, $\bar{\rho}$ is iteratively computed by applying the ideal-gas relationship $\bar{\rho} = \bar{p}/R\bar{T}$, where R is the gas constant, with $\bar{p}$ and $\bar{T}$ being among the results of a previous iteration or an initial guess. Also, the superscript $(*)$ refers to a submatrix or a subvector that is identical to its zeroth-order counterpart (as defined in the preceding section), with the exception that the Jacobian $| J_0 |$ is now replaced by $| J_1 |$, which is defined, on an element basis, as follows:

$$| J_1 | = \begin{vmatrix} \sum_{i=1}^{20} \frac{\partial N_i}{\partial \zeta} x_i & \sum_{i=1}^{20} \frac{\partial N_i}{\partial \zeta} \lambda_i & \sum_{i=1}^{20} \frac{\partial N_i}{\partial \zeta} z_i \\ \sum_{i=1}^{20} \frac{\partial N_i}{\partial \eta} x_i & \sum_{i=1}^{20} \frac{\partial N_i}{\partial \eta} \lambda_i & \sum_{i=1}^{20} \frac{\partial N_i}{\partial \eta} z_i \\ \sum_{i=1}^{20} \frac{\partial N_i}{\partial \xi} x_i & \sum_{i=1}^{20} \frac{\partial N_i}{\partial \xi} \lambda_i & \sum_{i=1}^{20} \frac{\partial N_i}{\partial \xi} z_i \end{vmatrix}$$

where λ_i is the fraction of the rotor eccentricity ϵ by which the typical node i is displaced, as depicted in Figure 16.13.

The remainder of the perturbation analysis and determination of the gas-induced force components is very much similar to the earlier segment in the main text ending with the two force-gradient expressions (16.28) and (16.29).

Comment

The challenge in this proposed analysis upgrade is in efficiently managing what is now a larger-size model, considering that the energy equation has become part of the flow-governing equations. Nevertheless, this upgrade paves the way to numerous applications in the gas-turbine engine area. These are not necessarily confined to the "hot" side of the engine, specifically the turbine section, but also include compressor applications such as that depicted in Figure 16.62. Moreover, there is clearly no obstacle in combining the cylindrical-whirl type of rotor-axis excitation (the topic of this section) with conical whirl in the manner explained earlier in the main text.

Index

absolute total pressure, 83
accuracy
  assessment, 128
  degradation, 128
  enhancement, 128
adaptation of axisymmetric flow solution, 253
adjustable parameters, 24
admission losses, 267
adverse pressure gradient, 176
aerodynamic loading, 119, 167
airframe structures, 1
airfoil cascade, 14
airfoil suction and pressure surfaces, 123
airfoil trailing edge, 114
algebraic equations, 68
alternate frame of reference, 252
analytical over-specification, 290
annular seal, 244
approximate solution, 44
angular eccentricity, 299
angular pitch, 130
approximation functions, 6
area coordinates, 13
arithmetic average, 63
aspect ratio, 165
assessment of centered-rotor flow, 249
auxiliary power unit, 162
axisymmetric flow, 153

Baldwin and Lomax, 169
  algebraic turbulence model, 169
  Benim and Zinser near-wall model of, 169
  vorticity-based, 169
biquadratic element, 191
Blade
  suction side, 91
  wake, 162
blade-to-blade spacing, 122, *See also* pitch
blockage, 153
boundary
  closed, 4
  open, 4
boundary conditions, 3, 61
boundary layer thickness, 164
boundary layer separation, 120
boundary values problems, 3
  linear, 23
  nonlinear, 23
boundaries, 123
  domain-splitting, 123
  periodic, 123
bulk flow model, 237
bulkiness, 131

calculus of variations, 35, 68
Cal. Tech's bench mark, 258
canned programs, 57
capacitance matrix, 99
cascade, 118
  centroid, 58
  geometry, 115
  of lifting bodies, 114
  rectilinear, 118
  rotating-blade, 167
centered-impeller domain, 190
centered-rotor flow field, 239, 275
centrifugal compressor, 130
centripetal acceleration, 167
characteristic length, 84
circulation, 6, 120
  around an airfoil, 123
  -like, 134
  rectilinear, 119
  spanwise variation of, 138
classification of viscous-flow problems, 324
column matrix, 58
compatibility requirement, 34
completeness requirement, 34
compressible-flow gas seals (adiabatic flow), 307

compressible-flow problems (rigorous adaptation), 355
gas turbine applications, 355
high-speed non-contact seals, 355
linearized convection terms of, 355
compressibility, 237
computational-domain mapping, 114
computer-based solution, 106
computer-oriented, 52
conduction applications, 66
conformal transformation, 114
conical whirl, 275, 298
continuity equation, 120
continuum problems, 3
convection
heat transfer coefficient, 84
resistance, 46
convection-dominated flow, 49, *See also* inertia-dominated flow
converging passage, 107
conversion, 34
coolant flow rate, 83
cooling, 66
air, 66
aspects, 85
blade, 86
effectiveness, 83, 87
external, 87
of a radial turbine rotor, 66
rotor disk, 86
coordinate system
global, 10
local, 17
rotating, 252
rotating-translating, 252
Coriolis acceleration, 167
CPU time consumption, 66
Crank-Nicolson method, 103, 210
creeping flow, 327
critical Mach number, 219
curvilinear coordinates, 149
cusp trailing edge, 116
cyclic fluctuations, 201
cylindrical frame of reference, 177

damping coefficient, 190
degree of freedom, 4, 246
destabilizing effect, 189
deswirl vanes, 274
deviation angle, 292
differential formulation, 68
diffuser
annular, 176
conical, 176
cubic, 73
exhaust, 176
plannar, 176
radial, 176
diffusion
excessive, 152
over the suction side, 152
Dirichlit boundary condition, 107
disk backside, 85
discharge velocity, 122
discontinuity, 120
discretization of the continuum, 71
discretized domain, 27
displaced-rotor operation, 344
displacement field, 27
domain, 123
with high degree of multi-connectivity, 135
splitting boundaries, 123
double
discharge, 191
entry, 191
doubly-connected domain, 118, 134
driven cavity, 237

eccentricity, 238
element, 10
biquadratic, 13
conformal, 34
curve-sided, 10
equations, 30
finite, 1
intensity, 123
isoparametric, 14
Lagrangian, 158, 161
matrices, 20
orientation of, 10
quadrilateral, 13
straight-sided, 10
subparametric, 19
superparametric, 19
tetrahedral, 74
triangular, 80
elliptic flow field, 150
equilateral, 33
error consistency criterion, 208
Euler-Lagrange equation, 37
Euler's method, 101
exact solution, 25
expanded rotordynamic analysis, 304
explicit algorithm, 210
external constraint, 114

face seal, 286
favorable pressure gradient, 204
field discretization model, 128
finite-difference technique, 1, 2, 24, 40, 64
finite-element
analyses, 29
approximation, 43
formulation, 35
fundamentals, 1
grid points, 2
mesh, 27
method, 1, 26
superiority of, 65

finite-volume computational technique, 114, 117
flat-plate cascade, 114
flow ellipticity, 180
flow-permeable boundary, 136
flow separation, 85
fluid-induced force components, 260, 280
fluid/rotor interaction system, 299
formulation, 40
  finite-element, 40
  variational, 27
forward whirl, 239
frame of reference
  rotating, 253, 255
  rotating-translating, 253
free-vortex boundary condition, 135
friction coefficient, 84
full guidedness, 116
functional minimization, 46

Galerkin's algorithm, 210
Galerkin's weighted-residual approach, 127
Gauss theorem, 127
Gaussian elimination method, 102
global
  coordinates, 14
  matrix, 79
  vector, 256
Gresho and Lee, findings of, 177
grid dependency, 184, 195, 264, 278
guide vanes, 135

heat balance equation, 86
heat conduction problems, 51
heat convection coefficient, 66
heat generation rate, 58
heat flux, 67
heat sink, 66
housing, 238
hydraulic diameter, 86
hydraulic pump, 258

impeller, 168
implicit algorithm, 210
implicit set of equations, 102
inclusion of shear stress perturbations, 309
independent functions, 42
independent variable, 6
  one, 6
  three, 7
  two, 7
inertia coefficient, 190
inertia-dominated flow, 181
influence coefficients, 204
instability, 189
integrals, 38
  approximation of, 38
  elemental, 45
integration
  by parts, 37
  limits of, 22
  numerical, 22
interaction effects, cylindrical/conical whirl of, 303
interelement continuity, 34
internal flow, 111
interpolation functions, 6, 8, 33, 72, 75
  deriving, 8
  order of, 75
interrelated rotordynamic coefficients, 306
inviscid, incompressible, 107
irregular mesh, 106
irrotational flow, 115
irrotationality condition, 120
isometric view, 85
isoparametric element, 16
iterative step, 121

Jacobian, determinant of, 21

Kutta condition, 116

labyrinth seal, 189, 273
lamiar flow, 153
Laplacian, 121
leading edge, 84
leakage flow, 189
leakage passage, 189, *See also* secondary passage
leakage suppressant, 197
length coordinates, 11
linear interpolation, 11
load vector, 31, 109
local, 14
  coordinates, 14
  frame of reference, 126

main flow, 67
mapping
  functions, 18
  matrix, 75
mass concentration, 3
master streamline, 291
meridional velocity, 195
mesh refinement, 33
metal creep, 88
metal temperature, 86
mixed-order interpolation, 207
monotonic convergence, 33
Monte Carlo method, 23
multi-connectivity, 136
multiply-connected domain, 106, 114
multi-valuedness, 114

natural boundary conditions, 124, 206
natural coordinates, 10
  in one dimension, 11
  in two dimensions, 12
natural coordinates for 3-D surface elements, 321

Navier-Stokes problems, 150, 324
near-wall zone, 176, 244
Neuman boundary condition, 107
Newtonian fluid, 155
nine-noded element, 156
no-penetration boundary condition, 123
no-slip boundary condition, 107, 156
nodal
  degree of freedom, 8
  points, 65
  positions, 57,59
  temperatures, 77
  unknowns, 41
  values, 27
nodeless degree of freedom, 114, 138
nodes, 4, *See also* nodal
  exterior, 4
  interior, 4
  locations, 44
  types of, 4
numerical
  convergence, 162
  fluctuations, 162
  oscillations, 162
  stability, 106

off-design performance, 169, 184
off-diagonal rows, 59
one-dimensional problems, 42

parabolized flow field, 150
parent element, 17, 164
partially-packed arrays, 128
periodic boundary, 123
periodicity conditions in radial cascades, 130
perturbation approach, 237
  finite-element–based, 237
  foundation of the, 238
  results of the, 294
  to unsteady-flow problems, 237
Petrov-Galerkin W.R. method, 194
pitch, 211
piecewise approximation, 2
plane-wall transient, 100
pointwise approximation, 2
polynomials, 19
potential flow solver, 112, 119
power series, 23
power systems, 177
Prandtl-Van Driest formulation, 178
preswirl velocity, 261
primary flow passage, 189
primary flow rate
primitive formulation, 158, 329
probability schemes, 23
propulsion applications, 131
purely implicit method, 103

quadratic element, 19
quadrilateral element, 19
quasi-three-dimensional flow model, 151

radial-airfoil cascade, 130
radial inflow turbine, 66
radial turbine scroll, 130
radius of curvature, 154
reaction forces, 256, 269, 294
  aggravating, 256
  restoring, 256
reattachment, 198
recirculation zone, 150
recovery
  characteristics, 184
  coefficient, 185
rectilinear cascade, 118
reference frame, 3
relative
  gas velocity, 84
  stability ratio, 308
  total temperature, 83
  velocity, 253
residual, 127
reversed lean, 165
Reynolds number, 153
Richardson quadrature, 211
Ritz method, 24
rotordynamic, 190
  coefficients, 240
  stability, 240, 282

scroll tongue, 134
seal
  face, 191
  resistance of, 301
  wear-ring, 191
secondary passage, 189
self-adjoint, 31
serendipity, 17
  family, 19
  interpolation, 19
shape functions, 6
shear stress, 163
shrouded-pump impeller, 190, 242, 283
similarity solution, 23
simply-connected domain, 118
simulation of the primary-impeller, 291
single-harmonic perturbation, 296
single-valued solution, 114
singularity, 110
solidity ratio, 211
solution
  domain, 24, 66
  exact, 25
  multi-valued, 120
specified-temperature nodes, 82
splitting boundaries, 122
stagnation streamline, 150

static pressure recovery, 186
stator/rotor
  flow interaction, 201
  interface, 203
stiffness coefficients, 31, 190
stiffness matrix, 128
stream-function formulation, 107, 109
streamline-curvature method, 114
streamlined object, 107
stream tube thickness, 152, 153
stress concentration, 88
subsonic flow, 121
superposition, flow around multiple bodies by, 111
surface tractions, 157
swirl brake, 286
swirl velocity, 275
swirling flow, 177
symmetric matrix, 59, 105
synchronous whirl, 262

temperature distribution, 45
tetrahedron, 15
time dependency, 201
time marching, 205
thermal conductivity, 55
thermal resistance, 46
thin-layer approximation, 150
thin rod problem, 48, 52
three-noded triangle, 33
through-flow velocity, 131
tolerance factor, 162
tooth-to-tooth chamber, 276
total pressure loss, 186
trailing edge condition, 116, *See also* Kutta condition
transformation
  Fourier, 23
  Laplace, 23
transient problem, 95
trapezoidal rule, 39
trial functions, 24
triangular prism, 73
tridiagonal, 102
turbomachinery component, 138
turbulence, 169
  closure, 169
  vorticity-based, 169
twenty-noded isoparametric element, 132
two-way exchange of boundary conditions, 220

underrelaxation factor, 214
universal law of the wall, 179
unsteady flow problems, 201
upwinding technique, 182, 244

variational
  approach, 3, 66
  formulation, 68
  interpolation, 2
  principles, 3
  problems, 3
  statement, 42, 44, 51, 52, 64
velocity potential
  discontinuity, 134
  formulation, 6
  virtual displacement, 251
viscosity coefficient
  dynamic, 84
  eddy, 180
  effective, 84
  kinematic, 84
volume coordinates, 14
vortex
  breakdown, 197
  shedding, 201
vorticity, 178

wake cutting, 201
wear-ring seal, 286
weight functions, 177, 190
weighted-residual method, 24, 117
whirling motion, 240
whirl-orbit radius, 312
wiggles, 177, 244
Wu's flow model, 151

Zeroth-order flow field, 240, 335

Printed in the United States
by Baker & Taylor Publisher Services